Penguins

Penguins

NATURAL HISTORY AND CONSERVATION

Edited by

PABLO GARCIA BORBOROGLU

and

P. DEE BOERSMA

A SAMUEL AND ALTHEA STROUM BOOK

UNIVERSITY OF WASHINGTON PRESS
Seattle & London

Penguins is published with the assistance of a grant from the Samuel and Althea Stroum Endowed Book Fund, and supports the work of the Global Penguin Society, www.globalpenguinsociety.org.

© 2013 by the University of Washington Press
Printed and bound in Korea
Design by Thomas Eykemans
Composed in Warnock, typeface designed by Robert Slimbach
Display type set in Bodoni, designed by Morris Fuller Benton
Headings set in Trade Gothic, designed by Jackson Burke
16 15 14 13   5 4 3 2 1

UNIVERSITY OF WASHINGTON PRESS
PO Box 50096, Seattle, WA 98145, USA
www.washington.edu/uwpress

The paper used in this publication is acid-free and meets the minimum requirements of American National Standard for Information Sciences—Permanence of Paper for Printed Library Materials, ANSI Z39.48–1984.∞

Complete cataloging information for this title
is available from the Library of Congress.
ISBN 978-0-295-99284-6

Many years of work, effort,
and persistence reside in this book,
which we hope will improve the world
for penguins and for people.

Contents

GPS

GLOBAL PENGUIN SOCIETY

The Global Penguin Society (GPS) is an international alliance-building organization that promotes the protection of the world's penguin species through science, management, and education. As an international forum for conservation, GPS helps NGOs, academic and research institutions, individual projects, local communities, and other partners work together for the conservation of penguins and oceans. Working synergistically, GPS accelerates and enhances penguin and ocean conservation efforts.

Penguins are conservation subjects as well as tools for ocean conservation. The Society fosters the science needed for conservation and adequate management of penguins and marine environments at local, regional, and global scales. GPS uses science to help decision makers improve management and educate local communities about the value of penguins.

The Global Penguin Society links local stakeholders to policy change using penguins. It provides opportunities for the public, scientists, and managers to develop and advocate solutions for sustainable activities and management for marine and coastal environments. When appropriate, GPS campaigns to educate people about how to improve the quality of life for both penguins and people. In brief, the Global Penguin Society encourages synergy, strategy, and integration and provides a unified vision to enhance the scope of penguin conservation achievements.

www.globalpenguinsociety.org

Penguins

Introduction

Penguins are Southern Hemisphere seabirds. They are athletic, interesting, and ancient. Some Magellanic penguins migrate more than 2,400 miles from the Strait of Magellan in Argentina to Rio de Janeiro, Brazil. Magellanic penguins can travel more than 170 kilometers in a day, and do so under their own power—no fossil fuel use for them. A breeding Magellanic likely covers 16,000 kilometers a year, the average distance a car is driven in the United States. Emperor penguins breed in Antarctica, where they keep their eggs warm by holding them on their feet while enduring temperatures as cold as –30 to –40°C with the wind blowing 40 meters per second. Humans couldn't survive for long in those conditions. If that doesn't take your breath away, imagine fasting in these conditions as a male emperor does for up to four months, without a bite to eat except snow and ice.

Penguins are remarkable creatures. They vary in size from the little penguin, which weighs just about 1 kilogram and is a shallow diver, to the emperor penguin, which weighs up to 40 kilograms and can dive 500 meters and hold its breath for 23 minutes. In comparison, the record dive for a human is 101 meters in 4.13 minutes. Penguin species live in environments ranging from the tropics of the Galápagos Islands to frozen Antarctica and across islands and continents in the Southern Hemisphere. They all are black and white, and some have yellow and orange crests or a bluish coloration. Some individuals in the wild live more than 30 years. They nest in deserts, in forests, on bare rock, in burrows, under bushes, under trees, and in the open.

Fossils suggest that penguins flourished between 10 and 40 million years ago, and the oldest fossil is about 55 million years old. Their origins, however, may be rooted in the Cretaceous period, 140 to 65 million years ago, when their ancestor was a flying seabird. Whether that ancestral form was a loon, an albatross, or a frigate bird is unclear. In this book, we organized the chapters by the relationships among penguins based on current morphological and molecular knowledge. The six genera of living penguins (*Aptenodytes*, *Eudyptes*, *Eudyptula*, *Megadyptes*, *Pygoscelis*, and *Spheniscus*) are clearly defined, and their classification has stood the test of time. The relationship between species within and outside their genera is not so fixed.

We start the book with the largest penguins, the king and emperor (*Aptenodytes*), and end with the smallest, the little (*Eudyptula*). Each species account provides the common and scientific names, description, taxonomic status, range and distribution, summary of population trends, International Union for Conservation of Nature (IUCN) status, natural history, population sizes and trends, map and size of colonies, general annual cycle, main threats, recommended research and conservation actions, and current conservation efforts for the species. We use the same order in each chapter. The material presented should inform human action, whether it is research, policy, or on-the-ground conservation. The editors hope this book will do more than inform you about penguins. First we must know, then understand, and, finally, we can act. We hope you will be moved to help penguins.

Even though penguins live in remote areas of the world, humans have a big impact on them. Early Antarctic explorers depended on them for food, businessmen harvested tens of thousands of them for their oil, and earlier settlers on islands ate them and collected their eggs. Now penguins are among the most endangered seabirds. About two-thirds of penguin species are on the IUCN Red List of Threatened Species. The threat of human harvest has decreased, but climate variation, fishing, habitat modification, disease, and even tourism loom as threats. Human impacts will likely continue to be largely negative on seabirds as human numbers grow beyond 7 billion and consumption increases. The fate of seabirds, from albatrosses to penguins, is linked with our own.

Penguins are among the most popular and best-loved birds. *March of the Penguins*, the highest-grossing documentary of all time, features emperor penguins. Whether a person turns to the movies or children's books, penguins figure significantly in human culture. This book details the lives of the world's penguins. A total of 49 researchers from 12 countries on 5 continents participated in this effort. This is the first book to bring the world's experts together to share what they know about each species.

The book is a milestone for the Global Penguin Society (GPS), a group dedicated to the well-being of penguins. Mitigating and managing the threats to penguins on local, regional, and global scales require funding and social networking. The Global Penguin Society helps researchers, landowners, politicians, and governments give penguins a voice.

Penguins are environmental sentinels. People love penguins but are unaware of their decline. Their natural charisma makes them the perfect ambassadors to advocate for the health of our oceans and coasts.

I

Large Penguins

Genus Aptenodytes

King Penguin

(Aptenodytes patagonicus)

Charles-André Bost, Karine Delord, Christophe Barbraud, Yves Cherel,
Klemens Pütz, Cédric Cotté, Clara Péron, and H. Weimerskirch

1. SPECIES (COMMON AND SCIENTIFIC NAMES)

King penguin, *Aptenodytes patagonicus*

2. DESCRIPTION OF THE SPECIES

The king penguin is one of two extant species in the genus *Aptenodytes*. Together with its close relative, the emperor, the king differs from other penguins in its large size, long and thin beak, and bright colors. Paleontological evidence suggests that about 3 million years ago, a third *Aptenodytes* species, the Ridgen's penguin (*A. ridgeni*), lived in New Zealand (Fordyce and Jones 1990; Williams 1995). The genus *Aptenodytes* split off around 40 million years ago from the branch that led to all other living penguin species (Jouventin 1982).

The king, the second-largest penguin, differs from the emperor in its thinner appearance and the marked auricular and brighter orange patches that extend as a narrow stripe around the neck (fig. 1). Unlike the emperor penguin, the king has an unfeathered tarsus. The head, chin, and throat are blackish. The lower beak mandibles are orange to bright pink. The eye is brown with a thin, dark iris. The flippers are proportionally larger than the emperor's and ventrally white with blue black at the tip. The dorsal side of the body, ranging from the neck over the back to the tail and the feathered legs, looks bright silver blue after the molt. The dorsal plumage is separated from the white ven-tral side by a thin black line and becomes a dull gray after the end of summer. The tail is rather short and dark gray. The robust feet are black. Within pairs, the male is most often larger than the female. The crèched chick is feathered with brown down before developing thick brown dunes that are maintained until around 10 months of age (fig. 2). The beak is blackish until the chick molts into immature plumage. The auricular patches of immature birds are much paler (yellow) than

FIG. 1 (*FACING PAGE*) All penguins are social. These king penguins are in lock step as they march showing their unfeathered tarsus, which distinguishes them from the larger but less colorful emperor penguins. (C. Sutter)

FIG. 2 Adult king penguin near a king penguin chick. The brown-feathered chicks were thought to be a separate species and were called "oakum boys" because they resembled sailors who became covered in tar and hemp when filling cracks between the boards on ships. (P. D. Boersma)

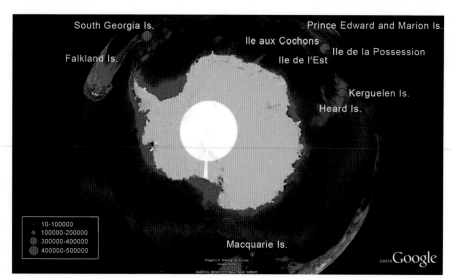

FIG. 3 Distribution and abundance of the king penguin, with counts based on pairs.

Legend:
- 10-100000
- 100000-200000
- 300000-400000
- 400000-500000

those of adults. The king acquires adult plumage at the age of three years (Williams 1995).

3. TAXONOMIC STATUS

The king penguin, first described in 1778 by the English naturalist John Frederick Miller, was distinguished from the emperor penguin in 1844 by George Robert Gray (Gray 1844). The generic name *Aptenodytes* is derived from the ancient Greek and means "without the capacity to fly," and its specific epithet *patagonicus* is derived from "Patagonia." Several subspecies have been suggested, but two are currently recognized: *A. p. patagonicus* breeds in the South Atlantic Ocean, and *A. p. halli* breeds in the southern Indian and Pacific Oceans. Some distinct variation in size occurs by locality, but accurate and comparable data are lacking. For example, the flipper lengths of kings at the Crozet Islands and Marion Island are about 5% longer than in South Georgia Island. Conversely, flipper lengths of birds from the Kerguelen Islands are about 16% smaller than on Crozet (table 1.1). The Crozet and Kerguelen archipelagoes lie only 1,400 kilometers apart, but penguin populations appear genetically isolated from one another (Viot 1987). Body mass also varies by locality. For example, birds from South Georgia are as much as 25% heavier than the Crozet birds during courtship.

4. RANGE AND DISTRIBUTION

King penguins breed on sub-Antarctic islands between latitude 45° south and 55° south, namely on the Falkland/Malvinas Islands and South Georgia Island in the South Atlantic Ocean, on Prince Edward, Crozet (including Cochons, Possession, and l'Est), Kerguelen, and Heard Islands in the southern Indian Ocean, and on Macquarie Island in the southwestern Pacific Ocean (table 1.2; fig. 3). One breeding pair has been reported from the South Sandwich Islands, but consistent breeding awaits confirmation. All breeding sites are generally located within a distance of 400 kilometers from the Antarctic Polar Front.

The king's foraging ecology, including their range and distribution in relation to marine features, is one of the most extensively studied among marine vertebrates. During summer, irrespective of the location of their breeding site, their at-sea distribution is strongly dependent on frontal zone features, particularly the Antarctic Polar Front. However, the Sub-Antarctic Front is also frequented by birds from some breeding sites (Crozet Islands [Bost et al. 2009], Falklands/Malvinas [Pütz 2002; Pütz and Cherel 2005]).

In autumn and winter, birds caring for chicks forage beyond the Antarctic Polar Front, in Antarctic waters, up to the limit of the pack ice (Moore et al. 1999; Pütz et al. 1999; Bost et al. 2004). Only birds from the small breeding colony of the Falklands/Malvinas make use of the slope of the Patagonian Shelf up to 38° south (Pütz 2002). In general, the kings' winter foraging area is encompassed by the Antarctic Polar Front to the north and the Antarctic Divergence (up to the limit of the sea ice) to the south. Breeding birds foraging very close to the colonies in winter, over slope waters, can return with almost-intact prey in their stomachs (Cherel et al. 1996).

FIG. 4 Adult king penguin with an egg threatens to peck another king penguin. (P. D. Boersma)

Where adult nonbreeders and immature birds disperse is unknown. Juveniles probably disperse off the Polar Frontal Zone as far as 60° south (Ainley et al. 1984), that is, in the same sectors used by juvenile emperor penguins (Kooyman et al. 1996).

5. SUMMARY OF POPULATION TRENDS

King populations throughout the sub-Antarctic have increased since the end of the last century following a drastic decline during the 19th and the beginning of the 20th century (Macquarie Island [Rounsevell and Copson 1982], Heard Island [van den Hoff et al. 1993], Kerguelen Islands [Weimerskirch et al. 1989], Crozet Islands [Delord et al. 2004], Marion Island [Williams et al. 1979], South Georgia Island [Lewis Smith and Tallowin 1979], Falklands/Malvinas [Otley et al. 2007]). The king population declined due to exploitation associated with the sealing industry (Conroy and White 1973; Conroy 1975; Rounsevell and Copson 1982). Most of the colonies grew rapidly between 1970 and 1990, sometimes at a rate especially high for a long-lived seabird (Weimerskirch et al. 1992).

During the past 15 years, population variability was high (table 1.2). Recent surveys indicate that several large colonies appear to have decreased and stabilized at a lower population level. The population has begun to stabilize over the past decade, and we estimate the global population at 1.6 million annual breeding pairs (range 1,584,320–1,728,320). The counts occur in January, at peak or just following peak egg laying (fig. 4). Therefore, the numbers of breeding pairs are under-

estimates, because they do not include the pairs that failed before the counts or late-nesting pairs that lay eggs until mid-March. The 18-month breeding-molt cycle of this species means the proportion of late breeders is regulated primarily by the breeding success of the preceding year.

6. IUCN STATUS

Since 2004, the International Union for Conservation of Nature (IUCN) has listed the king penguin in the Least Concern category on its Red List of Threatened Species (IUCN 2011). The population size is large, and colonies have recently increased and/or stabilized at all breeding sites throughout its range.

LEGAL STATUS. This species is protected throughout its range, with colonies located either in designated nature reserves or sites of special scientific interest administered by the respective national authorities and regulated by management plans.

7. NATURAL HISTORY

BREEDING BIOLOGY. Kings have a breeding-molt cycle of more than one year, the longest among seabirds because the chicks fast during the winter. The life cycle was studied in detail at South Georgia Island (Stonehouse 1960; Olsson 1996; Olsson and Brodin 1997) and the Crozet Islands (Barrat 1976; Weimerskirch et al. 1992; Gauthier-Clerc et al. 2001; Descamps et al. 2002) and investigated at Marion Island (du Plessis et al. 1994; van Heezik et al. 1994) and the Falklands/Malvinas (Otley et al. 2007).

FIG. 5 King penguins nesting on Salisbury Plain, South Georgia Island. (P. D. Boersma)

Kings often breed in large, dense colonies (mean density about two breeders per square meter, [Barrat 1976]) located on flat, sandy beaches or near-shore valleys (fig. 5). Protection against the dominant wind direction during winter is a determinant of colonization and colony breeding success (Weimerskirch et al. 1992).

The breeding cycle lasts about 14–15 months from courtship to chick fledging (but see Otley et al. 2007). This unusual extended cycle results from the large size of the bird and the halting of chick growth from autumn to early spring. Some chicks fast for up to five months during winter before being fed by their parents and fledging (Cherel et al. 1987). The extended breeding cycle means that colonies are continuously occupied with crèched chicks throughout the year. The reproductive cycle (n) begins at the end of winter and is initiated mostly by failed breeders of the current year (cycle n-1) returning to the colonies for the prenuptial molt (date of first arrival: Crozet, range 22 Sept.–1 Oct., n = 3 years; South Georgia, range 8 Sept.–6 Nov.) (Weimerskirch et al. 1992; Williams 1995). Males and females do not arrive to molt at the same time, which results in a low rate of

mate fidelity (28.6 % [Barrat 1976]). After their molt, they forage at sea for more than two weeks (Crozet, 17.1 ± 0.5 days, n = 3 years [Descamps et al. 2002]) before returning to shore between late October and December. Penguins come to the colonies with large body reserves, approximately four kilograms (Weimerskirch et al. 1992; Cherel et al. 1993; Gauthier-Clerc et al. 2001), as they have to fast during the first part of breeding (courtship and early incubation for males), and insufficient body reserves may induce a delay or even complete breeding failure (Gauthier-Clerc et al. 2001).

The laying period is asynchronous and extends over four months, from early November to mid-March, but with considerable variation between breeding sites and years (fig. 6). At Crozet, the mean laying date over three consecutive years ranged from 22 December ± 19.5 days to 14 January ± 15.9 days (Weimerskirch et al. 1992). A marked laying peak usually occurs between the second half of November and the second half of December, corresponding to the so-called early breeders (Gauthier-Clerc et al. 2001). The king penguin, like its close relative, the emperor penguin, has no nest and incubates its single large egg on its feet (table 1.3), sheltered by a feathered-skin pouch (fig. 7). Incubation lasts around 54 days. Re-laying is rare and occurs only when the egg is lost during incubation (Weimerskirch et al. 1992).

From courtship to hatching, the two parents exchange duties six times, one incubating while the other forages at sea (Descamps et al. 2002). Hatching may take 2–3 days, but the chick can be fed through a hole in the shell. Chicks have a thin covering of down and are brooded while balanced on their parents' feet (fig. 8). From hatching to crèching, both parents alternate brooding duties, each making four to five foraging trips usually 4–12 days in duration. This period lasts about one month before the chicks form crèches, by mid-February (fig. 9). During mid-April (15–26 April [Weimerskirch et al. 1992]), the almost fully grown chicks usually have gained extensive body reserves (Cherel et al. 1993) (fig. 10). At the same time, the adults start to desert the colony for extended periods because of the scarcity of prey (Charrassin and Bost 2001). During the winter period, the chicks lose weight, fast, and get an average of 4.7 feeds from their parents between 1 May and 31 August (Crozet, n = 3 years [Descamps et al. 2002]). Parents return to feed chicks usually by mid-September and continue to feed them until the chicks fledge. The chicks' molt occurs between the second half of November and early January.

Breeders usually lay eggs every year, even when failure is almost certain, late in the season. At best, successful breeding occurs every two years (Olsson 1996; Jiguet and Jouventin 1999). Birds that ended molt early (October) have their eggs between mid-November and early January (early breeders), thereby maximizing the likelihood that their chicks will have sufficient energy reserves to survive the winter fasting period. Conversely, kings that ended their molt later, that is, in December and January (late breeders), do not have eggs until late January to March, and thus their chicks are crèched with insufficient energy reserves for the winter fast. The heavier the crèched chick, the higher its probability of fledging (Cherel et al. 1987).

PREY AND FORAGING ECOLOGY. With the exception of the Falklands/Malvinas, all king breeding sites are located within the vicinity of the Antarctic Polar Front. Many aspects of the kings' foraging ecology have been intensively studied at various breeding sites, including their

FIG. 6 King penguins are a bill length away from their neighbors as they keep their eggs and young chicks on their feet. (P. Ryan)

FIG. 7 Adult king penguin incubating an egg at South Georgia Island. (P. D. Boersma)

FIG. 8 King penguin with chick between its feet. (P. D. Boersma)

FIG. 9 King penguin chicks fatten up for the winter and enjoy the close contact with other chicks as they wait for their parents. (P. Ryan)

FIG. 10 Adult king penguin is lost in a crowd of chicks that must fast in the winter when adults depart the colony, leaving the chicks without food until they return. (J. Weller)

diving behavior (Kooyman et al. 1992; Pütz et al. 1998; Charrassin et al. 1998, Charrassin et al. 1999; Moore et al. 1999; Charrassin and Bost 2001; Bost et al. 2002; Charrassin et al. 2002; Wienecke and Robertson 2006; Pütz and Cherel 2005; Bost et al. 2007) and diet (Adams and Klages 1987; Hindell 1988; Cherel and Ridoux 1992; Olsson and North 1997; Moore et al. 1999; Cherel et al. 2002). Except for the emperor penguin, no bird is able to dive deeper (up to 440 m [Charrassin et al. 2002]), although mean foraging depths are between 100 and 200 meters, increasing from incubating to crèching (Charrassin et al. 2002) and during winter (Moore et al. 1999). Regardless of the breeding location, kings are specialized predators that target mainly lanternfish (myctophids), which are small, schooling mesopelagic fish (review in Cherel et al. 2002). These fish constitute the bulk of mesopelagic fish in the Polar Frontal Zone (Sabourenkov 1991) and are caught by pursuit-diving, mostly during daylight hours (Pütz and Bost 1994). King penguins may dive synchronously (Pütz and Cherel 2005). During winter, the proportion of squid (e.g., Moore et al. 1998) in the diet increases up to 64% by biomass for breeders feeding their chicks (Cherel et al. 1996). When myctophid availability decreases, kings catch some other pelagic fish such as the ice fish (*Champsocephalus gunnari*) (Bost et al., unpubl. data).

PREDATORS. At some breeding sites, killer whales (*Orcinus orca*) are the major at-sea predator targeting both adults and fledglings during their departure from and arrival on the beach. Vagrant young leopard seals (*Hydrurga leptonyx*) also prey on king penguins during their winter sojourn along sub-Antarctic coasts. Some adult male Kerguelen fur seals (*Arctocephalus gazella*) at Marion Island (Hofmeyr and Bester 1993) and the Crozet Islands (Charbonnier et al. 2010) regularly take king penguins, and South American sea lion (*Otaria flavescens*) bulls prey on kings at the Falklands/Malvinas (Pütz, pers. obs.).

Ashore, the sub-Antarctic and Antarctic giant petrels *Macronectes halli* and *M. giganteus* prey on injured penguins or penguins in poor condition (Hunter 1983). Giant petrels also take chicks, mostly those in poor body condition, but can also catch and kill full-grown, healthy chicks, sometimes at night (Le Bohec et al. 2003). Predation by skuas (*Stercoraria* spp.) is mostly on deserted eggs or eggs robbed from inexperienced breeding birds when mates are exchanging nest duties. The lesser sheathbill (*Chionis minor*) takes mostly deserted eggs and unattended small chicks, but it also steals food when chicks are fed by their parents during winter and spring (Verheyden and Jouventin 1991). In the Falklands/Malvinas, turkey vultures and crested caracaras prey on abandoned eggs and chicks (Pütz, pers. obs.).

FIG. 11 A fat juvenile king penguin sleeps with its bill under its wing as it gets ready to molt at Neko Harbor, Antarctic Peninsula. Note the lack of yellow or orange plumage. (P. D. Boersma)

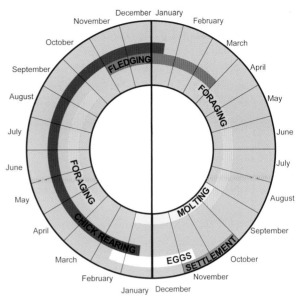

FIG. 12 The annual breeding-molt cycle of the king penguin is about 18 months, the longest of any penguin species.

MOLT. The first penguins coming ashore to molt are unsuccessful breeders. The peak of their molt is 17 October ± 4.6 days at Crozet (n = 3 years [Descamps et al. 2002]) and before November at the Falklands/Malvinas (Otley et al. 2007). Molting birds usually congregate along the banks of rivers, which they often use as a freshwater supply. The peak of molt for successful breeders is from late November to early January at both sites. Between fledging and the beginning of the molt, successful parents spend one month at sea. For early molters, molt duration is the same at the northern and southern limits of the range (Crozet, 31.0 + 3.9 days [Weimerskirch et al.1992]; South Georgia, 31.2 days, range 27–36 [Stonehouse 1960]). At Crozet, the molt duration of successful breeders is shorter (22.7 + 1.0 days [Weimerskirch et al. 1992]), indicating that molt duration depends on the outcome of the previous breeding season. During molt on land, the birds replace all their feathers and also their mandible plates (fig. 11). Feather growth begins at sea when penguins build up their energy reserves, including fat and protein, in anticipation of fasting and growing new feathers (Cherel et al. 1993; Cherel et al. 1994). Kings reach their peak body mass at the very beginning of the molting process and suffer their greatest mass loss during the molting fast, corresponding to about 51% and 45% of the initial weight of early and late breeders, respectively (Weimerskirch et al. 1992).

ANNUAL CYCLE. The king has one of the most unusual annual cycles of any of the penguins. It takes fourteen months to rear one chick, which means there is an overlap during the summer months when both eggs and large chicks are present. See figure 12.

8. POPULATION SIZES AND TRENDS

In recent decades, king populations have increased across their range (table 1.2). Exploitation of penguins for their oil associated with the sealing industry in the 19th and early 20th centuries resulted in their population decline. After commercial exploitation ceased, king populations quickly recovered during the second part of the 20th century at all breeding sites (Macquarie Island [Rounsevell and Copson 1982], Heard Island [van den Hoff et al. 1993], Kerguelen Islands [Weimerskirch et al. 1989], Crozet Islands [Delord et al. 2004], Marion Island [Williams et al. 1979], South Georgia Island [Lewis Smith and Tallowin 1979], Falklands/Malvinas [Otley et al. 2007]). The largest population occurs on the Crozet Islands (at least 612,000 pairs, i.e., about 50% of the global population). However, whereas the breeding colonies on Possession Island are counted annually, no recent data are available for the world's largest king colony at Cochons Island (500,000 pairs in 1988 [Guinet et al. 1995]). Kerguelen and Heard Island colonies were slower to recover and are still increasing. At Macquarie Island, the most recent estimate is of 150,000–170,000 breeding pairs (Parks and Wildlife Service 2006), representing an increase of 4–6% per year since 1984 (D. E. Rounsvell, unpubl. data), with recent recolonization of collapsed colony sites (Van Den Hoff et al. 2009). The South Georgia population is now approximately 450,000 pairs, an increase of 11% per year since 1985–86 (P. N. Trathan, pers. comm.). The Falklands/Malvinas population fledged the first chick in 1965 and is estimated at more than 1,000 pairs (Otley et al. 2007). The breeding pairs in the king colony at the Falklands/Malvinas has increased recently after declining slightly or being stable for a decade (Pütz, unpubl. data).

More king penguins are visiting the coasts of Tierra del Fuego to molt, and breeding is anticipated. The total breeding population is currently estimated at approximately 1.6 million pairs (table 1.2), and the population is expected to increase in at least some localities (e.g., Kerguelen Island, the Falklands/Malvinas). Recent surveys also indicate that several large colonies have stabilized or decreased compared to the beginning of the 21st century.

9. MAIN THREATS

Human exploitation probably caused the collapse of king populations during the 19th and early 20th centuries when penguins were killed for their oil by the sealing industry. At some breeding sites, populations were either exterminated (e.g., Heard Island) or at least substantially reduced. The cessation of sealing likely explains the exponential increase of subsisting populations and the recolonization of former breeding sites (Conroy and White 1973; Rounsevell and Copson 1982). Demographic studies at two breeding localities indicate very high adult survival at least since the mid-1970s (Crozet, 1975–90, 90.7–95.2% [Weimerskirch et al. 1992]; Crozet, 1998–2003, 90.6% [Le Bohec et al. 2007]; South Georgia, 1992–96, 83.0–97.7% [Olsson and van der Jeugd 2002]) (table 1.4). Juvenile and immature survival also appears to be relatively high (Olsson and van der Jeugd 2002; Gauthier-Clerc et al. 2004). Breeding success is highly variable (Crozet, 1986–88, 30.6%, range 0.7–53.2% [Weimerskirch et al. 1992]; Crozet, 1997–2001, range 20–34% [Descamps et al. 2002]; Gauthier-Clerc et al. 2004) and comparatively low, which is a direct consequence of the mix of early and late breeders; this in turn is affected by each individual's breeding success in the previous year. Breeding success of birds that did not breed or failed in a previous breeding attempt is much

higher (51–60%) than for birds that previously bred successfully (5% [Le Bohec et al. 2007]). Thus, the increase in size of the king population was probably caused by a combination of high survival rates for adults, juveniles, and immatures and the relatively high breeding frequency. Furthermore, an increased food supply (myctophid fish) made available by the reduction of whale stocks may have had a positive impact on king penguins (Laws 1977; Rounsevell and Copson 1982).

Some populations have stabilized since the mid-1990s (e.g., Crozet Islands). These populations were approaching carrying capacity during the early 1990s, and climatic factors, such as the Southern Oscillation Index, may have modified their population dynamics (Delord et al. 2004). The warm phase of the Southern Oscillation and warm sea surface temperatures in foraging areas may negatively affect kings' breeding success and adult survival (Le Bohec et al. 2008).

POTENTIAL THREATS. The king is one of the most fortunate species in term of conservation status, with either increasing or stable populations throughout its breeding range. In contrast to the majority of southern seabirds, introduced predators such as cats and rats have had no proven impact on the species.

Increasing tourism within the sub-Antarctic regions could induce disease outbreaks among king colonies. For example, unknown diseases killed at least 250 to 300 kings in 1992–93 at Marion Island (Cooper et al. 2009).

The most important potential threats may be a change in the availability of their main prey. Myctophid fish were commercially exploited by the end of the 1970s and the beginning of the 1990s, with more than 200,000 tons harvested in the South Georgia sector (Collins et al. 2008). Several attempts to develop new commercial fisheries on a limited scale within the Southern Ocean are ongoing. A large, uncontrolled development of myctophid fisheries close to key foraging areas, especially in the southern Indian Ocean sector (Kerguelen) or in the Scotia Sea (South Georgia) may have deleterious effects on the foraging success and populations of the king.

In the long term, projected temperature increase in the Southern Ocean may also drastically modify king colony distribution and populations (Peron et al. 2012). For example, Intergovernmental Panel on Climate Change (IPCC) models predict a linear increase of 2°C in sea surface temperature in the southern Indian Ocean by the end of the 21st century. The Crozet population of kings is of concern for the following reasons: (1) it depends on distant, dynamic frontal zones, (2) it holds more than half of the world's population (Delord et al. 2004), and (3) demographic parameters might be negatively affected by warm sea surface temperatures within the foraging areas (Le Bohec et al. 2008). Such an increase in sea surface temperature should lead to a decrease in sea-ice extent and a significant southward shift (around 2.5° in latitude) of the position of the Antarctic Polar Front, the key king foraging area (Bost et al. 1997, 2009; Peron et al. 2012). As a consequence, the minimal distance from the colony to the limit of the front is predicted to double by 2099 during the incubation and brood, with a mean distance of approximately 800 kilometers during summer.

Long-term tracking data for foraging king penguins breeding at Crozet indicate that during "warm" years, the predicted foraging zones are much farther south and hence penguins increase their foraging range (Bost et al., in prep.). A maximal foraging range of 750 kilometers during brooding increases the trip duration to more than three weeks. Breeding and foraging data indicate that the probability of partner desertion drastically increases after a threshold of 22 days without relief. Thus, given the energetic constraints of the chick during brooding, the projected increase in traveling distance in the southern Indian Ocean will prevent successful chick rearing (Peron et al. 2012) and hence negatively affect demographic parameters and population sizes (Le Bohec et al. 2008). The projected situation appears similar for kings on Marion Island, and it remains to be seen whether they can adapt to these changing scenarios. Satellite-tracked Magellanic penguins (*Spheniscus magellanicus*) that made trips farther from the nest had lower reproductive success (Boersma and Rebstock 2009). Thus, the warming of the Southern Ocean represents a major threat for king penguins, especially in the southern Indian Ocean, without major adaptations in foraging strategies or timing of reproduction. These environmental changes may occur at time scales too short to allow rapid adaptation of this long-lived species.

10. RECOMMENDED PRIORITY RESEARCH ACTIONS FOR CONSERVATION

The following research actions are a priority for conservation:

1. Determine the marine habitats used, especially during winter, in the first years of life and for nonbreeding adults. Extensive short- and long-term data sets on the foraging ecology and habitat use of breeding birds are already available during the summer months for some key localities (Crozet, Marion, Kerguelen, and South Georgia). More information is needed on breeding birds during the austral winter, when they disperse widely (Pütz 2002; Bost et al. 2004). Finally, place special emphasis on determining the trophic niche of nonbreeding birds. This group includes not only nonbreeding adults and failed breeders but also juveniles (during their first year at sea) and immature birds (three to four years old).

2. Model the foraging habitat to determine the most relevant physical and biotic parameters determining the foraging distribution. Pay special attention to the southernmost breeding locations, because of predicted rise in water temperature in the Southern Ocean.

3. Project foraging responses in marine habitats in light of future changes in sea surface temperature. Analysis of contrasted foraging responses during warm and cold years could inform a modeling approach for projecting penguins' at-sea distribution given variation in sea surface temperature.

4. Complete a census of all breeding populations at regular intervals in order to detect temporal trends and potential interactions with environmental changes.

5. Reconsider the taxonomic status of the different populations through genetic analysis. The results of combined genetic, morphological, and behavioral analysis may help clarify the taxonomic status of the different populations.

11. CURRENT CONSERVATION EFFORTS

Access to all king breeding sites is restricted, and permits from the respective national authorities are required for research and conservation purposes. Ecotourism, where it occurs, is strictly regulated. Procedures for preventing the outbreak of diseases at sub-Antarctic islands (Kerry et al. 1999) are generally applied. At Crozet and Kerguelen Islands, all the king colonies are under protection and included in the Réserve Naturelle Nationale des Terres Australes et Antarctiques Françaises, a natural reserve. At Macquarie Island, the colonies are included in the World Network of Biosphere Reserves under UNESCO's Man and the Biosphere Programme. The colonies of Marion Island and Prince Edward Island are part of a Special Nature Reserve under the South African Environmental Management: Protected Areas Act. South Georgia penguins live in a special protected area within the Environmental Management Plan for South Georgia. In the Falklands/Malvinas, all wildlife is protected under the Conservation of Wildlife and Nature Bill 1999.

12. RECOMMENDED PRIORITY CONSERVATION ACTIONS FOR INCREASING POPULATION RESILIENCE AND MINIMIZING THREATS AND IMPACTS

1. Implement marine protected areas (MPAs): Because of the imminent risk of the development of a commercial fishing industry targeting myctophid fish (the main king penguin prey), conservation efforts should focus on the development of marine sanctuaries or other MPA regimes in waters surrounding the respective breeding sites. Management of potential myctophid fisheries should take into account the periods when penguins' energy requirements are highest in order to limit any impact of prey depletion on the foraging range, namely the post-molt (November) and the brooding and early crèche (i.e., mid-January to mid-March) stages. At present, three king breeding sites are included in two recently established MPAs: Macquarie Island and Heard Island. Three other MPAs are planned around Marion and Prince Edward Islands, the Crozet and Kerguelen archipelagoes, and South Georgia Island.

2. Establish high-seas MPAs: Existing and planned MPAs encompass only part of the kings' oceanic habitat and do not include the main feeding areas. Establishing high-seas MPAs in key feeding areas outside national jurisdiction is key to protecting oceanic predators such as king penguins.

3. Restrict access to colonies: Ecotourism activities and access to the colonies must be controlled, especially within the context of possible disease outbreaks in sub-Antarctic localities.

ACKNOWLEDGMENTS

We are especially indebted to the Institut Polaire Français Paul Emile Victor (IPEV) for financial support of Antarctic research programs (IPEV Prog. 109: resp. H. Weimerskirch; IPEV Prog. 394: resp. C. A. Bost). The Terres Australes et Antarctiques Françaises also provided logistical support. We would like to thank P. N. Trathan for additional information. We thank the volunteers and colleagues who have enabled us to undertake this work and give our special thanks to all the colleagues and volunteers involved in research on the ecology of top predators in the Terres Australes et Antarctiques Françaises.

TABLE 1.1 Biometry of adult king penguins according to sex and locality

	MEAN	SD	RANGE	N	LOCALITY	REFERENCES
MALES						
Flipper length	360		347–373	23	Crozet	Barrat 1976
	343		321–379	70	South Georgia	Stonehouse 1960
Bill length	125[a]		117–132	23	Crozet	Barrat 1976
	137		123–149	70	South Georgia	Stonehouse 1960
	114.9[a]	3.4	110.5–118.8	5	Macquarie	Marchant and Higgins 1990
Foot length	185		170–200	70	South Georgia	Stonehouse 1960
Body mass						
start of courtship[1]	12.8		10.5–15.7	7	Crozet	Barrat 1976
start of courtship[2]	13.9		11.9–15.6	20	Crozet	Gauthier-Clerc et al. 2001
start of courtship[3]	11.6	0.2	9.3–13.1	6	Crozet	Gauthier-Clerc et al. 2001
start of courtship[1]	16.0	0.7	13.8–17.3	8	South Georgia	Stonehouse 1960
chick rearing*	10.8	1.1	7.8–12.8	26	Crozet	Bost, et al., unpubl.
chick rearing*	11.1	0.6	9.9–12.5	15	Kerguelen	Bost, et al., unpubl.
FEMALES						
flipper length	353		335–378	22	Crozet	Barrat 1976
	331		310–355	59	South Georgia	Stonehouse 1960
bill length	119[a]		109–127	22	Crozet	Barrat 1976
	129[a]		116–142	59	South Georgia	Stonehouse 1960
	114.1[a]	2.9	109.6–118.5	5	Macquarie	Marchant and Higgins 1990
Foot length	178		160–202	59	South Georgia	Stonehouse 1960
Body mass						
start of courtship[1]	11.5		9.3–12.5	10	Crozet	Barrat 1976
start of courtship[1]	14.3		13.2–6.2	11	South Georgia	Stonehouse 1960
chick rearing*	9.9	0.8	8.4–11.6	16	Crozet	Bost, et al., unpubl.
chick rearing*	10.3	0.6	9.3–11	6	Kerguelen	Bost, et al., unpubl.
UNSEXED BIRDS						
Flipper length	337	12		129	South Georgia	Stonehouse 1960
	325	10		51	Kerguelen	Viot 1987
	378	3		50	Crozet	Viot 1987
Bill length	92.3[β]	6.7		50	Crozet	Viot 1987
	87.6[β]	4.9		51	Kerguelen	Viot 1987

The measurements are in mm and the body mass in kg. Flippers were measured from the axilla to the body side; bill length from tip to mandibular plate excepted for mensurations indicated as α (from the base of mandibular plate to the tip) and β (culmen length). 1: unknown breeding success; 2: males successfully relieved during incubation; 3: males deserted during incubation; *: brooding period, relieved breeders (February–early March, 1 year).

TABLE 1.2 Estimated numbers of king penguin incubating and status of colonies monitored

AREA	LOCATION	ANNUAL BREEDING PAIRS	YEAR	STUDY PERIOD	LONG TERM TREND	POPULATION STATUS	SOURCES
Atlantic Ocean	South Georgia	c. 450,000	2002	1986–2002	Increasing	Increasing	Saudner 2006 Woehler and Croxall 1997 P.A. Prince and S. Poncet u/p
	Fakland Is.	c.1,000	2010	1980–2010	Increasing	Increasing	Clausen and Huin 2003 Pütz, unpubl. data
	South Sandwich Is.	1	1995	–	Unknown	Unknown	P. Harrison in Prince and Croxall 1996
Indian Ocean	Prince Edward & Marion Is.	65,000–70,000					
	Prince Edward Is.	2,000	2008	1952–2009	Increasing	Stable	Crawford et al. 2003 Crawford et al. 2009
	Marion Is.	65,000	2008	1952–2009	Increasing	Stable	Crawford et al. 2003 Crawford et al. 2009
	Crozet Is.	611,700–735,700					
	Île de la Possession	79,700	2009	1962–2003	Increasing	Increasing	Weimerskirch et al. 1992 Delord et al. 2004, Weimerskirch et al. unpublished data
	Île de l'Est	100,000	1984	1970, 1984	Stable	Unknown	Despin et al. 1972 Jouventin et al. 1984
	Île aux Cochons	432,000–556,000	1988	1962–1998	Increasing	Unknown	Guinet et al. 1995
	Kerguelen Is.	342,000*	1999			Unknown	See Chamaillé-James et al. 2000
	1. Courbet Peninsula (East)						
	Ratmanoff	52,000	2008	1963–2008	Increasing	Declining	Chamaillé-James et al. 2000 Weimerskirch et al. 1989 Weimerskirch et al. unpublished data
	Cap Digby	72,600	1999	1963–1999	Increasing	Unknown	Chamaillé-James et al. 2000
	2. Rallier du Baty Peninsula (South)						
	Baie Larose	21,400	1985	1973–1985	Increasing	Unknown	Chamaillé-James et al. 2000
	Feu de Joie	40,000	1987		Unknown	Unknown	Chamaillé-James et al. 2000
	Tellurometre	20,000	1982		Unknown	Unknown	Chamaillé-James et al. 2000
	Heard Is.	80,000	2003/04	1963–1993	Increasing	Increasing	Woehler 2006
	Macquarie Is.	150,000–170,000	2000	1930–2000	Increasing	Increasing	Parks and Wildlife Service 2006 Van Den Hoff et al. 2009

TABLE 1.3 King penguin egg size measurements according to the breeding locality

LOCALITY	EGG LENGTH	EGG BREADTH	EGG MASS	REFERENCES
South Georgia Is.	104.8(86–117) (75)	76.0(64–86) (291)	319 (205–440)	Stonehouse 1960, Murphy 1936, Williams 1995
Crozet Is.	104.1(92.9–124.0)(291)	73.9 (61.2–87.0) (291)	302 (235–380) (186)	Barrat 1976
Marion Is.	105.8 (93.8–112.5) (35)	74.0 (65.2–79.0)(35)	304 (243–351)(16)	From Rand, 1954; Barrat 1976
Macquarie Is.	109.7 (91–106)(15)	73.9 (69.5–78)(15)	–	From Barrat 1976

Sample size is the last value on the right side inside each block

TABLE 1.4 Main breeding and demographic parameters of king penguin

PARAMETERS	POSSESSION IS. (CROZET)	SOUTH GEORGIA	REFERENCE
Incubation (d)	53.8 +1.5	54–55	Barrat 1976; Stonehouse 1960; Williams 1995
Chick rearing (d)	1971 : 350 (n=3) 1997–2000: 324 ± 5d n=24)	300–390; 313	Barrat 1976; Stonehouse 1960; Croxall and Prince 1987; Descamps et al. 2002
Average reproductive success	1986–1988: 30.6% 1997–2001: 20–34%	84 % (n=1 year)	Stonehouse 1960; Weimerskirch et al. 1992; Descamps et al. 2002, Gauthier-Clerc et al. 2004
Age of first breeding (yr)	3	–	

REFERENCES

Adams, N. J., and N. T. Klages. 1987. Seasonal variation in the diet of the king penguin (*Aptenodytes patagonicus*) at sub-Antarctic Marion Island. *Journal of Zoology* (London) 212:303–24.

Ainley, D. G., E. F. O'Connor, and R. J. Boekelheide. 1984. *The Marine Ecology of Birds in the Ross Sea, Antarctic.* Ornithological Monograph No. 32. 109 pp.

Barrat, A. 1976. Quelques aspects de la biologie et de l'écologie du manchot royal (*Aptenodytes patagonicus*) des iles Crozet. *Comité National Français de la Recherche Antarctique* 40:9–51.

Boersma, P. D., and G. A. Rebstock. 2009. Foraging distance affects reproductive success in Magellanic penguins. *Marine Ecology Progress Series* 375:263–75.

Bost, C. A., J. B. Charrassin, Y. Clerquin, Y. Ropert-Coudert, Y. Le Maho. 2004. Exploitation of the marginal ice zone by king penguins during winter. *Marine Ecology Progress Series* 283:293–97.

Bost, C. A., C. Cotté, F. Bailleul, Y. Cherel, J. B. Charrassin, C. Guinet, D. G. Ainley, and H. Weimerskirch. 2009. Importance of Southern Ocean fronts for seabird and marine mammals. J. Marine Systems. Special Issue on Processes at Oceanic Fronts of the *Journal of Marine Systems* (JMS-SIOF) 79:363–76.

Bost, C.A., C. Cotté, P. Terray, C. Barbraud, K. Delord, C. Guinet, J. B. Charrassin, H. Weimerskirch. In prep. Tracking penguins reveals impacts of large-scale climate variability on food webs.

Bost, C. A., J. Y. Georges, C. Guinet, Y. Cherel, K. Pütz, J. B. Charrassin, Y. Handrich, T. Zorn, J. Lage, and Y. Le Maho. 1997. Foraging habitat and food intake of satellite-tracked king penguins during the austral summer at Crozet archipelago. *Marine Ecology Progress Series* 150:21–33.

Bost, C. A., Y. Handrich, P. J. Butler, A. Fahlman, L. G.Halsey, A. J. Woakes, and Y. Ropert-Coudert. 2007. Change in dive profiles as an indicator of feeding success in king and Adélie penguins. *Deep-Sea Research II*, 54:248-55.

Bost, C. A., T. Zorn, Y. Le Maho, and G. Duhamel. 2002. Feeding of diving predators and diel vertical migration of prey: King penguins' diet versus trawl sampling at Kerguelen Islands. *Marine Ecology Progress Series* 227:51–62.

Chamaillé-James, S., C. Guinet, F. Nicoleau, and M. Argentier. 2000. A method to assess population changes in king penguins: The use of a Geographical Information System to estimate area-population relationships. *Polar Biology* 23:545–49.

Charbonnier, Y., K. Delord, and J. B. Thiebot. 2010. King-size fast food for Antarctic fur seals. *Polar Biology* 33:721–24.

Charrassin, J. B., and C. A. Bost. 2001. Utilisation of the oceanic habitat by king penguins over the annual cycle. *Marine Ecology Progress Series* 22:285–97.

Charrassin, J. B., C. A. Bost, K. Pütz, J. Lage, T. Dahier, and Y. Le Maho. 1999. Changes in depth utilisation in relation to the breeding stage: A case study with the king penguin *Aptenodytes patagonicus*. *Marine Ornithology* 27:43–47.

Charrassin, J. B., C. A. Bost, K. Pütz, J. Lage, T. Dahier, T. Zorn, and Y. Le Maho. 1998. Foraging strategies of incubating and brooding king penguins *Aptenodytes patagonicus*. *Oecologia* 114:194–201.

Charrassin, J. B., Y. Le Maho, and C. A. Bost. 2002. Seasonal changes in the diving parameters of king penguins. *Marine Biology* 141:581–89.

Cherel, Y., J. B. Charrassin, and E. Challet. 1994. Energy and protein requirements for molt in the king penguin *Aptenodytes patagonicus*. *American Journal of Physiology* 266:R1182–88.

Cherel, Y., J. B. Charrassin, and Y. Handrich. 1993. Comparison of body reserve buildup in prefasting chicks and adults of king penguins (*Aptenodytes patagonicus*). *Physiological Zoology* 66:750–70.

Cherel, Y., K. Pütz, and K. A. Hobson. 2002. Summer diet of king penguins (*Aptenodytes patagonicus*) at the Falkland Islands, southern Atlantic Ocean. *Polar Biology* 25:898–906.

Cherel, Y., and V. Ridoux. 1992. Prey species and nutritive value of food fed during summer to king penguin *Aptenodytes patagonica* chicks at Possession Island, Crozet Archipelago. *Ibis* 134:118–27.

Cherel, Y., V. Ridoux, and P. G. Rodhouse. 1996. Fish and squid in the diet of king penguin chicks *Aptenodytes patagonicus* during winter at sub-Antarctic Crozet Islands. *Marine Biology* 126:559-70.

Cherel, Y., J. C. Stahl, and Y. Le Maho. 1987. Ecology and physiology of fasting in king penguin chicks. *The Auk* 104:254–62.

Clausen, A. P., and N. Huin. 2003. Status and numerical trends of king, gentoo and rockhopper penguins breeding in the Falkland Islands. *Waterbirds* 26(4):389–402.

Collins, M. A., J. C. Xavier, N. M. Johnston, A. W. North, P. Enderlein, G. A. Tarling, C. M. Waluda, E. J. Hawker, N. J. Cunningham. 2008. Patterns in the distribution of myctophid fish in the northern Scotia Sea ecosystem. *Polar Biology* 31:837–51.

Conroy, J. W. H. 1975. Recent increases in penguin populations in the Antarctic and the Sub-Antarctic. In *The Biology of Penguins*, ed. B. Stonehouse, 321–36. London: Macmillan.

Conroy, J. W. H., and M. G. White. 1973. The breeding status of the king penguin. *British Antarctic Survey Bulletin.* 32:31–40.

Cooper, J., R. J. M. Crawford, M. S. De Villiers, B. M. Dyer, G. J. G. Hofmeyr, and A. Jonker. 2009. Disease outbreaks among penguins at sub-Antarctic Marion Island: A conservation concern. *Marine Ornithology* 37:193–96.

Crawford, R. J. M., J. Cooper, B. M. Dyer, M. D. Greyling, N. T. W. Klages, P. G. Ryan, S. L. Petersen, L. G. Underhill, L. Upfold, W. Wilkinson, M. S. De Villiers, S. Du Plessis, M. Du Toit, T. M. Leshoro, A. B. Makhado, M. S. Mason, D. Merkle, D. Tshingana, V. L. Ward, and P. A. Whittington. 2003. Populations of surface-nesting seabirds at Marion Island, 1994/95–2002/03. *African Journal of Marine Science* 25:427–40.

Crawford, R. J. M., P. A. Whittington, L. Upfold, P. G. Ryan, S. L. Petersen, B. M. Dyer, and J. Cooper. 2009. Recent trends in numbers of four species of penguins at the Prince Edward Islands. *African Journal of Marine Science* 31(3):419–26.

Croxall, J. P., and P. A. Prince. 1987. Seabirds as predators on marine resources, especially krill, at South Georgia. In *Seabirds: Feeding Ecology and Role in Marine Ecosystems*, ed. J. P. Croxall, 347–68. Cambridge: Cambridge University Press.

Delord, K., C. Barbraud, and H. Weimerskirch. 2004. Long-term trends in the population size of king penguins at Crozet archipelago: Environmental variability and density dependence? *Polar Biology* 27:793–800.

Descamps, S., M. Gauthier-Clerc, J. P. Gendner and Y. Le Maho. 2002. The annual breeding cycle of unbanded king penguins *Aptenodytes patagonicus* on Possession Island (Crozet). *Avian Science* 2:87–98.

Despin, B., J. L. Mougin, and M. Segonzac. 1972. Oiseaux et mammifères de l'Île de l'Est. *Comité National Français de la Recherche Antarctique* 31:1–106.

Du Plessis, C. J., Y. M. van Heezik, and P. J. Seddon. 1994. Timing of king penguin breeding at Marion Island. *Emu* 94:216–19.

Fordyce, R. C., and C. M. Jones. 1990. Penguin history and new fossil material from New Zealand. In *Penguin Biology*, ed. L. S. Davis and J. T. Darby, 551–648. San Diego, CA: Academic Press.

Gauthier-Clerc, M., J. P. Gendner, C. A. Ribic, W. R. Fraser, E. J. Woehler, S. Descamps, C. Gilly, C. Le Bohec and Y. Le Maho. 2004. Long-term effects of flipper bands on penguins. *Proceedings of the Royal Society B* 271:423–26.

Gauthier-Clerc, M., Y. Le Maho, J. P. Gendner, J. Durant and Y. Handrich. 2001. State-dependent decisions in long-term fasting king penguins, Aptenodytes patagonicus, during courtship and incubation. *Animal Behaviour* 62:661–69.

Gray, G. R. 1844. Aptenodytes. *Annals and Magazine for Natural History* 13:315.

Guinet, C., P. Jouventin, and J. Malacamp. 1995. Satellite remote sensing in monitoring change of seabirds: Use of Spot Image in king penguin population increase at Ile aux Cochons, Crozet Archipelago. *Polar Biology* 15:511–15.

Hindell, M. A. 1988. The diet of the king penguin *Aptenodytes patagonicus* at Macquarie Island. *Ibis* 130:193–203.

Hofmeyr, G. J. G., and M. M. Bester. 1993. Predation on king penguins by Antarctic fur seals. *South African Journal of Antarctic Research* 23:71–74.

Hunter, S. 1983. The food and feeding ecology of the giant petrels Macronectes halli and M. giganteus at South Georgia. *Journal of Zoology* 200(4):521–38.

IUCN (International Union for Conservation of Nature) 2011. IUCN Red List of Threatened Species. Version 2011.2. www.iucnredlist.org (accessed 15 March 2012).

Jiguet, F., and P. Jouventin. 1999. Individual breeding decisions and long-term reproductive strategy in the king penguin *Aptenodytes patagonicus*. *Ibis* 141 (3):428–33.

Jouventin, P. 1982. *Visual and Vocal Signals in Penguins: Their Evolution and Adaptative Characters*. Berlin: Paul Parey.

Jouventin, P., J. C. Stahl, H. Weimerskirch, and J. L. Mougin. 1984. The seabirds of the French subantarctic islands and Adélie Land, their status and conservation. In *Status and Conservation of the World's Seabirds*, ed. J. P. Croxall, P. G. H. Evans, and R. W. Schreiber, 609–25. ICBP Technical Publication No. 2. Cambridge: International Council for Bird Preservation.

Kerry, K., M. Riddle, and J. Clarke. 1999. Diseases of Antarctic wildlife. Scientific Committee on Antarctic Research (SCAR) and the Council of Managers of National Antarctic Programs (COMNAP). Kingston: Australian Antarctic Division.

Kooyman, G. L., Y. Cherel, Y. Le Maho, J. P. Croxall, P. H. Thorson, V. Ridoux, and C. A. Kooyman. 1992. Diving behavior and energetics during foraging cycles in king penguins. *Ecological Monographs* 62:143–63.

Kooyman, G. L, T. G. Kooyman, M. Horning, and C. A. Kooyman. 1996. Penguin dispersal after fledging. *Nature* 383:397.

Laws, R. J. 1977. Seals and whales of the Southern Ocean. *Philosophical Transactions of the Royal Society B* 279 London B: 81–96.

Le Bohec, C., J. Durant, M. Gauthier-Clerc, N. C. Stenseth, Y. H. Park, R. Pradel, D. Grémillet, J. P. Gendner, and Y. Le Maho. 2008. King penguin population threatened by Southern Ocean warming. *Proceedings of the National Academy Sciences of the United States of America* 105: 2493–97.

Le Bohec, C., M. Gauthier-Clerc, J. P. Gendner, N. Chatelain, and Y. Le Maho. 2003. Nocturnal predation of king penguins by giant petrels on the Crozet Islands. *Polar Biology* 26(9):587–90.

Le Bohec, C., M. Gauthier-Clerc, D. Grémillet, R. Pradel, A. Béchet, J. P. Gendner, Y. Le Maho. 2007. Population dynamics in a long-lived seabird: 1. Impact of breeding activity on survival and breeding probability in unbanded king penguins. *Journal of Animal Ecology* 76:1149–60.

Lewis Smith, R. I., and J. R. B. Tallowin. 1979. The distribution and size of king penguin rookeries on South Georgia. *British Antarctic Survey Bulletin* 49:259–76.

Marchant, S., and P. J. Higgins. 1990. *Handbook of Australian, New Zealand and Antarctic Birds*. Melbourne, Australia: Oxford University Press.

Moore, G. J., G. Robertson, and B. Wienecke. 1999. Seasonal change in foraging areas and dive depths of breeding king penguins at Heard Island. *Polar Biology* 21:376–84.

Moore, G. J., B. Wienecke, and G. Robertson. 1998. Food requirements of breeding king penguins at Heard Island and potential overlap with commercial fisheries. *Polar Biology* 20:393–402.

Murphy, R. C. 1936. *Oceanic Birds of South America*. Vol. 1. New York: American Museum of Natural History.

Olsson, O. 1996. Seasonal effect of timing and reproduction in the king penguin: A unique breeding cycle. *Journal of Avian Biology* 27:7–14.

Olsson, O., and A. Brodin. 1997. Changes in king penguin breeding cycle in response to food availability. *Condor* 99:994–97.

Olsson, O., and H. P. van der Jeugd. 2002. Survival in king penguins *Aptenodytes patagonicus*: Temporal and sex-specific effects of environmental variability. *Oecologia* 132(4):509–16.

Olsson, O., and A. W. North. 1997. Diet of the king penguin *Aptenodytes patagonicus* during three austral summers at South Georgia. *Ibis* 139:504–13.

Otley, H., A. Clausen, D. Christie, N. Huin, and K. Pütz. 2007. Breeding patterns of king penguins on the Falkland Islands. *Emu* 107:156–64

Parks and Wildlife Service. 2006. Macquarie Island Nature Reserve and World Heritage Area Management Plan. Hobart, Tasmania: Parks and Wildlife Service. 189 pp.

Peron, C., H. Weimerskirch, and C. A. Bost. 2012. Projected poleward shift of king penguins' (*Aptenodytes patagonicus*) foraging range at the Crozet Islands, southern Indian Ocean. *Proceedings of the Royal Society B* 279:2515–23.

Prince, P. A., and J. P. Croxall. 1996. The birds of South Georgia. *Bulletin of the British Ornithologists' Club* 116:81–104.

Pütz, K. 2002. Spatial and temporal variability in the foraging areas of breeding king penguins. *Condor* 104(3):528–38.

Pütz, K., and C. A. Bost. 1994. Feeding behavior of free-ranging king penguins (*Aptenodytes patagonicus*). *Ecology* 75 (2):489–97.

Pütz, K., and Y. Cherel. 2005. The diving behaviour of brooding king penguins (*Aptenodytes patagonicus*) from the Falkland Islands: Variation in dive profiles and synchronous underwater swimming provide new insights into their foraging strategies. *Marine Biology* 147:281–90.

Pütz, K., Y. Ropert-Coudert, J. B. Charrassin, and R. P. Wilson. 1999. Foraging areas of king penguins (*Aptenodytes patagonicus*) breeding at Possession Island, southern Indian Ocean. *Marine Ornithology* 27:77–84.

Pütz, K., R. P. Wilson, J. B. Charrassin., T. Raclot, J. Lage, Y. Le Maho, M. A. M. Kierspel, B. Culik, and D. Adelung. 1998. Foraging strategy of king penguins (*Aptenodytes patagonicus*) during summer at the Crozet Islands. *Ecology* 79(6):1905–21.

Rand, R. W. 1954. Notes on the birds of Marion island. *Ostrich* 26(2):57–69.

Rounsevell, D. E., and G. R. Copson. 1982. Growth rate and recovery of a king penguin, *Aptenodytes patagonicus*, population after exploitation. *Australian Wildlife Research* 9:519–25.

Sabourenkov, E. N. 1991. Mesopelagic fish of the southern ocean—summary results of recent Soviet studies. Commission for the Conservation of Antarctic Marine Living Resources, Hobart, Australia, SC-CAMLR-SSP/7: 433–57.

Sanders, S. 2006. *Important Bird Areas in the United Kingdom Overseas Territories: Priority Sites for Conservation.* Sandy, Bedfordshire, England: Royal Society for the Protection of Birds.

Stonehouse, B. 1960. The king penguin *Aptenodytes patagonicus* at South Georgia. 1: Breeding behaviour and development. *Scientific Report of the Falkland Islands Dependent Survey* 23:1–81.

Trathan, P. N., C. Bishop, G. Maclean, P. Brown, A. Fleming, M. A. Collins. 2008. Linear tracks and restricted temperature ranges characterise penguin foraging pathways. *Marine Ecology Progress Series* 370:285–94.

Van Den Hoff, J., R. J. Kirkwood, and P. B. Copley. 1993. Aspects of the breeding cycle of king penguins *Aptenodytes patagonicus* at Heard Island. *Marine Ornithology* 21:49–55.

Van den Hoff, J., C. R. McMahon, and I. Field. 2009. Tipping back the balance: Recolonization of the Macquarie Island isthmus by king penguins (*Aptenodytes patagonicus*) following extermination for human gain. *Antarctic Science* 21 (3):237–41.

Van Heezlg, Y., P. J. Seddon, J. Cooper, and A. L. Plös. 1994. Interrelationships between breeding frequency, timing and outcome in king penguins *Aptenodytes patagonicus:* Are king penguins biennial breeders? *Ibis* 136(3):279–84.

Verheyden, C., and P. Jouventin. 1991. Over-wintering strategies of the Lesser Sheathbill *Chionis minor* in an impoverished and insular environment. *Oecologia* 86:132–39.

Viot, C. R. 1987. Différenciation et isolement entre populations chez le manchot royal (*Aptenodytes patagonicus*) et le manchot papou (*Pygoscelis papua*). *L'Oiseau et la Revue Française d'Ornithologie* 57:251–59.

Weimerskirch, H., J. C. Stahl, and P. Jouventin. 1992. The breeding biology and population dynamics of king penguins *Aptenodytes patagonicus* on the Crozet Islands. *Ibis* 134:107–17.

Weimerskirch, H., R. Zotier, and P. Jouventin. 1989. The avifauna of the Kerguelen Islands. *Emu* 89:15–29.

Wienecke, B., and G. Robertson. 2006. Comparison of foraging strategies of incubating king penguins *Aptenodytes patagonicus* from Macquarie and Heard Islands. *Polar Biology* 29:424–38.

Williams, A. J., W. R. Siegfried, A. E. Burger, and A. Berruti. 1979. The Prince Edward Islands: A sanctuary for seabirds in the Southern Ocean. *Biological Conservation* 15:59–71.

Williams, T. D. 1995. *The Penguins.* Oxford: Oxford University Press.

Woehler, E. J. 2006. Status and conservation of the seabirds of Heard Island. In *Heard Island–Southern Ocean Sentinel*, ed. K. Green and E. J. Woehler, 128–65. Chipping Norton, NSW, Australia: Surrey Beatty & Sons.

Woehler, E. J., and J. P. Croxall. 1997. The status and trends of Antarctic and Subantarctic seabirds. *Marine Ornithology* 25:43–66.

Emperor Penguin
(Aptenodytes forsteri)

Barbara Wienecke, Gerald Kooyman, and Yvon Le Maho

1. SPECIES (COMMON AND SCIENTIFIC NAMES)

Emperor penguin, *Aptenodytes forsteri*

2. DESCRIPTION OF THE SPECIES

Emperor penguins are the largest member of the family Spheniscidae both in body mass and in height.

ADULT. Males and females are similar in appearance, although males tend to be slightly larger than females. The auricular patches are yellow, and the color is almost orange near the ear, becoming nearly white where the patch meets the neck, with a wash of light yellow across the chest. The auricular patches are poorly defined compared to those of the king penguin, and the upper chest is washed in light yellow. The upper and lower bills are black, and the mandibular plates are yellow and, in mature adults in good condition, may be nearly orange. The iris is dark brown. The back from the nape to the tail and the upper part of the flippers are dark gray to black, and the underside of the body, including the underside of the flippers, is white. The legs are densely covered in feathers, and the feet are black (fig. 2).

JUVENILE. Head coloration changes gradually over the first five years of life. In first-year birds, the chin is white, the auricular patches are white-gray, and the beak is black. As the birds mature, the kidney-shaped auricular patches

FIG. 1 (*FACING PAGE*) Emperor penguins have only one chick per season and are the only species to incubate their eggs during the Antarctic winter. (J. Weller)

FIG. 2 The main diagnostic feature of the emperor penguin, the largest of the penguins, is head coloration, which distinguishes it from the smaller king penguin. (J. Weller)

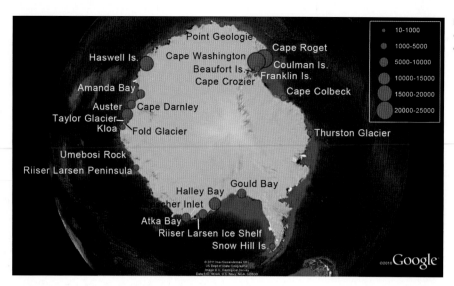

FIG. 3 Distribution and abundance of the emperor penguin, with counts based on pairs.

•	10-1000
●	1000-5000
●	5000-10000
●	10000-15000
●	15000-20000
●	20000-25000

become increasingly yellow. The mandibular plates on the lower bill gradually turn pink to orange.

CHICK. Chicks are covered in light gray down, but their heads are black with large white patches around the eyes that extend to the chin (fig. 1).

MORPHOMETRIC DATA. See table 2.1.

3. TAXONOMIC STATUS

Emperor penguins were probably first sighted on the second voyage of Captain James Cook (1773–75). On his mission to discover Terra Australis Incognita, or the Unknown Land of the South, Cook reached latitudes far enough south that he encountered emperor penguins on several occasions. The naturalist on board the *Resolution*, Johann Reinhold Forster, failed to notice that he had come across a new species of penguin. Instead, he described them as *Aptenodytes patachonica*, then the accepted name for king penguins.

It was only in 1844 that George Robert Gray, head of the ornithology section of the British Museum in London, separated emperors from kings (*A. patagonicus*), the closest relatives of emperor penguins, and classified them as a distinct species. Gray also reviewed the nomenclature of all penguin species known at the time and called the emperor *A. forsteri* in honor of Cook's naturalist. Incidentally, Gray also revised the name of kings from *A. patachonica* to *A. pennatii* in order to avoid further confusion (e.g., Wienecke 2010).

Recent phylogenetic work indicates a common ances-

try for penguins and albatrosses 62.4–77.3 million years ago. The genus *Aptenodytes* diverged around 40 million years ago, followed by *Pygoscelis* some 38 million years ago (Baker et al. 2006). There are no known subspecies of emperors.

4. RANGE AND DISTRIBUTION

Emperor colonies occur around the coast of Antarctica (fig. 3). While numerous sightings have been reported since the early 20th century, until 2012 the exact number of colonies was unknown; the status of most is uncertain (e.g., Wienecke 2010, 2009). Satellite imagery from the 2009 breeding season confirmed the existence of 37 out of 38 colonies found in the previous satellite survey and detected 4 new ones (Fretwell et al. 2012). Ground truthing is needed to learn more about the colonies, particularly in areas of Antarctica that are not frequently visited (Fretwell and Trathan 2009). Currently, 45 colonies are known to be occupied, 1 colony has been abandoned (Emporer Island, Dion Islands). The Ledda Bay colony was identified in satellite imagery in 1999 but was not found in 2009 because there was no fast ice at the site at the time the satellite images were taken (Fretwell et al. 2012). Locations of emperor colonies are listed in table 2.2.

Juvenile emperors regularly venture north far beyond their usual distribution (Kooyman and Ponganis 2008; Wienecke et al. 2010). In the past, juveniles were sighted on sub-Antarctic islands, such as Heard Island (Downes et al. 1959) and Macquarie Island (Copson and Brothers 2008). More recently, on June 21, 2011, a juvenile emperor penguin was spotted on Peka Peka Beach in

New Zealand (Miskelly 2011). The only other recorded sighting of an emperor penguin in New Zealand was in 1967 (Perry 2011; *New Zealand Herald* 2011).

5. SUMMARY OF POPULATION TRENDS

Due to the remoteness of most emperor colonies, information on population trends is available for only a few locations (see Wienecke 2011). The current global population estimate of about 238,000 breeding pairs (Fretwell et al. 2012) nearly doubled the previous population estimate, but this increase is due to better-informed survey methods rather than an actual increase in the number of breeding pairs. At the longest ongoing study at Pointe Géologie (66.65° S, 140.02° E), regular population estimates started with its discovery in November 1950 and continued with few interruptions. The first estimate, made in June 1951, put the population size at 5,000–5,500 individuals (Cendron 1952). The population remained stable from the 1950s to the late 1970s and then decreased significantly (Barbraud and Weimerskirch 2001). Since 1982, the population appears to have remained stable at about 3,000 pairs but has not recovered to its numbers at discovery (Jenouvrier et al. 2005; Barbraud et al. 2011).

The emperor colony at Haswell Island (66.55° S, 92.67° E) has been monitored intermittently. In the 1960s–70s, the population was estimated to comprise about 7,000–9,000 pairs (14,000–18,000 adults), but it appears to have decreased to approximately 2,900–4,500 pairs (5,700–9,000 adults) in the 1990s–early 2000s (Antarctic Treaty Secretariat 2006; Barbraud et al. 2011).

The Ross Sea region is unusual in that regular population counts exist for most of the known colonies there. Populations have fluctuated markedly, especially in the smaller colonies, but on a regional level, the population appears to be stable (Barber-Meyer et al. 2007).

One of the smallest known colonies in the western region of the Antarctic Peninsula was located at Emperor Island, Dion Islands (67.87° S, 68.72° W) (Stonehouse 1953). With 150–180 pairs, it appeared to be a marginal colony even when first sighted. Now it may have all but disappeared.

The colony at Taylor Glacier (67.47° S, 60.88° E) was monitored intermittently from the late 1950s until the early 1980s. Since 1988, annual censuses have shown that it has been relatively stable, with approximately 3,000 pairs (G. Robertson and B. Wienecke, unpubl. data).

6. IUCN STATUS

The International Union for Conservation of Nature lists the emperor in the Near Threatened category on its Red List of Threatened Species (BirdLife International 2012; IUCN 2012). This status is based on the penguins' large distribution range around the Antarctic continent, as well as the possibly large population size.

The species was uplisted because the population may experience moderately rapid declines caused by the effects of climate change, especially decreases in sea ice concentration and thickness.

LEGAL STATUS. Internationally, emperor penguins, like all native Antarctic fauna, and their entire breeding area are protected under Annex II to the Protocol on Environmental Protection to the Antarctic Treaty, adopted by the Antarctic Treaty nations during the XI-4 Special Antarctic Treaty Consultative Meeting in 1991 and recently revised (Antarctic Treaty Secretariat 1991 and 2009). However, while the annex sets out measures intended to minimize or avoid disturbance to concentrations of birds, it allows "taking" them for scientific purposes (with the proper permits) as long as "only small numbers of native mammals or birds are killed, and in no case more are killed from local populations than can, in combination with other permitted takings, normally be replaced by natural reproduction in the following season" (Antarctic Treaty Secretariat 2009). This holds true for populations in Antarctic Specially Protected Areas as well, as long as the taking is for compelling scientific purposes. The treaty does not mention individual species.

The U.S. Fish and Wildlife Service recently added a number of penguin species to the Endangered Species Act. Emperor penguins were not included because of their supposedly stable population and the insufficient accuracy of current climate models that predict significant changes to the emperors' habitat (USFWS 2008).

Discussions are currently under way in Australia about adding emperors to the species covered by the Environment Protection and Biodiversity Conservation Act 1999 (Commonwealth of Australia 2007).

7. NATURAL HISTORY

The first emperor penguin colony to be visited was at Cape Crozier (77°29′ S, 169°34′ E) during the Terra Nova Expedition (1911–13) under the command of Captain

James Clark Ross (Wilson 1907). An increasing number of emperor breeding colonies was discovered with the establishment of permanently occupied research stations on the Antarctic continent and the introduction of fixed-wing aircraft during the International Geophysical Year (1957–58). However, most colonies were too remote for ongoing studies. The noticeable exception was Pointe Géologie (66°39′ S, 140°01′ E), where the emperor colony is a short walk from the station. The number of penguins attending this colony in winter was first estimated in June 1951 (Cendron 1952), and the birds' breeding biology and ecology were investigated in 1952 (Prévost 1961). Much of what is known about the species is based on research at Dumont d'Urville, which commenced in 1956 (e.g., Prévost and Sapin-Jaloustre 1965; Mougin and van Beveren 1979). The transferability of the information to other colonies is unknown due to the lack of similar studies conducted elsewhere in Antarctica.

Most known colonies are located on landfast sea ice, but three are settled on land, and at least one is atop an iceberg (Wienecke 2012). Hence, the onset of breeding at most colonies is dependent on the formation of fast ice. Another determinant of the penguins' arrival time at their breeding location is the extent of fast ice, that is, the distance they have to traverse from the edge of the pack ice to their breeding sites (fig. 4). This distance is highly variable both spatially and temporally. The most expansive fast ice forms in the Weddell Sea, where it can stretch up to 400 kilometers north. In comparison, in large sections of East Antarctica, the fast ice is usually only about 100 kilometers wide; in the Ross Sea, the extent of fast ice in the area of colonies is only a few kilometers. At Pointe Géologie, most emperors reach the colony in March. Courtship ensues, and the majority of pairs form within 24 hours after the males return to the colony (Bried et al. 1999). Copulation commences in mid-April; the earliest observed was on 15 April (Prévost and Sapin-Jaloustre 1965).

Given the time constraints (extended incubation and equivalent absence of the females from the colony), relaying lost eggs or attempting to breed later in the season is physically impossible. The timing of laying, although less synchronized than one might expect (about five weeks from early May until about 10 June [Prévost and Sapin-Jaloustre 1965]), is crucial to the success of a pair (table 2.3). A penguin arriving late at the colony may

find that its partner from the previous season has paired with a new partner. Also, females compete for males, as they outnumber males by about 10%. The divorce rate among emperors, at about 85%, is significantly higher than in many other penguin species (Isenmann 1971). Thus, emperors are serial monogamists, having only one breeding partner per season but many partners over a lifetime.

Since most colonies form on the fast ice, there is no nesting material, nor do pairs occupy individual territories. The male alone incubates the single egg, which is nestled on his feet and tucked into the brood pouch, a fold of bare skin forming a pocket that covers the egg (fig. 5).

DIET. The emperors' diet comprises a variety of fish, cephalopods, and krill; the most important prey species are the Antarctic silverfish (*Pleuragramma antarcticum*), up to 95% by mass, and Antarctic krill (*Euphausia superba*), about 75% by number (Klages 1989; Pütz 1995; Wienecke and Robertson 1997; Cherel and Kooyman 1998). The arrow squid (*Psychroteuthis glacialis*) also features prominently in the emperors' diet (e.g., Piatkowski and Pütz 1994).

PREDATORS. Because of the emperors' winter breeding habit, their eggs and very young chicks are safe from predators, as both skuas and giant petrels return to Antarctica only in early spring (i.e., well after hatching). Southern giant petrels (*Macronectes giganteus*) are capable of attacking and killing chicks, even quite advanced, healthy ones. South polar skuas (*Catharacta maccormicki*) sometimes attack small chicks separated from their subcolonies (fig. 6).

Possibly the major threats at sea are leopard seals (*Hydrurga leptonyx*) and killer whales (*Orcinus orca*). Little is known about the diet of these seals and whales, and the proportion of penguins in their diet appears rather low (Green and Williams 1986; Casaux et al. 2009).

MOLT. Molting takes place from December to March (Prévost and Sapin-Jaloustre 1965). The early molters are immature birds, either birds that have not been successful in their breeding attempt in a given season or subadults that have not bred before. Occasionally, adults return to their breeding colonies to molt, but many postbreeders appear to travel vast distances to their molt

FIG. 4 (*FACING PAGE*) Emperor penguins often follow each other as they walk or toboggan to their breeding colony. (B. Wienecke)

FIG. 5 An emperor penguin stretches, showing the bare skin of the vertical brood patch where the egg and chick are kept warm. (B. Wienecke)

FIG. 6 Dead emperor penguin chick serves as dinner for a south polar skua. (J. Weller)

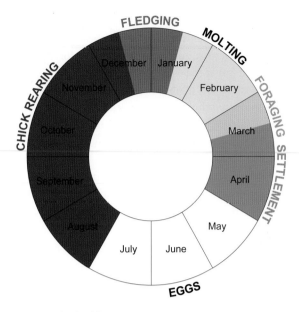

FIG. 7 Annual cycle of the emperor penguin.

locations (e.g., Kooyman et al. 2000; Wienecke et al. 2004). The molt lasts for 30–40 days (Groscolas 1978).

ANNUAL CYCLE. Emperors tend to return to their colonies in March when the sea ice is thick enough to support the colonies (fig. 7). Courtship and pair-bonding ensue over the following four to six weeks until the females lay their eggs in May to early June. Only the males incubate the eggs (53–55 days at Dion Islands [Stonehouse 1953]; 62–66 days at Pointe Géologie [Prévost 1961]), while the females depart the colony and forage in the pack-ice zone. In late July to early August, the chicks hatch and the females return, relieving the males from their long fast. Until the chicks are able to thermoregulate independently, they are brooded by their parents, who take turns caring for the chicks and hunting. From September onward, the energy requirements of the chicks are such that both parents need to supply them with food (fig. 8). In mid-December, the chicks start to fledge, and by early January, the overwhelming majority of emperors have abandoned their colonies. Most adults do not return to their breeding grounds for the annual molt. The juveniles fend for themselves and often travel immense distances away from their natal colonies. They are often not seen again until they are three to five years old.

8. POPULATION SIZES AND TRENDS

Based on a survey conducted using satellite imagery, Fretwell et al. (2012) estimate the total global population of emperor penguins to be about 238,000 breeding pairs. With the exception of the Pointe Géologie colony, access even to most known colonies is limited, and a number of breeding locations can be reached only relatively late in the chick-rearing period. Thus, it is difficult to estimate the number of pairs attending a colony in a given year. Satellite imagery can be used for remote counts.

The Ross Sea region is best studied in terms of the number of colonies that have been visited regularly over the past two decades (e.g., Barber-Meyer et al. 2007). Most other colonies have not been censused systematically or visited regularly. Table 2.4 summarizes the most recent census data available.

9. MAIN THREATS

The inaccessibility of many emperor colonies protects them from numerous human threats. Introduced predators and commercial fishing, for example, are not issues for this species. But the threats that juveniles face during their first few years at sea are generally still unknown. The large areas in which post-breeding adults in particular (Kooyman et al. 2000; Wienecke et al. 2004) and fledglings forage make the establishment of marine protected areas unsuitable as a conservation measure (Boersma and Parrish 1999) unless the entire Southern Ocean were to be declared an MPA.

The most serious threat—other than direct human disturbance and interference—is global warming. Increasing ocean temperatures are likely to change fast-ice extent and duration. A reduction in fast ice may be beneficial, as distances between emperor colonies and foraging areas are shortened, and it may reduce the time parent birds need to cross the ice (fig. 9) (e.g., Jenouvrier et al. 2009 and references therein). If prey becomes less abundant or shifts into different areas, foraging trips may be lengthened, reducing the penguins' reproductive success.

Changes in the timing of the breakout of the fast ice could seriously harm populations if it occurs before the chicks are ready to go to sea (fig. 10). A reduction in fast-ice extent also means that storm actions have a greater influence on the quality of the ice where the penguins breed. Hence, the fast ice may start to break out and disintegrate before the chicks are ready to fledge (Boersma 2008). This has disastrous consequences for the sea-ice-based colonies, as adults will continue to return to their traditional breeding grounds but will no longer be able

FIG. 8 Emperor penguin colonies are situated on ice. Here, young chicks wait for their parents to return to feed them on a sunny spring day. (B. Wienecke)

FIG. 9 Tracks on the ice show the arrival and departure of emperor penguins as they travel to feed their chicks. (B. Wienecke)

FIG. 10 By early December, chicks toboggan and walk to the sea. Their down wears away quickly once they are at sea. (J. Weller)

to rear their offspring to fledging. Male survival may also be affected by sea-ice extent (see Jenouvrier et al. 2005).

High carbon dioxide levels in the atmosphere cause marine waters to acidify. This changes the geochemical processes in the world's oceans, which in turn affect biological processes (e.g., McNeil and Matear 2008). An increase in pH reduces the ability of calcifying plankton to produce calcium carbonate (e.g., Riebesell et al. 2000). For example, Antarctic krill (*Euphausia superba*) require calcium carbonate for rebuilding their shells after the molt (e.g., Naczk et al. 1981; Nicol et al. 1992), but in an ocean that is becoming more acidic, calcium carbonate dissolves, putting at risk the long-term survival of krill and all species directly or indirectly dependent on them.

In terms of direct human impact, public interest in emperor penguins is strong, and many people wish to see them in their natural environment. Several tour operators fly visitors to emperor colonies, usually when the chicks are already advanced in their development and are relatively robust. There are no binding regulations on the tourism industry, which prefers to be self-regulating. In the absence of population data for most colonies, however, it is difficult to determine whether or not these visitations have an impact on the colonies. There is also no reporting requirement should unexpected events occur.

The international scientific community appears to have a growing interest in emperor penguins as well. Since scientists often interfere directly with their study subjects, they may do serious harm to individual penguins, as well as populations, if inadequate handling techniques and study methods are used.

FIG. 11 Emperor penguin chicks crèche for warmth and protection from predators such as giant petrels. Here, a pair of skuas flies overhead. (B. Wienecke)

10. RECOMMENDED PRIORITY RESEARCH ACTIONS FOR CONSERVATION

We recommend the following research actions for conservation of emperor penguins:

1. Develop international research and tourist visitation protocols. The social behavior of emperors—for example, huddling (fig. 11)—and the absence of territorial defense mean that these birds are sensitive to disturbance, and for such an emblematic species, it would be wise to develop an international protocol focused on reducing disturbance through new technologies.

2. Continue assessing the status of the global population of emperors using high-quality census data. As many colonies are inaccessible for most of the year, this calls for new methods and technologies. For example, automatic cameras (e.g., Newbery and Southwell 2009) and satellite technology might be useful in confirming that human disturbance is not a contributing factor. A number of colonies need to be selected that are considered representative of a region. The smaller colonies appear to experience much larger inter-annual changes in their population size than do larger colonies (see Barber-Meyer et al. 2007).

3. Avoid any invasive research during incubation and brooding. These are critical times in the breeding cycle of emperors, as disturbance can lead to major losses of eggs and young chicks and, hence, reduction in breeding success. Disturbance may also induce mortality in adult birds during the molt.

11. CURRENT CONSERVATION EFFORTS

The inaccessibility of many emperor colonies has protected this species. With the exception of Antarctic Specially Protected Areas, there are currently no special conservation activities under way designed to protect emperor penguins. Note that under Article 3 of Annex II of the Protocol on Environmental Protection to the Antarctic Treaty, research in these protected areas (including killing animals) is still permitted as long as the relevant national agency has issued a permit (Antarctic Treaty Secretariat 1991).

Reliable population data on a number of key colonies around the continent are needed, which will require the following actions:

1. Coordinate efforts and agree on how to conduct research as an increasing number of nations operating in Antarctica become interested in emperor penguin research.
2. Use remote monitoring (e.g., automated cameras, satellite imagery) wherever possible, to minimize disturbance. However, there are many practical difficulties, such as maintenance of cameras and the need to complement satellite imagery with ground observations.

TABLE 2.1 Measurements of emperor penguins and their eggs (showing mean, standard deviation or range, and sample size)

VARIABLE	MALES	FEMALES	EGG	REFERENCES
body mass (g)				
colony arrival	36.7, 35–40 (15)	28.4, 28–32 (14)		Prévost 1961
hatching	23.4±0.8 (25)	24.8±2.7 (25)		Wienecke unpubl.
	24.7, 21.9–27.7 (5)	27.4±4.1 (23)		Prévost 1961
early chick rearing	27.8±2.7 (23)			Wienecke unpubl.
flipper length (mm)	345, 320–380 (41)	347, 335–260 (6)		Prévost 1961
bill length (mm)	80.9±3.6 (25)	72.3±4.5 (25)		Wienecke unpubl.
	81.5, 81–82 (2)			Prévost 1961
tarsus length (mm)	153±9 (25)	150±6.3 (25)		Wienecke unpubl.
egg length (mm)			124.0, 110–135.7 (13)	Prévost 1961
			119.8±4.2 (100)	Wienecke unpubl.
egg breadth (mm)			84.2, 79–89 (13)	Prévost 1961
			83.4±4.2 (100)	Wienecke unpubl.

TABLE 2.2 List of emperor penguin colonies

NAME	LONGITUDE	LATITUDE
Amanda Bay	76.88	-69.28
Amundsen Bay	50.74	-66.77
Astrid Coast	8.31	-69.94
Atka Bay	-8.13	-70.62
Auster	64.00	-67.40
Bear Peninsula	-110.17	-74.37
Beaufort Island	167.04	-76.94
Bowman Island	120.80	-65.08
Brownson Islands	-103.70	-74.18
Cape Crozier	169.43	-77.51
Cape Darnley	69.70	-67.88
Cape Roget	170.56	-71.98
Cape Washington	165.38	-74.65
Coulman Ilsand	169.64	-73.34
Davies Bay	158.41	-69.33
Dawson Lambton	-26.56	-76.01
Dion Islands	-68.73	-67.90
Dolleman Island	-60.56	-70.68
Drescher Inlet	-19.12	-72.86
Edward VII Pen	-157.74	-77.13
Fold Glacier	59.38	-67.33
Franklin Island	168.40	-76.18
Gould Bay	-47.32	-77.74
Gunnerus Riiser Larsen Peninsula	34.39	-68.78
Halley Bay	-27.20	-75.52
Haswell Island	93.01	-66.53
Kloa Point	57.30	-66.64
Luitpold Coast	-33.65	-77.27
Mertz Glacier	146.45	-66.93
Noville Peninsula	-98.49	-71.75
Peterson Bank	110.20	-65.93
Point Geologie	140.01	-66.67
Ragnhild Coast	27.25	-69.97
Riiser Larsen	-15.13	-72.14
Sanae	-1.38	-70.05
Shackleton	96.00	-64.88
Smith Peninsula	-60.85	-74.38
Smyley Island	-78.75	-72.31
Snow Hill Island	-57.46	-64.52
Stancomb Wills	-23.02	-74.16
Taylor Glacier	60.88	-67.48
Thurston Glacier	-125.59	-73.43
Umebosi Rock	43.09	-68.06

Note that most colonies have not been surveyed and their population size is unknown.

TABLE 2.3 Breeding variables of emperor penguins

VARIABLE	DUMONT D'URVILLE	TAYLOR GLACIER	REFERENCE
Incubation (d)	62–67		Prévost 1961
Chick rearing (d)	~ 150		Williams 1995
Reproductive success	2–85% (40 yrs)	42–88 % (20)	Barbraud & Weimerskirch 2001 Wienecke unpubl.
Age of first breeding (yr)	males: 4–8 (25) females: 3–6 (16)		Mougin & van Beveren 1979
Adult survival rate	95.1 % (*based on banded birds*)		Mougin & van Beveren 1979
Maximum lifespan	unknown		

TABLE 2.4 Recent population estimates for various emperor penguin colonies

COLONY	LOCATION	ESTIMATED POPULATION SIZE	YEAR/DATE	REFERENCE
Cape Crozier	77.5°S,169.4°E	475 chicks	2004	Barber-Meyer et al. 2007
Beaufort Island	76.1°S,167.9°E	628 chicks	2005	Barber-Meyer et al. 2007
Franklin Island	76.1°S,168.3°E	1966 chicks	2005	Barber-Meyer et al. 2007
Cape Washington	74.7°S,165.4°E	23201 chicks	2005	Barber-Meyer et al. 2007
Coulman Island	73.3°S,169.6°E	24207 chicks	2005	Barber-Meyer et al. 2007
Cape Roget	71.9°S,170.5°E	7207 chicks	1996	Barber-Meyer et al. 2007
Pointe Géologie*	66.7°S,140.0°E	~3100 pairs	2005	Jenouvrier et al. 2009
Haswell Island				
Amanda Bay	69.3°S, 76.3°E	~6500 adults	2007	Wienecke and Pedersen 2009
Auster	67.4°S, 64.0°E	~ 13400 pairs	1994	Wienecke and Robertson 1997
Taylor Glacier*	67.5°S, 60.9°E	2939 adults	25 June 2008	Wienecke unpubl.
Fold Island	67.3°S, 59.3°E	238 chicks, 395 adults	26 Sep 2009	Wienecke unpubl.
Umebosi Rock	68.1°S, 43.1°E	~ 300 pairs	2000	Kato et al. 2004
Riiser-Larsen Peninsula	68.8°S, 34.4°E	~4000 pairs	2000	Kato et al. 2004
Snow Hill	64.5°S, 57.4°W	~ 1200 pairs	20 July 1997	Coria and Montalti 1999

* >20 years of data

REFERENCES

Antarctic Treaty Secretariat. 1991. Annex II to the Protocol on Environmental Protection to the Antarctic Treaty. XI-4 Special Antarctic Treaty Consultative Meeting XI-4, Madrid, Spain. http://www.ats.aq/documents/recatt/att009_e.pdf (accessed 11 January 2010).

———. 2006. Management plan for Antarctic Specially Protected Area No. 127, Haswell Island. 29th Antarctic Treaty Consultative Meeting. Committee for Environmental Protection IX, Edinburgh, Great Britain. http://www.ats.aq/documents/recatt/att335_e.pdf (accessed 11 January 2010).

———. 2009. Annex II to the Protocol on Environmental Protection to the Antarctic Treaty. 32nd Antarctic Treaty Consultative Meeting ,Committee for Environmental Protection XII, Baltimore, MD, USA. http://www.ats.aq/documents/recatt/att432 e.pdf (accessed 11 January 2010).

Baker, A. J., S. L Pereira, O. P. Haddrath, and K. A. Edge. 2006. Multiple gene evidence for expansion of extant penguins out of Antarctica due to global cooling. *Proceedings of the Royal Society B* 273:11–17.

Barber-Meyer, S. M., G. L. Kooyman, and P. J. Ponganis. 2007. Trends in western Ross Sea emperor penguin chick abundances and their relationships to climate. *Antarctic Science* 20:3–11.

Barbraud, C., M. Gavrilo, Y. Mizin, and H. Weimerskirch. 2011. Comparison of emperor penguin declines between Pointe Géologie and Haswell Island over the past 50 years. *Antarctic Science* 23(5):461–68.

Barbraud, C., and H. Weimerskirch. 2001. Emperor penguins and climate change. *Nature* 411:183–86.

BirdLife International. 2009. Species factsheet: *Aptenodytes forsteri*. http://www.birdlife.org (accessed 1 December 2010).

Boersma, P. D. 2008. Penguins as marine sentinels. *BioScience* 58:597–607.

Boersma, P. D., and J. Parrish. 1999. Limiting abuse: Marine protected areas, a limited solution. *Ecological Economics* 31:287–304.

Bried, J., F. Jiguet, and P. Jouventin. 1999. Why do *Aptenodytes* penguins have high divorce rates? *The Auk* 116:504–12.

Casaux, R., A. Baroni, A. Ramón, A. Carlini, M. Bertolin, and C. Y. DiPrinzio. 2009. Diet of the leopard seal *Hydrurga leptonyx* at the Danco Coast, Antarctic Peninsula. *Polar Biology* 32:307–10.

Cendron, J. 1952. Une visite hivernale à une rookerie de manchots empereurs. *La Terre et le Vie—Revue d'Écologie* 40:101–8.

Cherel, Y., and G. L. Kooyman. 1998. Food of emperor penguins (*Aptenodytes forsteri*) in the western Ross Sea, Antarctica. *Marine Biology* 130:335–44.

Commonwealth of Australia. 2007. Environment Protection and Biodiversity Conservation Act 1999. Canberra: Australian Government Department of the Environment and Water Resources.

Copson, G. R., and N. P. Brothers. 2008. Notes on rare, vagrant and exotic avifauna at Macquarie Island, 1901–2000. *Papers and Proceedings of the Royal Society of Tasmania* 142:105–15.

Coria, N., and D. Montalti. 2000. A newly discovered breeding colony of emperor penguins *Aptenodytes forsteri*. *Marine Ornithology* 28:119–20.

Downes M. C., E. H. M. Ealez, A. M. Gwynn, and P. S. Young. 1959. The birds of Heard Island. ANARE Reports, series B, vol. 1. Zoology, Australian National Antarctic Research Expeditions. 20 pp.

Fretwell, P. T., and P. N. Trathan. 2009. Penguins from space: Faecal stains reveal the location of emperor penguin colonies. *Global Ecology and Biogeography* 18:543–52.

Gray, G. R. 1844. *Aptenodytes*. *Annals and Magazine for Natural History* 13:315.

Green, K., and R. Williams. 1986. Observations on food remains in faeces of elephant, leopard and crabeater seals. *Polar Biology* 6:43–45.

Groscolas, R. 1978. Study of moult fasting followed by an experimental forced fasting in the emperor penguin *Aptenodytes forsteri*: Relationship between feather growth, body weight loss, body temperature and plasma fuel levels. *Comparative Biochemistry and Physiology A* 61:287–95.

Isenmann, P. 1971. Contribution à l'éthologie et à l'écologie du Manchot empereur (*Aptenodytes forsteri* Gray) à la colonie de Pointe Géologie (Terre Adélie). *L'Oiseau* 41:9–64.

IUCN (International Union for Conservation of Nature). 2012. IUCN Red List of Threatened Species. Version 2012.1. www.iucnredlist.org (accessed 21 August 2012).

Jenouvrier, S., C. Barbraud, and H. Weimerskirch. 2005. Long-term contrasted response to climate of two Antarctic seabird species. *Ecology* 86:2889–2903.

Jenouvrier, S., C. Barbraud, H. Weimerskirch, and H. Caswell. 2009. Limitation of population recovery: A stochastic approach to the case of the emperor penguin. *Oikos* 118:1292–98.

Kato, A., K. Watanabe, and Y. Naito. 2004. Population changes of Adélie and emperor penguins along the Prince Olaf Coast and on the Riiser-Larsen Peninsula. *Polar Bioscience* 17:117–22.

Klages, N. 1989. Food and feeding ecology of Emperor penguins in the Eastern Weddell Sea. *Polar Biology* 9:385–90.

Kooyman, G. L., E. C. Hunke, S. F. Ackley, R. P. van Dam, and G. Robertson. 2000. Moult of the emperor penguin: Travel, location, and habitat selection. *Marine Ecology Progress Series* 204:269–77.

Kooyman, G. L., and P. Ponganis. 2008. The initial journey of juvenile emperor penguins. *Aquatic Conservation: Marine and Freshwater Ecosystems* 17:S37-S43.

McNeil, B. I., and R. J. Matear. 2008. Southern Ocean acidification: A tipping point at 450-ppm atmospheric CO_2. *Proceedings of the National Academy of Science* 105:18860–64.

Miskelly, C. 2011. The global penguin—Part 1. How a lone emperor ventured into superstardom. http://blog.tepapa.govt.nz/2011/06/23/the-global-penguin---part-1-how-a-lone-Emperor-ventured-into-superstardom (accessed 30 July 2011).

Mougin, J. L., and M. van Beveren. 1979. Structure et dynamique de la population de manchots empereurs *Aptenodytes forsteri* de la colonie de l'archipel de Pointe Géologie, Terre Adélie. *Comptes Rendus de l'Académie des Sciences*, Paris 289:157–60

Naczk, M., J. Synowiecki, and Z. E. Sikorski. 1981. The gross chemical composition of Antarctic krill shell waste. *Food Chemistry* 7:175–79.

Newbery, K. B., and C. Southwell. 2009. An automated camera system for remote monitoring in polar environments. *Cold Regions Science and Technology* 55:47–51.

New Zealand Herald. 2011. http://www.pressdisplay.com/pressdisplay/services/OnlinePrintHandler (accessed 27 June 2011).

Nicol, S., M. Stolp, and O. Nordstrom. 1992. Change in the gross biochemistry and mineral content accompanying the moult cycle in the Antarctic krill *Euphausia superba*. *Marine Biology* 113:201–9.

Perry, Nick. 2011. A young Antarctic emperor penguin takes a rare wrong turn and ends up on New Zealand coast. Associated Press, 21 June.

Piatkowski, U., and K. Pütz. 1994. Squid diet of emperor penguins (*Aptenodytes forsteri*) in the eastern Weddell Sea, Antarctica during late summer. *Antarctic Science* 6:241–47.

Prévost, J. 1961. Écologie des manchots empereur. In *Actualités Scientifiques et Industrielles 1291*. Paris: Hermann.

Prévost, J., and J. Sapin-Jaloustre. 1965. Écologie des manchots antarctiques. In *Biogeography and Ecology in Antarctica*, ed. J. van Mieghem and P. van Oye, 551–648. The Hague: W. Junk.

Pütz, K. 1995. The post-moult diet of Emperor penguins (*Aptenodytes forsteri*) in the eastern Weddell Sea, Antarctica. *Polar Biology* 15:457–63.

Riebesell, U., I. Zondervan, B. Rost, P. D. Tortell, R. E. Zeebe, and F. M. M. Morel. 2000. Reduced calcification of marine plankton in response to increased atmospheric CO_2. *Nature* 407:364–67.

Stonehouse, B. 1953. The emperor penguin (*Aptenodytes forsteri*) Gray I. Breeding behaviour and development. Falkland Islands Dependency Survey, Scientific Report 6.

USFWS (U.S. Fish and Wildlife Service). 2008. Fish and Wildlife Service proposes addition of penguin species to Endangered Species List. http://www.fws.gov/news/NewsReleases/showNews.cfm?newsId=471BA860-0EA0-7F90-972F3B8DD9F6E185 (accessed 22 December 2009).

Wienecke, B. 2009. Emperor penguin colonies in the Australian Antarctic Territory: How many are there? *Polar Record* 45:304–12.

———. 2010. History of the discovery of emperor penguin colonies, 1902–2004. *Polar Record* 46:271–76.

———. 2011. Review of historical population information of emperor penguins. *Polar Biology* 34:153–67.

———. 2012. Emperor penguins at the West Ice Shelf. *Polar Biology* 35:1289–96.

Wienecke, B., R. Kirkwood, and G. Robertson. 2004. Pre-moult foraging trips and moult locations of emperor penguins at the Mawson Coast. *Polar Biology* 27:83–91.

Wienecke, B., and P. Pedersen. 2009. Population estimates of emperor penguins at Amanda Bay, Ingrid Christensen Coast, Antarctica. *Polar Record* 45:207–14.

Wienecke, B., B. Raymond, and G. Robertson. 2010. Maiden journey of fledgling emperor penguins from the Mawson Coast, East Antarctica. *Marine Ecology Progress Series* 410:269-82.

Wienecke, B., and G. Robertson. 1997. Foraging space of emperor penguins *Aptenodytes forsteri* in Antarctic shelf waters in winter. *Marine Ecology Progress Series* 159:249–63.

Williams, T. D. 1995. *The Penguins Spheniscidae: Bird Families of the World*. Oxford: Oxford University Press.

Wilson, E. A. 1907. *Natural History*. Vol. 2: *Zoology*, part 2: Aves. British National Antarctic Expedition 1901–1904. London, British Museum (Natural History).

II

Brush-Tailed Penguins

Genus *Pygoscelis*

3

Adélie Penguin
(Pygoscelis adeliae)

PHIL N. TRATHAN AND GRANT BALLARD

1. SPECIES (COMMON AND SCIENTIFIC NAMES)

Adélie penguin, *Pygoscelis adeliae* (Hombron and Jacquinot, 1841, Auckland Islands)

The Adélie penguin was first described under the name *Catarrhactes adeliae*, and the current name was adopted later. The bird's generic name, *Pygoscelis* (Wagler, 1832), refers to "rump legged," while its specific name derives from the wife of Jules Sébastien César Dumont d'Urville, the French explorer and naval officer who led the 1840 French expedition to the ice frontiers of Adélie Land (Terre Adélie), Antarctica, where the bird was discovered. All the common names are variations around the specific name: *Adéliepinguïn* (Dutch), *Adéliepinguin* (German), *manchot Adélie* (French), *pingüino de Adelia* or *de ojo blanco* (Spanish), *Adéliepikkewyne* (South African Dutch), and *pinguim-de-Adélia* (Portuguese).

2. DESCRIPTION OF THE SPECIES

The sexes are similar, though the female is usually smaller (Ainley and Emison 1972; Scolaro et al. 1991; Kerry et al. 1992); this is most apparent when both partners are together. There is no seasonal variation in plumage in either sex, but the feathers show increasing signs of wear throughout the breeding season. Juveniles up to the age of approximately one year old are distinguished from adults by their white chins and throats.

FIG. 1 (*FACING PAGE*) Male Adélie extending his neck and flippers in the ecstatic display. (P. D. Boersma)

FIG. 2 Adult Adélie penguin showing the bill feathered to the nares and the white-feather eye-ring. (P. D. Boersma)

ADULT. The head, chin, throat, upper parts of the body, back, and tail are blue black in a freshly molted bird, fading to brown with wear. The underparts and belly are pure white. The flippers are blue black on their dorsal surface with a narrow white trailing edge and white ventrally with a narrow black leading edge and a small black area near the tip. The tail is composed of stiff quills, forming a characteristic brush-tail (fig. 1). The eye is brown and ringed by a distinctive circle of white feathers. The bill is mostly black with some orange red near the base; it is covered in feathers for more than half its length and hence appears short (fig. 2). The legs are covered in

FIG. 3 Adult Adélie with young hatchling. Note the darker down growing on the head of the chick and the dense feathers of the adult. (J. Weller)

white feathers, showing pink to white skin at the base. The feet are also pink to white but with black soles. The dark face and absence of any crest, together with the absence of any breast-bands, separates adult Adélies from all other medium-size penguins.

JUVENILE. The chin and throat are white, and the eye-ring is black until the birds molt to adult plumage at age 14 months.

CHICK. The chicks are variable in color, with their first down when hatched ranging from pale silver to sooty black. By 10 days of age, the first down has been replaced by a woolly second down that is uniformly dark gray (fig. 3).

MORPHOMETRIC DATA. Adults can be sexed using the dimensions of the bill or by cloacal examination (Scolaro et al. 1991; Kerry et al. 1992). Males are generally larger than females (table 3.1). As with most species, Adélies vary markedly in mass throughout the breeding season (table 3.2) (see also Ballard et al. 2010a). Mean mass varies with age, as younger, less experienced Adélies are lighter in weight and younger birds generally arrive later in the season. Ainley (1975) describes penguins arriving in late October that weighed around 6,000 grams (males) and 5,400 grams (females), while those arriving in early January weighed only 4,100 grams (males) and 3,700 grams (females). Body mass decreases after first arrival and throughout the courtship period and egg laying as penguins fast. Adults recover mass after their alternating foraging trips during incubation; body mass then usually decreases once again during chick rearing as parents provision their chicks (Emmerson et

al. 2003; Ballard et al. 2010a). Wilson et al. (1991) report that adult mean body mass decreased from 4,350 grams at hatching to 4,050 grams at the end of the guard stage, increased to 4,370 grams over the first few days of the crèche period, and then declined over the remainder of the crèche period to 4,190 grams. After breeding is completed, birds return to sea where they regain weight before the annual molt, attaining their peak body mass as they come ashore, at which time they may weigh more than 6,770 grams (Penney and Lowry 1967).

EGGS AND NESTS. Eggs are laid in a rudimentary nest made out of rocks and pebbles. Birds constantly steal pebbles, which often leads to conflict and fighting. Adélie penguins lay only a single brood, generally of two eggs. When compared proportionally to the weight of the parent, the eggs are smaller than for other birds, which is also the case for all other penguin species (table 3.3). Adélie eggs differ in size, the first (A-egg) usually being slightly larger than the second (B-egg). The eggs are subspherical or broadly elliptical, with a chalky, white or greenish surface when laid that discolors and stains with increasing time in the nest.

3. TAXONOMIC STATUS

There are no subspecies of Adélies, and the global population has remarkable genetic homogeneity (Roeder et al. 2001). The genus *Pygoscelis* contains three living species collectively known as the "brush-tailed" or "elbow-legged" penguins. The two other living congeners are the gentoo penguin (*P. papua*) and the Chinstrap penguin (*P. antarctica*) (both J. R. Forster, 1781). All three species have their own distinctive plumage and are not easily confused with one another. From the fossil record, three other congeners are known to have once existed: *P. tyreei* (Simpson 1972), from New Zealand in the late Pliocene; *P. calderensis* (Walsh and Suárez 2006), from Chile in the late Miocene; and *P. grandis* (Walsh and Suárez 2006), also from Chile in the late Miocene.

4. RANGE AND DISTRIBUTION

Adélie penguins have a circumpolar range, with major breeding aggregations occurring on ice-free land in the Ross Sea, along the coast of the Antarctic continent, on the west coast of the Antarctic Peninsula, and on islands of the Scotia Arc (fig. 4). The most southerly breeding colonies are in the Ross Sea at Cape Royds (77° S) and

FIG. 4 Distribution and abundance of the Adélie penguin, with counts based on pairs.

the most northerly at Bouvet Island (Bouvetøya) (54° S). Outside the breeding season, the distribution is less well documented but is mainly pelagic and restricted to areas of pack ice (Ballard et al. 2010b; Dunn et al. 2010). Adélies occur as occasional vagrants in South America, Australia, New Zealand, and sub-Antarctic islands in the Atlantic, Indian, and Pacific Oceans.

The main breeding aggregations are in the Ross Sea region (1,000,000 pairs) (fig. 5), on the Antarctic Peninsula, throughout the Scotia Arc (727,000 pairs), and in East Antarctica (325,000) (Whitehead and Johnstone 1990); the largest known colony is at Cape Adare in the Ross Sea, where there were approximately 282,300 pairs in the mid-1980s (Marchant and Higgins 1990) but only 169,200 pairs in the mid-1990s (see Ainley 2002 for discussion). Marchant and Higgins (1990) and Ainley (2002) provide further details of other important

locations, with population sizes for individual colonies. The total breeding population has been estimated to be at least 2,610,000 pairs (Woehler 1993), or between 4,000,000 and 5,220,000 animals (BirdLife 2009). This figure is under review by scientists working with the Commission for the Conservation of Antarctic Marine Living Resources (CCAMLR). In addition to breeding birds, there are an estimated 10,000,000 immature birds (Croxall 1985).

5. SUMMARY OF POPULATION TRENDS

Globally, Adélie populations are thought to be stable (BirdLife 2009), though trends vary regionally.

6. IUCN STATUS

The threat status of Adélies, last assessed in 2012 by BirdLife International, is determined to be in the Near Threatened category on the International Union for Conservation of Nature's Red List of Threatened Species (IUCN 2012).

The species is listed as Near Threatened because the global population is expected to decline at a moderately rapid rate due to the effects of climate change. When assessing Adélies, BirdLife noted that the species has an extremely large range and hence does not approach the IUCN thresholds for Vulnerable under the range size criterion (extent of occurrence of less than 20,000 square kilometers combined with a declining or fluctuating range size, habitat extent or quality, or population size and a small number of locations or severe fragmentation). Further, the population trend appears to

FIG. 5 Penguins are well insulated. At Cape Bird, Ross Sea, Antarctica, snow sticks to the back of an Adélie penguin without melting. (P. D. Boersma)

be stable, and hence the species does not approach the IUCN thresholds for Vulnerable under the population trend criterion (greater than 30% decline over 10 years or three generations). Finally, the population size is large (see section 8 below) and hence does not approach the IUCN thresholds for Vulnerable under the population size criterion (less than 10,000 mature individuals with a continuing decline estimated to be greater than 10% in 10 years or three generations, or with a specified population structure).

The history of IUCN status is: 2008, Least Concern; 2004, Least Concern; 2000, Lower Risk/Least Concern; 1994, Lower Risk/Least Concern; 1988, Lower Risk/Least Concern. However, given the differences in regional population trend, the rapid warming over much of its range, and the species' dependence on sea ice, the IUCN status of the species should be reconsidered and upgraded to Vulnerable.

INTERNATIONAL AND LEGAL STATUS. The majority of the global population of Adélie penguins lives and breeds south of 60° south and so falls within the area covered by the Antarctic Treaty System. Only a small number of colonies occurs north of 60° south, mostly in the South Sandwich Islands and at Bouvet Island. In 1991, the Antarctic Treaty Consultative Meeting adopted the Protocol on Environmental Protection, which sets out environmental principles, procedures, and obligations for the comprehensive protection of the Antarctic environment and its dependent and associated ecosystems and species. The protocol designates Antarctica as a natural reserve devoted to peace and science and applies to both governmental and nongovernmental activities in the Antarctic Treaty area. The protocol seeks to ensure that all human activities, including science and tourism, do not have adverse impacts on the Antarctic environment or on its scientific and aesthetic values. Under the treaty, all activities must comply fully with national laws and regulations that implement the Antarctic Treaty System, as well as other national laws and regulations implementing international agreements on environmental protection, pollution, and safety that relate to the Antarctic Treaty area. All activities should also abide by the requirements imposed on organizers and operators under the protocol, insofar as they have not yet been implemented in national law. Adélie populations are therefore protected through national laws implementing the Antarctic Treaty.

Legal instruments afford specific protection to some colonies under the Antarctic Treaty. These are Antarctic Specially Managed Areas (ASMA) and Antarctic Specially Protected Areas (ASPA). A number of colonies, particularly in the West Antarctic Peninsula region and at the South Orkney Islands, receive such protection, principally because of the long-term scientific studies undertaken at these sites. Thus, there are currently 3 ASMAs and 34 ASPAs that protect Adélies across the whole of the Antarctic continent. However, these areas typically include only nesting areas (i.e., on land) and do not extend protection into the ocean, where Adélies spend most of their time.

Further protection is also available under CCAMLR. In 2009, the CCAMLR designated a high-seas marine protected area south of the South Orkney Islands. This was the first MPA anywhere in the World Ocean that is entirely within the high seas. It was designated partly because of the need to protect Adélie foraging grounds, particularly those used after the breeding season and before the molt.

7. NATURAL HISTORY

Major life-history studies for Adélie penguins in Sladen (1958, 1964) and Ainley (2002) are reviewed in Williams (1995). More recent work on Adélies focuses on their ecological niche and ecological interactions, rather than their breeding biology and natural history; consequently, this section draws heavily upon these early reviews for life-history details. Ratcliffe and Trathan (in press) review the Adélie diet, which also informs this section.

BREEDING BIOLOGY. The breeding biology of Adélies is well known, as a number of major studies were conducted in the mid-20th century. These include studies in the Ross Sea sector (Victoria Land) at Cape Bird (Spurr 1975b, 1975c; Davis 1982a, 1982b; Davis 1988; Davis and McCaffrey 1986), Cape Crozier (Sladen et al. 1968; Le Resche and Sladen 1970; Ainley and Schlatter 1972; Oelke 1975; Ainley et al. 1983), Cape Hallett (Reid 1964, 1965), and Cape Royds (Taylor 1962; Stonehouse 1963; Yeates 1968); in the South Orkney Islands at Signy Island (Sladen 1958; Lishman 1985a); in Wilkes Land (Penney 1968); and at Adélie Land (Sapin-Jaloustre 1960). Adélies are highly migratory, with birds at sea between May and August; they return to their colonies between September and October, when they commence breeding. They are highly colonial, often nesting in very large colonies and at high densities. Adélies (60–80%) usually remain with the same partner in successive years. They are also highly faithful to their nest sites, and 60–90% return to the same location. Egg laying begins in October and may last into November.

Adélies typically lay a two-egg clutch, with incubation lasting approximately 32–34 days. Both parents take turns incubating the eggs; there are usually two shifts of between 11 and 14 days, followed by one or more shorter shifts. On hatching, the chicks are immobile, lack down, and must be cared for by the adults (semi-altricial). They stay at their nests (nidicolous), where they are brooded or guarded for approximately 22 days after hatching. When they reach an age at which they are thermally independent and are able to avoid predation by raptorial seabirds, they begin to form small crèches of approximately 10 to 20 chicks. Chicks are fed every 1 to 2 days by both parents until they fledge at 50 to 60 days of age in February. Feeding chases are well developed and are a mechanism for dispersing the food among the chicks (Boersma and Davis 1997). After breeding, adults leave the colony to replenish body reserves lost over the preceding breeding season. Approximately four weeks after leaving the colony, adults undergo their postnuptial molt, usually starting in late February or early March and lasting between two and three weeks. For birds breeding at Signy Island and Ross Island, molt takes place on sea ice away from the colony (Dunn et al. 2010; Ballard et al. 2010b), while for birds breeding in East Antarctica, it may be at or close to the colony (Clarke et al. 2006).

WINTER DISTRIBUTION. The general seasonal distribution of Adélie penguins is poorly described, though the use of light-sensing geolocators is beginning to alter our understanding (Ballard et al. 2010b; Dunn et al. 2010; Erdmann et al. 2011). Birds generally disperse at the end of the breeding season and winter in the pack ice, although some birds have been seen close to their summer breeding sites when open water is available following winter storms (Parmalee et al. 1977). There are records of birds traveling a long distance from their breeding sites during the winter, with feeding grounds often a considerable distance from their breeding sites. Thus, during the winter months, Adélies can travel many hundreds or even thousands of kilometers away from their breeding

colonies (Kerry et al. 1995; Davis et al. 1996; Davis et al. 2001; Clarke et al. 2003; Ballard et al. 2010b) and are known to forage in areas with pack-ice up to about 80% cover, often several hundred kilometers south of the open water. They appear to be dependent on visible light for foraging or navigating during winter, which limits their distribution to north of about 73° south (the line of zero midwinter civil twilight) (Ballard et al. 2010b).

ARRIVAL AT THE BREEDING COLONY. The date of return to the breeding colony varies geographically and among years (fig. 6; table 3.5). At Signy Island in the South Orkney Islands, the arrival date can vary from 20 September to 8 October, though the mean date over 25 years was 27 September (Rootes 1988); at King George Island in the South Shetland Islands, arrival dates are generally similar, ranging from 28 September to 18 October (Jablonski 1987). In East Antarctica, birds arrive a little later, with dates varying between 4 and 17 October and a mean date of 12 October over an 11-year period (Johnstone et al. 1973). In the Ross Sea, at Cape Crozier, the dates are 18 to 25 October with a mean date of 20 October over an 8-year period. There is strong evidence to suggest that the date of arrival is related to latitude and is affected by local and/or regional sea-ice conditions, with arrival delayed in years of extensive sea ice (Ainley and LeRe-

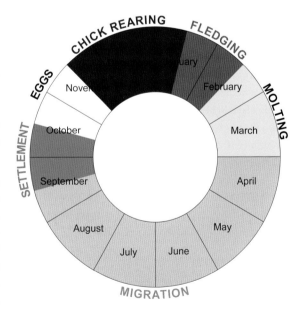

FIG. 6 Annual cycle of the Adélie penguin. Adélie penguins have such a broad geographic range that there are contrasting trends in initiation of egg laying on opposite sides of the Antarctic. (Emmerson et al. 2011)

sche 1973). Older, more experienced birds arrive at the breeding colony before younger birds; males also arrive slightly earlier than females at some locations (Spurr 1975c; Ainley et al. 1983; Trivelpiece and Trivelpiece 1990; Emmerson et al. 2003). After arrival, birds remain at the nest site until egg laying commences; the mean time between arrival and the first egg is approximately 21 days, though this does vary.

Colonies are variable in size, with only a few likely to have more than 200,000 pairs (Ainley 2002). Colonies may have satellite groups nesting nearby. The average distance between nests is generally about one meter (Trivelpiece and Volkman 1979).

NEST STRUCTURE. Nests consist of a shallow scrape lined with and surrounded by small stones and pebbles. Males initiate nest building by selecting the site, but both sexes build the nest together (Taylor 1962).

EGG LAYING. The date of the first egg varies geographically and between years, depending on latitude and sea-ice cover. The complex relationships between Adélie penguins and krill appear to vary regionally, with impacts on the initiation of egg laying depending on location in the Antarctic (Emmerson et al. 2011). At Signy Island in the South Orkney Islands, the date of laying varies between 22 October and 2 November, with a mean date of 27 October over 12 years. In the Ross Sea, egg laying starts later—at Cape Royds, it is generally between 9 and 22 November. Regardless of the date that the first egg is laid, laying is highly synchronous, with approximately 50% of clutches initiated within a six-day period. Young females lay slightly later than older, more experienced birds, although the age and experience of the male partner apparently does not affect the laying dates (Ainley et al. 1983).

CLUTCH SIZE. Adélies normally lay two eggs, though inexperienced three-year-olds tend to lay a greater proportion of single-egg clutches (Ainley et al. 1983). The mean clutch size increases from the edge of the colony toward the center; isolated nests average 1.45 eggs per nest, peripheral nests 1.78, and central nests 1.89, owing to the increasing proportion of two-egg clutches laid (45%, 68%, and 77%, respectively) (Spurr 1975c; Tenaza 1971). If the first egg is lost within 24 hours of laying, a third egg (C-egg) can be laid (Astheimer and Grau 1985). No replacement clutches are laid, possibly because of the short breeding season.

LAYING INTERVAL. The interval between the first and second eggs varies, with an average of three days (Spurr 1975c); for example, it may be a two-day interval at 13% of nests, three days at 68%, and four days at 19% (Taylor 1962).

INCUBATION. Incubation begins once the second egg is laid. Both sexes incubate the clutch in alternate shifts, though there is considerable variation recorded. At Signy Island (Lishman 1985b), the first shift is taken by the male and averages 13.7 days (range 9–18 days, n = 77); the second shift is taken by the female and averages 12.7 days (range 8–18 days, n = 73); and the third by the male, averaging 5.4 days (range 1–12, n = 74). Other studies show respective incubation shifts of 13, 15, and 8 days (Sladen 1958) and 11, 11, and 8 days (Taylor 1962). However, Davis (1982a) records only two long shifts of 16.6 days (range 9–25 days, n = 84) and 12.3 days (range 7–20 days, n = 84) before a period of alternating, shorter shifts. Females may take the first incubation shift, with males going to sea first. There is also inter-annual variation in the duration of the first shift, which may vary between 7.5 and 13.8 days (n = 5 years), and the second, which may vary between 8.0 and 12.5 days; shorter shifts usually occur in years of earlier sea-ice breakout (Yeates 1968).

INCUBATION PERIOD. The incubation period varies among years. The first egg generally hatches after 35.5 days, and the second egg hatches after 34.1 days, based on data from a large sample size of 100 first eggs and 100 second eggs (Reid 1965). Other studies report 34.7 days (range 32–38, n = 42) and 35.1 days (range 33–37 days, n = 9) for the first egg, and 33.2 days (range 31–35 days, n = 42) and 33.1 days (range 32–34 days, n = 9) for the second egg (Spurr 1975c), or 39.2 days (range 36–43) for the first egg and 37.9 days (range 33–42) for the second (Yeates 1968). See Ainley et al. (1983) for a more detailed review.

HATCHING. Hatching within a single nest is typically asynchronous, with the two eggs hatching on average 1.4 days apart. Within a given colony, however, hatching is generally synchronous between nests, reflecting the synchrony in egg laying. Some chicks hatch in less than 24 hours (17.7%), though most take up to 48 hours (74.5%) and a few longer than that (7.8%) (Taylor 1962).

HATCHING SUCCESS. Hatching success varies among years, between 57% and 82%; eggs may fail because they are addled or infertile, deserted, preyed upon (Spurr 1975a; Davis 1982a), or buried by heavy snow (Ballard 2010).

CHICK REARING. After hatching, the chick is guarded and brooded continuously by one or other of the parents, which exchange duties every day or so, though shifts may be as long as 4 days (Taylor 1962) (fig. 7). The parents share guard duty equally over the 22-day period. Early-hatching chicks and single chicks are guarded for longer than late-hatched ones (Taylor 1962; Lishman 1985a).

CRÈCHE. Crèches form slowly. In the Ross Sea, most chicks generally enter crèches when approximately 22 days (range 16–24) old (e.g., Davis 1982a). At Signy Island, chicks crèche when they are 16 and 19 days old (Lishman 1985b).

CHICK GROWTH. Chicks are fed one or more times per day (Ainley and Schlatter 1972; Oelke 1975; Volkman and Trivelpiece 1980; Trivelpiece et al. 1987). Meal size increases from 90–100 grams during guard stage to 200–300 grams during crèche stage. As chicks develop and become increasingly mobile, they often engage in feeding chases (Boersma and Davis 1997). Chicks reach a maximum weight of approximately 3,940 grams (equivalent to 74% of adult weight) before then losing weight and fledging at approximately 3,727 grams (70% of adult weight) (Volkman and Trivelpiece 1980). Some studies have shown similar growth rates for first and second chicks (Volkman and Trivelpiece 1980), but at some sites in some years, chicks differ in their rate of body mass increase (Davis and McCaffrey 1986).

FLEDGING PERIOD. The mean age at fledging varies, with reports of chicks fledging at 50.6 days (range 41–56 days, n = 113) of age and 48.4 days (range 42–57 days) of age (Taylor 1962; Ainley and Schlatter 1972) or at 60.3 days (range 54–64 days, n = 21) and 61.3 days (range 55–64, n = 8) of age, respectively, for first and second chicks (Lishman 1985a).

FLEDGING SUCCESS. Fledging success is markedly different among years and locations, with success varying between 63.3% and 83.2%. Skua predation can be considerable, and some sites experience very high rates

FIG. 7 Male (left) and female (right) Adélie penguins in the nest with a young chick. Note the depressed head feathers and raised neck feathers. (P. D. Boersma)

of chick loss due to predation in some years. Chicks in nests located at the periphery of colonies are more prone to predation than those in central locations. Starvation, nest desertion, and burial in heavy snow (with consequent freezing and/or suffocation) can also affect fledgling success.

BREEDING SUCCESS. The number of chicks fledged per breeding pair varies markedly among years and locations, with success between 0.68 to 1.39 chicks per nest, although in some years breeding failure within a colony is complete. Breeding success is lowest in years when sea ice persists, potentially affecting the abundance and distribution of prey or travel time between colonies and preferred feeding grounds (Ainley and LeResche 1973; Yeates 1975; Lishman 1985a; Whitehead 1991; Emmerson and Southwell 2008).

MOLT. Molt occurs after breeding and after the chicks have fledged. For birds from Signy Island, molt takes place at sea on sea ice (Dunn et al. 2010), but for birds from East Antarctica, molt occurs close to or at the colony (Clarke et al. 2006). For Signy Island birds, molting starts approximately four weeks after they leave the colony, with a mean start date of 19 February (range 7 February to 3 March, n = 19), and lasts between two and three weeks, on average 18.6 days (range 15–22 days). For others, it occurs at the breeding colony; for these, the average duration of the pre-molt foraging trip is approximately 9 days (range 1–15) for successful breeders, but 46 days (range 41–51) for unsuccessful breeders. The first adults come ashore to molt in early February, with the largest numbers molting in early March. Unsuccessful breeders begin molt on average slightly earlier than do successful breeders; immature birds complete molt before nonbreeders and breeding adults, with half of the presumed nonbreeders beginning molt before mid-February (Penney and Lowry 1967). For those molting on land, adults are ashore on average 19.8 days (range 15–23 days) and immature birds 18.6 days (range 15–21 days). Adults are on land for approximately 5.1 days before molt commences; the molt itself lasts 14.9 days, and birds remain for an additional 2.5 days after molt is complete. During molt, birds lose 45% of their initial body mass at a rate of 151–192 grams per day (Penney and Lowry 1967; Bougaeff 1975).

AGE AT FIRST BREEDING. Ainley et al. (1983) give the mean age at first breeding as 4.7–5.0 years for females and 6.2–6.8 years for males, with 19% of birds first breeding as 4-year-olds, 28% as 5-year-olds, 28% as 6-year-olds, and 18% as 7-year-olds. Immature birds return to the breeding site before recruitment into the breeding population; some return as 1-year-olds, usually in late January or February, approximately 30% return as 2-year-olds, and by age 4, 72% of males and 81% of females have visited the colony at least once.

SURVIVAL. Survival varies markedly among years and sites. It was approximately 80.9–97.0% for all birds over a 3-year period (Ainley et al. 1983) but fluctuated between 63% and 84% in an 11-year study, with a small colony having consistently lower apparent survival than larger ones (1996–2006 [Dugger et al. 2010]). Some reports (e.g., Davis and Speirs 1990) show that mortality can be higher in females, such that the sex ratio becomes male biased with age; thus, the ratio is 1.0:1.0 for 2-year-olds, but 1.0:0.4 for 14-year-olds (Ainley et al. 1983). Survival varies with age, with survival of 4-year-olds as high as 97.0% but decreasing to 93.3% for 7-year-olds. Most adult mortality occurs during winter, with approximately 82% of losses taking place after breeding birds have left the colony; winter conditions also affect the propensity of birds to breed in the following season (Spurr 1975c; Trathan et al. 1996). Survival rates of more successful breeders were found to be higher than those for unsuccessful breeders, but there was also indication of a survival penalty for those with the highest breeding effort (Lescroël et al. 2009).

PAIR FIDELITY. Pair fidelity varies among seasons and with breeding location. Penny (1968) records that up to 80% of birds retained the same pairing between two consecutive years. On average, over a four-year period, birds have approximately 2.0 partners, which may mean that multiple mating may be common in any given year.

SITE FIDELITY. Site fidelity is high, with 98.9% (range 98.1–100.0%, n = 4 years) of males and 65.3% (range 61.8–72.9%) of females retaining the same nest site in successive seasons; in 6 years, less than 0.1% of females and no males were found breeding at a colony different from that used in the previous year (Trivelpiece and Trivelpiece 1990). More recently, breeders have been found to move among colonies in response to adverse environmental conditions during 5 years in a 12-year study, with up to 3.5% of adults moving during that period (Dugger et al. 2010). Over a 4-year period, males used on average 1.4 nest sites, with 62% retaining the same nest for all 4 years; in contrast, females used on average 2.1 nest sites, and 29% retained the same site over this period; 35% used three nests in 4 years (Davis and Speirs 1990). For birds that survive to breeding age, 96% breed at their natal colony, the remaining 4% breed at adjacent colony sites. In fact, 77% of birds return to breed within 100 meters of their natal site (Ainley et al. 1983).

NATURAL LIFESPAN. Adélies have an estimated high survival rate (of 0.894 [Ainley 2002]) once they start to breed and so have a reasonable life expectancy of 15 years or more (possibly a maximum of 20) in the wild (Ainley 2002).

FIG. 8 An Adélie penguin foraging. The bird's hydrodynamic shape is adapted for swimming. (J. Weller)

PREY. Diet is dominated by euphausiid crustaceans and fish (see Ratcliffe and Trathan, in press), with squid and amphipods making relatively small contributions (table 3.6). Ratcliffe and Trathan (in press) explains that diet varies according to habitat around the colony: birds foraging over the Antarctic continental shelf feed mostly on Antarctic crystal krill (*Euphausia crystallorophias*) and fish, especially Antarctic silverfish (*Pleuragramma antarcticum*), while those feeding over the shelf slope or in oceanic waters feed on Antarctic krill (*Euphausia superba*). The relative contribution of fish to the diet varies among and within years in several studies and generally becomes more common during chick rearing (e.g., Ainley et al. 2003). One study of winter diet suggests that squid may become more important at this time, although sample sizes are small (Ainley et al. 1992). Adélies are thought to be visual predators and pursue their prey at depths of up to 45 meters (but dives of up to 170 meters are documented). Foraging is generally diurnal during chick rearing, particularly at northern latitudes; at southern latitudes, foraging trip arrivals and departures occur at all times of day (fig. 8).

Diet studies are based mostly on stomach flushing of adults returning to feed chicks. As such, they are likely to be unrepresentative of diet composition at other times of the year or of nonbreeding adults or, feasibly, even breeding adults, if they feed differentially for self and offspring during chick rearing. Two studies of diet using stable isotopes, one using fatty acid analysis and one using genetic signatures in feces, confirm these biases.

Studies have been carried out at colonies throughout the species range, though only a small proportion of the known colonies in each region have been sampled, and the variation in diet among sites is large. It may be possible to estimate diet composition at unsampled sites by modeling diet as a function of the percentage of the foraging range that is on the continental shelf, since this explains much of the geographic variation in diet (Ratcliffe and Trathan, in press).

Ainley et al. (2003) show that diet varies as a function of year, time of year, and percentage of foraging area covered by sea ice but not by colony location. These authors note that sea ice affects diet composition in neritic waters, that the quality of summer diet cannot explain different population growth rates among colonies, and that stable isotope analysis is a useful tool for synoptically describing diet in this species over a large area.

PREDATORS. On land, skuas are frequent predators of Adélie eggs and chicks. Sometimes a pair of skuas will work together to distract a brooding parent so that the egg or chick can be taken. Sickly, unguarded, or lost chicks are usually quickly identified, after which they are harassed until they leave the protection of the colony when they are finally killed and eaten. Giant petrels also regularly take eggs and chicks. Giant petrels remain a threat after fledging, though the main predators for chicks as they leave the colony are leopard seals. Leopard seals will grip birds by the feet and belly and vigorously beat them against the water to remove their skin in a process known

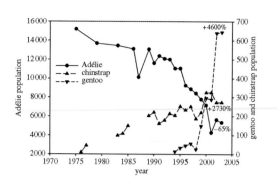

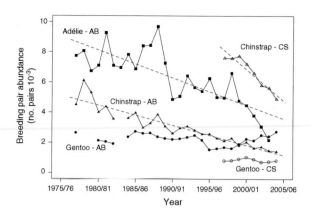

FIG. 9 Population trends for three penguin species in the Anvers Island vicinity, 1975–2003. The numbers on the graph indicate the percentage change from the initial sampling year for each species. Reproduced from Ducklow et al. 2007.

FIG. 10 Number of breeding pairs of Adélie (filled square), gentoo (filled and open circles), and chinstrap (filled and open triangles) penguins at Admiralty Bay (AB, filled symbols), King George Island, and Cape Shirreff (CS, open symbols), Livingston Island, South Shetland Islands, Antarctica, 1977–2004. Linear regressions with significant, non-zero slopes are plotted as dashed lines. Reproduced from Hinke et al. 2007.

as flailing. The birds are then dismembered and eaten, with the skin and feathers being discarded. Leopard seals also take adult birds and are often found patrolling shores close to penguin colonies so that they can catch incoming or outgoing birds, with several seals present at large colonies but few or none at smaller colonies (Ainley et al. 2005a). Queues of hesitant, nervous penguins are often found close to the edge of the water, waiting until they believe the water to be safe. Leopard seals can surface through thin ice, fracturing it and knocking penguins into the water. Killer whales probably at least occasionally take Adélie penguins, since they eat other pygoscelid penguins (Pitman and Durban 2010), although they more frequently coexist with no hostile interactions (Ballard and Ainley 2005).

8. POPULATION SIZES AND TRENDS

A number of authors (Fraser et al. 1992; Trathan et al. 1996; Croxall et al. 2002; Forcada et al. 2006; Forcada and Trathan 2009; Ballard et al. 2010b) have suggested that Adélie populations are influenced by climate variability and are therefore vulnerable to climate change. Potential impacts include poleward shifts in geographic distribution, range contraction, changes in the timing of biological events (phenology), changes in predator-prey interactions (Trathan et al. 2007), and changes to the availability of suitable wintering habitat (influenced by both sea-ice and daylight availability) (Ballard et al. 2010b). Indeed, recent poleward shifts and local

declines (Ducklow et al. 2007) have now been observed for those ice-dependent Adélie populations breeding in the West Antarctic Peninsula region and in the southern Scotia Sea (Trivelpiece et al. 2011). These colonies are some of the northernmost breeding locations for Adélie penguins, and at some of these sites, populations have almost disappeared since 1970. Thus, Adélies have declined by 70% on Anvers Island and the Antarctic Peninsula (Fraser et al. 1992; Ducklow et al. 2007) (fig. 9) and, to a lesser extent, at the South Shetland Islands (Hinke et al. 2007) (fig. 10) and the South Orkney Islands (Forcada et al. 2006; Forcada and Trathan 2009).

Away from the Antarctic Peninsula and the Scotia Sea, Adélie populations show strong evidence of increase (Croxall et al. 2002). In East Antarctica and the Ross Sea, where 38% of the total population lives (Woehler 1993), populations have increased as air temperatures have marginally decreased in some locations and seasonal sea ice became more extensive (Turner et al. 2009a; Turner et al. 2009b). Adélie penguins require ice- and snow-free nesting terrain, a limited resource in Antarctica, but one likely to increase under most climate-warming scenarios. Forcada et al. (2006) highlight the close interaction between penguin population processes and the marine environment and links these with prey availability. The regional consequences of global climate perturbations on the sea-ice phenology and penguin prey will affect the marine ecosystem, with repercussions for penguins (Trivelpiece et al. 2011).

Various population responses have been linked with climate change, particularly at the boundaries of the current geographic distribution, where penguins exist at the limits of their environmental tolerance. It is clear that the impacts of climate vary by regions, but they are most evident where rapid change has occurred; therefore some populations are faring better than others (Ainley 2002).

Potentially, these climate-related effects can be exacerbated by additional anthropogenic pressures such as habitat encroachment, fisheries, pollution, and tourism. These pressures confound a simple interpretation of population trends, and disentangling such complexity remains an important challenge (Fraser and Patterson 1997; Patterson et al. 2003; Smetacek and Nicol 2005; Ainley et al. 2007; Nicol et al. 2007; Trathan et al. 2007; Trathan and Reid 2009; Trivelpiece et al. 2011).

Population trends have been related to regional patterns in seasonal sea ice (fig. 11) and prey (Fraser et al. 1992; Trathan et al. 1996; Wilson et al. 2001; Croxall et al. 2002; Hinke et al. 2007; Ducklow et al. 2007; Forcada and Trathan 2009) (table 3.4), with trends varying depending on location. Westerly winds over the Southern Ocean that surrounds Antarctica have increased by around 15–20% since the 1970s (Turner et al. 2009a; Turner et al. 2009b). These stronger westerly winds bring warm, wet, oceanic air into the Amundsen and Bellingshausen Seas region. These winds have resulted in a clear dipole in seasonal sea ice with significant negative trends in the Amundsen and Bellingshausen Seas and significant positive trends in the Ross Sea (Turner et al. 2009a, 2009b). Current assessments suggest that the observed dipole will alter as the ozone hole heals in the second half of the 21st century (Turner et al. 2009b). This analysis is still being discussed (Sigmond and Fyfe 2010), but if correct, sea-ice trends will reverse, and the annual average sea-ice extent will diminish by 33%; most of this retreat will be in winter and spring (Turner et al. 2009a). Recent analyses (Ainley et al. 2010; Trivelpiece et al. 2011) suggest that with continued climate change and environmental warming, populations of Adélies are likely to experience marked declines. High-latitude areas, such as the Ross Sea, are therefore likely to become increasingly important as the environment changes.

Adélie population processes in the Amundsen and Bellingshausen Seas (Fraser and Hofmann 2003; Ducklow et al. 2007; Hinke et al. 2007) and at the South Orkney Islands (Forcada et al. 2006; Forcada and Trathan 2009)

FIG. 11 Adélie penguins often rest on ice. They spend their lives associated with ice. (J. Weller)

are also thought to be affected by strong teleconnections between El Niño, sea surface temperature (Trathan and Murphy 2002; Trathan et al. 2006), and sea ice (Quetin and Ross 2003; Trathan et al. 2006) that result in strong but periodic pulses of krill biomass (Murphy et al. 2007; Trivelpiece et al. 2011). Indeed, Fraser and Hofmann (2003) have shown that foraging-trip duration varies in a nonlinear manner but in accordance with sea-ice extent and changes in krill abundance; years with the lowest sea-ice extent are associated with the longest foraging-trip duration and the lowest measures of krill abundance. Years with intermediate or extensive sea ice are associated with shorter foraging-trip durations and higher levels of krill abundance. Changes in the population processes of pygoscelid penguins are now evident at Signy Island in the South Orkney Islands where congeneric Adélie, chinstrap, and gentoo penguins show population responses to the impact of sea-ice reduction (Trathan et al. 1996; Forcada et al. 2006). These three species breed sympatrically at Signy Island, where both Adélies and chinstraps have declined and the less ice-adapted gentoos have increased significantly in numbers over the past three decades. These trends have occurred in parallel with regional long-term warming and significant reductions in sea-ice extent. Periodic warm events, with teleconnections to the tropical El Niño region, have caused sea-ice cycles that have potentially led to reduced prey biomass and concurrent inter-annual population decreases in the three penguin species. With the loss of sea ice, Adélies are not as well buffered against the environment; consequently, their numbers have fluctuated greatly, such that their populations have shown a strong linear decline. Ultimately, variability in the population numbers of these three species reflects the local balance between penguin adaptation to sea-ice conditions and trophic-mediated changes cascading from global climate forcing (Forcada et al. 2006).

Estimates for the annual breeding population size of Adélie colonies on Ross Island in the Ross Sea (Ainley 2002) have the potential to form one of the most extensive long-term monitoring data sets anywhere in the Antarctic. Monitoring for these colonies first began in the late 1950s, with the recording of significant inter-annual variability. Wilson et al. (2001) suggest that changes in population numbers were related to natural physical environmental factors. Using available satellite imagery for sea ice between 1973 and 1997, they conclude that population growth (measured annually during the summer) is best explained by the inverse of the sea-ice extent five winters earlier and, further, that population growth is also related to the Southern Oscillation Index and, hence, the El Niño–Southern Oscillation (ENSO).

Using a demographic model to show how variation in survival of juveniles and subadults might account for the observed population variation, Wilson et al. (2001) propose that the five-year lag in population growth resulted from the interval that precedes recruitment of surviving juvenile and subadult birds into the breeding population. In the Ross Sea, extensive sea ice during winter appears to reduce subadult survival, but this phenomenon is expressed only when cohorts subsequently reach maturation. Thus, Wilson et al. suggest that extensive (more northerly) sea ice limits penguin access to productive waters that are known to occur south of the southern boundary of the Antarctic Circumpolar Current, with starvation or increased predation disproportionately affecting less experienced birds. They note that the observed patterns of penguin population change,

including those preceding the satellite era, result from differences in sea-ice extent and that this has changed significantly over recent decades.

In the western Pacific and Ross Sea sector, Ainley et al. (2005) show decadal-scale changes in population trajectories for Adélie penguins during the early 1970s and again during the late 1980s. These population changes correspond to changes in weather and sea-ice patterns, which were related to shifts in the Antarctic Oscillation (AAO) (Ainley et al. 2005b). The AAO is more usually known as the Southern Annular Mode (Hall and Visbeck 2002) and is based on the zonal mean sea-level pressure between 40° south and 65° south (Gong and Wang 1999); it is the leading pattern of variability of the extra-tropical winds (Kushner et al. 2001). The observed patterns of variability also show that the AAO and ENSO interact and that there is evidence of recent change (Carleton 2003). Both Wilson et al. (2001) and Ainley et al. (2005b) highlight the demographic lags inherent in the response of long-lived species to habitat or environmental variation. The proximate mechanisms responsible for the population changes are undoubtedly mediated through the prey field of the upper-trophic-level predator species but may involve shifts in coastal wind strength and air and sea temperature changes, which in turn affect the seasonal formation and decay of sea ice and polynyas and thus alter prey accessibility or availability (distribution or biomass). Environmental impacts on the prey field potentially influence predators through various vital rates, such as the proportion of adults breeding and consequently the reproductive output of populations.

9. MAIN THREATS

CLIMATE VARIATION. Adélie population processes have been strongly linked to climate variability and change (Ainley 2002; Forcada and Trathan 2009; Ainley et al. 2010), but the global population is not currently considered to be under any threat. Some additional consequences of climate change may become important in the future, exacerbating some of the more direct impacts of climate (fig. 12); these include increased accessibility of the Antarctic to humans, which may then lead to increased levels of illegal, unreported, or unregulated fishing, pollution, disease, and tourism (Trathan and Agnew 2010). Current levels of tourism are not thought to pose a threat (e.g., Trathan et al. 2008), though further studies are needed.

FIG. 12 Adélie penguins are a pack-ice obligate species, spending most of the year in and on the seasonal sea ice. (J. Weller)

KRILL HARVEST. Current levels of harvesting of Antarctic krill in the Southern Ocean are not thought to affect Adélie population processes, although future harvesting levels may have an impact. In recognition of this, the CCAMLR aspires to manage the krill fishery with ecosystem-based management methods that prevent any major impact on any krill-dependent predator species (CCAMLR 2000). Though the CCAMLR has been working to develop these methods for some years, progress has been slow, and no formal method has yet been adopted some 30 years after the Convention for the Conservation of Antarctic Marine Living Resources was signed. An agreed management protocol is now increasingly urgent, as recent interest in the krill fishery indicates that future catches may rise to a very high level (Nicol and Foster 2003).

HISTORICAL HARVEST. Historical levels of harvesting of fur seals, some penguin species, whales, and, more recently, some fish and crustaceans have been such that Southern Ocean ecosystem dynamics have been disrupted to a very great extent and community interactions greatly affected (e.g., the "krill surplus" hypothesis generally attributed to Laws 1977). This so-called krill surplus is hypothesized to have facilitated population growth in other kill predators, notably penguins (Sladen 1964; Laws 1977). Sladen (1964) acknowledges that increases in populations of Adélies were likely to be a direct result of declines in baleen whale stocks. Documented increases in other species, including chinstrap and gentoo penguins and Antarctic fur seals (Payne 1977), were also thought to be due to the much more abundant food supplies. Since this hypothesis was first proposed, however, the ecosystem consequences of such a large and rapid removal of upper-trophic-level species have been the subject of much scientific debate, conjecture, and disagreement (see, e.g., Ainley et al. 2007; Nicol et al. 2007; Trathan and Reid 2009 and references therein).

As the previously reduced populations of seals and whales recover and/or as climate change begins to affect other ecosystem components (Atkinson et al. 2004; Atkinson et al. 2008), the ensuing changes to ecosystem dynamics may cause penguin populations to decline (Trathan and Reid 2009).

Eggs of a number of species of penguin, including Adélies, were harvested by sealers and whalers well into the 1950s, when the practice ended (Bonner 1984). Its effects on bird populations are unknown (Bonner 1984).

10. RECOMMENDED PRIORITY RESEARCH ACTIONS FOR CONSERVATION

1. Continue the current CCAMLR initiative to collate regional population census data in order to develop adequate circumpolar and even regional breeding population estimates.
2. Develop an understanding of current and future regional population trends, which are critical for assessing changes in regional or global population size and guiding future monitoring and management actions. Many of the very largest Adélie colonies are poorly or infrequently surveyed, which makes it particularly challenging to assess changes in regional or global population size. Understanding current and future regional population trends is critical for future monitoring and management actions.
3. Examine the dependence of individual colonies on specific foraging areas at all times of year and increase understanding of the Adélies' winter diet and distribution.
4. Understand population genetic structure and phenotypic plasticity in order to understand how populations may respond to climate variability and change.

11. CURRENT CONSERVATION EFFORTS

Although there are a number of small-scale coastal Antarctic Specially Managed Areas and Antarctic Specially Protected Areas, CCAMLR members and Antarctic Treaty parties are actively engaged in developing a network of larger-scale marine protected areas across the Southern Ocean that will, among other things, include and protect important ecosystem processes (e.g., ocean frontal systems and areas of high primary productivity); representative areas; vulnerable areas; and scientific reference areas, in both the pelagic and benthic domains. Such a network would also contribute toward the conservation of biodiversity and increase the resilience of species facing the possible impact of climate change. The first CCAMLR MPA was adopted in 2009; it is located south of the South Orkney Islands and is just under 94,000 square kilometers in size. Fishing activities and discharge or refuse disposal from fishing vessels are prohibited. This area was designated because it has important oceanographic features and is important for benthic biodiversity; it is also an area where Adélies actively forage after the completion of breeding and before molt.

12. RECOMMENDED PRIORITY CONSERVATION ACTIONS FOR INCREASING POPULATION RESILIENCE AND MINIMIZING THREATS AND IMPACTS

1. Increase protection of marine Important Bird Areas (IBAs); such areas will allow life-critical actions to be maintained, specifically access for foraging and other important behavioral activities. Marine IBAs include seaward extensions to breeding colonies, including coastal foraging and maintenance areas; coastal congregations of nonbreeding seabirds; migration bottlenecks through or around which large numbers of seabirds pass regularly; and high-seas sites, often on very productive shelf-break areas, eddies, and upwellings, that are likely to be noncontiguous with breeding colonies, as they may lie hundreds of kilometers away.
2. Implement the CCAMLR's current initiative for developing a network of large-scale MPAs as it helps protect many key areas for Adélie penguins and other Antarctic predators (Ballard et al., in press).
3. Pay special attention to high-latitude areas such as the Ross Sea, as these areas may be important refugia for Adélies as climate change effects become increasingly apparent.

ACKNOWLEDGMENTS

We wish to acknowledge the extremely generous help of Tony Williams, who granted permission to use a great deal of information from his book *The Penguins*, without which this chapter would have been very much reduced. We wish to thank Heather Lynch, Louise Emmerson, and Colin Southwell for their careful reading of an earlier version of this chapter. We are also extremely grateful for the assistance of Penny Goodearl, who helped collate references and citations.

PRBO contribution #1893.

TABLE 3.1 Selected morphometric measurements for the Adélie penguin (showing mean, standard deviation and sample size [after Williams 1995])

	MALES	FEMALES	REFERENCE
Flipper length (mm)			
East Antarctica	192.7±7.7 (25)	189.3±7.5 (20)	Kerry et al. 1992
South Shetland Islands	190.1 (28)	185.8±6.4 (18)	Scolaro et al. 1991
Ross Sea	211.0 (28)	204.0 (16)	Ainley and Emison 1972
Bill length (mm)			
East Antarctica	40.0±2.2 (25)	36.7±1.8 (20)	Kerry et al. 1992
South Shetland Islands	41.8±2.5 (28)	38.4±2.4 (18)	Scolaro et al. 1991
Ross Sea	39.5 (31)	32.9 (18)	Ainley and Emison 1972
Bill depth (mm)			
East Antarctica	19.7±1.0 (25)	18.3±0.9 (20)	Kerry et al. 1992
South Shetland Islands	20.2±1.1 (28)	18.5±1.1 (18)	Scolaro et al. 1991
Toe length (mm)			
South Shetland Islands	78.4±3.5 (28)	75.3±2.5 (18)	Scolaro et al. 1991
Tarsus length (mm)			
Ross Sea	32.8 (23)	32.2 (13)	Ainley and Emison 1972

TABLE 3.2 Body mass of the Adélie penguin at different times of the breeding season (showing mean, standard error and sample size [after Williams 1995])

	MALES	FEMALES	REFERENCE
Arrival (g)			
South Shetland Islands	5350±100 (26)	4740±100 (21)	Trivelpiece and Trivelpiece 1990
Ross Sea	5200 (NA)	4610 (NA)	Davis and Spiers 1990
Clutch complete (g)			
South Shetland Islands	4340±100 (26)	3890±100 (21)	Trivelpiece and Trivelpiece 1990

TABLE 3.3 Adélie penguin egg size (with mean and sample size)

	EGG ORDER A	EGG ORDER B	REFERENCE
Length (mm)			
South Orkney Islands	69.2 (73)	68.4 (73)	Lishman 1985a
Ross Sea	69.8 (32)	68.7 (24)	Taylor 1962
Ross Sea	70.5 (15)	68.9 (15)	Reid 1965
Ross Sea	70.4 (86)	68.7 (86)	Yeates 1968
Breadth (mm)			
South Orkney Islands	55.3 (73)	55.2 (73)	Lishman 1985a
Ross Sea	55.8 (32)	54.5 (24)	Taylor 1962
Ross Sea	56.2 (15)	55.3 (15)	Reid 1965
Ross Sea	55.0 (86)	54.4 (86)	Yeates 1968
Mass (g)			
South Orkney Islands	120.8 (73)	113.2 (73)	Lishman 1985a
Ross Sea	124.0 (32)	115.0 (24)	Taylor 1962
Ross Sea	123.6 (15)	117.7 (15)	Reid 1965

TABLE 3.4 Decadal trends in breeding populations of Adélie penguins (after Croxall et al. 2002)

	1951–1960	1961–1970	1971–1980	1981–1990	1991–2000
Ross Sea		Stable	Increase	Increase/decrease	Fluctuate
East Antarctica		Increase		Increase	Increase
Anvers Island	Increase	Increase	Decrease	Stable	Decrease
King George Island	Increase	Increase	Stable	Stable/decrease	Stable
Signy Island	Increase	Increase	Stable	Increase/decrease	Fluctuate

TABLE 3.5 Annual cycle of the Adélie penguin (showing mean durations, and mean dates and ranges for two colonies)

ANNUAL CYCLE	STAGE DURATION	SOUTH ORKNEY ISLANDS	ROSS SEA
Return to colony	20 days	27 Sept (20 Sept–8 Oct)	21 Oct (18 Oct–25 Oct)
Egg laying	13 days	27 Oct (22 Oct–2 Nov)	13 Nov (9 Nov–22 Nov)
Incubation	1st egg-39 days (36–43 days) 2nd egg-38 days (33–42 days)	Start of clutch initiation on 3 Nov	Start of clutch initiation on 12 Nov
Hatching	Over 19 days; 24–68 hours	4 Dec (4 Dec–22 Dec)	17 Dec (5 Dec–30 Dec)
Brood/ guard	22 days (16–37 days)	4 Dec (4 Dec–9 Jan)	17 Dec (5 Dec–10 Jan)
Crèche	Up to 34 days	13 Jan (13 Jan–15 Feb)	3 Jan (29 Dec–10 Feb*)
Chick fledging	At age 50–60 days		3 Feb (31 Jan–10 Feb*)
Adult departure	Over 17 days	25 Jan (25 Jan–10 Feb)	25 Jan (20 Jan–10 Feb*)
Molt	19 days (18–22 days)	19 Feb (7 Feb–3 Mar)	10 Feb (1 Feb–???)
Reference		Lishman (1985a) BAS unpublished data	Ainley et al. 1983 Ainley & Ballard unpubl.

* = estimated

Note: For the east Antarctica, see Emmerson et al. 2011.

TABLE 3.6 Diet composition by mass of Adélie penguins (after Ratcliffe and Trathan, in press)

FAO SUB-AREA	N	CRUSTACEANS	EUPHAUSIIDAE	E. SUPERBA	E. CRYSTALLOROPHIAS	AMPHIPODS	FISH	CEPHALOPODS	UNIDENTIFIED	REFERENCE
48.1	48		99.6	98	1.6					Volkman et al. 1980
48.1	600		72.7	+++	+	3.9	1.9 – 24.0		11.0	Jablonski 1985
48.1	5			100.0						Nagy and Obst 1992
48.1	41			93.6 – 99.7		+				Coria et al. 1995
48.2	10			~100.0		+				White and Conroy 1975
48.2	28		98.3 – 99.1			+	0.9 – 1.7			Lishman 1985a
48.2	~200			99.5						Lynnes et al. 2002
48.2	~120			99.0		+	+			Libertelli et al. 2003
48.5	40	30.0		28.0			14.0	54.0		Ainley et al. 1992
58.4.1	105		79.0	41.0	38.0		18.0	3.0		Ridoux and Offredo 1989
58.4.1	41		48.3	39.7	8.5		48.7	+	3.0	Kent et al. 1998
58.4.1	28			75.0	4.0	1.0	17.0			Watanuki et al. 1997
58.4.1	30		55.0 – 80.0	27.0 – 38.0	22.0 – 39.0		20.0 – 45			Wienecke et al. 2000
58.4.1	10	93.6		4.4	93.9		6.2			Cherel 2008
58.4.2	574			24.0	42.0		31.8			Puddicome and Johnstone 1988
58.4.2	132		15.8 – 95.2	7.9 – 76.9	7.9	58.0	3.6 – 20.0		13.1	Green and Johnstone 1988
58.4.2	~500		0.0 – 85.0	20.0 – 90.0	0.0 – 90.0					Thomas and Green 1988
58.4.2	158			67.9	1.4	2.4	27.9			Watanuki et a. 1994
58.4.2	52			58.0	2.0		34.0			Watanuki et al. 1997
58.4.2	~200		+++				+++			Clarke et al. 1998
58.4.2	82			35.0 – 69.0	6.0 – 22.0		0.0 – 54.0			Kato et al. 2003; Watanuki et al. 2002
58.4.2	80			34.0 – 82.0			18.0 – 66.0			Nicol et al. 2008
58.4.2	83			25.5 – 66.7	1.5 – 2.5	1.0 – 1.8	23.9 – 66.7			Tierney et al. 2008a; Tierney et al. 2008b
88.1	201	91.5			90.8	1.2	7.8			Emison 1968
88.1	15				46.0	10.0	44.0			Paulin 1975
88.1	5			0.0 – 98.9	0.0 – 97.3		1.1 – 2.7			Ainley et al. 1984
88.1	16			+	77.8 – 94.0		+			van Heezik 1988

REFERENCES

Ainley, D. G. 1975. Displays of Adélie penguins: A reinterpretation. In *The Biology of Penguins*, ed. B. Stonehouse, 503–34. London: Macmillan.

———. 2002. *The Adélie Penguin: Bellwether of Climate Change.* New York: Columbia University Press.

Ainley, D., G. Ballard, S. Ackley, L. K. Blight, J. T. Eastman, S. D. Emslie, A. Lescroel, S. Olmastron, S. E. Townsend, C. T. Tynan, P. Wilson, and E. Woehler. 2007. Paradigm lost, or is top-down forcing no longer significant in the Antarctic marine ecosystem? *Antarctic Science* 19:283–90.

Ainley, D. G., G. Ballard, K. Barton, B. J. Karl, G. Rau, C. Ribic, and P. Wilson. 2003. Spatial and temporal variation of diet within a presumed metapopulation of Adélie penguins. *Condor* 105:95–106.

Ainley, D. G., G. Ballard, B. J. Karl, and K. M. Dugger. 2005a. Leopard seal predation rates at penguin colonies of different size. *Antarctic Science* 17:323–28.

Ainley, D. G., E. D. Clarke, K. Arrigo, W. R. Fraser, A. Kato, K. J. Barton, and P. R. Wilson. 2005b. Decadal-scale changes in the climate and biota of the Pacific sector of the Southern Ocean, 1950's to the 1990's. *Antarctic Science* 17:171–82.

Ainley, D. G., and W. B. Emison. 1972. Sexual size dimorphism in Adélie penguins. *Ibis* 114:267–71.

Ainley, D. G., and R. E. LeResche. 1973. The effects of weather and ice conditions on breeding in Adélie penguins. *Condor* 75:235–39.

Ainley, D. G., R. E. LeResche, and W. J. L. Sladen. 1983. *Breeding Biology of the Adélie Penguin.* Berkeley: University of California Press.

Ainley, D. G., E. F. O'Connor, and R. J. Boekelheide. 1984. The marine ecology of birds in the Ross Sea, Antarctica. *American Ornithological Union Monograph* 32:1–97.

Ainley, D. G., C. A. Rilbie, and W. R. Fraser. 1992. Does prey preference affect habitat choice in Antarctic seabirds? *Marine Ecology Progress Series* 90:207–21.

Ainley, D. G., J. Russell, S. Jenouvrier, E. Woehler, P. O. Lyver, W. R. Fraser, and G. L. Kooyman. 2010. Antarctic penguin response to habitat change as Earth's troposphere reaches 2 degrees C above preindustrial levels. *Ecological Monographs* 80:49–66.

Ainley, D. G., and R. P. Schlatter. 1972. Chick raising ability in Adélie penguins. *The Auk* 89:559–66.

Astheimer, L. B., and C. R. Grau. 1985. The timing and energetic consequences of egg formation in the Adélie penguin. *Condor* 87:256–67.

Atkinson, A., V. Siegel, E. Pakhamov, and P. Rothery. 2004. Long-term decline in krill stock and increase in salps within the Southern Ocean. *Nature* 432:100–103.

Atkinson, A., V. Siegel, E. A. Pakhomov, P. Rothery, V. Loeb, R. M. Ross, L. B. Quetin, K. Schmidt, P. Fretwell, E. J. Murphy, G. A. Tarling, and A. H. Fleming. 2008. Oceanic circumpolar habitats of Antarctic krill. *Marine Ecology Progress Series* 362:1–23.

Ballard, G. 2010. Biotic and physical forces as determinants of Adélie penguin population location and size. PhD diss., University of Auckland.

Ballard, G., and D. G. Ainley. 2005. Killer whale harassment of Adélie penguins at Ross Island. *Antarctic Science* 17:385–86.

Ballard, G., K. M. Dugger, N. Nur, D. G. Ainley. 2010a. Foraging strategies of Adélie penguins: Adjusting body condition to cope with environmental variability. *Marine Ecology Progress Series* 405:287–302.

Ballard, G., D. Jongsomjit, S. D. Veloz, and D. G. Ainley. In press. Coexistence of mesopredators in an intact polar ocean ecosystem: The basis for defining a Ross Sea marine protected area. *Biological Conservation.*

Ballard, G., V. Toniolo, D. G. Ainley, C. L. Parkinson, K. R. Arrigo, and P. N. Trathan. 2010b. Responding to climate change: Adélie penguins confront astronomical and ocean boundaries. *Ecology* 91:2056–69.

BirdLife International. 2009. Species factsheet: *Pygoscelis adeliae.* http://www.birdlife.org (accessed 10 March 2010).

Boersma, P. D., and L. S. Davis. 1997. Feeding chases and food allocation in Adélie penguins (*Pygoscelis adeliae*). *Animal Behaviour* 53:1047–52.

Bonner, W. N. 1984. Conservation in the Antarctic. In *Antarctic Ecology*, ed. R. M. Laws, vol. 2: 821–47. London: Academic Press.

Bougaeff, S. 1975. Variations pondérales et évaluation de la dépense énergétique chez le Manchot Adélie (*Pygoscelis adeliae*). *Compte Rendus Académie Science Paris Series D* 280:2373–76.

Carleton, A. M. 2003. Atmospheric teleconnections involving the Southern Ocean. *Journal Geophysical Research* 108:8080.

Cherel, Y. 2008. Isotopic niches of emperor and Adélie penguins in Adélie Land, Antarctica. *Marine Biology* 154:813–21.

CCAMLR (Commission for the Conservation of Antarctic Marine Living Resources). [[AUS: Per CMS, for acronym as author in in-text cit.]] 2000. *Understanding CCAMLR's Approach to Management*, ed. Karl-Hermann Kock. http://www.ccamlr.org/pu/e/e_pubs/am/am-all.pdf (accessed 10 March 2010).

Clarke, J., L. M. Emmerson, and P. Otahal. 2006. Environmental conditions and life history constraints determine foraging range in breeding Adélie penguins. *Marine Ecology Progress Series* 310:247–61.

Clarke, J., K. Kerry, C. Fowler, R. Lawless, S. Eberhard, and R. Murphy. 2003. Post-fledging and winter migration of Adélie penguins *Pygoscelis adeliae* in the Mawson region of East Antarctica. *Marine Ecology-Progress Series* 248:267–78.

Clarke, J., B. Manly, K. Kerry, H. Gardner, E. Franchi, S. Corsolini, and S. Forcadi. 1998. Sex differences in Adélie penguin foraging strategies. *Polar Biology* 20:248–58.

Coria, N. R., H. Spairani, S. Vivequin, and R. Fontana. 1995. Diet of Adélie penguins *Pygoscelis adeliae* during the posthatching period at Esperanza Bay, Antarctica, 1987/88. *Polar Biology* 15:415–18.

Croxall, J. P. 1985. The Adélie penguin (*Psygoscelis adeliae*). *Biologist* 32:165–70.

Croxall, J. P., P. N. Trathan, and E. J. Murphy. 2002. Environmental change and Antarctic seabird populations. *Science* 297:1510–14.

Davis, L. S. 1982a. Crèching behaviour of Adélie penguin chicks (*Pygoscelis adeliae*). *New Zealand Journal of Zoology* 9:279–86.

———. 1982b. Timing of nest relief and its effect of breeding success in the Adélie penguins (*Pygoscelis adeliae*). *Condor* 84:178–81.

———. 1988. Coordination of incubation routines and mate choice in Adélie penguins (*Pygoscelis adeliae*). *The Auk* 105:428–32.

Davis, L. S., P. D. Boersma, and G. S. Court. 1996. Satellite telemetry of the winter migration of Adélie penguins (*Pygoscelis adeliae*). *Polar Biology* 16:221–25.

Davis, L. S., R. G. Harcourt, and C. J. A. Bradshaw. 2001. The winter migration of Adélie penguins breeding in the Ross Sea sector of Antarctica. *Polar Biology* 24:593–97.

Davis, L. S., and F. T. McCaffrey. 1986. Survival analysis of eggs and chicks of Adélie penguins (*Pygoscelis adeliae*). *The Auk* 103:379–88.

Davis, L. S., and E. A. H. Speirs. 1990. Mate choice in penguins. In *Penguin Biology*, ed. L. S. Davis and J. T. Darby, 377–97. San Diego, CA: Academic Press.

Ducklow, H. W., K. Baker, G. G. Martinson, L. B. Quetin, R. M. Ross, R. C. Smith, S. E. Stammerjohn, M. Vernet, and W. R. Fraser. 2007. Marine pelagic ecosystems: The West Antarctic Peninsula. *Philosophical Transactions of the Royal Society B* 362: 67–94.

Dugger, K. M., D. G. Ainley, P. O'B Lyver, K. Barton, and G. Ballard. 2010. Survival differences and the effect of environmental instability on breeding dispersal in an Adélie penguin meta-population. *PNAS* 107(27):12375–80.

Dunn, M., J. Silk, and P. N. Trathan. 2010. Post-breeding distribution of Adélie penguins (*Pygoscelis adeliae*) nesting at Signy Island, South Orkney Islands. *Polar Biology* 34(2):205–14.

Emison, W. B. 1968. Feeding preferences of the Adélie penguin at Cape Crozier, Ross Island. *Antarctic Research Series* 12:191–212.

Emmerson, L., R. Pike, and C. Southwell. 2011. Reproductive consequences of environment-driven variation in Adélie penguin breeding phenology. *Marine Ecology Progress Series* 440:203–16.

Emmerson, L. M., J. Clarke, K. Kerry, and C. Southwell. 2003. Temporal variability and the interrelationships between CEMP parameters collected on Adélie penguins at Béchervaise Island. *CCAMLR Science* 10:75–93.

Emmerson, L. M., and C. J. Southwell. 2008. The effect of sea ice on Adélie penguin reproductive performance. *Ecology* 89:2096–2102.

Erdmann, E. S., C. A. Ribic, D. L. Patterson-Fraser, and W. R. Fraser. 2011. Characterization of winter foraging locations of Adélie penguins along the Western Antarctic Peninsula, 2001–2002. *Deep Sea Research Part II: Topical Studies in Oceanography* 58:13–16.

Forcada, J., and P. N. Trathan. 2009. Penguin responses to climate change in the Southern Ocean. *Global Change Biology* 15:1618–30.

Forcada, J., P. N. Trathan, K. Reid, and E. J. Murphy. 2006. Contrasting population changes in sympatric penguin species with climate warming. *Global Change Biology* 12:411–23.

Fraser, W. R., and E. E. Hofmann. 2003. A predator's perspective on causal links between climate change, physical forcing and ecosystem response. *Marine Ecology Progress Series* 265:1–15.

Fraser, W. R., and D. Patterson. 1997. Human disturbance and long-term changes in Adélie penguin populations: A natural experiment at Palmer Station, Antarctica. In *Antarctic Communities: Species, Structure and Survival*, ed. B. Bataglia, J. Valencia, and D. W. H. Walton, 445–52. Cambridge: Cambridge University Press.

Fraser, W. R., W. Z. Trivelpiece, D. G. Ainley, and S. G. Trivelpiece. 1992. Increases in Antarctic penguin populations: Reduced competition with whales of loss of sea ice due to environmental warming? *Polar Biology* 11:525–31.

Gong, D. Y., and S. W. Wang. 1999. Definition of Antarctic Oscillation Index. *Geophysical Research Letters* 26:459–62.

Green, K., and G. W. Johnstone. 1988. Changes in the diet of Adélie penguins breeding in East Antarctica. *Australian Wildlife Research* 15:103–10.

Hall, A., and M. Visbeck. 2002. Synchronous variability in the Southern Hemisphere atmosphere, sea-ice and ocean resulting from the annular mode. *Journal Climate* 15:3043–57.

Hinke, J. T., K. Salwicka, S. G. Trivelpiece, G. M. Watters, and W. Z. Trivelpiece. 2007. Divergent responses of Pygoscelis penguins reveal a common environmental driver. *Oecologia* 153:845–55.

IUCN (International Union for Conservation of Nature). 2012. IUCN Red List of Threatened Species. Version 2012.1. www.iucnredlist.org (accessed 22 August 2012).

Jablonski, B. 1985. The diet of penguins on King George Island, South Shetland Islands. *Acta Zoologica Cracoviensia* 29:117–86.

———. 1987. Diurnal pattern of changes in the number of penguins on land and the estimation of their abundance (Admiralty Bay, King George Island, South Shetland Islands). *Acta Zoologica Cracoviensia* 30:97–118.

Johnstone, G. W., D. J. Lugg, and D. A. Brown. 1973. The biology of the Vestfold Hills, Antarctica. *Australian National Antarctic Research Expedition Science Report B1* 123:1–62.

Kato, A., Y. Watanuki, and Y. Naito. 2003. Annual and seasonal changes in foraging site and diving behaviour in Adélie penguins. *Polar Biology* 26:389–95.

Kent, S., J. Seddon, G. Robertson, and B. Wienecke. 1998. Diet of Adélie penguins at Shirley Island, East Antarctica, January 1992. *Marine Ornithology* 26:7–10.

Kerry, K. R., D. J. Agnew, J. R. Clarke, and G. D. Else. 1992. Use of morphometric parameters for the determination of sex in Adélie penguins. *Wildlife Research* 19:657–64.

Kerry, K. R., J. R. Clarke, and G. D. Else. 1995. The foraging range of Adélie penguins at Béchervaise Island, MacRobertson Land, Antarctica, as determined by satellite telemetry. In *Penguins: Ecology and Management*,

ed. P. Dann, I. Norman, and P. Reilly, 216–43. Chipping Norton, NSW, Australia: Surrey Beatty and Sons.

Kushner, P. J., I. M. Held, and T. L. Delworth. 2001. Southern Hemisphere atmospheric circulation response to global warming. *Journal Climate* 14:2238–49.

Laws, R. M. 1977. Seals and whales of the Southern Ocean. *Philosophical Transactions of the Royal Society B* 279:81–96.

Le Resche, R. E., and W. J. L. Sladen. 1970. Establishment of pair and breeding site bonds by young known-age Adélie penguins (*Pygoscelis adeliae*). *Animal Behaviour* 18:517–26.

Lescroël, A., K. M. Dugger, G. Ballard, and D. G. Ainley. 2009. Effects of individual quality, reproductive success and environmental variability on survival of a long-lived seabird. *Journal of Animal Ecology* 78:798–806.

Libertelli, M., N. Coria, and G. Marateo. 2003. Diet of the Adélie penguin during three consecutive chick rearing periods at Laurie Island. *Polish Polar Research* 24:133–42.

Lishman, G. S. 1985a. The food and feeding ecology of Adélie penguins (*Pygoscelis adeliae*) and the chinstrap penguins (*Pygoscelis Antarctica*) at Signy Island, South Orkney Islands. *Journal of Zoology* (London) 205:245–63.

———. 1985b. The comparative breeding biology of Adélie and chinstrap penguins, *Pygoscelis adeliae* and *Pygoscelis antarctica*, at Signy Island, South Orkney Islands. *Ibis* 127:84–99.

Lynnes, A. S., K. Reid, J. P. Croxall, and P. N. Trathan. 2002. Conflict or coexistence? Foraging distribution and competition for prey between Adélie and chinstrap penguins. *Marine Biology* 141:1165–74.

Marchant, S., and P. J. Higgins. 1990. *Handbook of Australian, New Zealand and Antarctic Birds*. Vol. 1A. Melbourne: Oxford University Press.

Murphy, E. J., P. N. Trathan, J. L. Watkins, K. Reid, M. P. Meredith, J. Forcada, E. E. Thorpe, N. M. Johnston, and P. Rothery. 2007. Climatically driven fluctuations in Southern Ocean ecosystems. *Proceedings of the Royal Society of London Series B* 274:3057–67.

Nagy, K. A., and B. S. Obst. 1992. Food and energy requirements of Adélie penguins *Pygoscelis adeliae* on the Antarctic Peninsula. *Physiological Zoology* 65:1271–84.

Nicol, S., J. Clarke, S. J. Romaine, S. Kawaguchi, G. Williams, and G. W. Hosie. 2008. Krill *Euphausia superba* abundance and Adélie penguin *Pygoscelis adeliae* breeding performance in the waters of the Bechervaise Island colony, East Antarctica, in two years with contrasting ecological conditions. *Deep Sea Research Part II: Topical Studies in Oceanography* 55:540–57.

Nicol, S., J. P. Croxall, P. N. Trathan, N. Gales, and E. J. Murphy. 2007. Paradigm misplaced? Antarctic marine ecosystems are affected by climate change as well as biological processes and harvesting. *Antarctic Science* 19:291–95.

Nicol, S., and J. Foster. 2003. Recent trends in the fishery for Antarctic krill. *Aquatic Living Resources* 16:42–45.

Oelke, H. 1975. Breeding behaviour and success in a colony of Adélie penguins, *Pygoscelis adeliae*, at Cape Crozier, Antarctica. In *The Biology of Penguins*, ed. B. Stonehouse, 363–96. London: Macmillan.

Parmalee, D. F., W. R. Fraser, and D. R. Neilson. 1977. Birds of the Palmer Station area. *Antarctic Journal of the United States* 12:14–21.

Patterson, D. L., A. Easter-Pilcher, and W. R. Fraser. 2003. The effects of human activity and environmental variability on long-term changes in Adélie penguin populations at Palmer Station, Antarctica. In *Antarctic Biology in a Global Context*, ed. A. H. L. Huiskes, W. W. C. Gieskes, J. Rozema, R. M. L. Schorno, S. M. van der Vies, and W. J. Wolff, 301–7. Leiden: Backhuys Publishers.

Paulin, C. D. 1975. Feeding of the Adélie penguin *Pygoscelis adeliae*. *Mauri Ora* 3:27–30.

Payne, M. R. 1977. Growth of a fur seal population. *Philosophical Transactions of the Royal Society of London B* 279:67–79.

Penney, R. L. 1968. Territorial and social behaviour in the Adélie penguin. In *Antarctic Research Series*, vol. 12, ed. O. L. Austin, 83–131. Washington, DC: American Geophysical Union.

Penney, R. L., and G. Lowry. 1967. Leopard seal predation on Adélie penguins. *Ecology* 48: 878–82.

Pitman, R. L., and J. Durban. 2010. Killer whale predation on penguins in Antarctica. *Polar Biology.* Published online, 4 July.

Puddicome, R., and G. W. Johnstone. 1988. The breeding season diet of Adélie penguins at the Vestfold Hills, East Antarctica. *Hydrobiologia* 165:239–53.

Quetin, L. B., and R. M. Ross. 2003. Episodic recruitment in Antarctic krill *Euphausia superba* in the Palmer LTER study region. *Marine Ecology Progress Series* 259:185–200.

Ratcliffe, N., and P. N. Trathan. In press. A review of the diet and at-sea distribution of penguins breeding within the CCAMLR area. *CCAMLR Science.*

Reid, B. 1964. The Cape Hallett Adélie penguin rookery: Its size, composition, and structure. *Records of the Dominion Museum of New Zealand* 5:11–37.

———. 1965. The Adélie penguin (*Pygoscelis adeliae*) egg. *New Zealand Journal of Science* 8:503–14.

Ridoux, V., and C. Offredo. 1989. The diet of five summer breeding seabirds in Adélie Land, Antarctica. *Polar Biology* 9:137–253.

Roeder, A., R. K. Marshall, A. J. Mitchelson, T. Visagathilagar, P. A. Ritchie, D. R. Love, T. J. Pakai, H. C. McPartlan, N. D. Murray, N. A. Robinson, K. R. Kerry, and D. M. Lambert. 2001. Gene flow on the ice: Genetic differentiation among Adélie penguin colonies around Antarctica. *Molecular Ecology* 10:1645–56.

Rootes, D. M. 1988. The status of birds at Signy Island, South Orkney Islands. *British Antarctic Survey Bulletin* 80:87–119.

Sapin-Jaloustre, J. 1960. *Ecologie du manchot Adélie.* Paris: Hermann.

Scolaro, J. A., Z. B. Stanganelli, H. Gallelli, and D. F. Vergani. 1991. Sexing of adult Adélie penguins by discriminant analysis of morphometric measurements. *CCAMLR Scientific Papers 1990*, 543–50.

Sigmond, M., and J. C. Fyfe. 2010. Has the ozone hole contributed to increased Antarctic sea ice extent? *Geophysical Research Letters* 37:L18502.

Simpson, G. G. 1972. Pliocene penguins from North Canterbury, New Zealand. *Records of the Canterbury Museum* 9(2):159–82.

Sladen, W. J. L. 1958. The *Pygoscelis* penguins. 1: Methods of study. 2: The Adélie penguin. *Falkland Islands Dependency Survey Science Report* 17:1–97.

———. 1964. The distribution of the Adélie and chinstrap penguins. In *Biologie Antarctique*, ed. R. Carrick, M. W. Holdgate, and J. Prévost, 359–65. Paris: Hermann.

Sladen, W. J. L., R. E. LeResche, and R. C. Wood. 1968. Antarctic avian population studies, 1967–68. *Antarctic Journal of the United States* 3:247–49.

Smetacek, V., and S. Nicol. 2005. Polar ocean ecosystems in a changing world. *Nature* 437:362–68.

Spurr, E. B. 1975a. Breeding of the Adélie penguin, *Pygoscelis adeliae*, at Cape Bird. *Ibis* 117:324–38.

———. 1975b. Communication in the Adélie penguin. In *The Biology of Penguins*, ed. B. Stonehouse, 449–501. London: Macmillan.

———. 1975c. Behaviour of the Adélie penguin chick. *Condor* 77:272–80.

Stonehouse, B. 1963. Observations on Adélie penguins (*Pygoscelis adeliae*) at Cape Royds, Antarctica. *Proceedings of the XIII International Ornithological Congress*, 766–79.

Taylor, R. H. 1962. The Adélie penguin, *Pygoscelis adeliae*, at Cape Royds, Antarctica. *Ibis* 104:176–204.

Tenaza, R. 1971. Behaviour and nesting success relative to nest location in Adélie penguins (*Pygoscelis adeliae*). *Condor* 73:81–92.

Thomas, P. G., and K. Green. 1988. Distribution of *Euphasia crystallorophias* within Prydz Bay and its importance to the inshore ecosystem. *Polar Biology* 8:327–31.

Tierney, M., P. D. Nichols, K. E. Wheatley, and M. A. Hindell. 2008a. Blood fatty acids indicate inter- and intra-annual variation in the diet of Adélie penguins: Comparison with stomach content and stable isotope analysis. *Journal of Experimental Marine Biology and Ecology* 367:65–74.

Tierney, M., C. Southwell, L. M. Emmerson, and M. A. Hindell. 2008b. Evaluating and using stable-isotope analysis to infer diet composition and foraging ecology of Adélie penguins *Pygoscelis adeliae*. *Marine Ecology Progress Series* 355:297–307.

Trathan, P. N., and D. J. Agnew. 2010. Climate change and the Antarctic marine ecosystem: An essay on management implications. *Antarctic Science* 22(4):387–98.

Trathan, P., C. Bishop, G. Maclean, P. Brown, A. Fleming, and M. A. Collins. 2008. Linear tracks and restricted temperature ranges characterise penguin foraging pathways. *Marine Ecology Progress Series* 370:285–95.

Trathan, P. N., J. P. Croxall, and E. J. Murphy. 1996. Dynamics of Antarctic penguin populations in relation to interannual variability in sea ice distribution. *Polar Biology* 16:321–30.

Trathan, P. N., J. Forcada, and E. J. Murphy. 2007. Environmental forcing and Southern Ocean marine predator populations: Effects of climate change and variability. *Philosophical Transactions of the Royal Society B* 362:2351–65.

Trathan, P. N., and E. J. Murphy. 2003. Sea surface temperature anomalies near South Georgia: Relationships with the Pacific El Niño regions. *Journal Geophysical Research* 108:1–10.

Trathan, P. N., E. J. Murphy, J. Forcada, J. P. Croxall, K. Reid, and S. E. Thorpe. 2006. Physical forcing in the southwest Atlantic: Ecosystem control. In *Top Predators in Marine Ecosystems: Their Role in Monitoring and Management*, ed. I. L. Boyd, S. Wanless, and K. Camphuysen, 28–45. Cambridge: Cambridge University Press.

Trathan, P. N., and K. Reid. 2009. Exploitation of the marine ecosystem in the sub-Antarctic: Historical impacts and current consequences. *Papers and Proceedings of the Royal Society of Tasmania* 143:9–14.

Trivelpiece, S. G., G. R. Geupel, J. Kjelmyr, A. Myrcha, J. Sicinski, and W. Z. Trivelpiece. 1987. Rare bird sightings from Admiralty Bay, King George Island, South Shetland Islands, Antarctica, 1976–1987. *Cormorant* 15:59–66.

Trivelpiece, W. Z., J. T. Hinke, A. K. Miller, C. S. Reiss, S. G. Trivelpiece, and G. M. Watters. 2011. Variability in krill biomass links harvesting and climate warming to penguin population changes in Antarctica. *PNAS* 108(18):7625–28.

Trivelpiece, W. Z., and S. G. Trivelpiece. 1990. Courtship period of Adélie, gentoo and chinstrap penguins. In *Penguin Biology*, ed. L. S. Davis and J. T. Darby, 113–27. San Diego, CA: Academic Press.

Trivelpiece, W. Z., and N. J. Volkman. 1979. Nest site competition between Adélie and Chinstrap penguins: An ecological interpretation. *The Auk* 96:675–81.

Turner, J., R. Bindschadler, P. Convey, G. Di Prisco, E. Fahrbach, J. Gutt, D. Hodgson, P. Mayewski, and C. Summerhayes, eds. 2009a. *Antarctic Change and the Environment.* Cambridge: Scientific Committee for Antarctic Research.

Turner, J., J. C. Comiso, G. J. Marshall, T. A. Lachlan-Cope, T. Bracegirdle, T. Maksym, M. P. Meredith, Z. M. Wang, and A. Orr. 2009b. Non-annular atmospheric circulation change induced by stratospheric ozone depletion and its role in the recent increase of Antarctic sea ice extent. *Geophysical Research Letters* 36:L08502.

van Heezik, Y. 1988. Diet of Adélie penguins during the incubation period at Cape Bird, Ross Island, Antarctica. *Notornis* 35:23–26.

Volkman, N. J., K. Jazdzewski, W. Kittel, and W. Z. Trivelpiece. 1980. Diets of Pygoscelis penguins at King George Island, Antarctica. *Condor* 82:373–78.

Volkman, N. J., and W. Trivelpiece. 1980. Growth of *Pygoscelid* penguin chicks. *Journal of Zoology* (London) 191:521–30.

Walsh, S. A., and M. E. Suarez. 2006. New penguin remains from the Pliocene of Northern Chile. *Historical Biology* 18(2):115–26.

Watanuki, Y., A. Kato, Y. Naito, G. Robertson, and S. Robinson. 1997. Diving and foraging behaviour of Adélie penguins in areas with and without fast sea ice. *Polar Biology* 17:296–304.

Watanuki, Y., A. Kato, K. Sato, Y. Niizuma, C. A. Bost, Y. Le Maho, and Y. Naito. 2002. Parental mass change and food provisioning in Adélie penguins rearing chicks in colonies with contrasting sea-ice conditions. *Polar Biology* 25:672–81.

Watanuki, Y., Y. Mori, and Y. Naito. 1994. *Euphausia superba* dominates in the diet of Adélie penguins feeding under fast sea-ice in the shelf areas of Enderby Land in summer. *Polar Biology* 14:429–32.

White, M. G., and J. W. Conroy. 1975. Aspects of competition between psygoscelid penguins at Signy Island, South Orkney Islands. *Ibis* 117:371–73.

Whitehead, M. D. 1991. Food resource utilisation by seabirds breeding in Prydz Bay, Antarctica. *ACTA XX Congressus Internationalis Ornithologici Symposium* 23: 1384–92.

Whitehead, M. D., and G. W. Johnstone. 1990. The distribution and estimated abundance of Adélie penguins breeding in Prydz Bay, Antarctica. *Proceedings of the NIPR Symposium on Polar Biology* 3:91–98.

Wienecke, B. C., R. Lawless, D. Rodary, C. A. Bost, R. Thompson, T. Pauly, G. Robertson, K. R. Kerry, and Y. LeMaho. 2000. Adélie penguin foraging behaviour and krill abundance along the Wilkes and Adélie Land coasts, Antarctica. *Deep Sea Research Part II: Topical Studies in Oceanography* 47:2573–87.

Williams, T. D. 1995. *The Penguins.* Oxford: Oxford University Press.

Wilson, P. R., D. G. Ainley, N. Nur, S. S. Jacobs, K. J. Barton, G. Ballard, and J. C. Comiso. 2001. Adélie penguin population change in the Pacific sector of Antarctica: Relation to sea-ice extent and the Antarctic Circumpolar Current. *Marine Ecology Progress Series* 213:301–9.

Wilson, R. P., B. M. Culik, D. Adelung, H. J. Spairani, and N. R. Coria. 1991. Depth utilisation by breeding Adélie penguins, *Pygoscelis adeliae*, at Esperanza Bay, Antarctica. *Marine Biology* 190:181–89.

Woehler, E. J. 1993. The distribution and abundance of Antarctic and sub-Antarctic penguins. Hobart, Australia: Scientific Committee on Antarctic Research.

Yeates, G. W. 1968. Studies of Adélie penguin at Cape Royds, 1964–65 and 1965–66. *New Zealand Journal of Marine and Freshwater Research* 2:472–96.

4

Chinstrap Penguin

(Pygoscelis antarctica)

WAYNE TRIVELPIECE AND SUE TRIVELPIECE

1. SPECIES (COMMON AND SCIENTIFIC NAMES)

Chinstrap penguin, *Pygoscelis antarctica* (J. R. Forster, 1781)

2. DESCRIPTION OF THE SPECIES

The chinstrap is the smallest of the *Pygoscelis* species, with adult weights ranging from 3.5 to 5.5 kilograms. Sexes are similar, but males are larger than females.

MORPHOMETRIC DATA. Chinstraps are similar in size to Adélies and some of the temperate penguins (table 4.1).

ADULT. The crown, forehead, and upper parts of the body are covered in blue-black feathers following molt, which turn brown over time from wear. The long tail is black. This species is easily distinguished by the conspicuous ear-to-ear black strap under the chin. The flippers are black above and white below, with a small black tip and thin black leading edge. The bill is black and nearly 50 millimeters long and has a slight hook at the tip. The feet range in color from pinkish to orange with black soles. Chinstraps have reddish-brown irises, while congeners have dark irises (fig. 2).

JUVENILE. Juveniles (up to 14 months of age) are distinguishable from adults by the black flecking on their faces, especially around the eyes (Trivelpiece et al. 1985).

FIG. 1 (*FACING PAGE*) Chinstrap penguin parents guarding chicks of various ages. (P. D. Boersma)

FIG. 2 Female (*left*) and male (*right*) chinstrap penguins mutually calling show the chinstrap characteristic of the species. (P. Angiel)

CHICK. Down is a uniform silver to light gray at birth with a second, darker gray down appearing at two to three weeks of age.

3. TAXONOMIC STATUS

The chinstrap is most closely related to the Adélie (*P. adeliae*) and gentoo (*P. papua*) penguins with which it

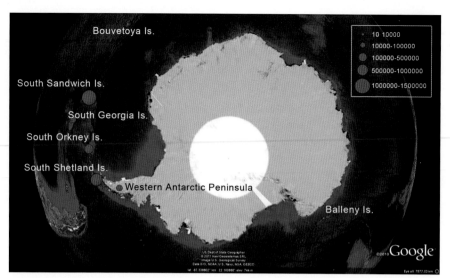

FIG. 3 Distribution and abundance of the chinstrap penguin, with counts based on pairs.

overlaps in breeding distribution throughout the core of its range in the Scotia Sea region (see section 4 for details). The *Pygoscelis* penguins branched off from the common ancestor of extant penguins about 38 million years ago and diversified into the Adélie penguin approximately 19 million years ago and the chinstrap and gentoo penguins around 14 million years ago (Baker et al. 2006). The *Pygoscelis* species compose one of only two extant penguin groups—with emperor and king penguins (*Aptenodytes*)—that stayed in Antarctica and adapted to the colder conditions that developed after the first major cooling event, which occurred between 34 million and 25 million years ago. There are no subspecies.

4. RANGE AND DISTRIBUTION

Chinstraps are located primarily on the Antarctic Peninsula, south to approximately 64° south, and on the South Shetland, South Orkney, and South Sandwich Islands (fig. 3). Small breeding populations are reported on South Georgia, Bouvet, Heard, and the Balleny Islands (Croxall and Kirkwood 1979). The nonbreeding range of the chinstrap is extensive; satellite-tracked penguins have been observed moving from the South Shetland Islands eastward 2,000 kilometers to the South Sandwich Islands and westward 4,000 kilometers to the boundary of the upper Ross (Trivelpiece et al. 2007; US AMLR 2010). In addition, a chinstrap from Bouvet Island (Bouvetøya) was tracked to the South Sandwich Islands, a winter migration of 3,600 kilometers to the southwest (Biuw et al. 2010).

5. SUMMARY OF POPULATION TRENDS

Population trends for chinstrap penguins, although complex and somewhat regional in extent and timing, suggest that this species experienced dramatic increases in numbers following the harvesting of fur seals and whales from the early 1800s to mid-1900s and the development and expansion of a finfish fishery that began in the 1960s. Chinstrap populations reached their peak in the late 1970s (Croxall and Kirkwood 1979) and have experienced significant declines at nearly all breeding sites since that time (Poncet 1997; Woehler et al. 2001; Forcada et al. 2006; Lynch et al. 2008; Forcada and Trathan 2009; Trivelpiece et al. 2011). A recent analysis of published data on chinstraps in the Antarctic Peninsula region reports an estimated population decline of 2.6 ± 0.7% per annum since 1979–80 (Lynch, pers. comm.).

6. IUCN STATUS

Chinstrap penguins are currently classified as Least Concern (BirdLife International 2010) on the International Union for Conservation of Nature's Red List of Threatened Species, based primarily on earlier census data suggesting increasing populations up through the 1970s (Croxall and Kirkwood 1979) and the assumption that this ice-avoiding species benefits from reduced winter sea ice due to climate warming (Fraser et al. 1992; Smith et al. 1999; Ducklow et al. 2007). However, recent analyses and insight from long-term research on this species in the South Shetland Islands (Hinke et al. 2007; Trivelpiece et al. 2011) and the South Orkney Islands (Forcada et al. 2006), and new regionwide census infor-

FIG. 4 A chinstrap penguin colony with unguarded chicks is pink from the krill pigment that the penguins excrete. (M. Lewis)

mation from throughout the Scotia Sea (Lynch et al. 2008), suggest that the number of chinstraps is declining significantly in all major breeding areas throughout their range. These declines appear linked to concurrent and significant reductions in krill, the chinstrap's primary food (Atkinson et al. 2004; Reiss et al. 2008; Trivelpiece et al. 2011).

Other threats to this species include disturbance by humans (primarily tourists) and interactions with fisheries. Some chinstrap breeding colonies are frequently visited (e.g., Half Moon Bay and Penguin Island [Naveen et al. 2000; Lynch et al. 2010a]) and/or are located in areas where fishing activity is concentrated (e.g., Cape Shirreff). At present, these activities do not appear to be affecting the chinstrap population as a whole but may become increasingly problematic as additional sources of stress should the climate-driven declines in krill biomass continue. Almost all chinstrap breeding populations are found south of 60° south and are thus protected under the Antarctic Treaty system, specifically, under Article 3 and Annex II of the Protocol on Environmental Protection to the Antarctic Treaty. In addition, several chinstrap breeding colonies are located in areas afforded special protection as Antarctic Specially Protected Areas and Antarctic Specially Managed Areas. Chinstraps are further protected under the Convention for the Conservation of Antarctic Marine Living Resources. However, given the magnitude of recent population declines among chinstraps throughout their breeding range, the links between climate change and reductions in krill biomass (Loeb et al. 1997; Miller and Trivelpiece 2008),

and predictions of increasing warming in this region (IPCC Report 2008), chinstrap populations should continue to be carefully monitored and their status reviewed following IUCN listing criteria (Trivelpiece et al. 2011).

7. NATURAL HISTORY

Key features of the life history of chinstraps are their highly migratory winter behavior, their nearly exclusive dependence on krill for food, and their distribution, which is confined largely to the South American quadrant of the Southern Ocean, where they overlap extensively with the developing krill fishery and tourist industries.

BREEDING BIOLOGY. Chinstraps are widely dispersed north of the seasonal pack ice during the nonbreeding winter period (April–Sept.). Breeders from the South Shetland Islands have been tracked eastward to South Georgia and the South Sandwich Islands and westward to the junction of the Amundsen and Ross Seas regions (Trivelpiece et al. 2007; US AMLR 2010). Breeders return to their colonies in early October through November and lay two eggs, following courtship fasting of two weeks for females and three weeks for males. Both sexes incubate, with initial 5- to 10-day shifts that shorten as hatching approaches, following a 33- to 36-day incubation period (table 4.2). Chicks are guarded for approximately four weeks following hatching, with parents alternating periods at the nest with foraging trips to sea that average 12–24 hours. After this guard phase, the chicks are left alone in the colony and form crèches, while both parents

FIG. 5 Chinstrap colony on a windblown slope at Bailey Head, Deception Island, Antarctica. (P. D. Boersma)

forage simultaneously and return to the nest to feed the chicks daily (fig. 4). Chicks fledge at approximately seven to eight weeks of age, after reaching 85% of adult weight.

Following chick fledging in late February, the adult penguins return to sea to feed and build up reserves for the three-week fast that accompanies their annual catastrophic molt (Trivelpiece et al. 1987, 1990; Jabłonski 1985; Lishman 1985). Chinstraps nest in dense colonies located predominantly on sloped habitat (fig. 5) and often in association with other *Pygoscelid* species (Volkman and Trivelpiece 1981). Pair-bonds are strong, with 82% of surviving pairs reuniting the following year, and site fidelity is 94% for males and 82% for females (Trivelpiece and Trivelpiece 1990).

PREY. Chinstraps are major consumers of Antarctic krill (*Euphausia superba*) (Volkman et al. 1980; Jabłonski 1985; Lishman 1985; Lynnes et al. 2004). A smaller percentage of their diet is composed of fish, primarily myctophiids (0–85% fish by frequency occurrence, less than 1–17% by weight [Jansen et al. 1998; Rombolá et al. 2003]). They generally feed pelagically at depths of less

than 40 meters and forage more frequently at night than do sympatrically breeding gentoo (*P. papua*) and Adélie (*P. adeliae*) penguins (Miller and Trivelpiece 2008). Typical dives at night are shallower than daytime dives, and overnight foragers consume a higher proportion of fish (4–35% by weight [Jansen et al. 1998]). Chinstraps appear to be making long trips at night to the shelf break, where more fish are found, and remain inshore feeding on krill during the day (Jansen et al. 1998; Ichii et al. 2007). The timing and duration of chinstrap foraging trips vary within and among breeding sites and seasons (Jansen et al. 2002), with penguins in the South Orkney Islands generally foraging farther from shore than those in the South Shetland Islands (Wilson and Peters 1999; Ichii et al. 2007; Lynnes et al. 2002; Miller and Trivelpiece 2008). Chinstraps at the South Orkney Islands have also been recorded feeding benthically, but, again, their prey was almost exclusively Antarctic krill (Takahashi et al. 2003).

The primary terrestrial predators of the chinstrap are the brown skua (*Catharacta antarctica lönnbergi*), the south polar skua (*C. maccormicki*), and the southern giant petrel (*Macronectes giganteus*). The brown skua is larger and more aggressive than the south polar skua, with which it overlaps in distribution on the Antarctic Peninsula and the South Shetland and South Orkney Islands. In areas of overlap, brown skuas control the penguin resources by defending feeding territories that encompass the penguin colonies, while south polar skuas feed primarily at sea (Trivelpiece et al. 1980; Trivelpiece and Volkman 1982; Pietz 1987). Predators spend more time searching for prey at large penguin colonies where they have a better chance of finding vulnerable targets. However, reproductive success is higher in these larger colonies, possibly because predation occurs more frequently at colony edges, where there are only a relatively small proportion of nests (Young 1994; Emslie et al. 1995; Hahn et al. 2005; Malzof and Quintana 2008). Skua predation on chinstraps is more prevalent during the incubation and early chick phases of the penguin reproductive cycle and decreases after the chicks reach crèche.

Giant petrel predation attempts on penguins increase coincident with the hatching of the first giant petrel chicks, usually in early to mid-January in the Scotia Sea region. Unlike skuas, giant petrels prey on penguins primarily after the chicks are left unattended by their parents and are in crèches. Most predation on penguins by skuas and nearly all by giant petrels occur on the ground. Only rarely do giant petrels take a chick on the wing, but Boersma (pers. comm.) saw a petrel come up a rise, grab a chick, and remove it from the small breeding colony on Aitcho Island, South Shetland Islands, without landing. The giant petrel landed at the colony earlier, but the chicks were surrounded by adults, so there was no land access. The primary marine predator is the leopard seal (*Hydrurga leptonyx*), which patrols beaches adjacent to penguin colonies and captures adult chinstrap and other penguins as they commute between the colony and the sea, where they forage over the four-month breeding season. Results from long-term studies in the South Shetland Islands suggest that leopard seals take 5% to 20% of the adult penguin breeding population per year (Mader 1998; Trivelpiece, unpubl. data).

MOLT Chinstrap penguins undergo a catastrophic molt each year after the completion of the breeding season, between February and April. Young penguins (i.e., nonbreeders) and failed breeders are the first to molt, followed by successful breeders. Chinstraps fatten up at sea for up to a month before coming ashore, usually at their breeding colonies, to molt (fig. 6). They spend approximately three weeks ashore during the molt, fasting throughout, and can lose up to 40% of their pre-molt mass during this period. Juvenile chinstraps retain their first plumage from fledging until the following February–March, when they undergo their first molt at 14 months of age. At this time, they acquire their first adult plumage, without the dark flecking on the face. Chinstraps depart the breeding colonies upon completion

FIG. 6 Chinstrap penguins arriving on the beach keep their balance by lifting their flippers. (M. Lewis)

of the annual molt and spend the winter ranging widely across the Southern Ocean (Trivelpiece et al. 2007; Biuw et al. 2010).

ANNUAL CYCLE. Chinstraps are highly migratory and are widely distributed throughout the Amundsen, Bellingshausen, and Scotia Seas in winter. They are found in the marginal ice zone (MIZ) and northward into the pelagic regions, often in association with the Antarctic Circumpolar Current (Trivelpiece et al. 2007; US AMLR 2010). The breeding season chronology of the chinstrap penguin is fairly uniform throughout its range. The annual cycle for chinstraps in the South Shetland Islands is shown in figure 7.

8. POPULATION SIZES AND TRENDS

Woehler (1993) summarized the chinstrap breeding population data throughout the Scotia Sea and estimated a total population of approximately 7,500,000 pairs. More complete censuses of the South Shetland Islands (Jabłonski 1984; Shuford and Spear 1988) reported several large colonies on Low and King George Islands that were not included in Woehler (1993). Chinstraps were distributed primarily in the South Sandwich Islands (5,000,000 pairs [Croxall et al. 1984]), South Shetland Islands (2,450,000 pairs [Shuford and Spear 1988]), and South Orkney Islands (600,000 pairs [Poncet and Poncet 1985]) and on the Antarctic Peninsula (76,000 [Croxall and Kirkwood 1979]) (table 4.3). Historical data on chinstrap populations, although complex, suggest a change in breeding population numbers over the past half century; populations were already increasing significantly but have been declining recently throughout the species' range (Lynch et al. 2008; Trivelpiece et al. 2011). Laws (1977) suggests that a large "krill surplus" resulted from the demise of the world's whale stocks during the historic whaling era. One of the possible consequences of this whale harvesting was that an estimated 150 million tons of krill was available to support other krill-consuming predators such as penguins.

Historical data on penguin populations and trends are few and largely anecdotal. The data that do exist are intriguing and tend to add support to the hypothesis that chinstrap populations increased severalfold at breeding colonies in the Scotia Sea region (Sladen 1964; Conroy 1975; Croxall and Kirkwood 1979; Croxall et al. 1984; Rootes 1988). Data from the South Orkney Islands

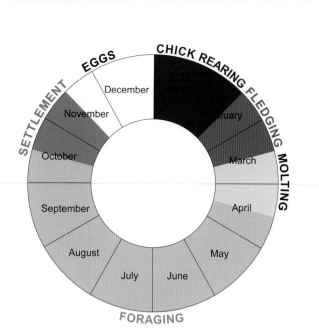

FIG. 7 Annual cycle of the chinstrap penguin.

report fivefold increases at chinstrap colonies between the 1930s and 1970 (Sladen 1964).

Recent data from the South Shetland Islands (Hinke et al. 2007) report 70% declines in breeding populations at Admiralty Bay, King George Island, since the late 1970s and declines of more than 50% from Cape Shirreff, Livingston Island, since the late 1990s. Breeding populations at Penguin Island and Chabrier Rocks have declined 75% since the early 1980s (Sander et al. 2007a; Sander et al. 2007b). Chinstrap penguins have also experienced significant population declines in the South Orkney Islands (Woehler et al. 2001; Forcada et al. 2006; Forcada and Trathan 2009), while in the South Sandwich Islands, long considered the heart of the chinstraps' distribution, populations have declined by more than 50% (Poncet 1997; Convey et al. 1999). Based on recent censuses of colonies throughout the Antarctic Peninsula, South Shetland Islands, and South Orkney Islands, Lynch et al. (2008) reports declines at all breeding sites for which data were available. Long thought to be ecological winners in the climate-warming scenario (Fraser et al. 1992; Trathan et al. 1996a; Trathan et al. 1996b; Smith et al. 1999; Ducklow et al. 2007), chinstraps may instead be among the most vulnerable species of penguins in Antarctica (Trivelpiece et al. 2011).

9. MAIN THREATS

CLIMATE VARIATION. Winter temperatures in the Antarctic Peninsula region have increased by 5–6°C over the past 60 years (Smith et al. 1996, 1999; Ducklow et al. 2007), affecting the periodicity of sea-ice coverage in the region (Murphy et al. 1995; Jacobs and Comiso 1997). This climatic change, through its associated reduction in winter sea-ice coverage, has had a profound impact on the Antarctic krill–based food web (Trathan et al. 2006; Hinke et al. 2007; Murphy et al. 2007a, Murphy et al. 2007b; Trathan et al. 2007; Forcada and Trathan 2009; Trivelpiece et al. 2011). Antarctic krill is a key species in the Antarctic marine ecosystem and accounts for nearly 100% of the prey eaten by chinstraps (Volkman et al. 1980; Lishman 1985; Trivelpiece et al. 1987, 1990; Jansen et al. 1998; Lynnes et al. 2002; Miller and Trivelpiece 2008). One of the factors that has emerged as a key link between climate variation and its impact on chinstrap populations is the relationship between sea ice and the over-winter survival of larval krill. The algal community associated with sea ice is an important winter food source for krill, particularly larval krill that, unlike adults, cannot fast for extended periods and depend on ice algae to survive over winter (Quetin and Ross 1991). Fecundity of adult krill is also enhanced following winters with extensive pack ice. The ice algae provide an important early-season food source that results in both a longer spawning season and a larger proportion of the adult population reproducing (Siegel and Loeb 1995).

The strong link between sea ice and krill fecundity and recruitment suggests that reduction in the frequency of winters with extensive sea ice has resulted in a decline in krill biomass (Loeb et al. 1997; Siegel et al. 1998; Nicol et al. 2000; Siegel 2005). Declines in krill biomass may have resulted in fewer chinstrap fledglings encountering sufficient food during their transition to independence, as fledgling survival has decreased (Hinke et al. 2007; Trivelpiece et al. 2011).

FISHERIES. The worldwide distribution of the chinstrap overlaps extensively with the distribution of the current krill fishery in the Scotia Sea region. And since chinstraps are largely confined to this region, they are potentially the species most likely to be affected by fishery activities. Current levels of krill harvesting are low (100,000–200,000 metric tons per year) compared to historical catch levels (400,000–500,000 metric tons annually in the 1980s) and represent only a small fraction of the estimated krill biomass in the region (CCAMLR 2010). There is concern that should the fishery effort increase in the future and/or be concentrated in a few select areas, it could adversely affect local chinstrap populations. To avoid this threat, the Commission for the Conservation of Antarctic Marine Living Resources' Working Group on Ecosystem Monitoring and Management proposed the establishment of small scale management units (SSMUs) (Hewitt et al. 2004) so that the allowable catch in the Scotia Sea region could be spread out across the region. The CCAMLR adopted these SSMUs as a conservation measure, but to date there has been no allocation of the krill catch quotas to individual management units.

HUMAN DISTURBANCE AND TOURISM. Tour ships began taking passengers to the Antarctic Peninsula region in the late 1960s when Lars Lindblad built the *Explorer*, nicknamed "the little red ship," and opened Antarctica to tourism. There were only two ships taking tourists to this region in the 1970s; however, since 1990, tourism activity has increased considerably on the Antarctic Peninsula, and recently more than 25 ships have begun plying the waters of this region annually (Lynch et al. 2010a). Although some authors have reported possible increased levels of stress in several penguin species exposed to visitation, including the closely related gentoo (Trathan et al. 2008), a recent analysis found no relationship between the current population declines of chinstraps and visitation levels (Lynch et al. 2010b). However, several chinstrap colonies near research stations have experienced greater declines, suggesting that some negative impacts on this species from prolonged exposure to human disturbance probably occurs (Trivelpiece, unpubl. data).

10. RECOMMENDED PRIORITY RESEARCH ACTIONS FOR CONSERVATION

Chinstraps are declining throughout their range, and thus any research that helps monitor their status and improves our understanding of factors important to their breeding biology is needed.

1. Combine censuses of chinstrap populations in the South Sandwich Islands, probably still the heart of this species' distribution, with monitoring studies of their diets, reproductive success, and foraging behaviors. Currently no research efforts are taking place in this archipelago.

2. Conduct demographic studies in the South Orkney Islands and at more southerly colonies along the Western Antarctic Peninsula for comparison with ongoing work in the South Shetlands to determine critical periods and risk factors for these populations.

3. Repeat the extensive survey of King George Island done in 1979–80 (Jabłonski 1984) with the goal of confirming trends in chinstrap population trajectories at numerous colonies of different sizes (large [50,000] to small [100s only]) on the Drakes Passage side of the South Shetland Islands where major chinstrap colonies occur but little research has been done to date.

4. Use time-energy budget analyses to help determine food needs for chinstraps so that krill fishery allotments may be established in SSMUs in these regions.

5. Collect winter distribution, diet, and survival data, which can provide a better understanding of where this species spends the nonbreeding season and the risks that may be associated with these areas.

11. CURRENT CONSERVATION EFFORTS

There are no special conservation efforts directed specifically at this species. Chinstraps benefit from several broader conservation measures designed to protect some of the sites where they breed and that require visitors and researchers to get permits to enter many of their breeding colonies. In addition, as soon as the CCAMLR has allocated fishery quotas in the krill SSMUs, chinstraps will be afforded some protection from overharvesting by the fishery in the vicinity of their breeding colonies throughout their range in the Scotia Sea).

12. RECOMMENDED PRIORITY CONSERVATION ACTIONS FOR INCREASING POPULATION RESILIENCE AND MINIMIZING THREATS AND IMPACTS

This species appears to be severely affected by the changes in krill populations that have been documented throughout the Scotia Sea region. Given their regional distribution, their heavy reliance on krill, and the extensive overlap with the krill fishery, chinstraps may be among the most vulnerable of all Antarctic penguins.

1. Initiate monitoring efforts for this species in regions where none currently exist. Monitoring studies of chinstrap populations must include annual censuses, dietary and foraging studies, reproductive research, and demographic data. Currently this information is available only from the South Shetland Islands (US AMLR 2010) and intermittently from the South Orkney Islands (UK-BAS 2007).

2. At a minimum, establish similar research efforts in the South Sandwich Islands and increase research efforts in the South Orkney Islands.

3. Gather data from the edge of the species range at South Georgia and along the Antarctic Peninsula near the southern extent of this species' distribution.

4. Collect information from all of these main breeding regions on distribution and diet during the nonbreeding winter period.

ACKNOWLEDGMENTS

We would like to thank the many field associates who have been part of our research teams in Antarctica over the past 30-plus years. In addition, we want to acknowledge and sincerely thank the long-term financial support and assistance of the U.S. Antarctic Marine Living Resources program, the National Science Foundation Office of Polar Programs, the Pew Fellows Program, and the Lenfest Program of the Pew Charitable Trusts.

TABLE 4.1 Morphometric data for the chinstrap penguin (with mean and standard deviation)

ATTRIBUTE	MALE	FEMALE	UNKNOWN	REFERENCES
Mass (g) at				
Arrival	4980±51.0 (26)	4470±45.8 (21)		Trivelpiece & Trivelpiece 1990
Clutch completion	3580±51.0 (26)	3430±45.8 (21)		Trivelpiece & Trivelpiece 1990
	4056±350.0 (17)	3384±301.0 (37)		Lishman 1983
Chick rearing	4440±318.2 (19)	3880±310.0 (25)		Croxall & Furse 1980
1st Egg	3918±314.2 (821)	3493±260.6 (821)		Trivelpiece unpubl. data
Flipper length (mm)	192.7±7.4 (27)	187.0±6.0 (28)		Amat et al. 1993
			191.0±6.8 (101)	Trivelpiece unpubl. data
Bill length (mm)	49.0±2.2 (27)	46.2±2.4 (28)		Amat et al. 1993
Egg weight (g)			both	
			113.6±8.7 (52)	Lishman 1983
			1st egg	
			104.9±10.6 (821)	Trivelpiece unpubl. data
Egg length (mm)			both	
			67.4±3.1 (51)	Croxall & Furse 1980
			1st egg	
			67.1±2.5 (51)	Lishman 1983
			67.7±3.0 (821)	Trivelpiece unpubl. data
Egg breadth (mm)			both	
			52.0±2.1 (51)	Croxall & Furse 1980
			1st egg	
			52.3±1.6 (51)	Lishman 1983
			52.2±2.4 (821)	Trivelpiece unpubl. data

TABLE 4.2 Key life history parameters for the chinstrap penguin

PARAMETER	VALUE	NOTES AND REFERENCES
Incubation period	33–37 days	Trivelpiece & Trivelpiece 1990
Chick rearing period	7–8 weeks	Volkman et al. 1980
Breeding success		
Admiralty Bay, KGI	0.81 chicks/pair	Trivelpiece, US AMLR CEMP data 1997–2010
Cape Shirreff, LI	0.92 chicks/pair	Trivelpiece, US AMLR CEMP data 1997–2010
Age at first breeding	3 years (3–22% of 3-year-olds breed)	Trivelpiece et al. 1990
Adult survival	N/A	
Juvenile recruitment		
Admiralty Bay, KGI	24%	Trivelpiece unpubl. (1984–1994)
Cape Shirreff, LI	18%	Trivelpiece unpubl. (1998–2007)
Maximum life span	16–20 years	Based on banding studies Trivelpiece unpubl. data

TABLE 4.3 Changes in populations of chinstrap penguins (by region in Antarctica)

REGION	HISTORICAL POPULATION SIZE	YEAR	REFERENCES	RECENT ESTIMATES	YEAR	NOTES/REFERENCES
South Shetland Islands	2,453,825	1987	Shuford & Spear, 1988	986,440*	2010*	*Estimate calculated using 2.6% per annum decrease in population size (Trivelpiece et al. 2011, Lynch pers.comm.)
South Sandwich Islands	5,000,000	1979	Croxall et al. 1984	1,285,000	1997	Poncet 1997
South Orkney Islands	600,000	1983	Poncet & Poncet 1985	405,600*	2010*	*Estimate calculated using 1.2% per annum decrease in population size for Signy Island, South Orkney Islands (Forcada & Trathan 2009)
Western Antarctic Peninsula	76,000	1979	Croxall & Kirkwood 1979	71,970	2010*	*Estimate calculated using 0.17% per annum decrease in population size (Trivelpiece et al. 2011)
Bouvet Island (Bou-vetøya)	200	1996	Biuw et al. 2010	40	2008	Biuw et al. 2010
South Georgia	1800	1979	Prince & Payne 1979	NA	—	Trathan pers. comm.
Balleny Islands	10	1980	Robertson et al. 1980	20	2000	Macdonald et al. 2002

REFERENCES

Amat, J. A., J. Vinuela, and M. Ferrer. 1993. Sexing chinstrap penguins (*Pygoscelis antarctica*) by morphological measurements. *Colonial Waterbirds* 16(2):213–15.

Atkinson, A., V. Siegel, E. Pakhomov, and P. Rothery. 2004. Long-term decline in krill stock and increase in salps within the Southern Ocean. *Nature* 432:100–103.

Baker, A. J., S. L. Pereira, O. P. Haddrath, and K.-A. Edge. 2006. Multiple gene evidence for expansion of extant penguins out of Antarctica due to global cooling. *Proceedings of the Royal Society B* 273:11–17.

BirdLife International. 2010. *Pygoscelis antarctica*. IUCN Red List of Threatened Species. Version 2010.1. www.iucnredlist.org (accessed 31 March 2010).

Biuw, M., C. Lydersen, P. J. Nico de Bruyn, A. Arriola, G. J. G. Hofmeyr, P. Kritzinger, and K. M. Kovacs. 2010. Long-range migration of a Chinstrap penguin from Bouvetøya to Montagu Island, South Sandwich Islands. *Antarctic Science* 22:157–62.

CCAMLR (Commission for the Conservation of Marine Living Resources). 2010. *Statistical Bulletin*, vol. 22 (2009–2010). CCAMLR, Hobart, Australia.

Conroy, J. W. H. 1975. Recent increases in penguin populations in Antarctica and Subantarctic. In *The Biology of Penguins*, ed. B. Stonehouse, 321–36. Baltimore, MD: University Park Press.

Convey, P., A. Morton, and J. Poncet. 1999. Survey of marine birds and mammals of the South Sandwich Islands. *Polar Record* 35(193):107–24.

Croxall, J. P., and J. R. Furse. 1980. Food of chinstrap penguins and macaroni penguins at Elephant Island, South Shetland Islands. *Ibis* 122:237–45.

Croxall, J. P., and E. D. Kirkwood. 1979. The distribution of penguins on the Antarctic Peninsula and islands of the Scotia Sea. Cambridge: British Antarctic Survey.

Croxall, J. P., P. A. Prince, I. Hunter, S. J. McInnes, and P. G. Copestake. 1984. The seabirds of the Antarctic Peninsula, islands of the Scotia Sea, and Antarctic continent between 80° W and 20° W: Their status and conservation. In *Status and Conservation of the World's Seabirds*, ed. J. P. Croxall, P. G. H. Evans, and R. W. Schreiber, 637–66. ICBP Technical Publication No. 2.

Ducklow, H. W., K. Baker, D. G. Martinson, L. B. Quetin, R. M. Ross, R. C. Smith, S. E. Stammerjohn, M. Vernet, and W. Fraser. 2007. Marine pelagic ecosystems: The West Antarctic Peninsula. *Philosophical Transactions of the Royal Society B* 362:67–94.

Emslie, S. D., N. Karnovsky, and W. Trivelpiece. 1995. Avian predation at penguin colonies on King George Island, Antarctica. *Wilson Bull* 107:317–27.

Forcada, J., and P. N. Trathan. 2009. Penguin responses to climate change in the Southern Ocean. *Global Change Biology* 15:1618–30.

Forcada, J., P. N. Trathan, K. Reid, E. J. Murphy, and J. P. Croxall. 2006. Contrasting population changes in sympatric penguin species in association with climate warming. *Global Change Biology* 12:411–23.

Fraser, W. R., W. Z. Trivelpiece, D. G. Ainley, and S. G. Trivelpiece. 1992. Increases in Antarctic penguin populations: Reduced competition with whales or a loss of sea ice due to environmental warming? *Polar Biology* 11:525–31.

Hahn, S., H.-U. Peter, and S. Bauer. 2005. Skuas at penguin carcass: Patch use and state-dependent leaving decisions in a top-predator. *Proceedings of the Royal Society B* 272:1449–54.

Hewitt, R. P., G. Watters, P. N. Trathan, J. P. Croxall, M. E. Goebel, D. Ramm, K. Reid, W. Z. Trivelpiece, and J. L. Watkins. 2004. Options for allocating the precautionary catch limits of krill among small-scale management units in the Scotia Sea. *CCAMLR Science* 11:81–97.

Hinke, J. T., K. Salwicka, S. G. Trivelpiece, G. M. Watters, and W. Z. Trivelpiece. 2007. Divergent responses of *Pygoscelis* penguins reveal common environmental driver. *Oecologia* 153: 845–55.

Ichii, T., J. L. Bengston, P. L. Boveng, Y. Takao, J. K. Jansen, L. M. Hiruki-Raring, M. F. Cameron, H. Okamura, T. Hayashi, and M. Naganobu. 2007. Provisioning strategies of Antarctic fur seals and chinstrap penguins produce different responses to distribution of common prey and habitat. *Marine Ecology Progress Series* 344:277–97.

IPCC (Intergovernmental Panel on Climate Change) Report. 2008.

Jabłonski, B. 1984. Distribution, numbers, and breeding preferences of penguins in the region of the Admiralty Bay (King George Island, South Shetland Islands) in the season 1979/1980. *Polish Polar Research* 5:5–16.

———. 1985. The diet of penguins on King George Island, South Shetland Islands. *Acta Zoologica Cracoviensia* 29(8):117–86.

Jacobs, S., and J. Comiso. 1997. Climate variability in the Amundsen and Bellingshausen Seas. *Journal of Climate* 10:697–709.

Jansen, J. K., P. L. Boveng, and J. L. Bengtson. 1998. Foraging modes of chinstrap penguins: Contrasts between day and night. *Marine Ecology Progress Series* 165:161–72.

Jansen, J. K., R.W. Russell, and W. R. Meyer. 2002. Seasonal shifts in the provisioning behavior of chinstrap penguins, *Pygoscelis antarctica*. *Oecologia* 131:306–18.

Laws, R. M. 1977. Seals and whales of the Southern Ocean. *Philosophical Transactions of the Royal Society of London B* 279:81–96.

Lishman, G. 1983. The comparative breeding biology, feeding ecology and bioenergetics of Adélie and Chinstrap penguins. PhD diss., University of Oxford.

———. 1985. The comparative breeding biology of Adélie and Chinstrap penguins at Signy Island, South Orkney Islands. *Ibis* 127(1):84–99.

Loeb, V., V. Siegel, O. Holm-Hansen, et al. 1997. Effects of sea-ice extent and krill or salp dominance on the Antarctic food web. *Nature* 387:897–900.

Lynch, H. J., K. Crosbie, W. F. Fagan, and R. Naveen. 2010a. Spatial patterns of tour ship traffic in the Antarctic Peninsula region. *Antarctic Science* 22(2):123–30.

Lynch, H. J., W. F. Fagan, and R. Naveen. 2010b. Population trends and reproductive success at a frequently visited penguin colony on the western Antarctic Peninsula. *Polar Biology* 33:493–503.

Lynch, H. J., R. Naveen, and W. F. Fagan. 2008. Censuses of penguins, blue-eyed shags, and southern giant petrel populations in the Antarctic Peninsula, 2001–2007. *Marine Ornithology* 36:83–97.

Lynnes, A. S., K. Reid, and J. P. Croxall. 2004. Diet and reproductive success of Adélie and Chinstrap penguins: Linking response of predators to prey population dynamics. *Polar Biology* 27:544–54.

Lynnes, A. S., K. Reid, J. P. Croxall, and P. N. Trathan. 2002. Conflict or coexistence? Foraging distribution and competition for prey between Adélie and Chinstrap penguins. *Marine Biology* 141:1165–74.

Macdonald, J. A., K. J. Barton, and P. Metcalf. 2002. Chinstrap penguins (*Pygoscelis antarctica*) nesting on Sabrina Islet, Balleny Islands, Antarctica. *Polar Biology* 25:442–47.

Mader, T. R. 1998. Temporal variation in leopard seal presence and predation near an Antarctic penguin rookery. MSc thesis, Montana State University.

Malzof, S. L., and R. D. Quintana. 2008. Diet of the south polar skua *Catharacta maccormicki* and the brown skua *C. lönnbergi* at Cierva Point, Antarctic Peninsula. *Polar Biology* 31:827–35.

Miller, A. K., and W. Z. Trivelpiece. 2008. Chinstrap penguins alter foraging and diving behavior in response to the size of their principle prey, Antarctic krill. *Marine Biology* 154:201–8.

Murphy, E. J., A. Clarke, C. Symon, and J. Priddle. 1995. Temporal variation in Antarctic sea-ice: Analysis of a long term fast-ice record from the South Orkney Islands. *Deep Sea Research* 42:1045–62.

Murphy, E. J., P. N. Trathan, J. L. Watkins, K. Reid, M. P. Meredith, J. Forcada, S. E. Thorpe, N. M. Johnston, and P. Rothery. 2007a. Climatically driven fluctuations in Southern Ocean ecosystems. *Proceedings of the Royal Society B*, 274:3057–67.

Murphy, E. J., J. L. Watkins, P. N. Trathan, K. Reid, M. P. Meredith, S. E. Thorpe, N. M. Johnston, A. Clarke, G. A. Tarling, M. A. Collins, J. Forcada, R. S. Shreeve, A. Atkinson, R. Korb, M. J. Whitehouse, P. Ward, P. G. Rodhouse, P. Enderlein, A. G. Hirst, A. R. Martin, S. L. Hill, I. J. Staniland, D. W. Pond, D. R. Briggs, N. J. Cunningham, and A. H. Fleming. 2007b. Spatial and temporal operation of the Scotia Sea ecosystem: A review of large-scale links in a krill centred food web. *Philosophical Transactions of the Royal Society B* 362:113–48.

Naveen, R., S. C. Forrest, R. G. Dagit, L. K. Blight, W. Z. Trivelpiece, and S. G. Trivelpiece. 2000. Censuses of penguin, blue-eyed shag, and southern giant petrel populations in the Antarctic Peninsula region, 1994–2000. *Polar Record* 36:323–34.

Nicol, S., T. Pauly, N. L. Bindoff, S. Wright, D. Thiele, G. W. Hosie, P. G. Strutton, and E. Woehler. 2000. Ocean circulation of East Antarctica affects ecosystem structure and sea-ice extent. *Nature* 406:504–7.

Pietz, P. J. 1987. Feeding and nesting ecology of sympatric south polar and brown skuas. *The Auk* 104:617–27.

Poncet, J. 1997. Report to Commissioner, South Georgia and South Sandwich Islands, Seabird species account. Stanley, Falkland Islands: Government Printing Office.

Poncet, S., and J. Poncet. 1985. A survey of penguin breeding populations at the South Orkney Islands. *British Antarctic Survey Bulletin* 68:71–81.

Prince, P. A., and M. R. Payne. 1979. Current status of birds at South Georgia. *British Antarctic Survey Bulletin* 48:103–18.

Quetin, L. B., and R. M. Ross. 1991. Behavioral and physiological characteristics of the Antarctic krill, *Euphausia superba*. *American Zoologist* 31:49–63.

Reiss, C. S., A. M. Cossio, V. Loeb, and D. A. Demer. 2008. Variations in the biomass of Antarctic krill (*Euphausia superba*) around the South Shetland Islands, 1996–2006, *ICES Journal of Marine Science* 65:497–508.

Robertson, C. J. R., J. R. Gilbert, and A. W. Erikson. 1980. Birds and seals of the Balleny Islands, Antarctica. *National Museum of New Zealand Records* 1:271–79.

Rombolá, E., V. E. Marscho, and N. Coria. 2003. Comparative study of the effects of late pack-ice break-off on chinstrap and Adélie penguins' diet and reproductive success at Laurie Island, South Orkney Islands, Antarctica. *Polar Biology* 26:41–48.

Rootes, D. M. 1988. The status of birds at Signy Island, South Orkney Islands. *British Antarctic Survey Bulletin* 80:87–119.

Sander, M., T. C. Balbão, E. S. Costa, C. R. dos Santos, and M. V. Petry. 2007a. Decline in the breeding population of *Pygoscelis antarctica* and *Pygoscelis adeliae* on Penguin Island, South Shetland, Antarctica. *Polar Biology* 30:651–54.

Sander, M., T. C. Balbão, M. J. Polito, E. S. Costa, and A. P. B. Carneiro. 2007b. Recent decrease in chinstrap penguin (*Pygoscelis antarctica*) populations at two of Admiralty Bay's islets on King George Island, South Shetland, Antarctica. *Polar Biology* 30:659–61.

Shuford, W. D., and L. B. Spear. 1988. Surveys of breeding penguins and other seabirds in the South Shetland Islands, Antarctica, January–February 1987. NOAA Technical Memorandum NMFS-F/NEC-59.

Siegel, V. 2005. Distribution and population dynamics of *Euphausia superba*: Summary of recent findings. *Polar Biology* 29:1–22.

Siegel, V., and V. Loeb. 1995. Recruitment of Antarctic krill (*Euphausia superba*) and possible causes for its variability. *Marine Ecology Progress Series* 123:45–56.

Siegel, V., V. Loeb, and J. Groger. 1998. Krill density, proportional and absolute recruitment and biomass in the Elephant Island region during the period 1977–1997. *Polar Biology* 19:393–98.

Sladen, W. J. L. 1964. The distribution of the Adélie and chinstrap penguins. In *Biologie Antarctique*, ed. R. Carrick, M. W. Holdgate, and J. Prevost, 359–65. Paris: Hermann.

Smith, R. C., D. Ainley, K. Baker, E. Domack, S. Emslie, B. Fraser, J. Kennett,

A. Leventer, E. Mosley-Thompson, S. Stammerjohn, and M. Vernet. 1999. Marine ecosystem sensitivity to climate change. *BioScience* 49:393–404.

Smith, R. C., S. E. Stammerjohn, and K. S. Baker. 1996. Surface air temperature variations in the Western Antarctic Peninsula region. In *Foundations for Ecological Research West of the Antarctic Peninsula*, vol. 70, ed. R. M. Ross, L. B. Quetin, and E. E. Hofmann, 105–21. Antarctic Research Series. Washington, DC: American Geophysical Union.

Takahashi, A., M. J. Dunn, P. N. Trathan, K. Sato, Y. Naito, and J. P. Croxall. 2003. Foraging strategies of Chinstrap penguins at Signy Island, Antarctica: Importance of benthic feeding on Antarctic krill. *Marine Ecology Progress Series* 250:279–89.

Trathan, P. N., J. P. Croxall, and E. J. Murphy. 1996a. Dynamics of Antarctic penguin populations in relation to the annual variability in sea ice distribution. *Polar Biology* 16:321–30.

Trathan, P. N., F. H. J. Daunt, and E. J. Murphy, eds. 1996b. *South Georgia: An Ecological Atlas*. Cambridge: British Antarctic Survey.

Trathan, P. N., E. J. Murphy, J. Forcada, J. P. Croxall, K. Reid, and S. E. Thorpe. 2006. Physical forcing in the southwest Atlantic: Ecosystem control. In *Top Predators in Marine Ecosystems: Their Role in Monitoring and Management*, ed. L. Boyd, S. Wanless, and C. J. Camphuysen, 28–45. Cambridge: Cambridge University Press.

Trathan, P. N., J. Forcada, and E. J. Murphy. 2007. Environmental forcing and Southern Ocean marine predator populations: Effects of climate change variability. *Philosophical Transactions of the Royal Society B* 362:2351–65.

Trathan, P. N., J. Forcada, R. Atkinson, R. H. Downie, and J. R. Shears. 2008. Population assessments of gentoo penguins (*Pygoscelis papua*) breeding at an important Antarctic tourist site, Goudier Island, Port Lockroy, Palmer Archipelago, Antarctica. *Biological Conservation* 141:3019–28.

Trivelpiece, W. Z., S. Buckelew, C. Reiss, and S. G. Trivelpiece. 2007. The winter distribution of Chinstrap penguins from two breeding sites in the South Shetland Islands of Antarctica. *Polar Biology* 30:1231–37.

Trivelpiece, W. Z., R. G. Butler, and N. J. Volkman. 1980. Feeding territories of brown skuas (*Catharacta lönnbergi*). *The Auk* 97:669–76.

Trivelpiece, W. Z., J. T. Hinke, A. K. Miller, C. S. Reiss, S. G. Trivelpiece, and G. M. Watters. 2011. Variability in krill biomass links harvesting and climate warming to penguin population changes in Antarctica. *Proceedings of the National Academy of Sciences* 108 (18):7625–28.

Trivelpiece, W. Z., and S. G. Trivelpiece. 1990. Courtship period of Adélie, gentoo and chinstrap penguins. In *Penguin Biology*, ed. L. S. Davis and J. T. Darby, 113–27. San Diego, CA: Academic Press.

Trivelpiece, W. Z., S. G. Trivelpiece, G. R. Geupel, J. Kjelmyr, and N. J. Volkman. 1990. Adélie and chinstrap penguins: Their potential as monitors of the Southern Ocean marine ecosystem. In *Ecological Change and the Conservation of Antarctic Ecosystems*, ed. K. R. Kerry and G. Hempel, 191–202. Proceedings of the Fifth SCAR Symposium on Antarctic Biology, Hobart. Berlin: Springer-Verlag.

Trivelpiece, S. G., W. Z. Trivelpiece, and N. J. Volkman. 1985. Plumage characteristics of juvenile pygoscelid penguins. *Ibis* 127:378–80.

Trivelpiece, W. Z., S. G. Trivelpiece, and N. J. Volkman. 1987. Ecological segregation of Adélie, gentoo, and chinstrap penguins at King George Island. *Antarctica: Ecology* 68(2):351–61.

Trivelpiece, W. Z., and N. J. Volkman. 1982. Feeding strategies of sympatric south polar *Catharacta maccormicki* and brown skuas *C. lönnbergi*. *Ibis* 124:50–54.

UK-BAS (U.K. British Antarctic Survey). 2007. Data and Collections. http://www.antarctica.ac.uk/bas_research/data/index.php (accessed 18 April 2012).

US AMLR (U.S. Antarctic Marine Living Resources). 2010. Overwinter movement patterns of Antarctic predators. http://swfsc.noaa.gov/ge.aspx?ParentMenuID=42&TopPG=16274&BottomPG=16272&Project=2010ChinstrapTrack (accessed 28 June 2011).

Volkman, N. J., P. Presler, and W. Trivelpiece. 1980. Diets of pygoscelid penguins at King George Island, Antarctica. *Condor* 82:373–78.

Volkman, N. J., and W. Trivelpiece. 1981. Nest-site selection among Adélie, Chinstrap and gentoo penguins in mixed species rookeries. *The Wilson Bulletin* 93:243–48.

Wilson, R. P., and G. Peters. 1999. Foraging behaviour of the Chinstrap penguin *Pygoscelis antarctica* at Ardley Island, Antarctica. *Marine Ornithology* 27:85–95.

Woehler, E. J. 1993. *The Distribution and Abundance of Antarctic and Subantarctic Penguins.* Scientific Committee on Antarctic Research, Cambridge. 76 pp.

Woehler, E. J., J. Cooper, J. P. Croxall, W. R. Fraser, G. L. Kooyman, G. D. Miller, D. C. Nel, D. L. Patterson, H.-U. Peter, C. A. Ribic, K. Salwicka, W. Z. Trivelpiece, and H. Weimerskirch. 2001. A statistical assessment of the status and trends of Antarctic and Subantarctic seabirds. SCAR report.

Young, E. C. 1994. *Skua and Penguin: Predator and Prey.* Cambridge: Cambridge University Press.

5

Gentoo Penguin

(Pygoscelis papua)

Heather J. Lynch

1. SPECIES (COMMON AND SCIENTIFIC NAMES)

Gentoo penguin, *Pygoscelis papua* (J. R. Forster, 1781)

2. DESCRIPTION OF THE SPECIES

The gentoo is the third-largest penguin. Adults weigh between five and eight kilograms.

ADULT. The throat, exterior portion of the wing, and back are dark (described variously as black, bluish black, or gray brown), and the ventral surface from the breast down to the vent are white. The undersides of the wings are white with a narrow black leading edge and a black tip. The face and head are black with the distinguishing exception of two white but variably sized patches above, and usually continuous with, white eye-rings and connected by a thin white band over the crown of the head (fig. 2). Additional scattered white feathers are also often evident on the head and nape, particularly at the trailing edge of the eye patch. The gentoo bill is red orange except along the upper mandible and at the tip, where it is black. The upper mandible may also have an elongated yellow-to-red spot (Metcheva et al. 2008). The feet range in color from pale pink to orange to red. Gentoos, like all other penguins, have a uropygial gland that they use to lubricate and protect their feathers (fig. 3). Males are typically larger than females, although the differences can be difficult to detect in the field (Renner et al. 1998). Body size and morphology are highly variable (table 5.1).

FIG. 1 (*FACING PAGE*) Adult (*background*) and juvenile (*foreground*) gentoo penguin. Note that the juvenile's bill is not as red, the chin is white, and the head patch is less distinct than in the adult. (P. Ryan)

FIG. 2 Gentoo penguins have orange bills and head patches of white feathers that are individually distinct. (P. Ryan)

Moreover, gentoos breeding at different localities within the same geographic area can vary in morphologies, possibly adaptive to local oceanographic conditions (Bost and Jouventin 1990b; Bost et al. 1992).

JUVENILE. Similar to the adult but smaller, with a smaller white eye patch that may or may not be continuous with the white eye-ring (Williams 1995) (see fig. 1).

suggests that it is most closely related to the chinstrap penguin, from which it diverged between 10.8 and 18.3 million years ago (Baker et al. 2006). There are two gentoo subspecies (*P. papua ellsworthi* and *P. papua papua*) (Stonehouse 1970). *P. papua ellsworthi* breeds on the Antarctic Peninsula and the South Shetland, South Orkney, and South Sandwich Islands; while the larger *P. papua papua* breeds farther north on the Falkland/Malvinas Islands, South Georgia Island, and other sub-Antarctic islands (Stonehouse 1970; Williams 1995).

4. RANGE AND DISTRIBUTION

The gentoo has a circumpolar breeding distribution that ranges in latitude from Cape Tuxen on the Antarctic Peninsula (65°16′ S) to the Crozet Islands (46°00′ S) (fig. 5). Although Petermann Island (65°10′ S, 64°10′ W) has historically been reported as the gentoo's southernmost breeding location (Croxall and Kirkwood 1979), the distribution of gentoo breeding has moved southward since 2000, concurrent with a rapid increase in the gentoo population at its southern extent (Lynch et al. 2008, 2012) (see also section 8).

Breeding populations are summarized in table 5.2. Gentoo colonies are small compared to Adélie and chinstrap colonies; the largest (e.g., Cuverville Island, Antarctic Peninsula [64°41′ S, 62°38′ W] and North Beach, New Island, Falkland/Malvinas Islands [51°43′ S, 61°18′ W]) include only around 6,000 breeding pairs (Lynch et al. 2008). The nonbreeding range of gentoo penguins is not as well known, although gentoos are considered more sedentary than the other pygoscelids (Williams 1995) and adults have been found as far south in the winter as they are found to breed during the summer (M. Chesalin, unpubl. data).

5. SUMMARY OF POPULATION TRENDS

The current estimate of 387,000 breeding pairs is 23% larger than the last estimate of 314,000 breeding pairs (Woehler 1993), and the global gentoo population appears to have increased over the past 20–30 years. The three largest populations of gentoos (Falkland/Malvinas Islands, South Georgia Island, and the Antarctic Peninsula) contain 80% of the world's gentoo penguins (fig. 6), and at these locations, populations are either stable (Falkland/Malvinas Islands, South Georgia Island) or increasing (Antarctic Peninsula) (see table 5.2).

FIG. 3 Penguins use the uropygial gland, shown here in detail, to oil their feathers. Note the long tail feathers characteristic of the brush-tailed penguins. (P. D. Boersma)

FIG. 4 A gentoo penguin chick covered in dense down has its juvenile plumage under the down. (P. Ryan)

CHICK. Gray dorsally from the crown to the tail and white ventrally including the throat (fig. 4).

MORPHOMETRIC DATA. See table 5.1.

3. TAXONOMIC STATUS

The gentoo shares the genus *Pygoscelis* with the chinstrap and Adélie penguins, although DNA evidence

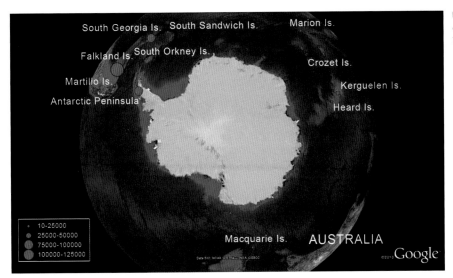

FIG. 5 Distribution and abundance of the gentoo penguin, with counts based on pairs.

South Georgia Is. South Sandwich Is. Marion Is.

Falkland Is. South Orkney Is.

Martillo Is. Crozet Is.

Antarctic Peninsula Kerguelen Is.

Heard Is.

- 10-25000
- 25000-50000
- 75000-100000
- 100000-125000

Macquarie Is. AUSTRALIA

Google

6. IUCN STATUS

The International Union for Conservation of Nature has listed the gentoo penguin as Near Threatened on its Red List of Threatened Species "because it is suspected to be undergoing a moderately rapid population decline, owing to rapid declines in some populations" (BirdLife International 2010). This status appears unjustified on the basis of the apparent increase in the global gentoo population over the past two decades and also because of the stable or increasing populations at most of the gentoo's breeding sites. Its classification based on current evidence should be Least Concern.

Gentoo penguins are far from qualifying as Vulnerable under any of the IUCN Red List criteria: population size reduction (global populations are increasing), small geographic range (gentoos are one of the most widespread of the penguin species [Bost and Jouventin 1990b]

FIG. 6 A male gentoo penguin vocalizes on Beaver Island, Falkland/Malvinas Islands. (P. D. Boersma)

and are expanding at their southern extreme [Lynch et al. 2008]), small populations (390,000 breeding pairs far exceed the threshold of 10,000 mature individuals), or evidence of likely extinction within the next 100 years (there is none). Threats identified alongside its status are disturbance by humans, local pollution, and potential interaction with fisheries, none of which appear to be significant negative factors for the majority of the population (see section 9).

LEGAL STATUS. Gentoos occupy a large geographic range and are protected under a variety of legal frameworks. Populations south of 60° are protected under the Antarctic Treaty system and, specifically, Article 3 and Annex II to the Protocol on Environmental Protection to the Antarctic Treaty. Within this region, several areas have been afforded special protection under the designations Antarctic Specially Protected Areas and Antarctic Specially Managed Areas, including several with resident gentoo populations. These populations are also subject to protections under the Convention for the Conservation of Antarctic Marine Living Resources, which extends protection north to the Antarctic Convergence. At South Georgia Island, a 22-kilometer fishery exclusion buffer around the coastline mitigates potential overlap with the foraging habitats of nearshore foraging species such as the gentoo penguin. Additionally, all tourist sites at South Georgia are strictly regulated and subject to visitor permitting.

Gentoos north of the Antarctic Convergence are generally subject to lower levels of environmental protec-

FIG. 7 Gentoo penguin brooding an egg and a chick. (P. D. Boersma)

tion. In response to concerns over the impact of fishing on seabirds and marine resources, the Falkland Islands Interim Conservation and Management Zone and the Falkland Outer Conservation Zone have been established. These areas, and additional measures such as the Falkland Islands Fisheries Ordinance, established fishing zones, a system of fishing licenses, and procedures for the calculation of catch limits, could mitigate impacts on the Falkland/Malvinas Islands gentoo population. Marine protected areas have been either designated or proposed for several of the sub-Antarctic islands containing gentoo populations, including Heard and Prince Edward Islands and the Crozet and Kerguelen Islands (ASOC 2008).

7. NATURAL HISTORY

Key features of the life-history characteristics of gentoos are the high degree of flexibility evident in almost all aspects of their breeding, diet, and phenology, a preference for inshore foraging, and a resident (nonmigratory) life history.

BREEDING BIOLOGY. Key life-history parameters are included in table 5.3, and the annual cycle of reproduction and molt are illustrated for two representative locations in table 5.4. As with all the pygoscelids, gentoos lay two eggs separated by a three- to four-day interval (Williams 1990, 1995; Williams and Croxall 1991). There is little size difference between the first and second eggs (see table 5.1). Intraclutch asynchrony in hatching is variable (mean 1.2 ± 0.7 day and a range of 0–5 days in one study [Lescroël et al. 2009]) (fig. 7); the extent to which gentoos are employing a brood reduction strategy is

equivocal and may depend on location and subsequent food availability (Williams and Croxall 1991; Croxall and Davis 1999; Lescroël et al. 2009). Although it had been thought that none of the penguin species participated in extended parental care, more recent evidence suggests that gentoo chicks receive some post-fledging provisioning by adults (Williams 1990; Bost and Jouventin 1991; Polito and Trivelpiece 2008). Gentoos can breed as early as two years of age (Williams and Rodwell 1992) and most are breeding by the time they are three or four years old (Williams 1995).

Gentoos in South Georgia Island and the Falkland/Malvinas Islands nest on flat beaches or in tussock grasses (fig. 8) (Croxall and Prince 1980a; Reilly and Kerle 1981), whereas those breeding farther south on the Antarctic Peninsula typically nest on low-lying gravel beaches and dry moraines (fig. 9) (Jablonski 1984; Volkman and Trivelpiece 1981).

Gentoos exhibit a high degree of plasticity in a range of life-history traits, including in their breeding biology (Bost and Jouventin 1990b). The relatively large year-to-year fluctuations in breeding population at a given site are probably related to a high degree of skipped breeding among mature individuals (Williams and Rodwell 1992; Croxall and Rothery 1995) and delayed recruitment of first-time breeders (Williams 1990). Breeding success is also highly variable and appears related to a suite of factors including food availability, latitude, individual age, and colony size (Williams 1995). Breeding success is generally higher in Antarctic populations than in sub-Antarctic populations (Bost and Jouventin 1990b; Lescroël et al. 2009), with most reports falling between 0.4 to 1.5 chicks crèched per nest (see table 5.3; fig. 10). Gentoos are capable of laying a replacement clutch upon egg or nest loss (Bost and Jouventin 1990b; Williams 1980), although clutch replacement is far more frequent at more northern locations (e.g., Crozet Island [Bost and Clobert 1992]) than on the Antarctic Peninsula, where clutch replacement appears possible only within the first week of incubation (Cordier et al. 1983).

Mate fidelity among gentoos is highly variable, both among breeding locations and also between years (Williams and Rodwell 1992). Estimates of mate fidelity range from 49% at Crozet Island (Bost and Jouventin 1991), to 72–89% at South Georgia Island (Williams and Rodwell 1992), to 90% in the South Shetland Islands (Trivelpiece and Trivelpiece 1990). Gentoos also display a relatively

FIGS. 8A (*ABOVE*) AND 8B Gentoo penguins defending their turf. (P. Ryan, P. D. Boersma)

FIG. 9 Gentoo penguins must lay their eggs on rock so that the eggs do not freeze. (J. Weller)

high degree of site fidelity (estimated at 89–100% at South Georgia [Williams and Rodwell 1992] and around 60% at King George Island [Williams 1995]), although they appear to be excellent colonizers of new breeding territory, especially at their southern range boundary (H. Lynch, pers. obs.). Gentoos are renowned for the mobility of their nesting sites despite a high level of phylopatry, particularly when located in vegetated areas (Bost and Jouventin 1990b).

PREY. The gentoo exhibits considerable variation in diet between years and among breeding locations (Bost and Jouventin 1990b; Pütz et al. 2001; Klages et al. 1990; Ridoux 1994; Lescroël et al. 2004; Lescroël and Bost 2005; Miller et al. 2009). At all locations, gentoos have a preference for foraging inshore close to where they breed (Croxall and Prince 1980a; Croxall and Prince 1980b; Trivelpiece et al. 1987; Klages et al. 1990; Wilson et al. 1998; Adams and Wilson 1987; Williams and Siegfried

FIG. 10 Adult gentoo calling to its two chicks. (P. D. Boersma)

FIG. 11 Gentoo penguin chick on a feeding chase with a skua looking for food at New Island, Falkland/Malvinas Islands. (P. D. Boersma)

1980; Tanton et al. 2004). Gentoos are opportunistic foragers feeding on species that are spatially and temporally patchy (Robinson and Hindell 1996; Adams and Klages 1989; Lescroël and Bost 2005; Miller et al. 2009). They feed predominantly on crustaceans, fish, and squid (see table 5.5) (Ratcliffe and Trathan, in review), although the relative mix of each component varies within their breeding range. Antarctic krill (*Euphausia superba*) are the dominant species eaten in the South Shetland Islands, and in the South Orkneys, the penguins are piscivorous. Diet can also vary between the breeding period and the nonbreeding period (Hindell 1989; Clausen and Pütz 2003). The general trend appears to be one of decreasing variability and increasing krill consumption at higher latitudes (Bost and Jouventin 1990b).

PREDATORS. The sub-Antarctic skua (*Catharacta antarctica*) is the primary predator of gentoo eggs and small chicks (fig. 11) (Bost and Jouventin 1990b, 1991; Quintana and Cirelli 2000; Crawford et al. 2003a; Hahn et al. 2005). Giant petrels (*Macronectes* spp.) also kill young chicks and even adults. Leopard seals (Walker et al. 1998; Forcada et al. 2009), sea lions, and killer whales all prey on adult gentoos.

MOLT. Following chick fledging, gentoos leave the colony and later reassemble at the breeding colony for molt.

Although it lasts only one to two weeks at higher latitudes (e.g., 10 days at South Georgia [Croxall 1982]), the pre-molt period is longer at lower latitudes (one to two months or longer), especially for those birds completing reproduction early in the breeding season (Bost and Jouventin 1991; Williams 1995). The molting process itself takes an average of 19.5 days (Adams and Brown 1990), during which individuals lose approximately 3% of body mass per day (Croxall 1982; Davis et al. 1989). A detailed description of the plumage replacement process in gentoos may be found in Reilly and Kerle (1981).

ANNUAL CYCLE. The annual cycle (fig. 12) is less protracted in northern populations (north of the Antarctic Convergence, e.g., Crozet and Macquarie Islands) than in southern populations (south of the Antarctic Convergence), as summarized in table 5.4. Northern populations of gentoo penguins start breeding in early winter and continue over an extended period of almost five months, whereas southern populations lay eggs for two to three weeks in early spring (Williams 1980; Bost and Jouventin 1990a, 1990b). The extended breeding period at northern location is due largely to re-laying after egg or chick loss. More than 80% of late egg laying (more than a month after breeding commenced) involved clutch replacement.

Compared with their congeners, gentoos exhibit

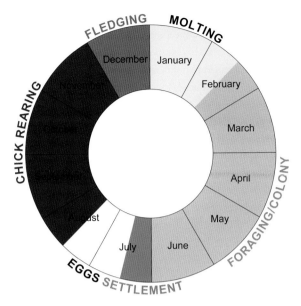

FIG. 12 Annual cycle of the gentoo penguin.

more phenotypic plasticity (Lescroël et al. 2009; Lynch et al. 2009), which may offer considerable advantage in an era of changing environmental conditions (Forcada and Trathan 2009). Such flexibility can lead to a high degree of year-to-year variability in mean clutch initiation, above and beyond any within-year asynchrony (Croxall and Prince 1979; Williams 1990).

8. POPULATION SIZES AND TRENDS

The current estimate of the total gentoo population is about 387,000 breeding pairs (see table 5.2), which is 23% larger than the last global estimate of 314,000 breeding pairs (Woehler 1993). Population trends are difficult to establish because of large year-to-year fluctuations in the size of the gentoo breeding population at a given location. Where possible, trends have been assessed on the 20- to 30-year time scale. Gentoo populations are increasing at most sites on the Antarctic Peninsula where they are monitored, particularly at those sites at the southern extent of their breeding range (Lynch et al. 2008, 2012). Populations also appear to be increasing on the South Orkney (Lynch et al. 2012; Forcada and Trathan 2009) and South Sandwich Islands (Convey et al. 1999).

Although highly variable, gentoo populations appear stable in the Falkland/Malvinas Islands (Bingham 2002; Clausen and Huin 2003; Baylis et al. 2012), on Prince Edward Island (Crawford et al. 2009), on Marion Island (Crawford et al. 2003a, 2003b, 2009), and on South

Georgia (Trathan et al. 1996; Forcada and Trathan 2009) and appear to be stable or slightly increasing on Crozet Island (Voisin 1984). Populations appear to be decreasing on Kerguelen Island (Lescroël and Bost 2006) and Heard Island (E. Woehler, pers. comm.). A tiny population on Martillo Island, Tierra del Fuego, appears to be increasing (Ghys et al. 2008). The population trajectory on Macquarie Island is unknown.

9. MAIN THREATS

Due to the latitudinal range over which gentoos breed, threats to the species are best divided into those that affect the northern population and those that affect the more remote southern population. Historically, threats to northern gentoo populations were egg collecting and potential competition with fisheries. Collection of gentoo eggs has decreased dramatically over the past few years in the Falkland/Malvinas Islands (Clausen and Pütz 2002), although some legal egg collection continues (Otley et al. 2004). The extent to which gentoo penguins compete with fisheries in the Falkland/Malvinas Islands has been a matter of considerable debate (Pütz et al. 2001; Clausen and Pütz 2002; Bingham 2002; Clausen and Pütz 2003), in part because the extent of overlap between the gentoo diet and commercial targeted species varies widely over the course of the year (Pütz et al. 2001; Clausen and Pütz 2002, 2003). More recent threats include oil spills from oil rig supply vessels operating in the vicinity of the Falkland/Malvinas Islands (Bingham 2002). Expanded oil exploration close to the Falkland/Malvinas Islands is a current concern given the potential negative impact of oil spills on gentoo populations, as has occurred with other species of penguins throughout the world (e.g., Underhill et al. 1999; Crawford et al. 2000; Garcia-Borboroglu et al. 2006; Boersma 2008). Heavy metals and other pollutants have been found in chicks (Jerez et al. 2011).

Tourism is a concern for breeding gentoos. Studies on other species of penguins show that human visitation results in physiological or behavioral responses indicating stress (Nimon et al. 1995; Walker et al. 2005, 2006; Wilson et al. 1991) and may change long-term population trends of colonies neighboring human settlements (Woehler et al. 1994). The evidence for decreased breeding productivity at frequently visited sites is equivocal, with some studies finding evidence of depressed breeding productivity (Trathan et al. 2008; Lynch et al. 2009) and some showing

FIG. 13 A gentoo penguin chick pants as it tries to dissipate heat on a hot day. Note the denticles on the tongue, which help both adults and chicks hold their prey. (P. D. Boersma)

FIG. 14 Gentoo penguins are expanding southward. Here, a gentoo examines recently exposed glacially striated rock. (H. Lynch)

no effect (Cobley and Shears 1999; Holmes et al. 2006). Cobley and Shears (1999) suggests that gentoos in disturbed colonies are not disadvantaged during particularly successful breeding seasons, consistent with the pattern of long-term studies (presumably covering a range of conditions) that find a more significant impact from tourism than single-season studies (Lynch et al. 2009).

Although the effect of tourism on penguins is most frequently couched in terms of the impact on breeding penguins resident at their terrestrial breeding colonies, by-products of marine traffic such as noise pollution and disturbance to penguins (particularly gentoos) foraging in inshore waters (Lynch et al. 2010a) should also be considered. Tourism activity has increased exponentially since 1990 on the Antarctic Peninsula, as has associated marine traffic. It is important to note that such traffic can affect even those sites that may be designated as off-limits for visitation, especially if the sites are located along frequently traveled ship routes or are in the vicinity of other, more popular tourist landing sites. Although typically not considered a major concern in Antarctic waters, oil spills (most famously, the *Bahia Paraiso* [Kennicutt et al. 1991]) and oiled penguins (Reid 1995) have been reported, highlighting the concerns accompanying greater marine traffic in otherwise pristine waters.

The Antarctic populations of gentoo penguins have been increasing over the past 30 years (Forcada et al. 2006; Lynch et al. 2008), a period during which the climate has changed significantly (fig. 13). The Antarctic Peninsula has experienced significant warming (Vaughan et al. 2003), particularly in mid-winter temperatures (Smith et al. 1999), a shorter period of sea-ice coverage (Smith and Stammerjohn 2001; Stammerjohn et al. 2008a; Stammerjohn et al. 2008b), and changing patterns of precipitation (Turner et al. 2005). These changes have had a significant impact on multiple facets of the Southern Ocean ecosystem in which much of the global gentoo population breeds (e.g., Loeb et al. 1997; Smith et al. 1999; Croxall et al. 2002; Clarke et al. 2007; Ducklow et al. 2007; Murphy et al. 2007; Trathan et al. 2007). The fastest rates of population growth are seen at newly established colonies clustered at the southern boundary of the gentoo's breeding range, where populations appear to be expanding as a result of more open water and increased snow-free breeding habitat (Lynch et al. 2012) (fig. 14). In all respects, the highly flexible nature of gentoos appears to make the bird a beneficiary

FIG. 15 Gentoo penguin incubating eggs on its stone nest at Brown Bluff, Antarctic Peninsula. Penguins may use more than 1,000 rocks to keep their nest platforms above runoff from melting snow and ice. (P. D. Boersma)

of climate change on the Antarctic Peninsula. Gentoos are less obligately colonial than either Adélie or chinstrap penguins (Volkman and Trivelpiece 1981) and are excellent colonizers of new breeding habitat that becomes available either through glacial retreat or snow melt. The capacity to re-lay a lost clutch buffers gentoos against nest loss resulting from extreme snow and rain events (fig. 15). Their sedentary nature allows them to initiate breeding according to local environmental conditions, and their phenotypic plasticity provides flexibility as the peninsular climate changes. Finally, the opportunistic foraging habits of the gentoo may allow for rapid adaptation as the marine food web shifts in response to climate change.

10. RECOMMENDED PRIORITY RESEARCH ACTIONS FOR CONSERVATION

Our understanding of gentoo populations is aided by the relative accessibility of their breeding areas and their small colony size, which make complete censuses feasible. Although sites along their southern boundary are becoming more hospitable for gentoos with climate change, it is important to assess whether breeding sites at the northern end of the gentoo range are becoming less suitable, particularly if environmental changes impact the marine food web.

1. Complete censuses of populations breeding on the sub-Antarctic islands, specifically, Crozet, Heard, and Macquarie, the sites with the least recent census counts.

2. Survey potential breeding areas south of Petermann Island to monitor southward range expansion.

11. CURRENT CONSERVATION EFFORTS

No special conservation efforts are currently being undertaken specifically for gentoo penguins, although they benefit from a number of broadly scoped legal protections outlined in section 6.

12. RECOMMENDED PRIORITY CONSERVATION ACTIONS FOR INCREASING POPULATION RESILIENCE AND MINIMIZING THREATS AND IMPACTS

Marine protections should focus on pollution, particularly on minimizing oil pollution, which could have significant consequences for a sedentary inshore forager like the gentoo. Terrestrial protections for gentoos should include the protection of breeding habitat and the minimization of colony disturbance during the breeding season. In the Antarctic, visitor site guidelines already specify minimum approach distances of 5 meters and off-limit areas. Gentoo colonies outside of the Antarctic may also benefit from such guidelines for minimizing the negative impact of an expanding ecotourism industry.

ACKNOWLEDGMENTS

I gratefully acknowledge assistance from the U.S. National Science Foundation Office of Polar Programs (Award No. NSF/OPP-0739515). I would also like to thank Ron Naveen and Evan H. C. Grant for their assistance with the preparation of this manuscript.

TABLE 5.1 Biometric data for the gentoo penguin (rounded to the tenth, mean ± standard deviation)

VARIABLE	MALES	FEMALES	UNSEXED	REFERENCES
Body mass (kg)[1]				
Crozet Islands			7.3±0.8 (18)[2]	Bost and Jouventin. 1991
Crozet Islands			6.7±0.8 (56)	Bost and Jouventin 1990b
Kerguelen Island			5.7±0.6 (70)	Bost and Jouventin 1990b
Kerguelen Island			4.6±0.6 (70)	Bost and Jouventin 1990b
Falkland/Malvinas Islands	7.0±0.7 (7)	6.6±0.6 (9)		Otley et al. 2004
Macquarie Island			5.7±0.6 (97)	Reilly and Kerle 1981
South Georgia			5.8	Croxall and Prince 1980a
South Georgia	5.6–8.0	4.9–7.5		Williams 1995[3]
South Georgia	6.6±0.09 (26)	6.0±0.1 (26)		Bost and Jouventin 1990b
King George Island	5.5	5.1		Volkman et al. 1980
Body length (mm)				
King George Island	652	622		Volkman et al. 1980
Bill length (mm)				
Crozet Islands	62.8±3.5 (42)	57.7±2.8 (48)		Williams 1995
Crozet Islands			60.7±4.0 (60)	Bost and Jouventin 1990b
Crozet Islands			61 (1)	Stonehouse 1970
Marion Island			59	Bost and Jouventin 1990b
Kerguelen Island	56.2±2.1 (49)	50.8±2.2 (40)		Williams 1995
Kerguelen Island			55.6±3.0 (42)	Bost and Jouventin 1990b
Kerguelen Island			53.2±3.0 (60)	Bost and Jouventin 1990b
Kerguelen Island			51.9±3.0 (55)	Bost and Jouventin 1990b
Kerguelen Island			55.7±3.0 (28)	Bost and Jouventin 1990b
Kerguelen Island			53.0±5.1 (19)	Stonehouse 1970
Falkland/Malvinas Islands	60.5±2.3 (12)	54.1±1.9 (10)		Otley et al. 2004
Falkland/Malvinas Islands			56±4.5 (15)	Stonehouse 1970
Heard Island			52 (2)	Stonehouse 1970
Bird Island	55.5±0.3 (56)	50.4±0.2 (56)		Williams 1990
South Georgia			51±5.0 (16)	Stonehouse 1970
South Georgia			52	Croxall and Prince 1980a
South Georgia	55±3 (66)	50±3 (56)		Bost and Jouventin 1990b
Macquarie Island			56.3±3.3 (136)	Reilly and Kerle 1981
Macquarie Island			53±2.8 (23)	Stonehouse 1970
King George Island	50.8	48.4		Volkman et al. 1980
Antarctic Peninsula			47.6 (20)	Bost and Jouventin 1990b
Antarctic Peninsula			47 (2)	Stonehouse 1970
Bill width (mm)				
Macquarie Island			10.0±0.8 (16)	Reilly and Kerle 1981
King George Island	18.0	16.1		Volkman et al. 1980
Bill depth (mm)				
Crozet Islands	19.5 (16)	17.3±1.5 (15)		Williams 1995
Kerguelen Island	18.4±1.3 (14)	17.9±1.1 (14)		Williams 1995
Falkland/Malvinas Islands	24.8±4.0 (12)	23.2±2.6 (10)		Otley et al. 2004
Bird Island	17.3±0.1 (56)	15.4±0.1 (56)		Williams 1990
Macquarie Island			18.3±1.6 (156)	Reilly and Kerle 1981
Flipper area (cm²)				
Crozet Islands			109.0 (1)	Stonehouse 1970
Kerguelen Island			93.9±7.8 (15)	Stonehouse 1970
Falkland/Malvinas Islands			104.4±8.4 (7)	Stonehouse 1970
Heard Island			103.0 (1)	Stonehouse 1970
South Georgia			100.8±4.1 (10)	Stonehouse 1970
South Georgia			103	Croxall and Prince 1980a
Macquarie Island			95.4±5.3 (19)	Stonehouse 1970
South Orkneys/ South Shetlands			86.8±5.7 (15)	Stonehouse 1970
Antarctic Peninsula			92.0 (2)	Stonehouse 1970

TABLE 5.1 (cont.)

VARIABLE	MALES	FEMALES	UNSEXED	REFERENCES
Flipper length (cm)				
Crozet Islands			24.0 (1)	Stonehouse 1970
Crozet Islands	25.7±0.8 (42)	24.9±0.9 (48)		Williams 1995
Crozet Islands			25.4±1.1 (40)	Bost and Jouventin 1990b
Marion Island			23.8 (2)	Bost and Jouventin 1990b
Kerguelen Island			21.1±1.5 (18)	Stonehouse 1970
Kerguelen Island	23.8±0.6 (49)	22.8±0.8 (35)		Williams 1995
Kerguelen Island			23.1±0.9 (57)	Bost and Jouventin 1990b
Kerguelen Island			22.2±0.9 (55)	Bost and Jouventin 1990b
Kerguelen Island			23.7±1.0 (29)	Bost and Jouventin 1990b
Kerguelen Island			24.0±1.2 (42)	Bost and Jouventin 1990b
Falkland/Malvinas Islands			22.5±1.2 (15)	Stonehouse 1970
Heard Island			21.1 (2)	Stonehouse 1970
South Georgia			21.7±1.0 (16)	Stonehouse 1970
South Georgia	23.4±1.3 (25)	22.2±1.1 (25)		Bost and Jouventin 1990b
Macquarie Island			21.5±1.0 (23)	Stonehouse 1970
Macquarie Island			23.0±0.1 (19)	Reilly and Kerle 1981
South Orkneys/ South Shetlands			19.7±0.7 (17)	Stonehouse 1970
Antarctic Peninsula			20.5 (2)	Stonehouse 1970
Antarctic Peninsula			20 (20)	Bost and Jouventin 1990b
Egg weight (g)				
Crozet Islands			First: 128.7±11.4 (20) Second: 124.1±10.8 (18)	Bost and Jouventin 1991
Marion Island			First: 128.6±8.3 (16) Second: 126.5±13.3 (13)	Williams 1980
Bird Island			First: 128.2±0.7 (201) Second: 130.0±0.8 (201)	Williams 1990
Egg length (mm)				
Marion Island			First: 68.0±2.4 (17) Second: 67.4±2.9 (17)	Williams 1980
Kerguelen Island			First: 67.7±2.1 (35) Second: 67.6±2.1 (35)	Lescroël et al. 2009
Egg width (mm)				
Marion Island			First: 58.3±1.4 (17) Second: 58.1±1.4 (17)	Williams 1980
Kerguelen Island			First: 57.6±2.3 (35) Second: 57.3±1.6 (35)	Lescroël et al. 2009

Note: Sample size is between parentheses. All measurements are from live adults except those from Stonehouse 1970, which represent museum specimens. Sites are ordered from north to south.

1 Body mass is highly variable among sites and years and over the course of the breeding season.

2 Before breeding.

3 Range in values reflects variation in body mass throughout the breeding season, with the highest values reflecting body condition just prior to molt.

TABLE 5.2 Population estimates for all main gentoo breeding locations (ordered by population size)

Falkland/Malvinas Islands	115,327	2000/01	Clausen and Huin 2003
South Georgia	98,867	1996	Trathan et al. 1996
Antarctic Peninsula (incl. S. Shetland Is.)	94,751	Various	Lynch et al. Unpublished
Kerguelen Islands	30,000–40,000	Unknown	Weimerskirch et al. 1988
Heard Island	16,574	1987	Woehler 1993
South Orkney Islands	10,760	Various	Lynch et al. Unpublished
Crozet Islands	9,000	Unknown	Jouventin 1994
Macquarie Island	3,800	Unknown	Robinson Unpublished[1]
S. Sandwich Islands	1,572	1996/97–1997/98	Convey et al. 1999
Marion Island	1,100	2008	Crawford et al. 2009
Prince Edward Island	40	2008	Crawford et al. 2009
Martillo Island, Tierra del Fuego	12	2006/07	Ghys et al. 2008
TOTAL	**C. 387,000**		

1   As cited in Holmes et al. 2006

TABLE 5.3 Key life history parameters for the gentoo penguin

PARAMETER	VALUE	REFERENCES
Age at first breeding	2–4 years	W95
Incubation period	34–37 days	BJ91, G08, L09, O04, W90
Brood period /	25–30 days	L09, O04, W90
Age at crèche		
Fledging period	80–105 days	W95
Reproductive success[a]	0.4–1.5 chicks/nest	BJ91, C97, C99, CP02, CS99, G08, H06, L09, L10, M09, P01, R86, RK81, T83, W90, WC91
Nest density	0.25–1.8 nests/m^2	CP80a, QC00, S75, VT81
Juvenile survival rates	0.27–0.59	W95
Adult survival rates	0.75–0.89	W95
Maximum lifespan	Unknown[b]	

a   This range encompasses most of the reported values for reproductive success in a given year, although very low reproductive success rates (as low as 0.0 chicks/nest) have also been reported.

b   Gentoos have been recorded breeding at least until 16 years of age (J. Hinke), although long-term studies have yet to establish the maximum longevity of gentoos in the wild.

Reference codes: BJ91 = Bost and Jouventin 1991, C97 = Croxall et al. 1997, C99 = Croxall et al. 1999, CP80a = Croxall and Prince 1980a, CP02 = Clausen and Pütz 2002, CS99 = Cobley and Shears 1999, G08 = Ghys et al. 2008, H06 = Holmes et al. 2006, L09 = Lescroël et al. 2009, L10 = Lynch et al. 2010b, M09 = Miller et al. 2009, O04 = Otley et al. 2004, P01 = Pütz et al. 2001, QC00 = Quintana and Cirelli 2000, R86 = Robertson 1986, RK81 = Reilly and Kerle 1981, S75 = Stonehouse 1975, T83 = Trivelpiece et al. 1983, VT81 = Volkman and Trivelpiece 1981, W90 = Williams 1990, W95 = Williams 1995, WC91 = Williams and Croxall 1991.

TABLE 5.4 Annual life cycle for the gentoo penguin

PARAMETER	BIRD ISLAND, SOUTH GEORGIA[a] (Williams 1990)	POSSESSION ISLAND, CROZET ISLANDS (Bost and Jouventin 1991)
Settlement	No defined period of settlement	28 June–10 July
Egg laying	Mean: 26.5 Oct. ± 3.7	Mean: 23 July ± 7 days / Range: Late July–late Nov.
Incubation	(2nd egg) 35.2± 0.1 days	(1st egg) 37.0 ± 1.7 days / (2nd egg) 35.6 ± 2.0 days
Crèching	25.4 ± 0.6 days	Late Sept.–early Oct.
Fledging	79.6 ± 0.4 days (1976 data)	89.3 ± 4 days
Molting		Mean completion: mid-December to mid-February

a   Unless otherwise indicated, data from 1988 is reported.

TABLE 5.5 Diet composition of gentoo penguins (as measured by stomach contents and by wet weight)

LOCATION	CRUSTACEANS	FISH	CEPHALOPODS	REFERENCES
Crozet Islands	54.2% [EV]	43.9% [M,N]	1.8%	Ridoux 1994[6]
Falkland/Malvinas Islands	36%	53%	11%	Pütz et al. 2001
Falkland/Malvinas Islands	30–35% [MG]	50–55%	15–20%	Clausen and Pütz 2002
Falkland/Malvinas Islands[1]	12% [MG]	35%	53% [LG]	Clausen and Pütz 2003
Heard Island	7.8% [EV]	90.5% [C,M,N]	1.7%	Klages et al. 1990
Kerguelen Islands ♂	2.9%	83.1%	13.3%	Lescroël et al. 2009
Kerguelen Islands ♀	41.2%	52.3%	5.9%	Lescroël et al. 2009
Kerguelen Islands[2]	84–86% [EV]	14–16%		Bost et al. 1994
King George Island	98.4% [ES]	1.5%		Volkman et al. 1980
King George Island	40.4% [ES]	48.6% [N]		Jablonski 1985
Livingston Island	67–97% [ES]	3–33% [C,M,N]		Miller et al. 2009
Macquarie Island	<1%	91.6% [M,N]	8.3%	Robinson and Hindell 1996
Macquarie Island	<1%	81.6%[M,N]	18.6%	Hindell 1989
Marion Island	44% [EV]	53% [C,M,N]	2%	Adams and Klages 1989
Marion Island	46.2% [EV]	53.2% [N]	0.6%	Adams and Wilson 1987
Marion Island	70%	30% [N]		La Cock et al. 1984
South Georgia[3]	68% [ES,EF]	32% [C,N]		Croxall and Prince 1980b
South Georgia	See note[4] [ES]	See note[3] [C,N]		Croxall et al. 1988
South Georgia	63.2%[5] [ES]	36.3% [C]		Croxall et al. 1999
South Georgia	36–95% [ES]	5–62% [C,N]		Berrow et al. 1999
South Orkney Islands		100%		White and Conroy 1975
South Orkney Islands	c. 15%-70%[ES]	c. 25%-81.1% [N]		Coria et al. 2000

Major diet constituents reported. Family- and species-specific information coded as: ES = *Euphausia superba*, EF = *E. frigid*, EV =*E. vallentini*, MG = *Munida gregaria*, LG = *Loligo gahi*, TG = *Themisto gaudichaudii*; C = Channichthyidae, M = Myctophidae, N = Nototheniidae.

1 This study reflects diet during the winter.

2 See also Lescroël et al. 2004

3   See also Croxall and Prince 1979, Croxall et al. 1988, and Croxall et al. 1997

4   Diet composition of gentoos in this study was strongly bimodal, with individual diets composed of either mostly fish (91% by weight) or mostly krill (87% by weight).

5 Average over 11 years for which data were available between 1977 and 1995 (1986 and 1994 excluded).

6   See also Ridoux et al. 1988

REFERENCES

Adams, N. J., and C. R. Brown. 1990. Energetics of molt in penguins. In *Penguin Biology*, ed. L. Davis, and J. Darby, 297–315. San Diego, CA: Academic Press.

Adams, N. J., and N. T. Klages. 1989. Temporal variation in the diet of the gentoo penguin *Pygoscelis papua* at sub-Antarctic Marion Island. *Colonial Waterbirds* 12(1):30–36.

Adams, N. J., and M. P. Wilson. 1987. Foraging parameters of gentoo penguins *Pygoscelis papua* at Marion Island. *Polar Biology* 7:51–56.

ASOC (Antarctic and Southern Ocean Coalition). 2008. Designation of marine protected areas within the Antarctic Treaty Area. Information Paper 119 presented to the 31st Antarctic Treaty Consultative Meeting, Kyiv, Ukraine.

Baker, A. J., S. L. Pereira, O. P. Haddrath, and K. A. Edge. 2006. Multiple gene evidence for expansion of extant penguins out of Antarctica due to global cooling. *Proceedings of the Royal Society B* 273:11–17.

Baylis, A. M. M., A. F. Zuur, P. Brickle, and P. A. Pistorius. 2012. Climate as a driver of population variability in breeding gentoo penguins *Pygoscelis papua* at the Falkland Islands. *Ibis* 154:30–41.

Berrow, S. D., R. I. Taylor, and A. W. A. Murray. 1999. Influence of sampling protocol on diet determination of gentoo penguins *Pygoscelis papua* and Antarctic fur seals *Arctocephalus gazella*. *Polar Biology* 22:156–63.

Bingham, M. 2002. The decline of Falkland Islands penguins in the presence of a commercial fishing industry. *Revista Chilena de Historia Natural* 75:805–18.

BirdLife International. 2010. *Pygoscelis papua*. IUCN Red List of Threatened Species. Version 2010.1. www.iucnredlist.org (accessed 31 March 2010).

Boersma, P. D. 2008. Penguins as marine sentinels. *BioScience* 58(7):597–607.

Bost, C. A., and J. Clobert. 1992. Gentoo penguin *Pygoscelis papua*: Factors affecting the process of laying a replacement clutch. *Acta Oecologica* 13(5):593–605.

Bost, C. A., and P. Jouventin. 1990a. Laying asynchrony in gentoo penguins on Crozet Islands: Causes and consequences. *Ornis Scandinavica* 21:63–70.

———. 1990b. Evolutionary ecology of the gentoo penguin *Pygoscelis papua*. In *Penguin Biology*, ed. L. Davis and J. Darby, 85–112. San Diego, CA: Academic Press.

———. 1991. The breeding performance of the gentoo penguin *Pygoscelis papua* at the northern edge of its range. *Ibis* 133:14–25.

Bost, C. A., P. Jouventin, and N. Pincson du Sel. 1992. Morphometric variability on a microgeographical scale in two inshore seabirds. *Journal of Zoology* (London) 226:135–49.

Bost, C. A., P. Koubbi, F. Genevois, L. Ruchon, and V. Ridoux. 1994. Gentoo penguin *Pygoscelis papua* diet as an indicator of planktonic availability in the Kerguelen Islands. *Polar Biology* 14:147–53.

Clarke, A., E. J. Murphy, M. P. Meredith, J. C. King, L. S. Peck, D. K. A. Barnes, and R. C. Smith. 2007. Climate change and the marine ecosystem of the western Antarctic Peninsula. *Philosophical Transactions of the Royal Society B* 362:149–66.

Clausen, A. P., and N. Huin. 2003. Status and numerical trends of king, gentoo, and rockhopper penguins breeding in the Falkland Islands. *Waterbirds* 26(4):389–402.

Clausen, A. P., and K. Pütz. 2002. Recent trends in diet composition and productivity of gentoo, Magellanic and rockhopper penguins in the Falkland Islands. *Aquatic Conservation: Marine and Freshwater Ecosystems* 12:51–61.

Clausen, A. P., and K. Pütz. 2003. Winter diet and foraging range of gentoo penguins (*Pygoscelis papua*) from Kidney Cove, Falkland Islands. *Polar Biology* 26:32–40.

Cobley, N. D., and J. R. Shears. 1999. Breeding performance of gentoo penguins (*Pygoscelis papua*) at a colony exposed to high levels of human disturbance. *Polar Biology* 21:355–60.

Convey, P., A. Morton, and J. Poncet. 1999. Survey of marine birds and mammals of the South Sandwich Islands. *Polar Record* 35(193):107–24.

Cordier, J. R., A. Mendez, J. L. Mougin, and G. Visbeek. 1983. Les oiseaux de la Baie de l'Espérance, Péninsule antarctique (63°24′ S, 56°59′ W) (à suivre). *L'Oiseau et la Revue Française d'Ornithologie* 53(2).

Coria, N., M. Libertelli, R. Casaux, and C. Darrieu. 2000. Inter-annual variation in the autumn diet of the gentoo penguin at Laurie Island, Antarctica. *Waterbirds* 23(3):511–17.

Crawford, R. J. M., J. Cooper, M. du Toit, M. D. Greyling, B. Hanise, C. L. Holness, D. G. Keith, J. L. Nel, S. L. Petersen, K. Spencer, D. Tshingana, and A. C. Wolfaardt. 2003a. Population and breeding of the gentoo penguin *Pygoscelis papua* at Marion Island, 1994/95–2002/03. *African Journal of Marine Science* 25:463–74.

Crawford, R. J. M., J. Cooper, B. M. Dyer, M. D. Greyling, N. T. W. Klages, P. G. Ryan, S. L. Petersen, L. G. Underhill, L. Upfold, W. Wilkinson, M. S. De Villiers, S. du Plessis, M. du Toit, T. M. Leshoro, A. B. Makhado, and M. S. Mason. 2003b. Populations of surface nesting seabirds at Marion Island, 1994/95–2002/03. *African Journal of Marine Science* 25:427–40.

Crawford, R. J. M., S. A. Davis, R. T. Harding, L. F. Jackson, T. M. Leshoro, M. A. Meyer, R. M. Randall, L. G. Underhill, L. Upfold, A. P. van Dalsen, E. van der Merwe, P. A. Whittington, A. J. Williams, and A. C. Wolfaardt. 2000. Initial impact of the Treasure oil spill on seabirds off western South Africa. *South African Journal of Marine Science* 22:157–76.

Crawford, R. J. M., P. A. Whittington, L. Upfold, P. G. Ryan, S. L. Petersen, B. M. Dyer, and J. Cooper. 2009. Recent trends in numbers of four species of penguins at the Prince Edward Islands. *African Journal of Marine Science* 31(3):419–26.

Croxall, J. P. 1982. Energy costs of incubation and moult in petrels and penguins. *Journal of Animal Ecology* 51(1):177–94.

Croxall, J. P., and L. S. Davis. 1999. Penguins: Paradoxes and patterns. *Marine Ornithology* 27:1–12.

Croxall, J. P., and E. D. Kirkwood. 1979. The distribution of penguins on the Antarctic Peninsula and islands of the Scotia Sea. Cambridge: British Antarctic Survey.

Croxall, J. P., T. S. McCann, P. A. Prince, and P. Rothery. 1988. Reproductive performance of seabirds and seals at South Georgia and Signy Islands, 1976–1987: Implications for Southern Ocean monitoring studies. In *Antarctic Ocean and Resources Variability*, ed. D. Sahrhage, 261–85. Berlin: Springer Verlag.

Croxall, J. P., and P. A. Prince. 1979. Antarctic seabird and seal monitoring studies. *Polar Record* 19:573–95.

———. 1980a. Food, feeding ecology and ecological segregation of seabirds at South Georgia. *Biological Journal of the Linnean Society* 14:103–31.

———. 1980b. The food of gentoo penguins *Pygoscelis papua* and macaroni penguins *Eudyptes chrysolophus* at South Georgia. *Ibis* 122:245–53.

Croxall, J. P., P. A. Prince, and K. Reid. 1997. Dietary segregation of krill-eating South Georgia seabirds. *Journal of Zoology* (London) 242:531–56.

Croxall, J. P., K. Reid, and P. A. Prince. 1999. Diet, provisioning and productivity responses of marine predators to differences in availability of Antarctic krill. *Marine Ecology Progress Series* 177:115–31.

Croxall, J. P., and P. Rothery. 1995. Population change in gentoo penguins *Pygoscelis papua* at Bird Island, South Georgia: Potential roles of adult survival, recruitment, and deferred breeding. In *The Penguins: Ecology and Management*, ed. P. Dann, I. Norman, and P. Reilly. Chipping Norton, NSW, Australia: Surrey Beatty and Sons.

Croxall, J. P., P. N. Trathan, and E. J. Murphy. 2002. Environmental change and Antarctic seabird populations. *Science* 297:1510–14.

Davis, R. W., J. P. Croxall, and M. J. O'Connell. 1989. The reproductive energetics of gentoo (*Pygoscelis papua*) and macaroni (*Eudyptes chrysolophus*) penguins at South Georgia. *Journal of Animal Ecology* 58(1):59–74.

Ducklow, H. W., K. Baker, D. G. Martinson, L. B. Quetin, R. M. Ross, R. C. Smith, S. E. Stammerjohn, M. Vernet, and W. Fraser. 2007. Marine Pelagic ecosystems: The West Antarctic Peninsula. *Philosophical Transactions of the Royal Society B* 362:67–94.

Forcada, J., D. Malone, J. Andrew Royle, and I. J. Staniland. 2009. Modelling predation by transient leopard seals for an ecosystem-based management of Southern Ocean fisheries. *Ecological Modelling* 220:1513–21.

Forcada, J., and P. N. Trathan. 2009. Penguin responses to climate change in the Southern Ocean. *Global Change Biology* 15:1618–30.

Forcada, J., P. N. Trathan, K. Reid, E. J. Murphy, and J. P. Croxall. 2006. Contrasting population changes in sympatric penguin species in association with climate warming. *Global Change Biology* 12:411–23.

Garcia-Borboroglu, P., P. D. Boersma, V. Ruoppolo, L. Reyes, G. A. Rebstock, K. Griot, S. R. Heredia, A. C. Adornes, and R. P. da Silva. 2006. Chronic oil pollution harms Magellanic penguins in the Southwest Atlantic. *Marine Pollution Bulletin* 52(2):193–98.

Ghys, M. I., A. R. Rey, and A. Schiavini. 2008. Population trend and breeding biology of gentoo penguin in Martillo Island, Tierra del Fuego, Argentina. *Waterbirds* 31(4):625–31.

Hahn, S., H. U. Peter, and S. Bauer. 2005. Skuas at penguin carcass: Patch use and state-dependent leaving decisions in a top-predator. *Proceedings of the Royal Society B* 272:1449–54.

Hindell, M. A. 1989. The diet of gentoo penguins *Pygoscelis papua* at Macquarie Island: Winter and early breeding season. *Emu* 89:71–78.

Holmes, N. D., M. Giese, H. Achurch, S. Robinson, and L. K. Kriwoken. 2006. Behavior and breeding success of gentoo penguins *Pygoscelis papua* in areas of low and high human activity. *Polar Biology* 29:399–412.

Jablonski, B. 1984. Distribution, numbers, and breeding preferences of penguins in the region of the Admiralty Bay (King George Island, South Shetland Islands) in the season 1979/1980. *Polish Polar Research* 5:5–16.

———. 1985. The diet of penguins on King George Island, South Shetland Islands. *Acta Zoologica Cracoviensia* 29(8):117–86.

Jerez, S. M. Motas, B. Fusaro, and A. Barbosa. 2011. Presence and distribution of heavy metals and other elements in chick individuals of gentoo penguin from King George Island, Antarctica. *Toxicology Letters* 205:S193–94.

Jouventin, P. 1994. Les populations d'oiseaux marins des T.A.A.F.: Résumé de 20 années de recherche. *Alauda* 62:44–47.

Kennicutt, M. C., II, S. T. Sweet, W. R. Fraser, W. L. Stockton, and M. Culver. [[AUS: Placement of 'II' is per CMS.]] 1991. Grounding of the *Bahia Paraiso* at Arthur Harbor, Antarctica. 1. Distribution and fate of oil spill related hydrocarbons. *Environmental Science & Technology* 25:509–18.

Klages, N. T. W., D. Pemberton, and R. P. Gales. 1990. The diets of king and gentoo penguins at Heard Island. *Australian Wildlife Research* 17:53–60.

La Cock, G. D., T. Hecht, and N. Klages. 1984. The winter diet of gentoo penguins at Marion Island. *Ostrich* 55:188–91.

Lescroël, A., C. Bajzak, and C. A. Bost. 2009. Breeding ecology of the gentoo penguin *Pygoscelis papua* at Kerguelen Archipelago. *Polar Biology* 32:1495–1505.

Lescroël, A., and C. A. Bost. 2005. Foraging under contrasting oceanographic conditions: The gentoo penguin at Kerguelen Archipelago. *Marine Ecology Progress Series* 302:245–61.

———. 2006. Recent decrease in gentoo penguin populations at Iles Kerguelen. *Antarctic Science* 18(2):171–74.

Lescroël, A., V. Ridoux, and C. A. Bost. 2004. Spatial and temporal variation in the diet of the gentoo penguin (*Pygoscelis papua*) at Kerguelen Islands. *Polar Biology* 27:206–16.

Loeb, V., V. Siegel, and O. Holm-Hansen. 1997. Effects of sea-ice extent and krill or salp dominance on the Antarctic food web. *Nature* 387:897–900.

Lynch, H. J., K. Crosbie, W. F. Fagan, and R. Naveen. 2010a. Spatial patterns of tour ship traffic in the Antarctic Peninsula region. *Antarctic Science* 22(2):123–30.

Lynch, H. J., W. F. Fagan, and R. Naveen. 2010b. Population trends and reproductive success at a frequently visited penguin colony on the western Antarctic Peninsula. *Polar Biology* 33:493–503.

Lynch, H. J., W. F. Fagan, R. Naveen, S. G. Trivelpiece, and W. Z. Trivelpiece. 2009. Timing of clutch initiation in *Pygoscelis* penguins on the Antarctic Peninsula: Towards an improved understanding of off-peak census correction factors. *CCAMLR Science* 16:149–65.

Lynch, H. J., R. Naveen, and W. F. Fagan. 2008. Censuses of penguins, blue-eyed shags, and southern giant petrel populations in the Antarctic Peninsula, 2001–2007. *Marine Ornithology* 36:83–97.

Lynch, H. J., R. Naveen, P. N. Trathan, and W. F. Fagan. 2012. Spatially integrated assessment reveals widespread changes in penguin populations on the Antarctic Peninsula. *Ecology.* http://www.esajournals.org/doi/pdf/10.1890/11–1588.1 (accessed 18 April 2012).

Metcheva, R., V. Beztukov, S. E. Teodorova, and Y. Yankov. 2008. "Yellow spot": A new trait of gentoo penguins *Pygoscelis papua ellsworthii* in Antarctica. *Marine Ornithology* 36:47–51.

Miller, A. K., N. J. Karnovsky, and W. Z. Trivelpiece. 2009. Flexible foraging strategies of gentoo penguins *Pygoscelis papua* over 5 years in the South Shetland Islands, Antarctica. *Marine Biology* 156:2527–37.

Murphy, E. J., J. L. Watkins, P. N. Trathan, K. Reid, M. P. Meredith, S. E. Thorpe, N. M. Johnston, A. Clarke, G. A. Tarling, M. A. Collins, J. Forcada, R. S. Shreeve, A. Atkinson, R. Korb, M. J. Whitehouse, P. Ward, P. G. Rodhouse, P. Enderlein, A. G. Hirst, A. R. Martin, S. L. Hill, I. J. Staniland, D. W. Pond, D. R. Briggs, N. J. Cunningham, and A. H. Fleming. 2007. Spatial and temporal operation of the Scotia Sea ecosystem: A review of large-scale links in a krill centred food web. *Philosophical Transactions of the Royal Society B* 362:113–48.

Nimon, A. J., R. C. Schroter, and B. Stonehouse. 1995. Heart rate of disturbed penguins. *Nature* 374:415.

Otley, H. M., A. P. Clausen, D. J. Christie, and K. Pütz. 2004. Aspects of the breeding biology of the gentoo penguin *Pygoscelis papua* at Volunteer Beach, Falkland Islands, 2001/02. *Marine Ornithology* 33:167–71.

Polito, M. J., and W. Z. Trivelpiece. 2008. Transition to independence and evidence of extended parental care in the gentoo penguin (*Pygoscelis papua*). *Marine Biology* 154:231–40.

Pütz, K., R. J. Ingham, J. G. Smith, and J. P. Croxall. 2001. Population trends, breeding success and diet composition of gentoo *Pygoscelis papua*, Magellanic *Spheniscus magellanicus* and rockhopper *Eudyptes chrysocome* penguins in the Falkland Islands. *Polar Biology* 24:793–807.

Quintana, R. D., and V. Cirelli. 2000. Breeding dynamics of a gentoo penguin *Pygoscelis papua* population at Cierva Point, Antarctic Peninsula. *Marine Ornithology* 28:29–35.

Ratcliffe, N., and P. Trathan. In review. A review of the diet and foraging movements of penguins breeding within the CCAMLR area.

Reid, K. 1995. Oiled penguins observed at Bird Island, South Georgia. *Marine Ornithology* 23(1):53–57.

Reilly, P. N., and J. A. Kerle. 1981. A study of the gentoo penguin. *Notornis* 28:189–202.

Renner, M., J. Valencia, L. S. Davis, D. Saez, and O. Cifuentes. 1998. Sexing of adult gentoo penguins in Antarctica using morphometrics. *Colonial Waterbirds* 21(3):444–49.

Ridoux, V. 1994. The diets and dietary segregation of seabirds at the subantarctic Crozet Islands. *Marine Ornithology* 22(1):1–192.

Ridoux, V., P. Jouventin, J. C. Stahl, and H. Weimerskirch. 1988. Écologie alimentaire comparée des manchotsnicheurs aux iles Crozet. *Revue d'Écologie (La Terre et la Vie)* 43:345–55.

Robertson, G. 1986. Population size and breeding success of the gentoo penguin, *Pygoscelis papua*, at Macquarie Island. *Australian Wildlife Research* 13:583–87.

Robinson, S. A., and M. A. Hindell. 1996. Foraging ecology of gentoo penguins *Pygoscelis papua* at Macquarie Island during the period of chick care. *Ibis* 138:722–31.

Smith, R. C., D. Ainley, K. Baker, E. Domack, S. Emslie, B. Fraser, J. Kennett, A. Leventer, E. Mosley-Thompson, S. Stammerjohn, and M. Vernet. 1999.

Marine ecosystem sensitivity to climate change. *BioScience* 49:393–404.

Smith, R. C., and S. E. Stammerjohn. 2001. Variations in surface air temperature and sea-ice extent in the western Antarctic Peninsula region. *Annals of Glaciology* 33:493–500.

Stammerjohn, S. E., D. G. Martinson, R. C. Smith, and R. A. Iannuzzi. 2008a. Sea ice in the western Antarctic Peninsula region: Spatio-temporal variability from ecological and climate change perspectives. *Deep Sea Research Part II: Topical Studies in Oceanography* 55(18–19):2041–58.

Stammerjohn, S. E., D. G. Martinson, R. C. Smith, X. Yuan, and D. Rind. 2008b. Trends in Antarctic annual sea ice retreat and advance and their relation to El Nino–Southern Oscillation and Southern Annular Mode variability. *Journal of Geophysical Research—Oceans* 113(C3):C03S90.

Stonehouse, B. 1970. Geographic variation in gentoo penguins, *Pygoscelis papua*. *Ibis* 112:52–57.

———. 1975. *The Biology of Penguins*. Baltimore, MD: University Park Press.

Tanton, J. L., K. Reid, J. P. Croxall, and P. N. Trathan. 2004. Winter distribution and behavior of gentoo penguins *Pygoscelis papua* at South Georgia. *Polar Biology* 27:299–303.

Trathan, P. N., F. H. J. Daunt, and E. J. Murphy, eds. 1996. *South Georgia: An Ecological Atlas*. Cambridge: British Antarctic Survey.

Trathan, P. N., J. Forcada, R. Atkinson, R. H. Downie, and J. R. Shears. 2008. Population assessments of gentoo penguins (*Pygoscelis papua*) breeding at an important Antarctic tourist site, Goudier Island, Port Lockroy, Palmer Archipelago, Antarctica. *Biological Conservation* 141:3019–28.

Trathan, P. N., J. Forcada, and E. J. Murphy. 2007. Environmental forcing and Southern Ocean marine predator populations: Effects of climate change variability. *Philosophical Transactions of the Royal Society B* 362:2351–65.

Trivelpiece, W. Z., and S. G. Trivelpiece. 1990. Courtship period of Adélie, gentoo and chinstrap penguins. In *Penguin Biology*, ed. L. S. Davis and J. T. Darby, 113–27. San Diego, CA: Academic Press.

Trivelpiece, W. Z., S. G. Trivelpiece, and N. J. Volkman. 1987. Ecological segregation of Adélie, gentoo, and chinstrap penguins at King George Island, Antarctica. *Ecology* 68(2):351–61.

Trivelpiece, W. Z., S. G. Trivelpiece, N. J. Volkman, and S. H. Ware. 1983. Breeding and feeding ecologies of pygoscelid penguins. *Antarctic Journal of the United States* 18:209–10.

Turner, J., T. Lachlan-Cope, S. Colwell, and G. J. Marshall. 2005. A positive trend in western Antarctic Peninsula precipitation over the last 50 years reflecting regional and Antarctic-wide atmospheric circulation changes. *Annals of Glaciology* 41:85–91.

Underhill, L. G., P. A. Bartlett, L. Baumann, R. J. M. Crawford, B. M. Dyer, A. Gildenhuys, D. C. Nel, T. B. Oatley, M. Thornton, L. Upfold, A. J. Williams, P. A. Whittington, and A. C. Wolfaardt. 1999. Mortality and survival of African penguins *Spheniscus demersus* involved in the *Apollo Sea* oil spill: An evaluation of rehabilitation efforts. *Ibis* 141:29–37.

Vaughan, D. G., G. J. Marshall, W. M. Connolley, C. Parkinson, R. Mulvaney, D. A. Hodgson, J. C. King, C. J. Pudsey, and J. Turner. 2003. Recent rapid regional climate warming on the Antarctic Peninsula. *Climatic Change* 60:243–74.

Voisin, J. F. 1984. Observations on the birds and mammals of île aux Cochons, Crozet Islands, in February 1982. *South African Journal of Antarctic Research* 14:11–17.

Volkman, N. J., P. Presler, and W. Trivelpiece. 1980. Diets of pygoscelid penguins at King George Island, Antarctica. *Condor* 82:373–78.

Volkman, N. J., and W. Trivelpiece. 1981. Nest-site selection among Adélie, chinstrap and gentoo penguins in mixed species rookeries. *The Wilson Bulletin* 93:243–48.

Walker, B. G., P. D. Boersma, and J. C. Wingfield. 2005. Physiological and behavioral differences in Magellanic penguin chicks in undisturbed and tourist-visited locations of a colony. *Conservation Biology* 19(5):1571–77.

———. 2006. Habituation of adult Magellanic penguins to human visitation as expressed through behavior and corticosterone secretion. *Conservation Biology* 20(1):146–54.

Walker, T. R., I. L. Boyd, D. J. McCafferty, N. Huin, R. I. Taylor, and K. Reid. 1998. Seasonal occurrence and diet of leopard seals (*Hydrurga leptonyx*) at Bird Island, South Georgia. *Antarctic Science* 10(1):75–81.

Weimerskirch, H., R. Zotier, and P. Jouventin. 1988. The avifauna of the Kerguelen Islands. *Emu* 89:15–29.

White, M. G., and J. W. H. Conroy. 1975. Aspects of competition between pygoscelid penguins at Signy Island, South Orkney Islands. *Ibis* 117:371–73.

Williams, A. J. 1980. Aspects of the breeding biology of the gentoo penguin *Pygoscelis papua*. *Le Gerfaut* 70:283–95.

Williams, A. J., and W. R. Siegfried. 1980. Foraging ranges of krill-eating penguins. *Polar Record* 20(125):159–75.

Williams, T. D. 1990. Annual variation in breeding biology of gentoo penguins, *Pygoscelis papua*, at Bird Island, South Georgia. *Journal of Zoology* (London) 222:247–58.

———. 1995. *Bird Families of the World: The Penguins*. Oxford: Oxford University Press.

Williams, T. D., and J. P. Croxall. 1991. Chick growth and survival in gentoo penguins (*Pygoscelis papua*): Effect of hatching asynchrony and variation in food supply. *Polar Biology* 11:197–202.

Williams, T. D., and S. Rodwell. 1992. Annual variation in return rate, mate and nest-site fidelity in breeding gentoo and macaroni penguins. *Condor* 94(3):636–45.

Wilson, R. P., B. Alvarrez, L. Latorre, D. Adelung, B. Culik, and R. Bannasch. 1998. The movements of gentoo penguins *Pygoscelis papua* from Ardley Island, Antarctica. *Polar Biology* 19:407–13.

Wilson, R. P., B. Culik, R. Danfeld, and D. Adelung. 1991. People in Antarctica—how much do Adélie penguins *Pygoscelis adeliae* care? *Polar Biology* 11:363–70.

Woehler, E. J. 1993. The distribution and abundance of Antarctic and Subantarctic penguins. Scientific Committee on Antarctic Research, Cambridge.

Woehler, E. J., R. L. Penney, S. M. Creet, and H. R. Burton. 1994. Impacts of human visitors on breeding success and long-term population trends in Adélie penguins at Casey, Antarctica. *Polar Biology* 14:269–74.

III

Yellow-Eyed Penguin

Genus *Megadyptes*

6

Yellow-Eyed Penguin

(Megadyptes antipodes)

PHILIP J. SEDDON, URSULA ELLENBERG, AND YOLANDA VAN HEEZIK

1. SPECIES (COMMON AND SCIENTIFIC NAMES)

Yellow-eyed penguin, *Megadyptes antipodes* ("big southern diver")

This penguin was first described as *Catarrhactes antipodes* (Hombron and Jacquinot, 1841, Auckland Islands). It is also known as *hoiho* (Maori), which means "noise-maker."

2. DESCRIPTION OF THE SPECIES

ADULT. The yellow-eyed penguin is tall (about 65 cm average standing height; table 6.1) and heavy, ranging between 4.2 and 8.5 kilograms (table 6.2). It has a distinctive pale yellow, uncrested band of feathers passing across the nape and around the eyes (fig. 2). The forecrown, chin, and cheeks are black-flecked yellow; the sides of the head and the foreneck are a light fawn-brown; and the back and tail are slate blue (paler and browner near molt). The chest, belly, front thighs, and the leading edge and underside of the flippers are white. The bill is long and relatively slender (compared to *Eudyptes*) and red-brown-pale cream. The eyes are yellow; the feet are pale to deeper pink dorsally and black-brown ventrally.

FIG. 1 (*FACING PAGE*) Traditionally, yellow-eyed penguins nested in lowland podocarp forests; however, following the destruction of most natural forest cover on the New Zealand mainland, penguins now breed in remnant shrub habitats to find shelter from heat stress. (U. Ellenberg)

FIG. 2 The adult yellow-eyed penguin has a bright yellow eye and an orange-and-pink bill. (H. Ratz)

FIG. 3. Yellow-eyed penguin chick, almost in juvenile plumage but still with chick down, has the yellow color beginning on the crown. (H. Ratz)

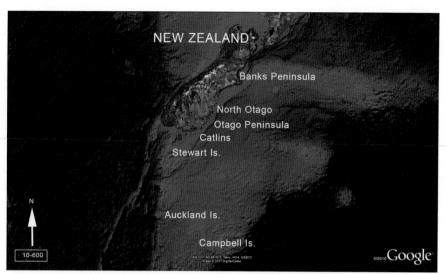

FIG. 4 Distribution and abundance of the yellow-eyed penguin, with counts based on pairs.

JUVENILE. Juveniles lack the pale yellow band and have a paler eye and paler dorsal head (fig. 3). Sexes look alike, with males being larger than females (Seddon and Seddon 1991). A combination of head and foot measurements will sex 93% of the adults correctly; using morphometrics to sex fledglings is less reliable (Setiawan et al. 2004).

MORPHOMETRIC DATA. See tables 6.1 and 6.2

3. TAXONOMIC STATUS

Recent research has overturned some long-held beliefs about the mono-generic status of yellow-eyed penguins. A combined analysis of ancient and modern DNA (Boessenkool et al. 2008) and ancient and modern penguin bone structure revealed the presence on the New Zealand South Island of a sister species, *Megadyptes waitaha* (Boessenkool 2009, Boessenkool et al. 2009a). Up until about 1500 AD yellow-eyeds were restricted to Auckland and Campbell Island groups, while the mainland was occupied by *M. waitaha*. It is thought that *M. waitaha* was harvested to extinction and that its disappearance opened the way for straggler yellow-eyed penguins from the sub-Antarctic to gain a foothold on the mainland, where a change in harvest patterns by humans and a loss of large marine mammals allowed them to rapidly expand into the population we see today. The endemic yellow-eyed penguin is now the only extant member of the genus *Megadyptes* remaining in and restricted to the New Zealand region.

4. RANGE AND DISTRIBUTION

Yellow-eyed penguins breed on the southeast coast of New Zealand's South Island (mainland), on Stewart Island and adjacent islands, and in the sub-Antarctic on Auckland and Campbell Islands (fig. 4).

On the New Zealand mainland, they breed in four distinct breeding regions: the Catlins, Otago Peninsula, North Otago, and Banks Peninsula. The few pairs on Banks Peninsula usually breed with little success, and recruitment appears to come from more successful breeding areas farther south. On two occasions a breeding attempt has been observed near Kaikoura (Mike Morrissey, DOC, pers. comm.).

On Stewart Island, yellow-eyeds breed along the northeastern (Anglem coast) and eastern shores, as well as on a range of island outliers, most importantly Codfish Island, around the exit of Paterson Inlet including Bench Island, and farther south in Port Pegasus.

Management of this species has proceeded as if it exists as one large management unit, stretching from Banks Peninsula to Campbell Island. Band recoveries (14,000 records) from 6,990 birds banded between 1981 and 1997 throughout their range, including 560 in the sub-Antarctic, indicated very little or no movement between the sub-Antarctic and mainland populations, or from north to south (Darby, unpubl. data). Early genetic studies by Triggs and Darby (1989) showed genetic separation between the South Island and Campbell and Auckland Islands.

More recent analyses by Boessenkool et al. (2009b, 2010a and b) revealed two genetically and geographically distinct populations: the South Island (including Rakiura/Stewart Island and surrounding islands) and the sub-Antarctic—confirming that there is no significant migration of yellow-eyeds between the sub-Antarctic and mainland New Zealand and that these populations should be managed as separate units (Boessenkool 2009; Boessenkool et al. 2009b, 2010a and b).

5. SUMMARY OF POPULATION TRENDS

There are estimated to be around 1,700 breeding yellow-eyed pairs, with the greatest proportion (~60%) in the sub-Antarctic. The South Island of New Zealand holds between 400 and 600 pairs (McKinlay 2001), while there are estimated to be at least 180 pairs on and around Rakiura / Stewart Island (Darby 2003; Massaro and Blair 2003). Monitoring of South Island yellow-eyeds is regular and intense, and there is a good understanding of population size and trends from year to year. However, the isolation of the sub-Antarctic yellow-eyed population means that the dynamics on the South Island are not necessarily mirrored farther south. Populations on the South Island mainland are generally stable but subject to extreme fluctuations. The populations on Stewart Island and surrounding islands are stable in some sites, but in decline for unknown reasons in others. The sub-Antarctic populations are insufficiently well surveyed to be able to determine trends.

6. IUCN STATUS

The World Conservation Union (IUCN) lists the yellow-eyed penguin as "Endangered" in the 2010 Red List, on the basis of its being confined to a small range when breeding and because its forest/scrub habitat is considered to have declined in quality and its populations to have undergone extreme fluctuations in numbers. The area occupied by breeding yellow-eyed penguins is considered to be sufficiently small to confer Endangered status. However, the decline in habitat quality is believed to have halted, and because of ongoing conservation measures the species could be downlisted in the future (IUCN 2010).

While the decline in breeding habitat quality may have been addressed over much of the South Island range, potential sea-based threats are not currently understood, appropriately monitored, or managed. Similarly, we have little idea about any potential impact of climate change. Premature downlisting this species could

hinder ongoing efforts for habitat restoration, predator control, and management of increasing tourism pressure. The yellow-eyed penguin is listed as Nationally Vulnerable under the New Zealand Threat Classification System (Miskelly et al. 2008).

LEGAL STATUS. The New Zealand Wildlife Act of 1953 (2008) confers absolute protection on yellow-eyed penguins throughout New Zealand and New Zealand fisheries waters. No person may kill or have in their possession any living or dead yellow-eyed unless authorized. However, it is a defense against prosecution if yellow-eyeds are accidentally caught as part of a fishing operation and the capture is reported to the authorities.

Under the Crown's settlement with the New Zealand Māori tribe Ngāi Tahu, yellow-eyeds are formally recognized as one of the *taonga* (natural treasure) species found within the Ngāi Tahu *takiwā* (the tribal region defined by the Te Rūnanga o Ngāi Tahu Act 1996 as being most of the South Island and the islands to the south including Stewart Island / Rakiura) and its outliers. This recognition does not change the absolute protection of the yellow-eyed penguin under the Wildlife Act, but does formalize Ngāi Tahu involvement in management of the species in partnership with the New Zealand Department of Conservation (DOC 2010).

The yellow-eyed penguin is listed as Threatened under the United States Endangered Species Act 1973 (ESA; USF and WS), administered by the U.S. Fish and Wildlife Service of the Department of the Interior, making import or export of the species without an ESA permit illegal (Rieben 2008).

7. NATURAL HISTORY

BREEDING BIOLOGY. The present-day distribution of yellow-eyed breeding sites corresponds to the pre-European distribution of cool coastal hardwood forest. Currently, breeding occurs in mature coastal forest, regenerating coastal scrub, but also in pasture, windbreaks of planted exotic trees, and relatively exposed cliffs (see fig. 1). Access to breeding and loafing sites may be via sandy or pebble/rocky beaches, or rocky platforms (Smith 1987, Darby and Seddon 1990, Williams 1995). The majority of nests are sufficiently spaced among dense vegetation to be visually isolated (Darby and Seddon 1990, Clark 2007).

This species of penguin selects well-concealed nest sites under dense vegetation (Seddon and Davis 1989), with both sexes contributing to nest building. The nest itself is an open shallow bowl made of twigs, grass, and leaves (Seddon and Davis 1989). Most birds return to the same general nest area each year, although fewer than 30% use the same nest site from year to year (Darby and Seddon 1990). Several nest sites can be occupied before laying, and nest sites can be different from year to year or be retained for up to seven successive seasons (Darby and Seddon 1990). Nearly three-quarters of breeding pairs remain together for more than one season (61% for 2–6 seasons, 12% for 7–13 seasons, compared to 27% for only one season, with divorce the cause of about 13% of pair-bond changes [Richdale 1957; see also Setiawan et al. 2005 cf. 6%]).

Females breed at an earlier age than males (45% and 96% of 2- and 3-year-old females, respectively), with an average age of 2.6 yrs, ranging between 2 and 6 years, compared to 7% and 35% of 2- and 3-year old males, with an average age of 4.3 yrs and ranging between 2 and 11 years (Richdale 1957; Darby, unpubl. data).

Little is known about pair-formation, which may take place at any time during the annual cycle. Pairs may be seen together at potential nest sites as early as July-August (fig. 5). The pre-egg phase is characterized by intermittent nest occupation, usually by lone males and pairs during the day and pairs overnight (Darby and Seddon 1990). During this time males and females show high testosterone (highest in males) and oestradiol (highest in females) levels, which drop when the eggs are laid (Cockrem and Seddon 1994). Two copulations have been observed, one of which took place 12 days before the first egg was laid (Seddon 1989a). Females tend to remain at the nest throughout the laying period but may leave for a single day between eggs (Darby and Seddon 1990). Males accompany the nest overnight and stay with the single egg during the female's absence (Darby and Seddon 1990). Egg dimensions are summarized in table 6.5.

The mean laying date is 24 September on the Otago Peninsula but can occur later at more southern locations, such as Nugget Point (van Heezik 1988), and for Campbell Island can be up to two weeks later than the mainland (Moore 1992a). A clutch of two eggs is usually laid 3–5 days apart (Richdale 1957). Some clutches of three and four eggs have been recorded at well-monitored sites on the Otago Peninsula; however, it is not known whether these additional eggs were laid by a third adult attending the nest. Young birds are more likely to lay single eggs (7.5% and 5% of all clutches contained only one egg) (Richdale 1957; Darby and Seddon 1990): 34% of clutches laid by 2-year-old females were single egg clutches compared to <1% laid by birds aged 3 years and older. Some individual females lay consistently early or late, regardless of their age, and egg size generally increases with the age of the female (Massaro et al. 2002). No replacement clutches are laid.

Incubation starts after the laying of the second egg (Darby and Seddon 1990). Both sexes incubate the eggs: the mean male incubation spell is 2.0 + 0.7 (+1 standard deviation) days and the mean female spell is 1.8 + 0.6 days (Seddon 1989b). The incubation period ranges between 39 and 51 days (mean = 43.5, Richdale 1957), and is the most variable among the penguins.

Hatching is mostly synchronous (Seddon 1990, 1991). In 94% of two-egg clutches, eggs hatched within one day of each other and 63% hatched on the same day (Darby and Seddon 1990). During six seasons, hatching success on the Otago Peninsula ranged between 81 and 87% for 1,215 eggs (Darby and Seddon 1990) and was 83% during one season on Campbell Island (Moore 1992a). Egg survival is also high (85%, Darby and Seddon 1990; 78%, Richdale 1957). The main cause of egg failure is infertility and early embryonic death, which accounted for 89% of all eggs that failed to hatch (Richdale 1957), although mustelids can take significant numbers of eggs.

FIG. 5 Male (*left*) and female (*right*) yellow-eyed penguins preening on the South Island, New Zealand. (H. Ratz)

Chick rearing can be divided into two phases: the guard phase when the chick is constantly attended for between 40 to 50 days at the nest, and the post-guard phase when chicks are left alone at the nest during the day and fed in the late afternoon or early evening when the adults come ashore, usually within 1–3 hours of each other (fig. 6) (Darby and Seddon 1990; Seddon 1990, 1991; Seddon and Darby 1990; Schuster and Darby 2000). By day 20, the sparse primary down has been replaced by thicker secondary down, and after 25 days, chicks no longer require brooding and may wander from the nest site (Darby and Seddon 1990), often congregating with chicks from adjacent nest sites to form small groups of 4–6, and up to 12 chicks, particularly in more open habitat and when nests are close together.

Seasonal and geographical variations in growth rates are small and mainly reflect variation in food supply (van Heezik 1990a); there is evidence that growth rates in the 1980s were faster than in the 1930s (van Heezik 1991). Growth rates of first- and second-hatched chicks are similar, even in years of poor food availability (van Heezik and Davis 1990). Hatching weight is about 108g and mean fledging weights from the Otago coast ranged between 4.1 kg and 6.2 kg during three years of variable food supply (van Heezik 1990a); the weight was 5.1 kg during one season at Campbell Island (Moore 1992a) and 5.3 kg from one season of poor reproductive success

FIG. 6 Adult yellow-eyed penguin with a chick in its nest. Note the chick's bluish-gray eye. (P. D. Boersma)

on Enderby Island (M. J. Young and L. S. Argilla 2009, unpubl. data). Variation in food availability can result in considerable inter-annual variation in fledging weights and chick mortality (van Heezik and Davis 1990).

Accident, predation and starvation are the major causes of chick death (Darby and Seddon 1990), with King (2007) commenting that disease may also be a concern. Early deaths are usually accidental, such as when small chicks are crushed, or can be due to the failure of the parents to feed the chicks (Darby and Seddon 1990). Chicks are most likely to be killed by predators between 5 and 20 days after hatching, but remain vulnerable until 40–50 days after hatching (Darby and Seddon 1990). Starvation is uncommon but may occur after 9 weeks (van Heezik 1988; Darby and Seddon 1990). When food availability was sufficiently poor to result in starvation of chicks, both chicks lost weight at similar rates and no brood reduction mechanisms were evident (van Heezik and Davis 1990).

Fledging of chicks occurs after about 106 days (Marchant and Higgins 1990). Little is known of the movements of newly fledged juveniles other than that in most years they move north and may spend most of the winter at sea, although in good food years, they will remain in the vicinity of their natal area and can be seen onshore from time to time (Darby, unpubl. data).

Breeding success of young (2–3 year-old) birds was lower (63%) than that of older birds (89%, Williams 1995). Richdale (1957) recorded overall fledging success at 76% (45–92%), a mean of 1.16 chicks per nest for birds on the Otago Peninsula. Fledging success for 1,029 chicks hatched over 6 seasons at 8 breeding areas along the Otago Peninsula and the Catlins coast was 69.3%, ranging between 27.9% and 86.6% (Darby and Seddon 1990). The overall mean number of chicks fledged per nest was 1.1 on the South Island (range: 0.57–1.39; Darby and Seddon 1990—622 nests at 8 localities during 6 seasons; Moore and Wakelin 1997—306 nests at 4 localities during 4 seasons); 1.3 at the Otago Peninsula (range 0.97—1.94; Ratz et al. 2004—285 nests at 2 localities during 6 seasons); 1.2 at Codfish Island (range: 0.96–1.51; King 2008—81 nests during 4 seasons); 0.70 at Stewart Island (range: 0–1.35; King 2008—233 nests at 2 localities during 4 seasons); 1.4 at Campbell Island (Moore 1992a—33 nests at one location during one season); and 0.6 at Enderby Island (Young 2009a; L. S. Argilla, unpubl. 2009 data—48 nests during one season). Overall, only 18.8% of fledglings will survive to maturity (Stein 2012).

Between 0% and 13% of males, and 0% and 17% of females skipped one breeding season, but skipping two consecutive seasons was rare (Ratz et al. 2004). Among 138 birds that deferred breeding, 17% did so following divorce, 29% as a matter of choice (birds bred as a pair later), and 54% after the death of their partner. The oldest known bird is at least 25 years of age (Darby, unpubl. data).

PREY. Yellow-eyed penguins (n=512) ate 26 prey species across seven mainland localities between 1984 and 1986, with fish (mostly six species) comprising 87% of the weight and 91% of the total number of prey items, and arrow squid (*Nototodarus* spp.) making up the remainder. Sprat (*Sprattus antipodum*) was most frequently taken, but also red cod (*Pseudophychis bachus*), opalfish (*Hemerocoetes* spp.), ahuru (*Auchenoceros punctatusi*), silversides (*Argentina elongata*), and blue cod (*Parapercis colias*) (van Heezik 1990b). Between 1991 and 1993, Moore and Wakelin (1997) identified 43 prey types from 198 samples from the Otago coast, including 37 fish, 4 cephalopod, and several crustaceans. The same species

made up the majority of biomass, but proportions of blue cod and opalfish were higher and sprat lower than in the previous study. Most prey are <200mm in length (van Heezik 1990b, 1990c; Moore and Wakelin 1997). Considerable variation in species composition and meal sizes occurs between years and different localities (van Heezik 1990b, 1990c; van Heezik and Davis 1990; Moore and Wakelin 1997). Prey species are both pelagic and demersal in habit (van Heezik 1990b, 1990c; Moore and Wakelin 1997). Juvenile penguins ate more squid and less fish than did adults (van Heezik 1990b). Stable isotope analysis of yellow-eyeds on Stewart Island indicates adults selectively provision chicks with opalfish while feeding mainly on blue cod themselves (fig. 7) (Browne et al. 2010).

FORAGING. Most yellow-eyeds forage in the mid-shelf region between 2 and 25 km offshore, diving to depths of up to 40–120 m to obtain bottom-dwelling (demersal) fish species (Moore 1999; Mattern et al. 2007a). Breeding yellow-eyeds make two kinds of foraging trips: day trips that ranged between 12 and 20km from the coast

FIG. 7 Adult yellow-eyed penguin is followed by its fluffy, down-covered chick begging for food. (H. Ratz)

and shorter evening trips of less than 7 km (Mattern et al. 2007a). Diving behavior is mainly benthic (87% of dives, Mattern et al. 2007a). While foraging trips are very consistent, considerable inter-individual variation exists in how far offshore birds feed: most are mid-shelf foragers (5–16km off the coast); some foraged <5 km from the coast; while others were active >16km from the coast (Moore 1999; Mattern et al. 2007a). Failed breeders and nonbreeders traveled farther (females) and for longer (especially males) than breeding birds (Moore 1999).

PREDATORS. One of the most significant impacts on yellow-eyed productivity is predation of chicks (particularly chicks between 5 and 20 days of age) by a suite of introduced mammalian predators, including feral cats (*Felis catus*), ferrets (*Mustela furo*), and stoats (*M. erminea*) (Alterio and Moller 1997; Clapperton 2001; Darby and Seddon 1990; Moller and Alterio 1999; Ratz 1997, 2000; Ratz et al. 1999, 2004). Annual chick mortality ranges as high as 63% in the absence of effective predator control (Darby and Seddon 1990). Rats (*Rattus norvegicus*) may prey on small chicks (Massaro and Blair 2003): there are two known instances of chicks wounded by rats on Campbell Island (Amey and Moore 1995). On the main Auckland Island, pigs (*Sus scrofa*) probably kill both chicks and adults (Taylor 2000). Otherwise adult penguins appear not to be at great risk from exotic terrestrial predators, except for dogs (Hocken 2005). Mustelids may also take eggs (53 lost from 27 nests at Double Bay in 1996; Darby, unpubl. data).

Sharks, barracouta (*Thyrsites atun*), fur seals (*Arctocephalus forsteri*), and New Zealand sea lions (*Phocarctos hookeri*) take or injure yellow-eyeds at sea (Hocken 2005, Schweigman and Darby 1997). The population-level significance of these causes of mortality is not known. As fur seal and New Zealand sea lion numbers are increasing around the New Zealand mainland coastline, at-sea predation of penguins likely will increase (fig. 8). An individual sea lion can take yellow-eyed penguins at rates that could extirpate small local populations, but this habit is not widespread in the sea lion population (Lalas and Ratz 2008; Lalas et al. 2007). Predation of yellow-eyeds by sea lions has been observed at Campbell Island in 1988, but predation levels were considered to be low and due to some individuals adopting this predatory behavior (Moore and Moffat 1992).

FIG. 8 The New Zealand sea lion is increasing in number and preys on yellow-eyed penguins given the opportunity. (P. D. Boersma)

MOLT. Molt takes place between late February and late March, on average 22 days after chicks fledge (Williams 1995), but more recently has extended into April. Birds remain onshore for about 24 days. Juveniles and failed and nonbreeders commence molting before successful breeders. Though most breeding birds molt on or near their nest sites, many move north, some as far as away as Canterbury, Kaikoura, Cape Campbell, and, rarely, the lower North Island.

ANNUAL CYCLE. See figure 9.

8. POPULATION SIZES AND TRENDS

Population trends of yellow-eyed penguins are variable among sites. We have detailed them by their management unit.

SOUTHEASTERN COAST OF SOUTH ISLAND, NEW ZEALAND. An estimated 400–600 breeding pairs, with some evidence of a decline in breeding pairs on Otago Peninsula over the last 20 years, but other breeding regions probably stable with extreme fluctuations.

The South Island breeding population has shown marked inter-annual fluctuations but no clear overall trends over ~30 years since the late 1970s, as indicated by extensive nest searches on Otago Peninsula and selected sites in North Otago and the Catlins (Darby 1985; Darby and Seddon 1990), the first detailed nest counts since the population study of Lance Richdale (1936–48, Richdale 1957) (fig. 10). There is some anecdotal evidence that yellow-eyeds were more numerous in some areas during the period of monitoring by Richdale (Darby and Seddon 1990); however, for other sites, available records suggest that the number of breeding birds may actu-

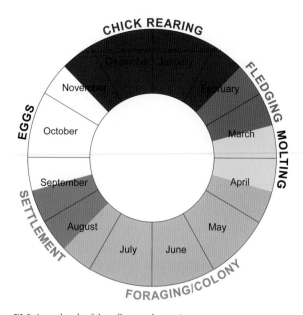

FIG. 9 Annual cycle of the yellow-eyed penguin.

ally have increased between the 1930s and the 1980s (Moore 2001). It is also apparent that while some breeding sites may be occupied for many decades, others may be abandoned, perhaps due to demographic stochasticity (McKinlay 1997), and new breeding sites can develop. In at least one case considerable set-net bycatch in the immediate vicinity is thought to be responsible for the failure of a breeding site (Darby and Dawson 2000).

Various factors combine to potentially give a spurious impression of historical population-level change. These include the use of nesting numbers as a proxy for adult population size, marked inter-annual variation in the numbers of birds attempting to breed, and individual breeding site extinctions and colonizations, along with incomplete monitoring in the past and the imprecision of historical estimates of abundance, including the use of extrapolations from partial nest counts. However, a detailed analysis of 10 of the most intensively monitored sites on Otago Peninsula indicates a significant decline in the number of breeding pairs, at 9–10 sites since 1992 (Ellenberg and Mattern 2012).

After a major population crash in 1990 (Gill and Darby 1993), more regular monitoring of virtually all known mainland (South Island) breeding areas was implemented by DOC, and currently more than half (~34) of the 53 known breeding areas are assessed each year (fig. 11). Mainland monitoring sites receive 3–5 (sometimes more) visits for estimating reproductive output.

Following an initial rapid increase in breeding pairs in 1991, when previously known breeders returned the following season, the following six years of low adult mortality and good recruitment allowed full recovery (see fig. 4; Efford et al. 1994, 1996). However, despite the compilation of a substantial data set, the drivers of extreme fluctuations in yellow-eyed breeding numbers remain little understood. The most recent nest counts at monitored sites indicate a total of 454 breeding pairs on the mainland, over a third of which are on the Otago Peninsula (table 6.3).

Population sizes have been assessed through annual counts of active nests. However, because of the dispersed nature of yellow-eyed breeding areas, the relative isolation of individual nests, and the dense vegetation that characterizes many sites (Clark 2007; Clark et al. 2008; Poole 2005; Seddon and Davis 1989), even intensive searches may fail to locate all active nests within a given area. Nest counts should therefore be used with caution and considered minimum counts of unknown accuracy (Hegg et al. 2012).

Considerable fluctuations in year-to-year numbers of breeding pairs likely depend also on variations in marine productivity, along with factors such as habitat change, disease outbreaks, variable intensity of predator control, and different levels of human disturbance. A project classifying mainland breeding sites and analyzing the effects of a range of natural and human derived factors on mainland population trends has been recommended by the Research Advisory Group appointed by the Yellow-eyed Penguin Recovery Group in 2009 (see section 10, below).

STEWART ISLAND AND ADJACENT ISLANDS, NEW ZEALAND. About 180 breeding pairs are declining on the Anglem coast and Codfish Island, probably stable in other areas.

A total of 178 active nests were found during the last comprehensive survey (1999–2001) covering Stewart Island (79 pairs) and adjacent islands (99 pairs, Massaro and Blair 2003). The most recent survey (2008–9) found 107 nests on Stewart Island (77 pairs) and smaller outliers (30 pairs) (table 6.3); however, larger islands such as Bench and Codfish were not surveyed (King 2009). Findings suggest that this subpopulation is currently in decline along the Anglem Coast (King 2009) and recently at Codfish Island / Whenua Hou (Houston and Nelson 2012). Numbers at other areas on and around Stewart Island appear stable (King 2009).

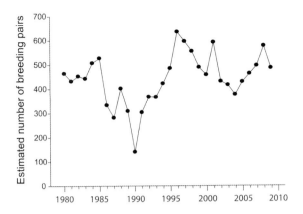

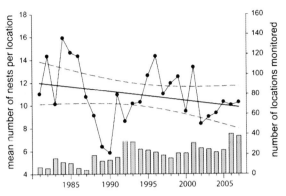

FIG. 10 Estimated number of yellow-eyed penguin breeding pairs on Otago Peninsula, New Zealand. Data provided by J. Darby (1981–97), D. Houston, and M. Young of the Department of Conservation, which maintains a spreadsheet (DOCDM-28226) estimating the number of breeding pairs, including those from sites that are not searched every year.

FIG. 11 Mean number of yellow-eyed penguin nests per mainland breeding location (South Island, New Zealand, black dots) and number of locations searched each breeding season (gray bars). A linear trend line including 95% confidence intervals illustrates large fluctuations in breeding pairs among seasons. Data from the current yellow-eyed penguin database managed by the Department of Conservation, New Zealand.

Darby (2003) summarized historical data and earlier surveys (1984–85, 1989–1991, 1999–2001; see also Blair 2000) in an attempt to provide baseline numbers for future monitoring; 22 breeding areas were identified on Stewart Island and a further 16 breeding areas on island outliers, including Bench, Codfish, and Ruapuke (Blair 2000; Darby 2003). Using ground searches as well as beach counts, Darby (2003) estimated 88–168 breeding pairs for Stewart Island, 84–146 breeding pairs for the smaller island outliers, and 48–78 breeding pairs for Codfish Island, an estimated total of 220–400 breeding pairs for this region.

The number of known breeding areas recorded for Stewart Island increased from 24 (1999–2000) to 29 (2008–2009) (King 2009). However, some previously known breeding sites had reduced numbers of breeding pairs, which led to the suspicion that increased search effort and experience could have obscured a potential decline. In any case, apart from the well-documented Anglem coast (nests monitored yearly 2003–2008; King 2008) data currently available for breeding areas on Stewart Island are insufficient to allow reliable conclusions about population trends.

Considering the available habitat, the overall number of birds observed in fragmented and small breeding areas on Stewart Island is surprisingly low compared to the large breeding aggregations on Codfish Island (Darby 2003; King 2008, 2009). A recent study (2003–8) initiated by the Yellow-eyed Penguin Trust concludes that predation (as suggested by Massaro and Blair 2003) is unlikely the cause for the observed difference between Stewart Island and predator-free Codfish Island. Starvation resulting from changes in the marine environment and disease may be important factors (King 2008), and differences in habitat quality caused by depletion of the understorey by herbivores on Stewart Island (Darby 2003) may also play a role. However, today many apparently suitable nesting areas on Stewart Island remain unoccupied.

Reasons for the observed population decline along the Anglem coast (north-eastern Stewart Island) are currently unknown, but have resulted in catastrophic reproductive failure in the seasons monitored. Commercial fisheries possibly had an important impact via intensive oyster dredging in Foveaux Strait that appears to affect prey availability and quality (Mattern 2008; Browne et al. 2010). Fisheries bycatch may also play an important role in the demise of yellow-eyed penguins in this area (Rowe 2009, 2010; Ramm 2010; Housten and Nelson 2012). Disease, such as the hemoparasite *Leucocytozoon* may affect chick survival (Hill et al. 2007; Hill 2008; King 2008), although it is currently unclear if disease directly causes chick deaths or makes chicks more vulnerable to other factors, such as starvation, or if depletion of foraging habitat and thus of food quality makes chicks more vulnerable to disease.

AUCKLAND ISLAND, NEW ZEALAND. An estimated 520–570 breeding pairs, based on a 1989 census of the northern part of the Auckland Island group and a reconnaissance of the extensive but more sparsely populated eastern shores and part of Adams Island (Moore 1992b); trend unknown.

In 1989 penguins were counted on Enderby Island and on adjacent islands, as well as the northernmost part of the main Auckland Island. The survey found 115 landing sites used by a total of 934 penguins, which may have represented an estimated 420–470 breeding pairs, of which 260–290 pairs (63%) were based on Enderby Island (Moore 1992b). Based on a brief reconnaissance of some sparsely populated eastern bays of the Auckland Islands, Moore (1992b) estimates a minimum of 520–570 breeding pairs for the entire Auckland Island group (table 6.4). This number was subsequently adjusted to 520–680 pairs. There are no recent estimates of population size for the Auckland Islands, but an indication of relative abundance comes from a total of 306 landing sites identified during a 2009 survey of the main island coastline (Beer 2010; table 6.4). Predator-free Enderby Island is thought to be a major population center in the north of the Auckland Island group based on beach count figures during the guard stage of chick rearing (Young 2009a).

CAMPBELL ISLAND, NEW ZEALAND. About 350–540 breeding pairs (last census in 1990/91; Moore et al. 2001); trend unknown.

On Campbell Island, yellow-eyeds are found in the more sheltered harbors with population centers at Perseverance Harbour, Northeast Harbour, Northwest Bay, and Southeast Harbour. Via beach counts at 172 landing sites, using mark-recapture data from one area, Moore and Moffat (1990) estimated 2050–2550 adults (ca. 610–890 breeding pairs) in 1987–88. By 1991–92, the population had decreased by 41% to 1200–1550 adults, i.e., 490–600 pairs if 60% were breeders (Richdale 1957), or 560–700 breeding pairs if 70% were breeders (Efford et al. 1994), with only 140 active landing sites found (Moore et al. 2001) (table 6.4).

Between 1987 and 1998 a selection of 11 landing sites, 8 in Northwest Bay and 3 in Southeast Harbour, was monitored opportunistically via counts (mostly in November). A dramatic decline between 1987 and 1992 coincided with a population crash on the mainland (South Island, NZ); however, reasons remain unclear (Efford et al. 1996). Most index sites began to recover by 1995 (Moore et al. 2001). In November 2008 index counts were repeated (two counts per landing site). Numbers at Northwest Bay (on average 131 birds) were higher than the previous November count in 1996 but have not yet reached previous peak records. At Southeast Harbour, on average 21.5 birds were observed, the lowest record ever and considerably lower than the previous count in 1997 (Hiscock 2008). However, it is unclear if this is caused by lower adult survival or merely a shift in use of landing sites potentially to avoid sea lion harassment (Hiscock 2008; see also Moore et al. 2001). The last comprehensive population survey in 1991–92 (Amey and Moore 1995) has not been repeated to date.

9. MAIN THREATS

TERRESTRIAL THREATS. On land, habitat degradation has the greatest impact on the species. Breeding habitat changes and loss and fragmentation of habitat reduce reproductive success through predation, disease, human disturbance, heat stress, and hypothermia if chicks get wet.

Loss of mature coastal forest may be responsible for historic declines in yellow-eyed populations and may force penguins to nest in possibly sub-optimal habitats (Roberts and Roberts 1973, Smith 1987), where they may face greater heat stress on land during breeding (Darby and Seddon 1990; McKinlay 2001; Clark 2007; Seddon 1988; Seddon and Davis 1989; Seddon and Darby 1990). However, there is evidence that penguins are able to breed and fledge chicks in even very modified pasture habitats (McKay et al. 1999) (fig. 12). The effects of predation were thought to be ameliorated (but not eliminated) through the creation of areas of rank vegetation that could support fewer rabbits (primary prey of cats, ferrets, and stoats) and hinder access of these predators into and through penguin breeding sites (Darby and Seddon 1990; Seddon 1988). Quantification of predator movements and the relative abundance of predators and prey do not support this "vegetation buffer" hypothesis. The evidence suggests buffer zones may attract predators, possibly through increased abundance of other prey species such as mice and small birds (Bruce 1991; Alterio 1991, 1994; Dymond 1991; Alterio and Moller 1997; Alterio et al. 1998; Moller and Alterio 1999; Moller et al. 1998). Inadequately controlled dogs continue to cause problems on penguin landing beaches (Hocken 2000, 2005).

DISEASE. Yellow-eyed adults and chicks appear periodically to succumb to outbreaks of disease. The major

FIG. 12 Yellow-eyed penguin walking through pasture to its nest shows the black sole of its foot. (H. Ratz)

die-off of adults (estimated at 31% of the mainland population) in 1990, in which a number of agents were suggested, including avian malaria, was not confirmed to be caused by a specific agent (Gill and Darby 1993; Duignan 2001; Graczyk et al. 1995a and b; McDonald 2003; Sturrock 2007; Sturrock and Tompkins 2007, 2008).

In the 2002–3, 2004–5, 2006–7 and 2008–9 breeding seasons, young chicks in mainland breeding sites became infected with a disease later described as diphtheritic stomatitis, caused by a Corynebacterium (*Corynebacterium amycolatum*) (Alley et al. 2004; Houston 2005). This bacterium is naturally present in healthy penguins. Despite repeated disease outbreaks, the trigger that causes diphtheritic stomatitis in some years and in some areas is still little understood even though the outbreaks thus far have been recorded biennially. Any stressor that weakens the immune system, ranging from starvation to an unknown virus, could potentially be responsible. Although some chicks responded to antibiotic treatment, mortality in some breeding areas was as high as 86% (DOC, unpubl. data); renal failure due to dehydration was suggested as the primary cause of death.

Corynebacterium also affected chicks on Stewart Island, where it may have subsequently contributed to chick mortality due to Leucocytozoon in 2006–8 (King 2007). There has been some recent evidence of disease-related mortality of adults since 1990, and some evidence of disease affecting adults and chicks on the sub-Antarctic islands (Argilla et al. 2010).

HUMAN DISTURBANCE. This species is vulnerable to disturbance by humans, including tourists, researchers, and managers (Wright 1998; Seddon et al. 2004; Ellenberg et al. 2007, 2009; Seddon and Ellenberg 2008). Available evidence suggests that carefully regulated tourist viewing of penguins does not carry measurable impact (Ratz and Thompson 1999). However, unregulated and relatively high intensity disturbance by tourists is associated with reduced breeding success and lower chick weights at fledging, which results in lower first-year survival and recruitment probabilities (McClung et al. 2004; Ellenberg et al. 2007). Contrary to what could be expected, birds that continued to breed despite frequent disturbance do not habituate to human proximity but appear to be sensitized and show stronger heart rate and hormonal stress responses to human disturbance (Ellenberg et al. 2007, 2009). Hence, penguins exposed to unregulated tourism are not only disturbed more often, each disturbance event is more costly for the affected birds and disturbance effects accumulate, potentially leading to poorer breeding performance.

Human disturbance of yellow-eyeds in the form of well-intentioned research manipulations, has included stomach flushing (Browne 2007; Moore and Wakelin 1997; van Heezik 1988, 1990 a and c; van Heezik and Seddon 1989), blood sampling (Boessenkool 2009; Cockrem and Seddon 1994; Ellenberg et al. 2007; Massaro et al., 2007; Seddon 1988; Setiawan 2004; Setiawan et al. 2006, 2007), artificial brood reduction and other nest-related manipulations of eggs or chicks (Farner 1958; van Heezik 1988, 1990b; Nordin 1991; Edge 1998; Efford and Edge 1998; Edge et al. 1999; Schuster and Darby 2003; Massaro 2004; Massaro and Davis 2004a and b; Massaro et al. 2003, 2004, 2006; Ellenberg 2009; Kudo 2009), and the attachment of external devices to measure foraging behavior (Mattern 2005, 2006, 2008; Mattern et al. 2007a; Moore 1999; Seddon and van Heezik 1990). However, with the exception of one blood sampling study (and double banding), no individual fitness or population-level implications of research manipulations have been identified (Stein 2012).

Another area of concern is the impact of flipper banding on yellow-eyed survival (van Heezik et al. 2005). Studies in other species indicate that in some cases flipper banding can be associated with higher energetic costs due to impairment of swimming ability, and thus lower survival (Fallow et al. 2009, and references therein), but this is not true of all species (Boersma and Rebstock 2010). Stein (2012) did show there was a measurable negative impact on yellow-eyeds from double banding. A recent study found that individuals with previous exposure to the more intrusive research manipulations are less likely to habituate to short and consistent experimental disturbance (Ellenberg et al. 2009). Both initial stress responses and habituation potential also depend on the sex and character of affected individuals, and the observed differences in behavioral and physiological stress responses are related to fitness parameters. Thus, human disturbance may drive contemporary evolutionary change in the composition of breeding populations with as yet unrealized consequences for conservation.

MARINE THREATS. Although some threats from disease and humans are more common on land, they can also occur in the marine environment.

FISHERIES. The *National Plan of Action to Reduce Incidental Catch of Seabirds in New Zealand Fisheries* (MOF and DOC 2004) lists yellow-eyed penguins as by-catch in inshore set nets (set gillnets). However, historically there has been very low observer coverage (~2%) of the inshore fishing fleet; hence, there is little available information on the numbers of yellow-eyed penguins killed at sea each year due to fisheries activities. The absence of these data makes reliable impact assessment and the development of mitigation measures impossible (Maunder et al. 2007; Ellenberg and Mattern 2012). Between 1979 and 1997 there were a total of 72 confirmed deaths of yellow-eyeds in set-net entanglements, most at or near Otago Peninsula, and set-net by-catch was thus considered to pose a significant threat (Darby and Dawson 2000). More recently fisheries observers have recorded the capture of four yellow-eyeds over three years (2006–09) of observer coverage of commercial set-net fisheries (Rowe 2009). There is also some evidence that indirect effects of commercial fisheries may have an impact on yellow-eyed marine habitat degradation (Mattern 2008, Browne et al. 2010). The most recent inshore fishing fleet observer program reported five yellow-eyeds caught during a two month period (Jan/Feb 2009) covering only 13% of the set-net effort during this time (Ramm 2010).

OCEANOGRAPHIC CONDITIONS AND FOOD SUPPLY. There can be marked inter-annual variation in food supply for yellow-eyed penguins with intermittent poor food years being marked by reduced numbers of breeding attempts, slow chick growth, higher pre-fledging chick mortality, low chick fledging weights and lower survival of juveniles and adults (Darby and Seddon 1990; van Heezik and Davis 1990). Causes for poor seasons have not been conclusively identified but could be related to climate (Peacock et al. 2000), including both periodic events such as the El Niño Southern Oscillation (ENSO event) and the inter-decadal Southern Oscillation, and long-term climate change.

POLLUTION. The wide geographical distribution of yellow-eyeds, the low intensity of shipping activity over much of the species range, and the low number of recorded spill incidents and proposals for their containment under the *NZ Marine Oil Spill Response Strategy* (Maritime New Zealand 2006), means oil and chemical spills are not considered a significant threat to yellow-eyed populations at this time. However, there are proposals to permit oil exploitation in the western Foveaux Strait and to develop Dunedin as the major oil harbor of the South Island in the near future, with enlargement of shipping channels being planned. If these developments were to proceed, the entire mainland penguin population would be exposed to a higher risk of oil pollution and spills.

10. RECOMMENDED PRIORITY RESEARCH ACTIONS FOR CONSERVATION

In 2009 New Zealand's Department of Conservation (DOC) circulated a draft document titled "Research and Management Priorities for Yellow-eyed Penguins" which called for, among other things the "development of a research agenda with an emphasis on identifying important gaps in our knowledge of yellow-eyed penguins" (DOC, unpubl. draft doc.). An immediate outcome of this document saw an ad hoc Yellow-eyed Penguin Research Advisory Group (YEP RAG) formed in order "to provide a research framework and to make associated recommendations for methods and process to address current research priorities with regard to YEP." DOC has affirmed that the annual recording of the reproductive performance of marked individuals at selected sites must be sustained so as to contribute to an electronic relational database (Efford et al. 1994) maintained by DOC and currently containing more than 30 years of data (DOC, unpubl. data). A key approach for future research will entail the interrogation of this unique research resource.

Research priorities include the following:

1. Identify determinants of inter-annual and regional variation in yellow-eyed penguin productivity.
2. Systematically collate data on breeding site characteristics, annual reproductive performance, management interventions, and climatic and oceanographic indicators at local and regional scales across selected sites through the penguin's range (YEP RAG 2010).
3. Unify and direct what have been enthusiastic but disjointed regional data-gathering exercises.
4. Identify specific projects that may be suitable for ongoing monitoring or one-time studies.

11. CURRENT CONSERVATION EFFORTS

Current conservation efforts fall into three categories.

HABITAT PROTECTION AND RESTORATION. Habitat protection and restoration (including predator control) is managed on a site-by-site basis by the Department of Conservation, the Yellow-eyed Penguin Trust, community groups, and private landowners. The Yellow-eyed Penguin Trust currently administers 7 breeding areas, 3 on the Otago Peninsula totalling over 250 ha, 1 in North Otago, and 3 in the Catlins. On Trust sites, active habitat restoration includes trapping exotic mammalian pests (primarily mustelids and cats) and planting native vegetation to convert former grazed pasture to coastal shrubland. Selected DOC sites have been revegetated, are subject to seasonal predator trapping of mustelids, and may limit public access at sensitive periods during the yellow-eyed breeding season. Several privately owned sites on the Otago Peninsula are a focus of nature-based tourism operations and undertake regular predator trapping and habitat enhancement, including controlled grazing by livestock, revegetation, or the provision of nest boxes. Nest boxes have been used successfully as a temporary measure to provide concealment until revegetation is sufficiently advanced (Lalas et al. 1999).

TOURISM MANAGEMENT. The yellow-eyed penguin is a major nature-based tourism attraction on mainland South Island. Commercial wildlife tourism takes place on both private land and under DOC concession on DOC-administered reserves. Unregulated visitor access is facilitated at some 20 key mainland sites with the intention of taking visitor pressure off more sensitive areas (McKinlay 2001). At unregulated sites, signage, viewing hides, and, in some cases, a volunteer warden scheme is used to mitigate impacts (Stein et al. 2010). There is some well-regulated seasonal tourist presence at yellow-eyed landing and nesting areas on Enderby Island, in the Auckland Island group. However, much of this must rely on capped tourist numbers, goodwill, and adherence to DOC's Subantarctic Minimum Impact Code and landing permit schedules (DOC 2008; Young 2009b).

REHABILITATION. Under DOC authority there are three approved rehabilitation centers for yellow-eyeds. These privately funded centers serve as collection points for sick, injured, or orphaned penguins, and for penguins that have not molted in good condition. They also act as focal points for management responses to periodic disease outbreaks during the breeding season (Busch and Cullen 2009). A recent evaluation of rehabilitation outcomes found that although rehabilitation of resident breeders did not generate a significant increase in mean annual survival, it can increase the local number of nesting attempts at sites where anthropogenic threats to the species are adequately managed (Ratz and Lalas 2010).

12. RECOMMENDED PRIORITY CONSERVATION ACTIONS FOR INCREASING RESILIENCE AND MINIMIZING THREATS AND IMPACTS

The long-term goal of the 2000–2025 Recovery Plan for the yellow-eyed penguin is to increase the population. As of 2010, effective seasonal predator control was in place for 22 of a total of 38 principal mainland sites, thus meeting the goal of protection of chicks from predators at 50% of all South Island nests (McKinlay 2001). Fisheries activities (via by-catch and alteration of foraging habitat and prey availability) have been postulated to have an impact on penguins breeding along the northern coasts of Stewart Island (Mattern 2008; Browne et al. 2010) and may negatively affect the mainland population (Darby and Dawson 2000; Ellenberg and Mattern 2012; Mattern 2005; Mattern et al. 2007b; Rose Grindley, Ministry of Fisheries NZ, pers. comm.). Recommendations are as follows:

1. Sustain and expand protection of chicks from predators at mainland sites.
2. Manage other anthropogenic threats, particularly human disturbance, at mainland sites.
3. Evaluate numbers and trends for sub-Antarctic populations and identify possible limiting factors.
4. Quantify the impact of commercial fisheries via by-catch, marine habitat degradation, and overfishing of spawning stock throughout the yellow-eyed range.

ACKNOWLEDGMENTS

This account was expanded and greatly improved by comments from David Agnew, John Darby, Bruce McKinlay, Thomas Mattern, Sue Murray, and Melanie Young. We are grateful to John Darby, Dave Houston, Hiltrun Ratz, Anita Spencer, and Chris Lalas for facilitating access to relevant literature.

TABLE 6.1 Summary of flipper, head, and foot biometric measurements (mm) for yellow-eyed adults

MEASUREMENT	SEX	MEAN	+1SD	RANGE	SAMPLE SIZE
Flipper length	Male[1]	215	3.7	207–223	66
	Female[1]	206	5.1	197–215	70
	Male[2]	209	2.5	204–215	39
	Female[2]	204	3.2	197–213	39
Bill length	Male[1]	55.1	1.52	51–58.8	66
	Female[1]	53.8	1.79	49.3–57.8	70
	Male[2]	54.4	1.3	51.6–56.8	39
	Female[2]	52.8	1.2	50.7–55.9	39
Total head	Male[3]	143.5	2.19	–	50
	Female[3]	137.1	3.21	–	52
Tarsus length	Male[2]	60.1	1.3	58.5–62.3	7
	Female[2]	58.1	1.2	56.7–60.0	7
Tarsus – toe	Male[2]	134.2	2.9	127.4–140.2	39
	Female[2]	130.4	2.7	125.0–135.8	39
	Male[3]	131.2	2.80	–	50
	Female[3]	126.1	2.95	–	52

1 Richdale (1951), Otago Peninsula; recalculated in Marchant & Higgins (1990).
2 Moore, Campbell Island; presented in Marchant & Higgins (1990).
3 Setiawan et al. 2004, Otago Peninsula.

TABLE 6.2 Body mass of adult yellow-eyed penguins (showing mean, range, and sample size)

STAGE / LOCATION SEX	PRE-MOULT OTAGO PENINSULA[1]	POST-MOULT OTAGO PENINSULA[1]	INCUBATION-GUARD OTAGO PENINSULA[2]	INCUBATION-GUARD CAMPBELL ISLAND[3]
Male	8.5; 7.3–8.9; (25)	4.4; 3.9–4.9; (25)	5.53; 4.88–6.35; (80)	5.4; 5.0–5.6; (8)
Female	7.5; 6.6–8.4; (25)	4.2; 3.6–4.8; (25)	5.26; 4.53–6.24; (80)	4.5; 4.0–4.7; (7)

1 J. T. Darby, P. J. Seddon and Y. van Heezik; in Marchant & Higgins (1990).

2 Richdale 1951; recalculated in Marchant & Higgins (1990).

3 P. J. Moore; presented in Marchant & Higgins (1990).

TABLE 6.3 Mainland yellow-eyed penguin management unit

REGION	NUMBER OF SITES	BREEDING PAIRS	SURVEY
Mainland NZ			
Banks Peninsula[1]	3	6	2011–12
North Otago[2]	8	50	2011–12
Otago Peninsula[2]	17	184	2011–12
Catlins[3]	9[†]	214	2011–12
Stewart Island			
Anglem Coast[4]	8	22	2008–9
East Coast[4]	11	37	2008–9
Port Pegasus[4]	4	18	2008–9
Smaller Outliers[4]	8	30	2008–9
Codfish Island[5]	4	46	2009–10
Bench Island[6]		14	1999–2000

† 9 localities are searched and monitored and the remaining 15 are searched every 5 years.

1 Anita Spencer and Stephen Parker (DOC)

2 Melanie Young (DOC)

3 Melanie Young and Cheryl Pullar (DOC)

4 King (2009)

5 Sandy King (YEP Trust) and Dave Houston (DOC)

6 Massaro and Blair (2003)

Notes: The "Anglem Coast" refers to the northeastern coast of Stewart Island between Halfmoon Bay and East Ruggedy.

Summary of confirmed penguin breeding sites and (minimum) number of breeding pairs encountered during ground searches on the South Island mainland (2009–10), and on Stewart Island (2008–9), Codfish Island/Whenua Hou (2009–10), and Bench Island (1999–2000) (see notes below for location details).

Geographic location (distribution best reflected by continuous line along the coast between the positions given for each breeding region):

Mainland NZ

Banks Peninsula: between 43°48'16S, 173°06'35E and 43°53'00S, 173°01'12E

North Otago: between 45°07'22S, 170°58'56E and 45°31'47S, 170°46'02E

Otago Peninsula: between 45°45'29S, 170°40'34E and 45°54'14S, 170°35'33E (outer coastline only not in harbour)
    and on Green Island 45°57'16S, 170°23'17E

Catlins: between 46°26'52S, 169°49'02E and 46°40'29S, 169°00'19E

TABLE 6.4 Sub-Antarctic Yellow-eyed penguin management unit

REGION	NUMBER OF SITES			BREEDING PAIRS
	Moore (1992)[1]	Young (2009a)	Beer (2010)[2]	Moore (1992)
AUCKLAND ISLAND				520-570
Northern part	115		100	420-470
Enderby Island	25	28	-	260-290
Frenchs Island	-		2	
Ewing Island	21		30	
Rose Island	13		9	
Ocean Island	1		3	
Shoe Island	-		2	
Davis Island	-		0	
Northern main island	34		54	
Southern part			179	
East Coast	-		75	
Carnley	-		10	
Adams Island	19		94	
	Moore et al. (2001)			Moore et al. (2001)
CAMPBELL ISLAND	140			350-540
Northwest Bay	13			
North East Harbour	51			
Perseverance Harbour	43			
South East Harbour	14			
Monument Harbour	6			
Shag Point	4			
Antarctic Bay	2			
Smoothwater Bay	2			
Rocky Bay	3			
Col Coast	2			

1 Moore (1992) found 19 landing sites on Adams Island in 1989, but methodological differences make direct comparison with the 2009 findings unreliable.

2 Enderby Island was surveyed separately in 2008–9.

3 The 2009 survey did not survey Enderby, Disappointment, Figure of Eight, Dundas, Green, or Friday islands; in addition a total of 27 previously unrecorded landing sites were recorded during beach counts, bringing the total number of recorded landings sites to 306.

Note: Summary of confirmed penguin landing sites and number of breeding pairs estimated from beach counts; Auckland Island: Moore (1992) based on 1989 survey; Beer (2010) based on 2009 survey; Campbell Island: Moore et al. 2001) based on 1992 survey.

TABLE 6.5 Egg biometrics and (sample size) for the yellow-eyed penguin (measurements presented as mean; standard deviation; range)

SOURCE	WEIGHT	LENGTH	BREADTH
Richdale (1957); recalculated in Marchant and Higgins (1990)	137.8; 7.5; 111–158; (446)	76.95; 2.3; 68–83; (592)	57.5; 1.35; 51.26–62.5; (592)
J. Darby; in Marchant and Higgins (1990)	128.4; 7.84; 100.65–134.8; (26)	75.62; 2.23; 71.5–80; (26)	57.06; 1.97; 51.8–60.7; (26)

Note: There is significant variation in the breadth (and volume) of eggs with female age (Massaro et al. 2002).

REFERENCES

Alley M. R., K. J. Morgan, J. M. Gill, and A. G. Hocken. 2004. Diseases and causes of mortality in yellow-eyed penguins, *Megadyptes antipodes*. *Kokako* 11(2):18–23.

Alterio, N. 1991. Feral cat (*Felis catus*) use of vegetation buffer zones surrounding yellow-eyed penguin (*Megadyptes antipodes*) breeding areas. Wildlife Management Report no. 11, University of Otago, New Zealand.

———. 1994. Diet and movements of carnivores and the distribution of their prey in grassland around yellow-eyed penguin (*Megadyptes antipodes*) breeding colonies. MSc thesis, University of Otago, New Zealand.

Alterio, N., and H. Moller. 1997. Diet of feral house cats *Felis catus*, ferrets *Mustela furo* and stoats *M. erminea* in grassland surrounding yellow-eyed penguin *Megadyptes antipodes* breeding areas, South Island, New Zealand. *Journal of Zoology* 243:869–77.

Alterio, N., H. Moller, and H. Ratz. 1998. Movements and habitat use of feral house cats *Felis catus*, stoats *Mustela erminea* and ferrets *Mustela furo*, in grassland surrounding yellow-eyed penguin *Megadyptes antipodes* breeding area in spring. *Biological Conservation* 83:187–94.

Amey, J., and P. J. Moore. 1995. Yellow-eyed penguin on Campbell Island. Department of Conservation, Invercargill, New Zealand.

Argilla, L.S., B. D. Gartrell, L. Howe, M. R. Alley, and K. J. Morgan. 2010. Investigation into the prevalence of Leucocytozoon in the endangered yellow-eyed penguin (*Megadyptes antipodes*) on Enderby Island. Abstracts of oral and poster presentations, 7th International Penguin Conference (30 Aug.-3 Sept.), Boston, MA.

Beer, K. J. 2010. Distribution of yellow-eyed penguins (*Megadyptes antipodes*) on the Auckland Islands: November–December 2009. Wildlife Management Report no. 235, Department of Zoology, University of Otago, Dunedin, New Zealand.

Blair, D. 2000. How many yellow-eyed penguins on Stewart Island? *Yellow-eyed Penguin News* 22:1–2.

Boersma, P. D., and G. A. Rebstock. 2010. Effects of double bands on magellanic penguins. *Journal of Field Ornithology* 81:195–205.

Boessenkool, S. 2009. Spatial and temporal genetic structuring in yellow-eyed penguins. PhD thesis, University of Otago, New Zealand.

Boessenkool, S., T. M. King, P. J. Seddon, and J. M. Waters. 2008. Isolation and characterization of microsatellite loci from the yellow-eyed penguin (*Megadyptes antipodes*). *Molecular Ecology Resources* 8:1043–45.

Boessenkool, S., J. J. Austin, T. H. Worthy, P. Scofield, A. Cooper, P. J. Seddon, and J. M. Waters. 2009a. Relict or colonizer? Extinction and range expansion of penguins in southern New Zealand. *Proceedings of the Royal Society B* 276:815–21.

Boessenkool, S., J. Waters, and P. J. Seddon. 2009b. Multilocus assignment analyses reveal multiple units and rare migration events in the recently expanded yellow-eyed penguin (*Megadyptes antipodes*). *Molecular Ecology* 18:2390–2400.

Boessenkool, S., B. Star, P. Scofield, P. J. Seddon, and J. M. Waters. 2010a. Lost in translation or deliberate falsification? Genetic analyses reveal erroneous museum data for historic penguin specimens. *Proceedings of the Royal Society B* 277:1057–64

———. 2010b. Temporal genetic samples indicate small effective population size of the endangered yellow-eyed penguin. *Conservation Genetics* 11:539–46.

Browne, T. 2007. Diet of yellow-eyed penguins on Stewart and Codfish Islands: Is diet responsible for poor yellow-eyed penguin chick survival on Stewart Island? MSc thesis, University of Otago, New Zealand.

Browne, T., C. Lalas, T. Mattern, and Y. van Heezik. 2010. Chick starvation In yellow-eyed penguins: Evidence for poor diet quality and selective provisioning of chicks from conventional diet analysis and stable isotopes. *Austral Ecology* (doi: 10.1111/j.1442–9993.2010.02125.x).

Bruce, L. 1991. Lagomorph abundance and distribution around yellow-eyed penguin (*Megadyptes antipodes*) breeding colonies on the Otago coast of New Zealand. Wildlife Management Report no. 13, University of Otago, New Zealand.

Busch, J., and R. Cullen. 2009. Effectiveness and cost-effectiveness of yellow-eyed penguin recovery. *Ecological Economics* 68:762–76.

Clapperton, B. K. 2001. Advances in New Zealand mammalogy, 1990–2000: Feral ferret. *Journal of the Royal Society of New Zealand* 31:185–203.

Clark, R. D. 2007. The spatial ecology of yellow-eyed penguin nest site selection at breeding areas with different habitat types on the South Island of New Zealand. MSc thesis, University of Otago, New Zealand.

Clark, R.D., R. Mathieu, and P. J. Seddon. 2008. Geographic information systems in wildlife management: A case study using yellow-eyed penguin nest site data. DOC Research and Development Series no.30, Department of Conservation, Wellington.

Cockrem, J. F., and P. J. Seddon. 1994. Annual cycle of sex steroids in the yellow-eyed penguin (*Megadyptes antipodes*) on South Island, New Zealand. *General and Comparative Endocrinology* 94:113–21.

Darby, J. T. 1985. The great yellow-eyed penguin count. *Forest and Bird* 16:16–18.

———. 2003. The yellow-eyed penguin (*Megadyptes antipodes*) on Stewart and Codfish Islands. *Notornis* 50(3):148–54.

Darby, J. T., and S. M. Dawson. 2000. By catch of yellow-eyed penguins (*Megadyptes antipodes*) in gill nets in New Zealand waters, 1979–1997. *Biological Conservation* 93(3):327–32.

Darby, J. T., and P. J. Seddon. 1990. Breeding biology of the yellow-eyed penguin. In *Penguin Biology*, edited by L. S. Davis and J. T. Darby, 45–62. Orlando, FL: Academic Press.

Department of Conservation. 2008. New Zealand's Subantarctic Islands: Minimum Impact Code. Southland Conservancy, Invercargill, New Zealand.

DOC. 2010. http://www.doc.govt.nz/about-doc/role/maori/settlements/ngai-tahu/species (accessed 8 June 2010)

Duignan, P. J. 2001. Diseases of penguins. *Surveillance* 28: 5–11.

Dymond, S. 1991. Winter use by ferrets (*Mustelo furo*) of vegetation buffer zones surrounding yellow-eyed penguin (*Megadyptes antipodes*) breeding areas in winter. Wildlife Management Report no. 15, University of Otago, New Zealand.

Edge, K. A. 1998. Parental investment in penguins: A phylogenetic and experimental approach. PhD thesis, University of Otago, New Zealand.

Edge, K. A., I. G. Jamieson, and J. T. Darby. 1999. Parental investment and the management of an endangered penguin. *Biological Conservation* 88(3):367–78.

Efford, M. G., N. Spencer, and J. T. Darby. 1994. A relational database for yellow-eyed penguin banding and breeding records. Manuscript. Landcare Research, Dunedin.

Efford, M. G., N. J. Spencer, and J. T. Darby. 1996. Population studies of yellow-eyed penguins, 1993–94 progress report. *Science for Conservation* 22, Department of Conservation, Wellington, New Zealand.

Efford, M. G., and K. A. Edge. 1998. Can artificial brood reduction assist the conservation of the yellow-eyed penguin (*Megadyptes antipodes*). *Animal Conservation* 1:263–71.

Ellenberg, U. 2010. Assessing the impact of human disturbance on penguins. PhD thesis, University of Otago, New Zealand.

Ellenberg, U., A. Setiawan, A. Cree, D. M. Houston, and P. J. Seddon. 2007. Elevated hormonal stress response and reduced reproductive output in yellow-eyed penguins exposed to unregulated tourism. *General and Comparative Endocrinology* 152:54–63.

Ellenberg, U., T. Mattern, and P. J. Seddon. 2009. Habituation potential of yellow-eyed penguins depend on sex, character and previous experience with humans. *Animal Behaviour* 77:289–96.

Ellenberg, U., and T. Mattern. 2012. Yellow-eyed penguin: Review of population information. Marine Conservation Services Programme, Depart-

ment of Conservation, Wellington. Available at http://www.doc.govt.
nz/mcs.

Fallow, P. M., A. Chiaradia, Y. Ropert-Coudet, A. Kato, and R. D. Reina. 2009.
Flipper bands modify the short-term diving behaviour of little penguins.
Journal of Wildlife Management 73:1348–54.

Farner, D. S. 1958. Incubation and body temperatures in the yellow-eyed
penguin. *Auk* 75:249–62.

Gill, J. M., and J. T. Darby. 1993. Deaths in yellow-eyed penguins (*Megadyptes
antipodes*) on the Otago Peninsula during the summer of 1990. *New
Zealand Veterinary Journal* 41(1):39–42.

Graczyk, T. K., J. F. Cockrem, M. R. Cranfield, J. T. Darby, and P. Moore.
1995a. Avian malaria seroprevalence in wild New Zealand penguins.
Journal de la Societie Francais de Parasitologie 2(4):401–5.

Graczyk, T. K., M. R. Cranfield, J. J. Brossy, J. F. Cockrem, P. Jouventin, and
P. J. Seddon. 1995b. Detection of avian malaria infections in wild and
captive penguins. *Journal of the Helminthological Society of Washington*
62:135–41.

Hegg, D., T. Giroir, U. Ellenberg, and P. J. Seddon. 2012. Yellow-eyed penguin
(*Megadyptes antipodes*) as a case study to assess the reliability of nest
counts. *Journal of Ornithology* 153:457–66.

Hill, A. 2008. An investigation of *Leucocytozoon* in the endangered yellow-
eyed penguin (*Megadyptes antipodes*). MSc thesis, Massey University,
New Zealand.

Hill, A. G., M. Alley, B. Gartrell, L. Howe, and R. Norman. 2007. Leucocytozoon in the yellow-eyed penguin (*Megadyptes antipodes*). In *Abstracts
of Oral and Poster Presentations*, 6th International Penguin Conference
(3–7 Sept.), edited by E. J. Woehler. Hobart, Australia.

Hiscock, J. 2008. Yellow-eyed penguin counts Northwest Bay and Southeast Harbour, Campbell Island (November). Manuscript. Department of
Conservation, Invercargill.

Hocken, A. G. 2005. Necropsy findings in yellow-eyed penguins (*Megadyptes antipodes*) from Otago, New Zealand. *New Zealand Journal of
Zoology* 32(1): 1–8.

Hombron, J. B., and H. Jacquinot. 1841. Description de plusieurs oiseaux
nouveaux ou peu connus, provenant de l'expédition autour du monde
faite par les corvettes *L'Astrolabe* et *La Zélée* . . . 2e série XVI: 312–20
(cited in Voisin, J-F., and J-L. Mougin. 2002. Liste des types d'oiseaux
des collections du Muséum national d'Histoire naturelle de Paris, 11:
Manchots (Spheniscidae). *Zoosystema* 24:187–90.

Houston, D. M. 2005. Diphtheritic stomatitis in yellow-eyed penguins. *New
Zealand Journal of Zoology* 32:267.

Houston, D., and D. Nelson. 2012. A survey of yellow-eyed penguins on Codfish Island/Whenua Hou during 28 October–3 November 2011. Report
of the Yellow-eyed Penguin Trust and the Department of Conservation,
Wellington, New Zealand.

IUCN. 2010. IUCN Red List of Threatened Species; http://www.iucnredlist.
org (accessed 2 June 2010).

King, S. 2007. Breeding success of yellow-eyed penguins on Stewart Island
and offshore islands, year four 2006/2007. Report prepared by Sandy
King for the Yellow-eyed Penguin Trust and the Department of Conservation (Southland Conservancy).

———. 2008. Breeding success of yellow-eyed penguins on Stewart Island
and off-shore islands, 2003–2008. Report of the Yellow-eyed Penguin
Trust and the Department of Conservation (Southland Conservancy).

———. 2009. A survey of yellow-eyed penguins on Stewart Island/Rakiura.
Manuscript. Yellow-eyed Penguin Trust, in association with the Department of Conservation (Southland Conservancy).

Kudo, F. 2009. Investigation of sexual selection in relation to post-orbital
ornamentation in yellow-eyed penguins (*Megadyptes antipodes*) and
Snares penguins (*Eudyptes robustus*). MSc thesis, University of Otago,
New Zealand.

Lalas, C., P. R. Jones, and J. Jones. 1999. The design and use of a nest box for

yellow-eyed penguins *Megadyptes antipodes*: A response to a conservation need. *Marine Ornithology* 27:199–204.

Lalas, C., and H. Ratz. 2004. Impact of predation by New Zealand sealions
on yellow-eyed penguins on Otago Penisula. *New Zealand Journal of
Zoology* 35:298.

Lalas, C., H. Ratz, K. McEwan, and S. D. McConkey. 2007. Predation by
New Zealand sea lions (*Phocarctos hookeri*) as a threat to the viability of
yellow-eyed penguins (*Megadyptes antipodes*) at Otago Peninsula, New
Zealand. *Biological Conservation* 135:235–46.

Marchant, S., and P. J. Higgins, co-ordinators. 1990. *Handbook of Australian, New Zealand, and Antarctic Birds*. Vol. 1, Part A: *Ratites to Petrels*.
Melbourne: Oxford University Press.

Maritime New Zealand. 2006. NZ Marine Oil Spill Response Strategy.
Available at http://www.maritimenz.govt.nz/Publications-and-forms/
Environmental-protection/Oil-spill-response-strategy.pdf (accessed 15
June 2010).

Massaro, M. 2004. Ecological, behavioural and physiological mechanisms
that underlie mate choice, egg laying and incubation in yellow-eyed penguins (*Megadyptes antipodes*) and Snares penguins (*Eudyptes robustus*).
PhD thesis, University of Otago, Dunedin.

Massaro, M., J. T. Darby, L. S. Davis, K. A. Edge, and M. J. Hazel. 2002.
Investigation of the interacting effects of female age, laying dates and egg
size in yellow-eyed penguins (*Megadyptes antipodes*). *Auk* 112:982–93.

Massaro, M., and D. Blair. 2003. Comparison of population numbers of
yellow-eyed penguins, *Megadyptes antipodes*, on Stewart Island and on
adjacent cat-free islands. *New Zealand Journal of Ecology* 27(2):107–13.

Massaro, M., L. S. Davis, and J. T. Darby. 2003. Carotenoid-derived orna-
ments reflect parental quality in male and female yellow-eyed penguins (*Megadyptes antipodes*). *Behavioural Ecology and Sociobiology*
55(2):169–75.

Massaro, M., and L. S. Davis. 2004a. Influence of laying date and maternal
age on eggshell thickness and pore density in yellow-eyed penguins.
Condor 106(3): 496–505.

———. 2004b. Preferential incubation positions for different sized eggs
and their influence on incubation period and hatching asynchrony in
Snares crested (*Eudyptes robustus*) and yellow-eyed penguins (*Megadyptes antipodes*). *Behavioural Ecology and Sociobiology* 56(5):426–34.

Massaro, M., L. S. Davis, J. T. Darby, G. J. Robertson, and A. N. Setiawan.
2004. Intraspecific variation of incubation periods in yellow-eyed penguins *Megadyptes antipodes*: Testing the influence of age, laying date and
egg size. *Ibis* 146:526–30.

Massaro, M., L. S. Davis, and R. S. Davidson. 2006. Plasticity of brood patch
development and its influence on incubation periods in the yellow-eyed
penguin *Megadyptes antipodes*: An experimental approach. *Journal of
Avian Biology* 37:497–506.

Massaro, M., A. N. Setiawan, and L. S. Davis. 2007. Effects of artificial eggs on
prolactin secretion, steroid levels, brood patch development, incubation
onset and clutch size in the yellow-eyed penguin (Megadyptes antipodes).
General and Comparative Endocrinology 151:220–29.

Mattern, T. 2005. Fish and ships? Indications of substantial fisheries interaction of yellow-eyed penguins (*Megadyptes antipodes*). *New Zealand
Journal of Zoology* 32:270.

———. 2006. Marine ecology of offshore and inshore foraging penguins: The
Snares penguin *Eudyptes robustus* and yellow-eyed penguin *Megadyptes
antipodes*. PhD thesis, University of Otago, New Zealand.

———. 2008. The tip of the iceberg? The decline of Stewart Island yellow-
eyed penguins indicates serious problems at sea. *New Zealand Journal
of Zoology* 35:298.

Mattern, T., U. Ellenberg, D. M. Houston, and L. S. Davis. 2007a. Consistent
foraging routes and benthic foraging behaviour in yellow-eyed penguins.
Marine Ecology Progress Series 343:295–306.

Mattern, T., U. Ellenberg, and L. S. Davis. 2007b. Decline for a delicacy:

Are decreased numbers of yellow-eyed penguins on Stewart Island a result of commercial oyster dredging? In *Abstracts of Oral and Poster Presentations, 6th International Penguin Conference* (3–7 Sept.) edited by E. J. Woehler. Hobart, Australia.

Maunder, M. N., A. Dunn, D. M. Houston, P. J. Seddon, and T. Kendrick. 2007. Evaluating fishery impact on a yellow-eyed penguin population using mark-recapture data within a population dynamics model. Contract report submitted to the Ministry of Fisheries, ENV2005–13.

McClung, M. R., P. J. Seddon, M. Massaro, and A. N. Setiawan. 2004. Nature-based tourism impacts on yellow-eyed penguins *Megadyptes antipodes*: Does unregulated visitor access affect fledging weight and juvenile survival? *Biological Conservation* 119:279–85.

McDonald, S. P. 2003. Parasitology of the yellow-eyed penguin (*Megadyptes antipodes*). MSc thesis, University of Otago, New Zealand.

McKay, R., C. Lalas, D. McKay, and S. McConkey. 1999. Nest-site selection by yellow-eyed penguins *Megadyptes antipodes* on grazed farmland. *Marine Ornithology* 27:29–25.

McKinlay, B. 1997. The conservation of yellow-eyed penguins (*Megadyptes antipodes*): Use of a PVA model to guide policy development for future conservation management direction. Wildlife Management Report no. 97, University of Otago, New Zealand.

———. 2001. Hoiho (*Megadyptes antipodes*) recovery plan, 2000–2025. Threatened Species Recovery Plan no. 35, Department of Conservation, Wellington.

Miskelly C. M., J. E. Dowding, G. P. Elliott, R. A. Hitchmough, R. G. Powlesland, H. A. Robertson, P. M. Sagar, R. P. Scofield, and G. A. Taylor. 2008. Conservation status of New Zealand birds. *Notornis* 55:117–35.

Moller, H., and N. Alterio. 1999. Home range and spatial organization of stoats (*Mustela erminea*), ferrets (*Mustela furo*), and feral house cats (*Felis catus*) on coastal grasslands, Otago Peninsula, New Zealand: Implications for yellow-eyed penguin (*Megadyptes antipodes*) conservation. *New Zealand Journal of Zoology* 26(3):165–74.

Moller, H., R. Keedwell, H. Ratz, and L. Bruce. 1998. Lagomorph abundance around yellow-eyed penguin (*Megadyptes antipodes*) colonies, South Island, New Zealand. *New Zealand Journal of Ecology* 22(1):65–70.

Moore, C. T., and R. D. Moffat. 1990. Yellow-eyed penguins on Campbell Island. Science and Research Internal Report no. 58, Department of Conservation, Wellington.

Moore, P. J. 1992a. Breeding biology of the yellow-eyed penguin *Megadyptes antipodes* on Campbell Island. *Emu* 92:157–62.

———. 1992b. Population estimates of yellow-eyed penguins (*Megadyptes antipodes*) on Campbell and Auckland Islands. *Notornis* 39:1–15.

———. 1999. Foraging range of the yellow-eyed penguin *Megadyptes antipodes*. *Marine Ornithology* 27:49–58

———. 2001. Historical records of yellow-eyed penguin (*Megadyptes antipodes*) in southern New Zealand. *Notornis* 48:145–56.

Moore, P. J., and R. D. Moffat. 1992. Predation of yellow-eyed penguin by Hooker's sealion. *Notornis* 39:68–69.

Moore, P. J., and M. D. Wakelin. 1997. Diet of the yellow-eyed penguin *Megadyptes antipodes*, South Island, New Zealand, 1991–1993. *Marine Ornithology* 25:17–29.

Moore, P. J., D. Fletcher, and J. Amey. 2001. Population estimates of yellow-eyed penguins, *Megadyptes antipodes*, on Campbell Island, 1987–98. *Emu* 101(3):225–36

Nordin, K. 1991. Individual recognition by vocalization of the yellow-eyed penguin (*Megadyptes antipodes*). MSc thesis, University of Otago, New Zealand.

Peacock, L., M. Paulin, and J. Darby. 2000. Investigations into climatic influence on population dynamics of yellow-eyed penguins *Megadyptes antipodes*. *New Zealand Journal of Zoology* 27(4):317–25.

Poole, J. 2005. The development and application of a GIS for monitoring hoiho (*Megadyptes antipodes*) nesting sites on the Otago Peninsula.

BAppSci diss., University of Otago, New Zealand.

Ramm, K. 2010. Conservation Services Programme Observer Report: 1 July 2008 to 30 June 2009. Final draft report, Department of Conservation, Wellington. Available at http://www.doc.govt.nz/mcs.

Ratz, H. 1997. Introduced predators of yellow-eyed penguins. PhD thesis. University of Otago, New Zealand.

———. 2000. Movements by stoats (*Mustela erminea*) and ferrets (*M-furo*) through rank grass of yellow-eyed penguin (*Megadyptes antipodes*) breeding areas. *New Zealand Journal of Zoology* 27(1):57–69.

Ratz, H., H. Moller, and D. Fletcher. 1999. Predator identification from bite marks on penguin and albatross chicks. *Marine Ornithology* 27:149–56.

Ratz, H., and C. Thompson. 1999. Who is watching whom? Checks for impacts of tourists on yellow-eyed penguins *Megadyptes antipodes*. *Marine Ornithology* 27:205–10.

Ratz, H., J. T. Darby, K. A. Edge, and C. Thompson. 2004. Survival and breeding of yellow-eyed penguins (*Megadyptes antipodes*), at two locations on the Otago Peninsula, South Island, New Zealand, 1991–96. *New Zealand Journal of Zoology* 31(2):133–47.

Ratz, H., and C. Lalas. 2010. An assessment of rehabilitation as a tool to increase population size of an endangered seabird, the yellow-eyed penguin (*Megadyptes antipodes*). *Journal of Wildlife Rehabilitation* 30:13–20.

Richdale, L. E. 1951. *Sexual Behaviour of Penguins*. Lawrence: University of Kansas Press.

———. 1957. *A Population Study of Penguins*. Oxford: Oxford University Press.

Rieben, C. 2008. Fish and Wildlife Service proposes addition of penguin species to Endangered Species List. http://www.fws.gov/news/newsreleases/showNews.cfm?newsId=471BA860-0EA0-7F90-972F3B8DD9F6E185 (accessed 8 June 2010).

Roberts, C. L., and S. L. Roberts. 1973. Survival rate of yellow-eyed penguin eggs and chicks on the Otago Peninsula. *Notornis* 20:1–5.

Rowe, S. 2008. Monitoring penguin bycatch in commercial set net fisheries. *New Zealand Journal of Zoology* 35:299

———. 2010. Level 1 Risk Assessment for incidental seabird mortality associated with New Zealand fisheries in the NZ-EEZ. Marine Conservation Services, Department of Conservation, Wellington. http://www.doc.govt.nz/mcs.

Rowe, S. 2009. Conservation Services Programme Observer Report, 1 July 2004 to 30 June 2007. DOC Marine Conservation Services Series no. 1. Department of Conservation, Wellington. http://www.doc.govt.nz/mcs.

Schuster, K., and J. T. Darby. 2000. Observations on the chick-rearing strategy of yellow-eyed penguins (*Megadyptes antipodes*) on Otago Peninsula, New Zealand. *Notornis* 47:141–47.

Schweigman, P., and J. T. Darby. 1997. Predation of yellow-eyed penguins (*Megadyptes antipodes*) on mainland New Zealand by Hooker's sealion (*Phocarctos hookeri*). *Notornis* 44:265–66.

Seddon, P. J. 1988. Patterns of behaviour and nest-site selection in the yellow-eyed penguin (*Megadyptes antipodes*). PhD thesis, University of Otago, Dunedin.

———. 1989a. Copulation in the yellow-eyed penguin. *Notornis* 36:50–51.

———. 1989b. Patterns of nest relief during incubation, and incubation span variability in the yellow-eyed penguin. *New Zealand Journal of Zoology* 16:393–400.

———. 1990. Behaviour of the yellow-eyed penguin chick. *Journal of Zoology (London)* 220:333–43.

———. 1991. An ethogram for the yellow-eyed penguin. *Marine Ornithology* 19:109–15.

Seddon, P. J., A. Smith, E. Dunlop, and R. Mathieu. 2004. Assessing the impact of unregulated nature-based tourism in Coastal Otago: Pilot study to quantify visitor numbers, attitudes and activities at the Sandfly Bay Wildlife Refuge, Otago Peninsula, Summer 2002/03. Wildlife Management Report no. 185, University of Otago, New Zealand.

Seddon, P. J., and J. T. Darby. 1990. Activity budget for breeding yellow-eyed penguins. *New Zealand Journal of Zoology* 17:527–32.

Seddon, P. J., and L. S. Davis. 1989. Nest site selection by yellow-eyed penguins. *Condor* 91:653–59.

Seddon, P. J., and U. Ellenberg. 2008. Effects of human disturbance on penguins. In *Marine Wildlife and Tourism Management*, edited by J. Higham and M. Lück, 163–81. Oxford: CABI Publishing.

Seddon, P.J., and R. J. Seddon. 1991. Chromosome analysis and sex identification of yellow-eyed penguins. *Marine Ornithology* 19:144–47.

Seddon, P. J., and Y. van Heezik. 1990. Diving depths of the yellow-eyed penguin. *Emu* 90:53–57.

Setiawan, A. N. 2004. Life history consequences of sociality in the yellow-eyed penguin (*Megadyptes antipodes*) in relation to social facilitation, vocal recognition and fidelity towards mates and nest sites. PhD thesis, University of Otago, New Zealand.

Setiawan, A. N., J. T. Darby, and D. M. Lambert. 2004. The use of morphometric measurements to sex yellow-eyed penguins. *Waterbirds* 27(1):96–101.

Setiawan A. N., M. Massaro, J. T. Darby, and L. S. Davis. 2005. Mate and territory retention in yellow-eyed penguins. *The Condor* 107(3):703–9.

Setiawan, A. N., L. S. Davis, J. T. Darby, P. M. Lokman, G. Young, M. A. Blackberry, B. L. Cannell, and G. B. Martin. 2006. Hormonal correlates of parental behavior in yellow-eyed penguins (*Megadyptes antipodes*). *Comparative Biochemistry and Physiology a-Molecular and Integrative Physiology* 145:357–62.

———. 2007. Effects of artificial social stimuli on the reproductive schedule and hormone levels of yellow-eyed penguins (Megadyptes antipodes). *Hormones and Behavior* 51:46–53.

Smith, R. 1987. Biogeography of a rare species: The yellow-eyed penguin (*Megadyptes antipodes*). BSc thesis. University of Otago, New Zealand.

Stein, A., K. Beer, and P. J. Seddon. 2010. Sandfly Bay revisited: A report on visitor attitudes, awareness, and activities at the Sandfly Bay Wildlife Refuge, Otago Peninsula. Wildlife Management Report no. 237, University of Otago, New Zealand.

Stein, A. 2012. Lifetime reproductive success in yellow-eyed penguins: Influence of life-history parameters and investigator disturbance. MSc thesis, University of Otago, New Zealand.

Sturrock, H. J. W. 2007. Epidemiology of avian malaria in yellow-eyed penguins (*Megadyptes antipodes*). MSc thesis, University of Otago, New Zealand.

Sturrock, H. J. W., and D. M. Tompkins. 2007. Avian malaria (*Plasmodium* spp.) in yellow-eyed penguins: Investigating the cause of high seroprevalence but low observed infection. *New Zealand Veterinary Journal* 55:158–60.

Sturrock, H. J. W., and D. M. Tompkins. 2008. Avian malaria parasites (*Plasmodium* spp.) in Dunedin and on the Otago Peninsula, southern New Zealand. *New Zealand Journal of Ecology* 32:98–102.

Taylor, G. A. 2000. Action plan for seabird conservation in New Zealand. Part A: Threatened seabirds. Threatened Species Occasional Publication no. 16, New Zealand Department of Conservation, Wellington.

Triggs, S. J., and J. Darby. 1989. Genetics and conservation of yellow-eyed penguins: An interim report. Science and Research Internal Report no. 43, Department of Conservation, Wellington.

United States Fish and Wildlife Service. 1973. Endangered Species Act of 1973. http://www.fws.gov/endangered/pdfs/ESAall.pdf (accessed 8 June 2010)

van Heezik, Y. 1988. The growth and diet of the yellow-eyed penguin. PhD thesis, University of Otago, New Zealand.

———. 1990a. Seasonal, geographical, and age-related variations in the diet of the yellow-eyed penguin (*Megadyptes antipodes*). *New Zealand Journal of Zoology* 17:201 12.

———. 1990b. Patterns and variability of growth in the yellow-eyed penguin. *Condor* 92:904–12.

———. 1990c. Diets of yellow-eyed, Fiordland crested, and little blue penguins breeding sympatrically on Codfish Island, New Zealand. *New Zealand Journal of Zoology* 17:543–48.

———. 1991. A comparison of yellow-eyed penguin growth rates across fifty years: Richdale revisited. *Notornis* 38:117–23.

van Heezik, Y., and L. S. Davis. 1990. Effects of food variability on growth rates, fledging sizes and reproductive success in the yellow-eyed penguin, *Megadyptes antipodes*. *Ibis* 132:354–65.

van Heezik, Y., and P. J. Seddon. 1989. Stomach sampling in the yellow-eyed penguin: Erosion of otoliths and squid beaks. *Journal of Field Ornithology* 60:451–58.

van Heezik, Y, P. J. Seddon, and J. T. Darby. 2005. Is flipper banding bad for yellow-eyed penguins? What can Richdale tell us? *New Zealand Journal of Zoology* 32:265–66.

YEP RAG (Yellow-eyed Penguin Research Advisory Group). 2010. Yellow-eyed Penguin Research Strategy, 2011–2016. Draft document, Department of Conservation, Dunedin, New Zealand.

Young, M. J. 2009a. File Note: Beach counts (total counts) of yellow-eyed penguins (*Megadyptes antipodes*) on Enderby Island, Auckland Islands, December 2008–February 2009. DOCDM-459794. Internal Report prepared by M. J. Young for Southern Islands Area Office, Department of Conservation.

Young, M. 2009b. Beach behaviour of yellow-eyed penguins on Enderby Island, Auckland Island Group, New Zealand. Wildlife Management Report no. 225, University of Otago, Dunedin.

Wildlife Act. 2008. Wildlife Act 1953 no. 31, as of 1 August 2008. http://www.legislation.govt.nz/act/public/1953/0031/latest/DLM277090.html (accessed 8 June 2010).

Williams, T. D. 1995. The Penguins. Oxford: Oxford University Press, Oxford.

Wright, M. 1998. Ecotourism on Otago Peninsula: preliminary studies of hoiho (*Megadyptes antipodes*) and Hookers sealion (*Phocartos hookerii*). Science for Conservation no. 68, Department of Conservation, Wellington.

IV

Crested Penguins

Genus *Eudyptes*

7

Southern Rockhopper Penguin
(Eudyptes chrysocome)

KLEMENS PÜTZ, ANDREA RAYA REY, AND HELEN OTLEY

1. SPECIES (COMMON AND SCIENTIFIC NAMES)

Southern rockhopper penguin, *Eudyptes chrysocome* (J. R. Foster, 1781)

Other common names are *gorfou sauteur* (French), *pingüino penacho amarillo* (Spanish), *pinguim-saltador-da-rocha* (Portuguese), *Südlicher Felsenpinguin* (German).

2. DESCRIPTION OF THE SPECIES

The southern rockhopper is the smallest of the crested penguins with a length of 45–55 centimeters and a body weight of 2–3.8 kilograms, depending on the season. It is thus similar in size to the Galápagos penguin and larger only than the little penguin.

ADULT. Southern rockhoppers are characterized by bright yellow superciliary stripes above each eye extending back horizontally from one to two centimeters behind the upper mandible and becoming laterally projecting yellow plumes (fig. 1). The strongly built bill is orange red and the iris is bright red. Plumage on the upperparts, head, and tail is grayish blue after the molt, turning brownish with time. The underparts are silky white. The flippers are blue black dorsally with white trailing edges and white with black markings ventrally. The legs and black-soled feet are pink. Sexes are similar, but females are smaller (fig. 2).

FIG. 1 (*FACING PAGE*) Long yellow head plumes may look like an unruly feather day for this southern rockhopper, but the length of the feathers is important in attracting a mate. Immatures have only short yellow feathers and are not nearly as striking. (P. Ryan)

FIG. 2 Male (*right*) and female (*left*) southern rockhopper penguins showing the deep orange bill and red eye. (K. Pütz)

JUVENILE. Immature birds can be distinguished from adults until the juveniles are two years of age. They are generally smaller than adults, with grayer plumage, and the superciliary stripe is pale yellow with much shorter plumes.

CHICK. Chicks in first and second down are dark grayish brown at the head, neck, and back and white at the front.

MORPHOMETRIC DATA. See table 7.1.

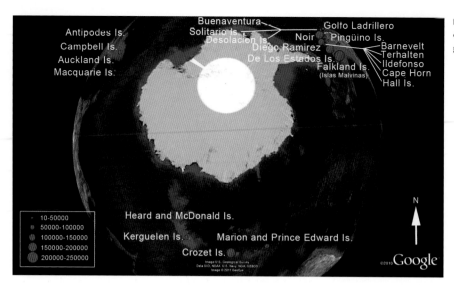

FIG. 3 Distribution and abundance of the southern rockhopper penguin, with counts based on pairs.

3. TAXONOMIC STATUS

The *Megadyptes-Eudyptes* clade occurs at similar latitudes and apparently diverged in the Middle Miocene (Langhian, roughly 15–14 million years ago), but the living species of *Eudyptes* are the product of a later radiation, stretching from about the Late Miocene (Tortonian, 8 million years ago) to the end of the Pliocene (Baker et al. 2006). Three rockhopper subspecies were recognized: the southern rockhopper (*Eudyptes chrysocome chrysocome*), the northern rockhopper (*E. c. moseleyi*), and the eastern rockhopper (*E. c. filholi*). Following Jouventin et al. (2006), *E. chrysocome* and *E. moseleyi* were split based on distinct morphological, vocal, and genetic differences (BirdLife International 2011). However, Banks et al. (2006) suggest a third species, *E. filholi* in the Indian Ocean, to be split from *E. chrysocome* in the Atlantic Ocean, but there are only small morphological differences and the sample size used in the study for the detection of genetic differences was low. Dinechin et al. (2009), using a larger sample size, propose the existence of three distinct species, and BirdLife International has yet to consider this new evidence. In total, three genetic studies, all using mitochondrial DNA, support dividing the species. Interbreeding occurs in a number of crested penguins, for example, between southern rockhoppers and macaronis (Woehler and Gilbert 1990; White and Clausen 2002) and between erect-cresteds (*E. sclateri*) (Napier 1968) and royals (*E. schlegeli*) (Hull and Wiltshire 1999).

4. RANGE AND DISTRIBUTION

The southern rockhopper is circumpolar in its distribution (fig. 3), breeding on sub-Antarctic and temperate islands from 46° south to 54° south in the Atlantic Ocean (Falkland/Malvinas Islands, Isla Pingüino, and Isla de los Estados), in the Indian Ocean (Marion and Prince Edward Islands, Crozet Islands, Kerguelen Islands, Heard and McDonald Islands), and in the Pacific Ocean (Macquarie Island, Campbell Island, Auckland Island, Antipodes Islands, and Barnevelt, Terhalten, Buenaventura, Ildefonso, Noir, and Diego Ramirez, seven islands off the coast of Chile).

The foraging ranges of southern rockhoppers have been investigated in detail in the southwestern Atlantic and the Indian Ocean. Satellite tracking revealed that birds from the Falklands/Malvinas (fig. 4) disperse after their molt over an area ranging from the Burdwood Bank to the south over coastal areas off southern South America and along the shelf break to the north up to 36° south (Pütz et al. 2002). These migration habits were confirmed by at-sea sightings (White et al. 1999) and stable isotope analysis (Dehnhard et al. 2011). In contrast, rockhoppers from Isla de los Estados disperse over an area ranging from the east coast of Tierra del Fuego to the South Shetland Islands, with some birds migrating also westward into the Pacific (Pütz et al. 2006b; Raya Rey et al. 2007a). Rockhoppers from the Crozet and Kerguelen Islands in the Indian Ocean wintered in similar habitats and at the same trophic level between the sub-Antarctic and Polar Fronts, but in different areas ranging from 30° to 70° east for rockhoppers from Crozet and

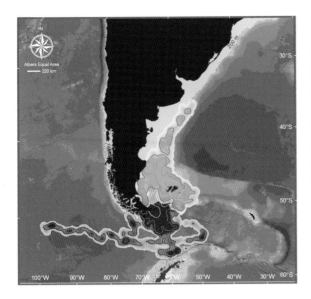

FIG. 4 Kernel distribution of the foraging ranges of southern rockhopper penguins from Isla de los Estados (red) and the Falkland/Malvinas Islands (green) in winter. (Pütz et al., unpubl. data)

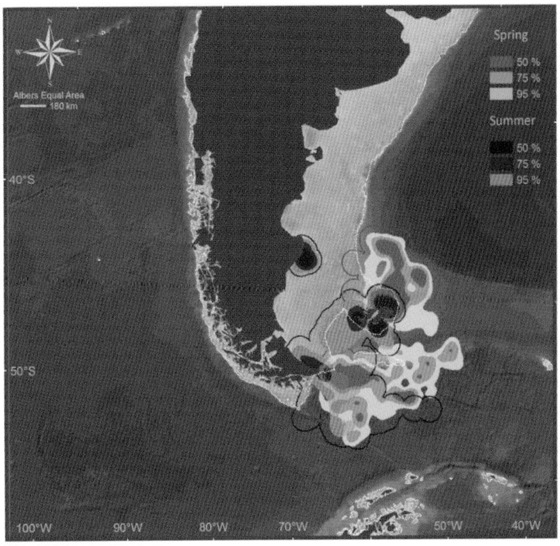

FIG. 5 Kernel distribution of the foraging ranges of southern rockhopper penguins from Isla de los Estados and Seal Bay, Falkland/Malvinas Islands, in spring (green) and from Isla Pingüino, Isla de los Estados, and Bird Island, Falkland/Malvinas Islands, in summer (red). (Pütz et al., unpubl. data)

from 70° to 125° east for birds from Kerguelen (Thiebot et al. 2012).

During the incubation period, males are foraging up to 500 kilometers from the breeding site (Pütz et al. 2003b, 2006a), whereas during the breeding season, foraging ranges are much more restricted due to time constraints caused by the need to provision the offspring with food (Boersma et al. 2002; Schiavini and Raya Rey 2004; Pütz et al. 2006a; Masello et al. 2010) (fig. 5). Other breeding sites where foraging ranges are known include Antipodes Island (Sagar et al. 2005) and Macquarie Island (Hull 1999b), where incubating birds foraged within the Antarctic Polar Frontal Zone up to 500 kilometers away from the breeding site (calculated from Hull 1999b).

Vagrants are recorded from South America, South Africa, Australia, and New Zealand and south from Antarctica and South Georgia Island (Harrison 1983; Marchant and Higgins 1990; Shirihai 2002).

5. SUMMARY OF POPULATION TRENDS

Southern rockhoppers have declined by 34% over the past 37 years (BirdLife International 2010). One of the biggest declines occurred in the Falkland/Malvinas Islands, where the population decreased from about 1.5 million breeding pairs in the mid-1930s to less than 300,000 in the 1990s (Pütz et al. 2003a), presumably caused, at least partly, by El Niño events (Boersma 1987). Similar population declines were observed at Campbell Island (Cunningham and Moors 1994). Counts were made at only a few breeding sites within the past three years. For most sites, the most recent population estimates were done in the 1980s or 1990s. Nevertheless, the best estimate of the current total population for southern rockhoppers is about 1.2 million pairs (BirdLife International 2010).

6. IUCN STATUS

The International Union for Conservation of Nature lists the southern rockhopper as Vulnerable on its Red List of Threatened Species (IUCN 2011). Rockhoppers have undergone rapid population declines, which, although ongoing for perhaps a century, appear to have worsened recently. The extent of these declines varies among locations, with some breeding sites showing slight population increases over recent years. The major threats are presumed to be associated with oceanographic changes caused by global warming, which largely influence food availability. Commercial fishing, hydrocarbon exploration and exploitation, and ecotourism pose severe threats, at least in some regions (BirdLife International 2010).

7. NATURAL HISTORY

BREEDING BIOLOGY. The onset of the breeding season can vary by up to six weeks, depending on year and location (Indian Ocean colonies are later than others). Generally, males return from their winter migration several days before females (Strange 1982). Breeding colonies are located from sea level to cliff tops, sometimes inland, usually in rocky areas stripped of vegetation (fig. 6); however, the penguins also nest in dense tussock (fig. 7), which is considered to provide shelter from predation (St. Clair and St. Clair 1996) and heavy precipitation (Demongin et al. 2010b). On Macquarie Island, interannual fidelity to nest sites and mates was about 50% (Hull et al. 2004). Nest shapes vary from small, circular scrape lines to more elaborate constructions incorporating vegetation, twigs, and bones (Strange 1982). In the Falklands/Malvinas, southern rockhoppers also occupy abandoned black-browed albatross and king shag nests.

Body mass of the birds on arrival can have a significant impact on subsequent breeding success (Raya Rey et al. 2007b; Crawford et al. 2008). After courtship, the female lays two eggs within 4–5 days in early November, and the eggs are incubated for 32–34 days. After

FIG. 6 Southern rockhoppers nest among rocks and sometimes with other seabirds on New Island, Falkland/Malvinas Islands. (P. D. Boersma)

FIG. 7 Southern rockhopper penguins nesting in tussock habitat, Falkland/Malvinas Islands. (P. D. Boersma)

egg laying, both partners remain at the nest for some days before the male leaves for two to four weeks (fig. 8). When he returns, the female takes to the sea for one to two weeks, to restore body condition. The length of foraging-trips may vary depending on food availability and foraging success (Williams 1995; BirdLife International 2010).

After hatching and during the guard stage, which lasts about three weeks, the female alone provisions the offspring with food while the male guards. Males leave to regain body condition for some days after the chicks have crèched. Thereafter, both partners share provisioning duties until the chick fledges by the end of January or early February when it is around 70 days old. Then both adults forage at sea for three to four weeks before returning to the nest site to molt, which takes another three to four weeks (data compiled from Pettingill 1960; Warham 1972, 1975; Strange 1982; Wilson 1995; and Bird-Life International 2010).

The first-laid egg (A-egg) is usually smaller and lighter than the second egg (B-egg) (Strange 1982; Lamey 1990). Hatching success is disproportionate but less pronounced than in other crested penguin species, and in the Falkland/Malvinas Islands, more than 80% of A-eggs are retained until at least one chick hatches successfully (Lamey 1993). Furthermore, there is evidence that in the Falklands/Malvinas, the probability of the smaller chicks hatching from A-eggs (which are larger than elsewhere) is higher and that both chicks may survive to fledging under favorable conditions (Poisbleau et al. 2008; Demongin et al. 2010a).

At Isla de los Estados, Marion Island, and Macquarie Island, the breeding success of southern rockhoppers ranged from 0.23 to 0.49 chicks fledged per nest. In the Falklands/Malvinas, it was much higher, ranging from 0.35 to 0.61 chicks fledged per nest (Pütz et al. 2001; Clausen and Pütz 2002; Hull et al. 2004; Raya Rey et al. 2007b; Crawford et al. 2008; Poisbleau et al. 2008).

Breeding frequency has been studied only on Macquarie Island (Hull et al. 2004), where a small proportion of sexually mature rockhopper adults returned to the breeding site each year. Population estimates for this species might be too low if this is common for all nesting sites.

Little is known about breeding success and juvenile and adult survival of the southern rockhopper. No long-term survival data exists for the species in the wild, and

FIG. 8 Southern rockhopper penguin guarding its egg on its nest. (K. Pütz)

FIG. 9 Adult southern rockhopper penguins greeting each other on New Island, Falkland/Malvinas Islands. (P. D. Boersma)

they can live for up to 30 years in captivity (BirdLife International 2010).

PREY AND FORAGING BEHAVIOR. In general, rockhopper penguins are opportunistic feeders, preying on a mixture of fish, crustaceans, and cephalopods (table 7.2). Stomach content analyses at various locations show large temporal and spatial variation in the prey of southern rockhoppers. Birds in the Pacific and Indian Ocean sectors appear to focus on crustaceans, whereas cephalopods play an important part in the diet of southern rockhoppers in the Chilean and Patagonian regions (fig. 11). In the 2003–4 breeding season in the Falklands/Malvinas, incubating rockhoppers from Sea Lion Island in the south of the archipelago preyed exclusively on fish, while their conspecifics from Steeple Jason in the northwest concentrated on crustaceans (Huin 2005). Stable isotope analysis also showed that even signatures from southern rockhoppers from Isla de los Estados and the Falklands/ Malvinas did not overlap with each other, indicating that they use separate foraging areas. Furthermore, the bimodal distribution in d^{13}C (delta C-13 isotopic ratio) signatures indicated feeding over the Patagonian Shelf and also pelagically (Hilton et al. 2006). This clearly demonstrates the large variation in the southern rockhopper diet and makes it difficult to link their diet to human activities or any changes in their environment.

The spatial and temporal variation in diet is reflected in the southern rockhopper's diving behavior, which differs according to breeding stage and location (Tremblay and Cherel 2003; Pütz et al. 2006a; Raya Rey et al. 2009). Since the early 1990s, the diving behavior of southern

rockhoppers has been investigated using sophisticated data loggers that record relevant dive parameters, at first at Crozet on a single bird (Wilson et al. 1997). Hull (2000) presents a larger number of data sets, but results were pooled over sexes and breeding stages. Afterward, diving parameters were considered according to breeding site and location, and a number of studies focused on temporal and spatial variation (table 7.4).

Overall, the maximum dive depth recorded so far from southern rockhoppers is 113 meters, and the maximum dive duration is 252 seconds. In general, trip duration decreased from more than 12 days during incubation to less than 3 days during brooding, with the percentage of overnight trips during brooding varying spatially between 0% on Kerguelen and 60% on Isla Noir. Penguins spending the night at sea tend to rest for longer periods; thus, they spent less time underwater and made fewer dives on overnight foraging trips compared with day trips. Accordingly, time spent underwater and number of dives increased from less than 40% to more than 60%, and from 19–21 to 26–39 for the incubation and brooding periods, respectively. Mean dive depths ranged between 20 and 30 meters throughout the breeding stages and locations, except for Crozet during brooding (40 m) and the Falklands/Malvinas during late brooding (14 m). Generally, it appears that mean dive depths are highest during early brooding and lowest during late brooding. As dive duration is closely related to dive depth (e.g., Wilson 1995), mean dive durations were highest during early brooding (79–101 seconds), intermediate during incubation (62–72 seconds), and lowest during late brooding (59–64 seconds).

FIG. 10 Southern rockhopper crèche. Some chicks are starting to lose their down and show their yellow feathers. (P. D. Boersma)

FIG. 11 Adult southern rockhopper feeding squid to a chick. (P. D. Boersma)

The differences in diving performance observed during the same breeding stages at the various locations were attributed to spatial and temporal differences in prey abundance and access to more productive shelf and slope areas in the vicinity of the breeding site (Tremblay and Cherel 2003; Pütz et al. 2006a; Raya Rey et al. 2009), which is indicated by higher foraging efforts and shorter trip durations when foraging over shelf regions.

PREDATORS. A number of native predators such as skuas (*Catharacta* spp.), gulls (*Larus* spp.), giant petrels (*Macronectes* spp.), striated caracaras (*Phalcoboenus australis*), fur seals (*Arctocephalus* spp.), and sea lions (*Otaria* spp.) prey on southern rockhoppers (Strange 1982; Boswall 1972; Catry et al. 2008; BirdLife International 2010; Raya Rey et al. 2011). For avian predators, the shape of the colony, the proportion of nests on the periphery, and the presence of vegetation determine the relative impact of predation (Jackson et al. 2005; Liljesthröm et al. 2008). Therefore, in already declining populations, avian predation can accelerate the fragmentation of colonies, leading to increasing predation and potentially resulting in the total loss of some breeding sites.

Giant petrels, fur seals, and sea lions mainly prey on southern rockhoppers in waters adjacent to colony areas. Their predation on penguins can be intense (e.g., Guinard et al. 1998; Charbonnier et al. 2010; Van Buren, pers. obs.), but more specific studies are necessary to elucidate the degree of predation rates.

Invasive alien predators, such as feral cats, rats, mice, and pigs (on the Auckland Islands), appear to have little, if any, impact on southern rockhopper populations (BirdLife International 2010); however, the extent of potential damage is unknown. The same applies to vegetation damage caused by introduced livestock such as sheep, goats, deer, and cattle.

MOLT. The postnuptial molt occurs from February onward; it usually lasts between three and four weeks but can be delayed for several weeks in years of reduced food availability as, for example, in 1985–86 in the Falklands/Malvinas (Keymer et al. 2001). Molting birds lose about 40% of their body mass during the molting period (Brown 1986). By mid-May at the latest, the last southern rockhoppers have completed the molt and spend the winter months at sea regaining body and breeding condition (BirdLife International 2010).

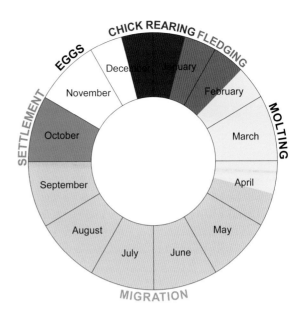

FIG. 12 Annual cycle of the southern rockhopper penguin.

ANNUAL CYCLE. The timing varies depending on year and location; for example, breeding season starts two weeks earlier in the Falklands/Malvinas compared to Isla de los Estados and Noir Island (Raya Rey et al. 2009) (fig. 12).

8. POPULATION SIZES AND TRENDS

Population sizes and trends are presented in table 7.3. For some sites, the most recent counts were made in the 1970s and 1980s. Determining population trends at breeding sites within each region is difficult because few sites are regularly monitored and/or are monitored using comparative count and estimate methods. Few robust population estimates are known before the 1970s, although some indications of population or colony size are known from memories, egging collection records kept by governments, and historical photographs. While there is general agreement that populations were much larger, the timing of the decline cannot be determined accurately for any populations before the 1970s.

The rate of decline is also difficult to determine for most breeding colonies. In the Falklands/Malvinas, annual monitoring of a subset of sites since 1987 indicates a strong stepwise decline (Huin 2007b). There was a major decline event in 1987, the proximate cause apparently being starvation of molting birds (Boersma 1987; Keymer et al. 2001), followed by stability until 2002–3, when there was a further reduction attributed to a harmful algal bloom (Uhart et al. 2004). Between these events, there was minimal recovery. Conversely, at Prince Edward Island, where a similar data set runs from 1994–95 to 2007–8, the rate of decline has been smooth (Crawford et al. 2009).

9. MAIN THREATS

None of the threats discussed below is likely to explain the widespread decline of rockhopper penguins over recent decades. Rather, it appears that the impact of these factors differs from site to site, and it is likely that a cumulative effect of several factors may persist. It is not yet clear whether the ecological and demographic drivers of current rockhopper trends are the same as those that drove trends through the 20th century.

CLIMATE VARIATION. Climate changes can have multiple effects on rockhopper populations. Severe rainfall can cause substantial mortality in crèched chicks (Demongin et al. 2010b). At sea, fluctuations in sea surface temperature (SST) resulted in a switch to prey of lower quality or a reduction in food availability (Cunningham and Moors 1994; Thompson and Sagar 2002; Jaeger and Cherel 2011). In contrast, isotopic evidence indicated no major shift in rockhopper diet (Hilton et al. 2006). However, our understanding of the specific relationships between biotic and abiotic factors, including historical changes, is still too incomplete to support a conclusion.

HARVEST. Historically, egg collection was common in some colonies until the 1950s but is now prohibited. Rockhoppers were also taken for use as bait in crab pots at a number of sites. The disappearance of the rockhopper colony from Isla Recalada in Chile indicates that human depredation, in this case, the collection of penguins for crab pots and as zoological specimens, is still threatening colonies that are not adequately protected (Oehler et al. 2007).

FISHERIES. Evidence of direct or indirect interactions between commercial fisheries and rockhopper penguins is lacking. Probably the most serious threat is seining, a fishing method that is no longer used in the range of rockhopper penguins. There is also no conclusive evidence of direct competition for target species or that fisheries have a negative impact at the population level or affect individual breeding success. Although the rockhopper diet is known for most breeding sites, the degree of overlap in the species and size of prey eaten

FIG. 13 Southern rockhopper penguins are well named, as they have no fear of launching into the air and hopping from rock to rock. (P. D. Boersma)

by rockhoppers and taken by commercial fishing fleets is limited, preventing conclusions of potential competition (e.g., Pütz et al. 2001; Bingham 2002; Clausen and Pütz 2002; Raya Rey and Schiavini 2005). Identifying potential detrimental effects, especially for indirect interactions, requires a greater understanding of the marine food web.

HABITAT DEGRADATION. Rockhoppers usually breed in rocky cliffs with some vegetation (fig. 13). At some sites, introduced grazing animals have significantly reduced vegetation. At Macquarie, for example, overgrazing by rabbits resulted in serious landslides that destroyed potential nesting habitat. On the Falklands/Malvinas, sheep, cattle, and horses destroyed the tussock fringe that provided cover for crèched chicks, thereby increasing chick mortality during heavy rainfalls (Demongin et al. 2010b). Therefore, the loss of the tussock fringe occurring on the Falklands/Malvinas, the Isla de los Estados, and some islands in the Indian Ocean should be further investigated. The expanding populations of fur seals and sea lions at some sites also can destroy the vegetation when they are resting ashore (BirdLife International 2010).

POLLUTION. Chemical pollution does not appear to be involved in the decline of southern rockhoppers. Karesh et al. (1999) notes that chlorinated pesticides and poly-chlorinated biphenyl levels in southern rockhoppers from Argentina were below detectable levels. Also, levels of organochlorines (e.g., DDT pesticides) and polychlorinated biphenyls in seabird eggs from the Falklands/Malvinas were a magnitude less than reported for comparable German seabird species (Hoerschelmann et al. 1979). Most southern rockhopper colonies and foraging areas are in the remoter parts of the world, but those in South America and adjacent areas are likely to be affected by oil pollution from existing oil platforms and internationally important shipping routes along the coastal waters of Brazil, Uruguay, and Argentina (Pütz et al. 2002, 2006b; Garcia Borboroglu et al. 2008).

Furthermore, test drilling for oil took place in 1998 in the Falklands/Malvinas and recommenced in 2010, when oil reservoirs in commercial quantities were found. Further exploratory drillings are scheduled for the near future. Apart from potential spills from these sources, chronic oil pollution is a threat for southern rockhoppers. In May 2008, a fishing vessel sank in Berkeley Sound in the Falklands/Malvinas Islands, discharging heavy crude oil in the vicinity of several southern rockhopper colonies (Bluff Cove Falkland Penguins Tour, n.d.). The effects evaluated by Falklands Conservation showed that rockhoppers were affected. Historically, small numbers of oiled rockhoppers have been seen ashore in the Falklands/Malvinas and South America (Bingham 2002), but real numbers are unknown and are likely to be significantly higher than thought because of the penguins' wide dispersal, especially in winter. Many oiled rockhoppers may die at sea, and, hence, their condition would be undetected.

Marine debris is another potential threat to penguins. In the Falklands/Malvinas, monthly beach surveys revealed an accumulation of 17.3 kilograms of large debris per kilometer, most of which was discarded fishing equipment and household waste from fishing vessels (Otley and Ingham 2003). Some penguins are killed when they become entangled in plastic debris or swallow small plastic items, but reports suggest that only a very small proportion of the breeding population is affected.

There is no provision under any Falklands/Malvinas fishing license legislation that sets specific standards of waste management.

DISEASE. The health of southern rockhoppers in the Falklands/Malvinas was evaluated in 1986–87 following a mass mortality (Keymer et al. 2001) and again in 2003 (Uhart et al. 2004). Both studies showed that southern rockhoppers have had very little exposure to diseases, and, hence, accidental introduction of diseases via tourism and research activities may be a risk. Another health survey, conducted in 1994 on southern rockhoppers in Santa Cruz Province, Argentina, found antibodies against *Chlamydia* spp., avian adenovirus, avian encephalomyelitis virus, infectious bronchitis virus, avian retrovirus, and paramyxoviruses (Karesh et al. 1999). The only known instance of poisoning occurred in the Falklands/Malvinas in November 2002, when one harmful algal bloom caused paralytic shellfish poisoning and the subsequent deaths of a large number of southern rockhoppers and other seabird species (Uhart et al. 2004). If the frequency of harmful algal blooms increases, it will likely become a more important problem for penguins and other seabirds (Shumway et al. 2003).

TOURISM. Due to the difficulty of accessing southern rockhopper colonies, most are spared disturbance by human visitors (fig. 14). Most visitors (i.e., 5,000–10,000 cruise ship passengers) go ashore at only two colonies, West Point Island and New Island in the Falklands/Malvinas (BirdLife International 2010). All visiting cruise ships are operating under standards set by the International Association of Antarctica Tour Operators, including maximum approach distances and restriction on the number of visitors ashore at the same time. The Falklands/Malvinas countryside code also contains these visitor guidelines. On New Island, the effects of visitors on the breeding success of southern rockhoppers have been monitored, and the results show that tourism within these guidelines does not have any significant effect (Catry et al. 2007). Small-scale tourism is permitted on Isla de los Estados and Isla Pingüino in Argentina.

10. RECOMMENDED PRIORITY RESEARCH ACTIONS FOR CONSERVATION

The International Rockhopper Penguin Workshop, held in June 2008 in Edinburgh, Scotland, identified the fol-

FIG. 14 Southern rockhopper penguins landing in rough surf on New Island, Falkland/Malvinas Islands. (P. D. Boersma)

lowing priority research actions on a global and regional scale (BirdLife International 2010):

1. Reexamine morphologic, behavioral, and genetic differences between populations to revise the taxonomic and, subsequently, conservation status.
2. Produce accurate population estimates throughout the species' range that can be used to address potential conservation needs.
3. Conduct long-term demographic studies assessing survival, age at first breeding, and breeding frequency, as lack of this knowledge is a significant impediment to understanding conservation status and potential reasons for the observed declines.
4. Investigate spatial and temporal links between rockhopper population trends and oceanographic parameters like sea surface temperature and primary productivity.
5. Research foraging behavior, particularly outside the breeding season.
6. Monitor for pandemic diseases and poisoning (e.g., through harmful algal blooms).
7. Lobby for vessels to uphold international pollution regulations.
8. Investigate potential competition between commercial fisheries and rockhopper penguins.
9. Research pinniped and rockhopper penguin interactions against the background of historical changes.

11. CURRENT CONSERVATION EFFORTS

Most southern rockhopper breeding sites are well protected, not least by their remoteness. In fact, nearly all breeding sites of this species in the Indian and Pacific Oceans are either national nature reserves or UNESCO World Heritage sites with the appropriate national management plans in place (BirdLife International 2010). In accordance, all land-based activities have to undergo a licensing process. Tourism is permitted only at selected sites and in low numbers (e.g., up to 1,000 tourists each year on Macquarie Island).

In contrast, less legislation is in place for the Chilean, Argentine, and Falklands/Malvinas populations (BirdLife International 2010). Furthermore, extensive fisheries and oil exploration and exploitation are occurring in these areas. In Chile, permission for any visit must be sought from the Chilean authorities, although most breeding sites are effectively protected by their inaccessibility. However, the most accessible colony on Isla Recalada, numbering 10,000 breeding pairs in 1993, vanished by 2005, presumably due, at least in part, to the collection of zoological specimens and the use of penguins as crab-pot bait (Oehler et al. 2007). Isla de los Estados in Argentina was declared an ecological, historical, and tourist provincial reserve, and some limited tourism is allowed, although no trips were made in recent years. Recently, the national government of Argentina and Santa Cruz Province designated Isla Pingüino as a new protected park. The island currently has a management plan that regulates ecotourism, but numbers did not exceed 300 tourists in recent years.

In the Falklands/Malvinas, southern rockhoppers are fully protected under the Conservation of Wildlife and Nature Ordinance 1999. Egging, which historically took place extensively, is now prohibited. There is also a Falklands/Malvinas countryside code in place, regulating the distances humans and vehicles must maintain from seabirds and mammals. Tourism is a growing sector, but sea-borne tourist visits to rockhopper colonies are limited mainly to colonies on New Island and West Point Island. Initial studies on the impact of tourists on the breeding success of rockhoppers on New Island revealed no significant human impact (Catry et al. 2007).

The at-sea foraging areas of southern rockhoppers around South America are subject not only to regulated and intense commercial fisheries but also to oil exploration and exploitation. Several studies so far have failed to find significant competition or mortality linked to fishing operations (e.g., Pütz et al. 2001; Clausen and Pütz 2002). Also, numbers of oiled rockhoppers found ashore have been negligible so far. Much more research is needed to identify potential threats that originate from these activities.

In view of these regulations and gaps in knowledge, the South Atlantic populations of southern rockhoppers are monitored. Censuses are conducted annually at Isla de los Estados and at selected sites in the Falklands/Malvinas, where, in addition, entire counts are conducted every five years. Several research programs also are investigating various aspects of the southern rockhoppers' breeding biology and foraging ecology at various sites in the Falklands/Malvinas and at Isla de los Estados. Demographic studies and, in some instances, investigations into foraging behavior have been initiated or will commence at other sites in Chile and the Indian and Pacific Oceans.

Research organizations involved are the Antarctic Research Trust (ART), the Consejo Nacional de Investigaciones Científicas y Técnicas (CONICET-CADIC) in Argentina, Falklands Conservation (FC), Feather Link in the United States, the Max Planck Institute for Ornithology (MPI) in Germany, the National Institute for Water and Atmosphere in New Zealand, the New Island Conservation Trust (NICT) in the Falkland/Malvinas Islands and the Centre d'Etudes Biologiques de Chizé (CEBC-CNRS) in France.

12. RECOMMENDED PRIORITY CONSERVATION ACTIONS FOR INCREASING POPULATION RESILIENCE AND MINIMIZING THREATS AND IMPACTS

There is an important need to conduct studies and increase our knowledge on a wide array of aspects related to the southern rockhopper. Many priority conservation actions will depend on the results of those studies, which should allow the design of short- and long-term conservation actions under the appropriate spatial framework.

1. Establish marine protected areas in the seasonal foraging areas and their migration pathways.
2. Protect breeding colonies from disturbance and introduction of predators.
3. Design and effectively implement adequate management plans where anthropogenic activities occur close to breeding colonies.

ACKNOWLEDGMENTS

We are grateful to all participants of and contributors to the International Rockhopper Penguin Workshop held in Edinburgh, Scotland, 3–5 June 2008, and all colleagues involved in rockhopper penguin research. We would like to thank the Global Penguin Society and the Antarctic Research Trust for their help in many different ways. Valeria Falabella, WCS, kindly prepared the kernel distribution maps.

TABLE 7.1 Biometric measurements of southern rockhoppers

Weight

| LOCATION | SEX | BODY MASS (G) | | | | SOURCE |
		PRE-EGG LAYING	CHICK-REARING	PRE-MOLT	POST-MOLT	
Isla de los Estados	M		2504 ± 379 (1600–3450; 62)		2054 ± 168 (1700–2250; 13)	Schiavini, Pütz, Raya Rey, unpubl. data
	F		2396 ± 247 (1800–3300; 141)		1808 ± 257 (1600–2500; 12)	
Falklands/ Malvinas	M		2440 ± 313 (2100–3200; 10)		2259 ± 274 (1800–2900; 17)	Pütz, unpubl. data
	F		2500 ± 223 (2200–2900; 9)		2033 ± 246 (1700–2400; 12)	
Marion Island				3800 ± 200 (3630–4320; 8)	2130 ± 110 (2050–2300; 8)	Brown 1986
Heard Island	M	2880 ± 290 (10)				Marchant & Higgins 1990
	F	3080 ± 250 (10)				
Campbell Island	M		2500 ± 260 (2000–2950; 14)			Marchant & Higgins 1990
	F		2440 ± 160 (2150–2700; 15)			
Antipodes Island	M	2425 ± 230 (49)				Warham 1972
	F	2225 ± 140 (33)				
Macquarie Island	M			2700 (2100–3200; 16)		Warham 1963
	F			2500 (2000–3200; 16)		

Body Morphometrics

LOCATION	SEX	BILL LENGTH (MM)	BILL DEPTH (MM)	FLIPPER LENGTH (MM)	HEAD LENGTH (MM)	SOURCE
Isla de los Estados	M	43.3 ± 1.8 (39.8–46.6; 12)	21.6 ± 0.8 (20.0–22.6; 12)			Schiavini, Pütz, Raya Rey, unpubl. data
	F	38.6 ± 2.9 (35.8–45.2; 12)	18.5 ± 0.9 (16.8–20.7; 12)			
Falklands/ Malvinas	M		26.0 ± 1.2 (24.1–27.6; 27)	175.9 (170–185; 15)		Pütz, unpubl. data
	F		22.3 ± 0.9 (20.6–24.2; 21)	167.9 (165–175; 10)		
Marion Island	M	45.7 ± 1.5 (5)		165.3 ± 7.5 (5)		Williams 1980
	F	40.6 ± 1.7 (6)		161.5 ± 5.6 (6)		
Heard Island	M	44.8 ± 1.2 (10)	21.4 ± 0.8 (10)			Marchant & Higgins 1990
	F	39.7 ± 2.5 (10)	17.9 ± 0.8 (10)			
Campbell Island	M	46.5 ± 1.9 (43.4–49.6; 15)	21.3 ± 1.2 (19.2–23.0; 15)	165 ± 4.9 (158–176; 15)		Marchant & Higgins 1990
	F	41.1 ± 1.6 (36.7–43.3; 15)	17.8 ± 0.8 (15.3–18.8; 15)	161 ± 4.4 (152–167; 15)		
Macquarie Island	M	46.4 ± 2.1 (60)	21.0 ± 1.0 (60)		115.6 ± 3.0 (60)	Hull 1996
	F	41.9 ± 2.1 (57)	18.7 ± 0.9 (57)		109.6 ± 3.6 (57)	

TABLE 7.1 (cont.)

Egg Sizes (Egg Sequence)

LOCATION	SEQ.	EGG LENGTH (MM)	EGG BREADTH (MM)	EGG WEIGHT (G)	SOURCE
Isla de los Estados	A	63.8 ± 3.7 (14)	49.4 ± 2.5 (14)		Demongin et al. 2010b
	B	69.5 ± 2.8 (14)	54.2 ± 1.9 (14)		
Falklands/ Malvinas	A	65.7 ± 2.6 (59.6–71.2; 100)	49.8 ± 1.8 (45.5–54.7; 100)	90.7 ± 8.8 (76.6–115.3; 100)	Poisbleau et al. 2008
	B	70.6 ± 2.3 (66.1–75.6; 100)	54.3 ± 1.6 (50.2–58.2; 100)	116.2 ± 9.4 (92.8–144.7; 100)	
Falklands/ Malvinas	A	66.1 ± 2.7 (545)	50.2 ± 1.7 (545)	92.5 ± 8.7 (545)	Demongin et al. 2010b
	B	71.1 ± 2.6 (394)	54.6 ± 1.7 (394)	118.4 ± 10.0 (394)	
Marion Island	A	62.3 (57.1–70.5; 122)	46.8 (43.6–50.5; 122)	76.0 ± 6.9 (122)	Williams 1995
	B	70.2 (64.4–78.6; 119)	52.9 (48.1–56.7; 119)	109.1 ± 9.1 (119)	
Crozet Islands	A	63.5 (57.6–70.7; 20)	47.3 (44.2–50.2; 20)	78.7 (62.1–88.9; 20)	Williams 1995
	B	70.4 (66.9–76.0; 20)	52.4 (49.7–54.4; 20)	107.0 (91.3–120.9; 20)	
Heard Island	A	63.9 (59.0–68.0; 11)	46.4 (42.2–51.5; 11)		Williams 1995
	B	71.9 (68.0–73.7; 11)	52.9 (49.0–56.7; 11)		
Campbell Island	A	63.5 (57.3–72.8; 37)	48.0 (45.2–53.4; 37)	79.6 (64.5–91.0; 37)	Williams 1995
	B	70.5 (65.3–78.6; 37)	53.7 (51.0–56.3; 37)	112.0 (92.0–125.0; 37)	
Macquarie Island	A	65 ± 4 (141)	49 ± 2 (141)	88.2 ± 13.8 (139)	Hull et al. 2004
	B	71 ± 3 (130)	53 ± 3 (130)	115.2 ± 14.8 (129)	

TABLE 7.2 Summary of diet data (by wet mass) collected from southern rockhoppers

SITE	DIET COMPOSITION BY WET MASS	SOURCE
Isla de los Estados	74%, 87% & 67% cephalopods, 26%, 14% & 33% crustaceans (1999, 2000 & 2001), Fish not estimated by mass	Raya Rey & Schiavini 2005
	Mixture of cephalopods, fish and crustaceans (no % by mass given)	Raya Rey et al. 2007b
Falklands/ Malvinas	Incubation: 61% crustaceans, 39% cephalopods and < 1% fish Chick-rearing: 51% crustaceans, 46% cephalopods and 2% fish Pre-moult: 8% crustaceans, 73% cephalopods and 14% fish	Thompson 1989
	53% cephalopods, 45% crustaceans, 2 % fish	Croxall et al. 1985
	53% crustaceans, 29 % cephalopod and 18% fish, large temporal & spatial variation	Pütz et al. 2001
	Mixture of fish, crustaceans cephalopods, large temporal & spatial variation, ranging from 100% fish at Sea Lion Island to 98& crustaceans at Steeple Jason	Huin 2005
Marion	100% & 91% euphausiids, 0% & 6% fish, 0% & 3% cephalopods	Brown & Klages 1987
Crozet	71% crustaceans, 17% cephalopods, 11% fish	Ridoux 1994
	95% crustaceans (E. vallentini), 4% fish, 1% squid	Tremblay & Cherel 2003
Kerguelen	97% crustaceans (E. vallentini), 3% fish	Tremblay & Cherel 2003
Heard	91% crustaceans, 8% fish, 1% cephalopods	Klages et al. 1989
Macquarie	70% euphausiids, 17% fish (myctophids)	Horne 1985
	69% euphausiids (E. vallentini), 17% fish (K. anderssoni)	Hindell 1988
	60% euphausiids, 29% fish (myctophids), 1% cephalopods	Hull 1999b
Campbell	mainly fish, but also cephalopods and crustaceans	Marchant & Higgins 1990

TABLE 7.3 Current population estimates and trends for southern rockhoppers (trends from BirdLife International 2010)

LOCATION		YEAR	COUNT	TREND	SOURCE
	Heard	2003	10,000	?	Woehler 2006
	Kerguelen	1985	85,480	?	Weimerskirch et al. 1989
	Crozet	1982	152,800	?	Woehler 1993
	Marion	2008	42,000	80% decline over 35 years	Crawford et al. 2009
	Prince Edward	2008	38,000	9% increase over 31 years	Crawford et al. 2009
Indian Ocean			328,280		
	Antipodes	1995	4,000	92% decline over 22 years	Tennyson et al. 2002
	Auckland	1990	3,000	Increase	Cooper 1992
	Campbell	1985	51,000	94% decline over 40 years	Cunningham and Moors 1994
	Macquarie	2007	37,500	Decrease	R. Gales in BirdLife International 2010
Pacific Ocean			95,500		
	Solitario	1978	35	?	Venegas and Jory 1979
	Golfo Ladrillero	1977	300	?	Venegas 1978
	Buenaventura	1984	500	?	Woehler 1993
	Desolación	1978	3,000	?	Venegas 1978
	Recalada	2005	0	100% decline	Oehler et al. 2007
	Noir	2005	158,200	15% increase over 25 years	Oehler et al. 2008
	Terhalten	2008	3,000	?	Oehler and Marin, pers. comm.
	Barnevelt	1984	10,800	?	Woehler 1993
	Ildefonso	2002	86,400	?	Kirkwood et al. 2007
	Cape Horn	1984	600	?	Clark et al. 1992
	Hall	1984	500	?	Clark et al. 1992
	Diego Ramirez	2002	132,721	15% increase over 21 years	Kirkwood et al. 2007
Chile			396,056		
	Falklands/Malvinas	2005	210,418	30% decline over 10 years	Huin 2006
	Pingüino	2007	501	295% increase over 13 years	E. Frere in BirdLife International 2010
	Estados	1998	173,793	?	Schiavini 2000
Atlantic Ocean			384,712		
Total			1,204,548		

TABLE 7.4 Summary of dive parameters recorded from southern rockhoppers at various breeding sites

LOCATION	MACQUARIE[1]	FALKLANDS/ MALVINAS[2]	ISLA DE LOS ESTADOS[2]	ISLA DE LOS ESTADOS[2]	KERGUELEN[3]	CROZET[3]	ISLA DE LOS ESTADOS[4]	FALKLANDS/ MALVINAS[2]	ISLA NOIR[5]
Study period	Incubation & brooding[a]	Incubation			Early brooding		Late brooding		
Sex studied	F & M[a]	M	M	F	F	F	F	F	F
Duration of day trips (days)					0.50 ± 0.08	0.47 ± 0.15	0.65 ± 0.21	0.61 ± 0.04	1.77 ± 1.67[b]
Duration of overnight trips (days)		13.3 ± 4.4	22.2 ± 5.5	12.3 ± 2.1		1.19 ± 0.21	1.45 ± 0.36		
Overnight trips (%)						10	53	1.6	60
Time spent underwater (%)	37	40 ± 3	38 ± 4	32 ± 3	64	64	69	63 ±3	61 ± 10
Dives per hour (n) day trips	14.8 ± 9.4[b]				29.6 ± 8.9	27.1 ± 5.9	32.7 ±5.7	39.4 ± 6.5	38 ± 14[b]
Dives per hour (n) overnight trips		20.7 ± 3.0	18.8 ± 2.7	18.6 ± 2.8		26.2 ± 3.9	28 ± 2.0		
Mean dive depth (m)	27.3 ± 20.3	21.4 ± 3.9	27.1 ± 5.7	25.1 ± 3.4	29.1 ± 14.1	40.4 ± 17.6	28.9 ± 24.6	14 ± 2.2	20.6 ± 19
Deepest dive (m)	104	93	100.5	79	85	104	113	66.5	101
Vertical distance (m) per hour under-water on day trips					1738 ± 468	2006 ± 747	2780 ± 1424	1716 ± 168	1260 ± 449
Vertical distance (m) per hour under-water on overnight trips		2157 ± 334	2691± 490	2909 ± 16		3950 ± 883	5996 ± 1783		
Bottom time (min) per hour underwater		22.7 ± 1.6	16.3 ± 1.6	20.4 ± 1.1				23.6 ± 1.0	17.0 ± 3.0
Mean dive duration (s)	72 ± 42	70 ± 6	72 ± 6	62 ± 9	87 ± 26	101 ± 27	79 ± 30	59 ± 9	64 ± 36
Longest dive (s)		153	184	140	177	190	164	189	252

Sources: 1: Hull 2000; 2: Pütz et al. 2006b; 3: Tremblay and Cherel 2003; 4: Schiavini and Raya Rey 2004, and 5: Raya Rey et al. 2009; M = Male, F = Female

a: Data for Macquarie Island were pooled over breeding stages and sexes; b: data for daily and overnight foraging trips pooled.

REFERENCES

Baker, A., S. L. Pereira, O. P. Haddrath, and K. A. Edge. 2006. Multiple gene evidence for expansion of extant penguins out of Antarctica due to global cooling. *Proceedings of the Royal Society B* 273:11–17.

Banks, R., A. van Buren, Y. Cherel, and J. Whitfield. 2006. Genetic evidence for three species of Rockhopper penguins *Eudyptes chrysocome. Polar Biology* 30:61–67.

Bingham, M. 2002. The decline of Falkland Islands penguins in the presence of a commercial fishing industry. *Revista Chilena de Historia Natural* 75:805–18.

BirdLife International. 2010. Rockhopper penguins: A plan for research and conservation action to investigate and address population changes. Proceedings of an International Workshop, Edinburgh, Scotland, 3–5 June 2008. http://www.birdlife.org/downloads/news/birdlife_rockhopper_penguin_report.pdf.

———. 2011. Southern rockhopper penguin *Eudyptes chrysocome*: BirdLife species factsheet. http://wwwbirdlife.org (accessed 7 July 2011).

Bluff Cove Lagoon. n.d. www.falklandpenguins.com/penguin-news-ocean8.asp (accessed 14 June 2011).

Boersma, P. D. 1987. Penguin deaths in the South Atlantic. *Nature* 327:96.

Boersma, P. D., D. L. Stokes, and I. Strange. 2002. Applying ecology to conservation: Tracking breeding penguins at New Island South Reserve, Falkland Islands. *Aquatic Conservation: Marine and Freshwater Ecosystems* 12:1–11.

Boswall, J. 1972. The South American sea lion *Otaria byronia* as a predator on penguins. *Bulletin of the British Ornithologists' Club* 92:129–32.

Brown, C. R. 1986. Feather growth, mass loss and duration of moult in macaroni and rockhopper penguins. *Ostrich* 57:180–84.

Brown, C. R., and N. T. Klages. 1987. Seasonal and annual variation in diets of macaroni (*Eudyptes c. chrysolophus*) and southern rockhopper (*E. chrysocome chrysocome*) penguins at sub-Antarctic Marion Island. *Journal of Zoology* (London) 212:7–28.

Catry, P., M. Lecoq, and I. Strange. 2008. Population growth and density, diet and breeding success of striated caracaras *Phalcoboenus australis* on New Island, Falkland Islands. *Polar Biology* 31:1167–74.

Catry, P., R. Matias, and R. van Noordwijk. 2007. A South Atlantic wildlife sanctuary for conservation management. New Island Conservation Trust, Falkland Islands.

Charbonnier, Y., K. Delord, and J. B. Thiebot. 2010. King-size fast food for Antarctic fur seals. *Polar Biology* 33:721–24.

Clark, G. S., A. Cowan, P. Harrison, and W. R. P. Bourne. 1992. Notes on the seabirds of the Cape Horn Islands. *Notornis* 39:133–44.

Clausen, A., and K. Pütz. 2002. Recent trends in diet composition and productivity of gentoo, Magellanic and rockhopper penguins in the Falkland Islands. *Aquatic Conservation: Marine and Freshwater Ecosystems* 12:51–61.

Cooper, W. 1992. Rockhopper penguins at the Auckland Islands. *Notornis* 39:66–67.

Crawford, R. J. M., A. B. Makhado, L. Upfold, and B. M. Dyer. 2008. Mass on arrival of rockhopper penguins at Marion Island influences breeding success. *African Journal of Marine Science* 30:185–88.

Crawford, R. J. M., P. A. Whittington, L. Upfold, P. G. Ryan, S. L. Petersen, B. M. Dyer, and J. Cooper. 2009. Recent trends in numbers of four species of penguins at the Prince Edward Islands. *African Journal of Marine Science* 31:419–26.

Croxall, J. P., P. A. Prince, A. Baird, and P. Ward. 1985. The diet of the southern rockhopper penguin *Eudyptes chrysocome chrysocome* at Beauchêne Island, Falkland Islands. *Journal of Zoology* (London) 206:485–96.

Cunningham, D. M., and P. J. Moors. 1994. The decline of rockhopper penguins *Eudyptes chrysocome* at Campbell Island, Southern Ocean, and the influence of rising sea temperatures. *Emu* 94:27–36.

Dehnhard, N., C. C. Voigt, M. Poisbleau, L. Demongin, and P. Quillfeldt. 2011. Stable isotopes in southern rockhopper penguins: Foraging areas and sexual differences in the non-breeding period. *Polar Biology*. doi:10.1007/s00300-011-1026-x.

Demongin, L., M. Poisbleau, A. Raya Rey, A. Schiavini, P. Quillfeldt, M. Eens, and I. J. Strange. 2010a. Geographical variation in egg size dimorphism in rockhopper penguins. *Polar Biology* 33:469–76.

Demongin, L., M. Poisbleau, I. J. Strange, and P. Quillfeldt. 2010b. Effects of severe rains on the mortality of southern rockhopper penguin (*Eudyptes chrysocome*) chicks and its impact on breeding success. *Ornitologia neotropical* 21:430–43.

Dinechin, M. de, R. Ottvall, P. Quillfeldt, and P. Jouventin. 2009. Speciation chronology of rockhopper penguins inferred from molecular, geological, and palaeoceanographic data. *Journal of Biogeography* 36:693–702.

Garcia Borboroglu, P., P. D. Boersma, L. Reyes, and E. Skewgar. 2008. Petroleum pollution and penguins: Marine conservation tools to reduce the problem. In *Marine Pollution: New Research*, ed. T. N. Hofer, 339–56. New York: Nova Science Publishers.

Guinard, E., H. Weimerskirch, and P. Jouventin. 1998. Population changes and demography of the northern rockhopper penguin on Amsterdam and Saint Paul islands. *Colonial Waterbirds* 21:222–28.

Harrison, P. 1983. *Seabirds: An Identification Guide.* Beckenham, United Kingdom: Croom Helm.

Hilton, G. M., D. R. Thompson, P. M. Sagar, R. J. Cuthbert, Y. Cherel, and S. J. Bury. 2006. A stable isotopic investigation into the causes of decline in a sub-Antarctic predator, the rockhopper penguin *Eudyptes chrysocome*. *Global Change Biology* 12:611–25.

Hindell, M. A. 1988. The diet of rockhopper penguins *Eudyptes chrysocome* at Macquarie Island. *Emu* 88:227–33.

Hoerschelmann, H., K. Polzhafer, K. Figge, and K. Ballschmiter. 1979. Organochlorine pesticides and polychlorinated biphenyls in bird eggs from the Falkland Islands and north Germany. *Environmental Pollution* 13:247–69.

Horne, R. S. C. 1985. Diet of royal and rockhopper penguins at Macquarie Islands. *Emu* 85:150–56.

Huin, N. 2005. Falkland Islands seabird monitoring programme. Annual Report 2003/04 and 2004/05. Falklands Conservation, Stanley, Falkland Islands.

———. 2007a. Falkland Islands penguin census 2005/2006. Falklands Conservation, Stanley, Falkland Islands.

———. 2007b. Falkland Island seabird monitoring programme. Annual Report 2006/2007 Falkland Conservation, Stanley, Falkland Islands.

Hull, C. L. 1996. Morphometric indices for sexing adult royal *Eudyptes schlegeli* and rockhopper *E. chrysocome* penguins at Macquarie Island. *Marine Ornithology* 24:23–27.

———. 1999a. Comparison of the diets of breeding royal (*Eudyptes schlegeli*) and rockhopper (*Eudyptes chrysocome*) penguins on Macquarie Island over three years. *Journal of Zoology* (London) 247:507–29.

———. 1999b. The foraging zones of breeding royal (*Eudyptes schlegeli*) and rockhopper (*E. chrysocome*) penguins: An assessment of techniques and species comparison. *Wildlife Research* 26:789–803.

———. 2000. Comparative diving behaviour and segregation of the marine habitat by breeding royal penguins, *Eudyptes schlegeli,* and eastern rockhopper penguins, *E. chrysocome filholi,* at Macquarie Island. *Canadian Journal of Zoology* 78:333–45.

Hull, C. L., M. Hindell, K. Le Mar, P. Scofield, J. Wilson, and M. A. Lea. 2004. The breeding biology and factors affecting reproductive success in rockhopper penguins *Eudyptes chrysocome* at Macquarie Island. *Polar Biology* 27:711–20.

Hull, C. L., and A. Wilthshire. 1999. An apparent hybrid royal × rockhop-

per penguin at Macquarie Island. *Australian Bird Watcher* 183:95–100.

IUCN (International Union for Conservation of Nature). 2011. IUCN Red List of Threatened Species. Version 2011.2. www.iucnredlist.org (accessed 15 March 2012).

Jackson, A. L., S. Bearhop, and D. R. Thompson. 2005. Shape can influence the rate of colony fragmentation in ground nesting seabirds. *Oikos* 111:473–78.

Jaeger, A., and Y. Cherel. 2011. Isotopic investigation of contemporary and historic changes in penguin trophic niches and carrying capacity of the southern Indian Ocean. *PLoS ONE* 6(2):e16484.

Jouventin, P., R. J. Cuthbert, and R. Ottval. 2006. Genetic isolation and divergence in sexual traits: Evidence for the northern rockhopper penguin *Eudyptes moseleyi* being a sibling species. *Molecular Ecology* 15:3413–23.

Karesh, W. B., M. M. Uhart, E. Frere, P. Gandini, W. E. Braselton, H. Puche, and R. A. Cook. 1999. Health evaluation of free-ranging rockhopper penguins (*Eudyptes chrysocome*) in Argentina. *Journal of Zoo and Wildlife Medicine* 30:25–31.

Keymer, I. F., H. M. Malcolm, A. Hunt, and D. T. Horsley. 2001. Health evaluation of penguins (Sphenisciformes) following mortality in the Falklands (South Atlantic). *Diseases of Aquatic Organisms* 45:159–69.

Kirkwood, R., K. Lawton, C. Moreno, J. Valencia, R. Schlatter, and G. Robertson. 2007. Estimates of southern rockhopper and macaroni penguin numbers at the Ildefonso and Diego Ramirez Archipelagos, Chile, using quadrat and distance-sampling techniques. *Waterbirds* 30:259–67.

Klages, N. T. W., R. P. Gales, and D. Pemberton. 1989. Dietary segregation of macaroni and rockhopper penguins at Heard Island. *Australian Wildlife Research* 16:599–604.

Lamey, T. C. 1990. Hatch asynchrony and brood reduction in penguins. In *Penguin Biology*, ed. L. S. Davis and J. T. Darby, 399–416. San Diego, CA: Academic Press.

———. 1993. Territorial aggression, timing of egg loss, and egg-size differences in rockhopper penguins, *Eudyptes c. chrysocome,* on New Island, Falkland Islands. *Oikos* 66:293–97.

Liljesthröm, M., S. D. Emslie, D. Frierson, and A. Schiavini. 2008. Avian predation at a southern rockhopper penguin colony on Isla de los Estados, Argentina. *Polar Biology* 31:465–74.

Marchant, S., and P. J. Higgins. 1990. *Handbook of Australian, New Zealand and Antarctic Birds.* Vol. 1A. Melbourne: Oxford University Press.

Masello, J. F., R. Mundry, M. Poisbleau, L. Demongin, C. C. Voigt, M. Wilkelski, and P. Quillfeldt. 2010. Diving seabirds share foraging space and time within and among species. *Ecosphere* 1:19.

Napier, R. B. 1968. Erect-crested and rockhopper penguins interbreeding in the Falkland Islands. *British Antarctic Survey Bulletin* 16:71–72.

Oehler, D. A., W. R. Fry, L. A. Weakley, and M. Marin. 2007. Rockhopper and macaroni penguin colonies absent from Isla Recalada, Chile. *Wilson Journal of Ornithology* 119:502–6.

Oehler, D. A., S. Pelikan, W. R. Fry, L. Weakley Jr., A. Kusch, and M. Marin. 2008. Status of crested penguin (*Eudyptes* ssp.) populations on three islands in southern Chile. *The Wilson Journal of Ornithology* 120:575–81.

Otley, H., and R. Ingham. 2003. Marine debris surveys at Volunteer Beach, Falkland Islands, during the summer of 2001/02. *Marine Pollution Bulletin* 46:1534–39.

Pettingill, O. S., Jr. 1960. Creche behavior and individual recognition in a colony of rockhopper penguins. *Wilson Bulletin* 72:209–21.

Poisbleau, M., L. Demongin, I. J. Strange, H. Otley, and P. Quillfeldt. 2008. Aspects of the breeding biology of the southern rockhopper penguin *Eudyptes c. chrysocome* and new consideration on the intrinsic capacity of the A-egg. *Polar Biology* 31:925–32.

Pütz, K., R. J. Ingham, J. G. Smith, and J. P. Croxall. 2001. Population trends, breeding success and diet composition of gentoo *Pygoscelis papua*, Magellanic *Spheniscus magellanicus* and rockhopper *Eudyptes chrysocome* penguins in the Falkland Islands. *Polar Biology* 4:793–807.

Pütz, K., A. P. Clausen, N. Huin, and J. P. Croxall. 2003a. Re-evaluation of historical rockhopper penguin population data in the Falkland Islands. *Waterbirds* 26:169–75.

Pütz, K., A. Raya Rey, N. Huin, A. Schiavini, A. Pütz, and B. H. Lüthi. 2006a. Diving characteristics of southern rockhopper penguins (*Eudyptes c. chrysocome*) in the southwest Atlantic. *Marine Biology* 149:125–37.

Pütz, K., A. Raya Rey, A. Schiavini, A. P. Clausen, and B. H. Lüthi. 2006b. Winter migration of rockhopper penguins (*Eudyptes c. chrysocome*) breeding in the Southwest Atlantic: Is utilisation of different foraging areas reflected in opposing population trends? *Polar Biology* 29:735–44.

Pütz, K., J. G. Smith, R. J. Ingham, and B. H. Lüthi. 2002. Winter dispersal of rockhopper penguins *Eudyptes chrysocome* from the Falkland Islands and its implications for conservation. *Marine Ecology Progress Series* 240:273–84.

———. 2003b. Satellite tracking of male rockhopper penguins *Eudyptes chrysocome* during the incubation period at the Falkland Islands. *Journal of Avian Biology* 34:139–44.

Raya Rey, A., K. Pütz, G. Luna-Jorquera, B. Lüthi, and A. Schiavini. 2009. Diving patterns of breeding female rockhopper penguins (*Eudyptes chrysocome*): Noir Island, Chile. *Polar Biology* 32:561–68.

Raya Rey, A., R. S. Samaniego, and P. F. Petracci. 2011. New records of South American sea lion *Otaria flavescens* predation on southern rockhopper penguins *Eudyptes chrysocome* at Staten Island, Argentina. *Polar Biology* 35:319–322.

Raya Rey, A., and A. Schiavini. 2005. Inter-annual variation in the diet of female southern rockhopper penguins (*Eudyptes chrysocome chrysocome*) at Tierra del Fuego. *Polar Biology* 28:132–41.

Raya Rey, A., P. Trathan, K. Pütz, and A. Schiavini. 2007a. Effect of oceanographic conditions on the winter movements of rockhopper penguins *Eudyptes chrysocome chrysocome* from Staten Island, Argentina. *Marine Ecology Progress Series* 330:285–95.

Raya Rey, A., P. Trathan, and A. Schiavini. 2007b. Inter-annual variation in provisioning behaviour of southern rockhopper penguins *Eudyptes chrysocome chrysocome* from Staten Island. *Ibis* 149: 826–35.

Ridoux, V. 1994. The diets and dietary segregation of seabirds at the subantarctic Crozet Islands. *Marine Ornithology* 22:1–192.

Sagar, P. M., R. Murdoch, M. W. Sagar, and D. R. Thompson. 2005. Rockhopper penguin (*Eudyptes chrysocome filholi*) foraging at Antipodes Islands. *Notornis* 52:75–80.

Schiavini, A. C. M. 2000. Isla de los Estados, Tierra del Fuego: The largest breeding ground for southern rockhopper penguins? *Waterbirds* 23:286–91.

Schiavini, A. C. M., and A. Raya Rey. 2004. Long days, long trips: Foraging ecology of female rockhopper penguins *Eudyptes chrysocome chrysocome* at Tierra del Fuego. *Marine Ecology Progress Series* 275:251–62.

Shirihai, H. 2002. *The Complete Guide to Antarctic Wildlife.* Princeton, NJ: Princeton University Press.

Shumway, S. E., S. M. Allen, and P. D. Boersma. 2003. Marine birds and harmful algal blooms: Sporadic victims or underreported events? *Harmful Algae* 2:1–17.

St. Clair, C. C., and R. C. St. Clair. 1996. Causes and consequences of egg loss in rockhopper penguins, *Eudyptes chrysocome. Oikos* 77:459–66.

Strange, I. 1982. Breeding ecology of rockhopper penguin (*Eudyptes crestatus*) in the Falkland Islands. *Le Gerfaut* 72:137–88.

Tennyson, A., G. Taylor, M. Imber, and T. Greene. 2002. Unusual bird records from the Antipodes Islands in 1978–1995, with a summary of other species recorded at the island group. *Notornis* 49:241–45.

Thiebot, J. B., Y. Cherel, P. N. Trathan, and C. A. Bost. 2012. Coexistence of oceanic predators on wintering areas explained by population-scale foraging segregation in space or time. *Ecology.* 93(1):122–30

Thompson, K. R. 1989. An assessment of the potential for competition between seabirds and fisheries in the Falkland Islands. Falkland Islands Foundation Project Report. Stanley, Falkland Islands. 94 pp.

Thompson, D., and P. Sagar. 2002. Declining rockhopper penguin populations in New Zealand. *Water & Atmosphere* 10:10–12.

Tremblay, T., and Y. Cherel. 2003. Geographic variation in the foraging behaviour, diet and chick growth of rockhopper penguins. *Marine Ecology Progress Series* 251:279–97.

Uhart, M., W. Karesh, R. Cook, N. Huin, K. Lawrence, L. Guzman, H. Pacheco, G. Pizarro, R. Mattsson, and T. Mörner. 2004. Paralytic shellfish poisoning in gentoo penguins (*Pygoscelis papua*) from the Falkland (Malvinas) Islands. In *Proceedings AAZV/AAWV/WDA Joint Conference*, San Diego, California, 28 August–3 September, 481–86.

Van Buren, A. 2010. Avian predation patterns in a Falkland Islands mixed seabird colony. 1st World Seabird Conference, Victoria, Canada, 7–11 September. Poster session abstracts P2–150.

Venegas, C. 1978. Pingüinos de Barbijo (*Pygoscelis antarctica*) y macaroni (*Eudyptes chrysolophus*) en Magallanes. *Anales Instituto Patagonia* 9:179–83.

Venegas, C., and J. Jory. 1979. Guía de campo para las aves de Magallanes. Series Monografías No. 11. Punta Arenas, Chile.

Warham, J. 1963. The rockhopper penguin *Eudyptes chrysocome* at Macquarie Island. *The Auk* 80:229–56.

———. 1972. Breeding season and sexual dimorphism in rockhopper penguins. *The Auk* 89:86–105.

———. 1975. The crested penguins. In *The Biology of Penguins*, ed. B. Stonehouse. London: Macmillan.

Weimerskirch, H., R. Zotier, and P. Jouventin. 1989. The avifauna of the Kerguelen Islands. *Emu* 89:15–29.

White, R. W., and A. P. Clausen. 2002. Rockhopper *Eudyptes chrysocome chrysocome* × macaroni *E. chrysolophus* penguin hybrids apparently breeding in the Falkland Islands. *Marine Ornithology* 30:40–42.

White, R. W., J. B. Reid, A. D. Black, and K. W. Gillon. 1999. Seabirds and marine mammal dispersion in the waters around the Falkland Islands, 1998–1999. Joint Nature Conservation Committee, Peterborough.

Williams, A. J. 1980. Rockhopper penguins *Eudyptes chrysocome* at Gough Island. *Bulletin of the British Ornithological Club* 100:208–12.

Williams, T. D. 1995. *Bird Families of the World: The Penguins.* Oxford: Oxford University Press.

Wilson, R. P. 1995. Foraging ecology. In *Bird Families of the World: The Penguins*, ed. T. D. Williams. Oxford: Oxford University Press.

Wilson, R. P., C. A. Bost, K. Pütz, J. B. Charrassin, B. M. Culik, and D. Adelung. 1997. Southern rockhopper penguin *Eudyptes chrysocome chrysocome* foraging at Possession Island. *Polar Biology* 17:323–29.

Woehler, E. J. 1993. The distribution and abundance of Antarctic and subantarctic penguins. Scientific Committee on Antarctic Research, Cambridge.

———. 2006. Status and conservation of the seabirds of Heard Island and the McDonald Islands. In *Heard Island, Southern Ocean Sentinel*, ed. K. Green and E. J. Woehler, 128–65. Chipping Norton, NSW, Australia: Surrey Beatty and Sons.

Woehler, E. J., and C. A. Gilbert. 1990. Hybrid rockhopper-macaroni penguins, interbreeding and mixed species pairs at Heard and Marion Islands. *Emu* 90:198–201.

8

Northern Rockhopper Penguin

(Eudyptes moseleyi)

Richard J. Cuthbert

1. SPECIES (COMMON AND SCIENTIFIC NAMES)

Northern rockhopper penguin, *Eudyptes moseleyi*

On Tristan da Cunha, northern rockhopper penguins are known as pinnamins.

2. DESCRIPTION OF THE SPECIES

The northern rockhopper is among the smallest of the *Eudyptes* group of crested penguins, with a body length of 55–65 centimeters and average mass of around three kilograms at the start of the breeding season. Like other crested penguins, northern rockhoppers have black plumage on the head and dorsal surface with a white chest and belly. Their bright red eye color and orange-red bill contrasts with the long and vivid yellow head plumes, which are longer than those of other rockhopper species (fig. 2). The flippers are black with a white trailing edge on the dorsal side and predominantly white on the underside with more black plumage on the leading edge and at the tip (Williams 1995). Male and female birds can be distinguished visually when in pairs by the deeper and heavier bill of the male. Individual birds need to be measured to determine sex. Stiff tail feathers and large pink feet with black or gray claws provide grip and stability on the boulders that the birds must clamber through when returning from and departing for the sea (see table 8.1).

Immature birds are less brilliantly plumaged, with dark gray feathers on their dorsal surface and a duller-red eye and bill color. Juveniles lack the long feather crest

FIG. 1 (*FACING PAGE*) Northern rockhopper penguin hopping among the rocks. (P. Ryan)

FIG. 2 The northern rockhopper penguin. (P. Ryan)

of adults (fig. 3) (Ryan 2007). Plumage abnormalities have been observed in the northern rockhopper, including partially leucistic birds, as well as birds with black spotted underparts. In comparison to the various forms of southern rockhoppers, the northern rockhopper is slightly larger, with longer and much more luxuriant head plumes (fig. 4), more extensive black plumage on the under flipper, and a black stripe at the base of the bill in contrast to the pale pink strip found in *Eudyptes chrysocome filholi* (Williams 1995). Their nuptial calls are also distinctly lower pitched in comparison to the southern forms (Jouventin 1982). Rockhoppers are well named, as they hop among the rocks.

FIG. 3 A juvenile northern rockhopper penguin lacks the long feather crest of adults (P. Ryan).

FIG. 4 The northern rockhopper penguin's crest is much longer than the southern rockhopper penguin's. (P. Ryan)

3. TAXONOMIC STATUS

Until recently, the taxonomic status of the northern rockhopper was confused, with this species being grouped with one or two other subspecies of rockhopper penguins. These forms were lumped together as the rockhopper *senso lato*, with three recognized separate subspecies consisting of the nominate southern form (*Eudyptes chrysocome chrysocome*), the eastern rockhopper (*Eudyptes chrysocome filholi*), and the northern rockhopper (*Eudyptes chrysocome moseleyi*). Clear differences between the morphology of the northern rockhopper, which has a longer (79–81 mm) and more luxuriant crest, and differences in voice, with northern rockhoppers having a lower-pitched nuptial song, were recognized from early on (Jouventin 1982).

Analysis of mitochondrial DNA from northern rockhoppers collected at Gough and Amsterdam Islands and eastern rockhoppers collected at Crozet and Kerguelen Islands revealed a deep split between the two forms, estimated to have occurred more than 680,000 years ago (Jouventin et al. 2006). Based on this early split,

differences in the timing of breeding, crest morphology, and voice, the northern rockhopper was recommended as a distinct species in Jouventin et al. (2006), and this split has been recognized by BirdLife International in its conservation status for this species. Less evidence has so far been gathered to justify a split between the two southern forms of rockhopper penguin. Although Banks et al. (2006) argue the case for these to be full species, most authorities, including BirdLife International, still recognize them as two subspecies.

4. RANGE AND DISTRIBUTION

Northern rockhoppers breed at the Tristan da Cunha island group and Gough Island in the central South Atlantic Ocean and at Amsterdam and Saint Paul Islands in the Indian Ocean (table 8.2, fig. 5). Tristan da Cunha and Gough Island are part of the United Kingdom's Overseas Territories, and Amsterdam and Saint Paul form part of the French Southern Territories. The location of these islands places the breeding distribution of most northern rockhoppers to the north of the Sub-

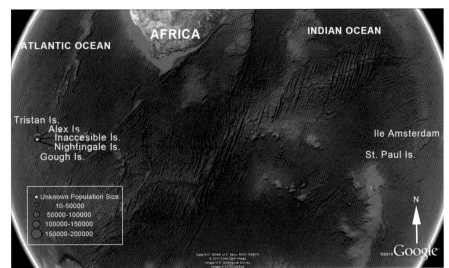

FIG. 5 Distribution and abundance of the northern rockhopper penguin, with counts based on pairs.

tropical Convergence Zone, with the exception of Gough Island, which is just to the south of this ocean front. The Subtropical Convergence Zone is one of three major oceanic frontal systems encircling Antarctica, dividing subtropical waters from nutrient-rich sub-Antarctic waters, where all other crested penguin species are located. After breeding and molting are complete, all birds depart the breeding sites and spend up to six months at sea before returning to breeding sites in the following season (fig. 6). The wintering distribution of the species is currently unknown. Some vagrant birds have been recorded from South Africa (Ryan 2007).

5. SUMMARY OF POPULATION TRENDS

The most recent population estimates indicate historical declines in excess of 96% for Gough Island and 98% for the main island of Tristan da Cunha that have occurred over at least 45 and 130 years, respectively (Cuthbert et al. 2009). Numbers have declined at Amsterdam Island and increased at Saint Paul Island in the Indian Ocean since the 1970s (Guinard et al. 1998); however, the overall trend for both islands suggests a decline of around 2% a year (Cuthbert et al. 2009). More recent comparisons of trends over the past 30 years for populations in the Atlantic (where more than 80% of the species occur) indicate a declining population at Gough Island and relatively stable populations at Tristan da Cunha and associated islands (Cuthbert et al. 2009; Robson et al. 2011). Overall, the population trend for the species is a decline.

6. IUCN STATUS

The International Union for Conservation of Nature classifies the northern rockhopper penguin as Endangered on its Red List of Threatened Species (IUCN 2011). This threat status is based on the recent split of this species from the other more abundant southern forms of rockhopper penguin and the rapid population decreases observed in northern rockhoppers over the past 30 years (in excess of 50% over three generations).

At Tristan da Cunha, penguins are protected under the conservation ordinance. Penguins at Nightingale, Stoltenhoff, and Middle Islands (all in the Tristan da Cunha island group) remain legally unprotected, and Tristan Islanders are still allowed a traditional but limited take of eggs and guano from Nightingale Island. Penguins became protected on Gough Island in 1976 when it was declared a wildlife reserve and subsequently designated a UNESCO World Heritage site, including its 22 kilometers of territorial waters (Cooper and Ryan 1994). Inaccessible Island was classified as a nature reserve in 1997, granting formal protection to penguins on and within 22 kilometers miles of the island. Inclusion of Inaccessible Island in an expanded Gough and Inaccessible Islands World Heritage site provided further legal protection in 2004. Amsterdam and Saint Paul Islands have been incorporated into the French Southern Territories Important Bird Area and are managed as nature reserves (BirdLife International 2010), although it is unclear if this protection includes marine areas as well as terrestrial sites.

FIG. 6 Northern rockhopper penguins rest on the beach. (P. Ryan)

7. NATURAL HISTORY

BREEDING BIOLOGY. Northern rockhoppers breed in a variety of habitats ranging from open boulder beaches at Gough Island and Tristan da Cunha to dense, high stands of tussock grass (mainly *Spartina arundinacea*) on Nightingale and Inaccessible Islands. Adults return from sea in late July and August and immediately pair up (usually with the same mate from previous years) within the breeding colonies. Pairs construct simple cup nests, with the female generally making the nest scrape and the male bringing grass, sticks, feathers, and stones with which the female constructs the nest (fig. 7). Courtship displays consist of a loud braying duet performed with raised flippers, heads held back, and shaking of head plumes. The resident pair aggressively shoos off other penguins that encroach within reach of its nest through a combination of pecking and rapid beating with a flipper. Such aggressive displays are often followed by a display duet after the intruder has moved on. Table 8.3 summarizes the breeding parameters reported for the species.

The female lays two white eggs, with the first (A-egg) being considerably smaller (63 by 50 mm, with a mass averaging 85 grams) in comparison to the second (B-egg) (70 by 54 mm, averaging 113 grams) (Wilson et al. 2010). Records from Tristan da Cunha suggest that some nests contain three eggs, with 4 of 353 nests (1.1%) observed to hold three eggs at one subcolony on Tristan da Cunha (C. Stone, pers. comm.) and similar reports from Nightingale Island (T. Glass, pers. comm.). Whether these three-egg clutches are a result of egg dumping or of birds laying three eggs is unclear.

Most of the A-eggs are lost in the first few days of incubation and usually only the B-egg is incubated (fig. 8). In around 6% of nests, two eggs are hatched, although only one chick is likely to be reared, as the B-egg hatches first and this larger chick dominates feeding opportunities (fig. 9). Incubation takes around 32–34 days, with male and female birds sharing incubation shifts. On hatching, the chick is brooded by the male for a further 20–26 days, while the female undertakes all foraging. After this period, chicks gather into crèches, and both male and female parents feed the chick (fig. 10). Feeding continues until chicks have molted out of their downy plumage and go to sea approximately 9–10 weeks after

FIG. 7 Northern rockhopper penguin incubating a pipped egg. (P. Ryan)

FIG. 8 Northern rockhopper pair with one egg. Parents often lose the first of their two eggs. (P. Ryan)

FIG. 9 Northern rockhoppers lay two eggs, with the second egg hatching first. The bill, flipper, and wet gray down of the chick hatching from the second egg are visible. (P. Ryan)

FIG. 10 Northern rockhopper chicks form a crèche near an adult. (P. Ryan)

hatching. Estimates of breeding success indicate that around 51% of clutches on Tristan da Cunha fledged one chick (Williams and Stone 1981), with a lower estimated breeding success of around 36% on Gough Island (Cuthbert and Sommer 2004). The majority of breeding failures occur during the incubation and early chick periods, with higher failure rates in smaller colonies, probably due to increased rates of predation by sub-Antarctic skuas (*Catharacta antarctica*) (Wilson et al. 2010).

PREY. Relatively few dietary studies have been undertaken on northern rockhoppers. One study at Gough Island indicated that crustaceans, in particular euphausids, make up 90% of prey items. The remaining prey items were fish and a very few cephalopods (Klages et al. 1988). Studies from Amsterdam and Saint Paul Islands suggest a lower proportion of euphausids in the diet, with one study indicating that cephalopods, euphausids, and fish constituted around 50%, 40%, and 10%, respectively, of dietary contents by mass (Cooper et al. 1990). Studies at Amsterdam Island also indicate seasonal changes in diet, with cephalopods (squid) dominating the diet (44% by mass) during the early chick-crèching stage in November, and fish (64% by mass) being the main prey item in later stages of chick rearing in December. These dietary differences were associated with deeper diving depths in the later period (average maximum depths of 83 vs. 57 m) (Tremblay et al. 1997).

PREDATORS. Sub-Antarctic skuas are a major predator of eggs and chicks at Gough Island and Tristan da Cunha and are likely to cause similar losses at Amsterdam and Saint Paul. Both southern giant petrels (*Macronectes halli*) and northern giant petrels (*M. giganteus*) have been observed preying on adult penguins in the sea at Nightingale Island, in the Tristan da Cunha island group (Ryan et al. 2008). Southern giant petrels are likely to take chicks at the breeding colonies, although this has not been documented. Sub-Antarctic fur seals (*Arctocephalus tropicalis*) occasionally hunt rockhoppers, as observed at Amsterdam Island (Guinard et al. 1998). Fur seals are experiencing a rapid and continuing increase at all islands where the northern rockhopper occurs, following historical exploitation. Although penguins must represent a rare component of their diet, this increase has the potential to raise predation pressure on penguin populations (Guinard et al. 1998; Cuthbert et al. 2009). Introduced house mice (*Mus musculus*) are present at Gough Island, and mice and ship rats (*Rattus rattus*) are found on Tristan da Cunha. While both species are known to be a serious predator of breeding seabirds—mice prey on large chicks of the Tristan albatross (*Diomedea dabbenena*) and other species at Gough (Cuthbert and Hilton 2004; Wanless et al. 2007)—there is no evidence that mice or rats prey on penguin eggs or chicks at either island. At Amsterdam and Saint Paul, a range of introduced mammalian predators and omnivores, including feral cats (*Felis domesticus*), rats, and feral pigs (*Sus scrofa*), are likely to have preyed on eggs and chicks in the past. The current impact of these predators on penguin populations is not recorded, although rats have now been eradicated from Saint Paul.

MOLT. Following the completion of breeding, adults embark on a pre-molt foraging trip for around a month, during which they lay down the fat reserves necessary for molting. Breeding birds return to molt in February–March at Tristan da Cunha, and around three to four weeks later at Gough. Failed breeders and immature birds molt earlier, returning as early as November to begin molting. Birds lay down and use considerable reserves of energy during molting. On Gough Island, they return from their pre-molt exodus at an average mass of 4.36 ± 0.53 kilograms, which contrasts with a mean mass before egg laying of 3.04 ± 0.36 kilograms. After the completion of molting, their average mass is around 2.58 ± 0.30 kilograms (R. J. Cuthbert, unpubl. data).

ANNUAL CYCLE. See figure 11.

8. POPULATION SIZE AND TRENDS

More than 80% of the word's northern rockhopper population is found at Tristan da Cunha (an estimated 160,000–175,000 pairs) and Gough Island (an estimated 32,000–65,000 pairs (Cuthbert et al. 2009; Robson et al. 2011). In the mid-1990s, numbers at Amsterdam and Saint Paul Islands were estimated to be around 25,000 and 9,000 pairs, respectively. Based on these estimates, the total population of this species is around 190,000–230,000 pairs.

Population trends for different islands are relatively poorly known due to the difficulty of accurately sur-

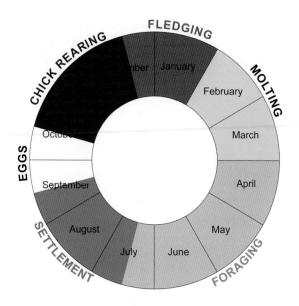

FIG. 11 Annual cycle of the northern rockhopper penguin.

veying these species and high inter-annual variation in nesting numbers. Estimating numbers of northern rockhopper penguins at Nightingale and Inaccessible Islands (Tristan da Cunha island group) is particularly difficult, as nesting colonies are located underneath dense two-meter-high stands of tussock grass, and landing on Inaccessible is, as the name implies, often difficult because of its steep cliffs and the rough seas. Two recent reviews of penguin numbers for Tristan da Cunha and Gough Island, based on relatively infrequent counts from the past 30 years, indicate that numbers are declining at Gough at an average rate of 3–4% a year (Cuthbert et al. 2009) but are relatively stable at Tristan da Cunha (Robson et al. 2011). A recent count from Middle Island, a small rock stack off Nightingale Island, recorded an estimated 80,000 pairs of penguins (Robson et al. 2011), which is not dissimilar to the estimate of 100,000 pairs recorded in 1973, the only other time that this site has been counted (Richardson 1984). Contrasting trends in numbers were recorded from Amsterdam and Saint Paul Islands in the Indian Ocean for the period 1971–73, with numbers at the most important breeding site of Amsterdam Island (25,000 pairs) declining over this period at 2.7% per year. Over the same period, the smaller Saint Paul population (9,000 pairs) increased at an annual rate of 5.5% (Guinard et al. 1998).

These estimated population trends have a great deal of uncertainty attached to them, due to the difficulty of accurately counting northern rockhoppers, their remote and difficult-to-access breeding sites, inter-annual variation in breeding numbers, and the relative infrequency of counts. Moreover, the shape of the declines is also highly uncertain, and it is unknown if populations declined at a steady rate of 2–3% a year or instead experienced periods of relative stability and then sudden and rapid population declines.

9. MAIN THREATS

Comparatively little information is available on the demographic parameters of this species, and even less is known on the demographic causes of the population declines. The only detailed demographic study is from Amsterdam Island, where annual adult survival was estimated to be 84.0 ± 1.1%, survival in the first year after fledging was around 39%, and age at first breeding was 4.7 ± 1.7 years (Guinard et al. 1998). Given the time taken to reach sexual maturity, relatively low reproductive output (a maximum of one chick per pair per year), and low values of breeding success observed (discussed below), a survival rate of around 84% is on the low side, and the use of flipper bands in this study may have negatively affected the penguins' survival probability, as has been recorded for other penguin species (Jackson and Wilson 2002). Values of breeding success (percentage of pairs fledging a chick) varied, with 28%, 35%, and 52% recorded for three consecutive years at Amsterdam Island (Guinard et al. 1995), 51% at Tristan da Cunha (Williams and Stone 1981), and around 36% and 53% at Gough Island (Cuthbert and Sommer 2004; R. J. Cuthbert, unpubl. data). These estimated values are likely to be relatively crude, due to the difficulty of documenting the breeding success of individual pairs once the guard stage has passed and chicks form large crèches.

The link between what little is known about the demographic parameters of northern rockhoppers and observed population declines is currently speculative, and more detailed and long-term studies are required for a better understanding of demographic rates and how these may change with time. If the estimated value of adult survival of around 84% is accurate (rather than an artifact of flipper banding), then such a low survival rate would result in a decline in population of around 6% a year, close to the observed rate of decline for this population at Amsterdam Island (Guinard et al. 1998). Values of breeding success also appear to be on the low side, and smaller colony sizes may result in further reduced breeding success due to increased predation rates at small colonies (Wilson et al. 2010).

Specific past, current or potential threats are discussed below.

CLIMATE VARIATION. Guinard et al. (1998) note a possible link between climatic variation and penguin numbers at Amsterdam Island, with the major decrease (by 57%) in penguin numbers from 1971 to 1993 occurring at the same time as a 0.8°C decrease in sea surface temperature around Amsterdam and Saint Paul Islands from 1982 to 1993. Such a change in sea temperatures may have altered the distribution of prey around these breeding sites, resulting in the observed decrease in breeding numbers at Amsterdam Island (Guinard et al. 1998). A major longitudinal study of stable isotope signatures from feathers of both northern and southern rockhopper species, including current and historical samples from Gough Island and Tristan da Cunha, was undertaken to investigate patterns of change in nitrogen and carbon signatures indicative of primary productivity in the ecosystem and prey trophic levels (Hilton et al. 2006). This study indicated no major systematic change in trophic levels or marine productivity for populations at Gough and Tristan da Cunha, suggesting that northern rockhoppers at these two sites have been feeding in a fairly constant marine food web in recent decades (Hilton et al. 2006; BirdLife International 2010). Given our lack of knowledge on the likely effect of climate variation on the foraging ecology and population trends of northern rockhoppers, trying to assess the putative impact of climate change is extremely difficult. Given the relatively small foraging range of penguins, particularly during the breeding season, they are likely to be more vulnerable to changes in the marine environment compared to seabirds that forage over a much larger range from the breeding site. Further study of the diet and foraging ecology of northern rockhoppers and the responses of key prey to climate change will enable more accurate predictions to be made on the potential negative or positive impacts of climate change.

HISTORICAL HARVESTS. Historically, the main threat to the northern rockhopper has come from human exploitation. Early accounts from both Gough Island and Tristan da Cunha report the presence of millions of birds, and these undoubtedly would have been used by early settlers on Tristan and sealing gangs at Gough and other islands. Vast numbers of birds and products from penguins were collected, with more than 25,200 eggs said to have been taken in one year at Tristan da Cunha in the 1920s (Rogers 1927), oil extracted from molting birds, feathers collected for stuffing pillows and mattresses, and even head plumes being used for making ornamental table "tossel" mats on Tristan. These practices had largely ceased by 1955 (Hagen 1952; Wace and Holdgate 1976; Richardson 1984). The current distribution of penguins on the main island of Tristan da Cunha is likely to reflect this past exploitation, with extant colonies all found on the far side of the island, away from the sole human settlement. Gough has never been permanently settled; the small number of sealers based ashore and the difficulty of landing on and accessing the steep boulder beach where most birds breed mean that only localized losses are likely to have occurred. Tristan Islanders are still allowed to take eggs at Nightingale and Middle Islands; however, the number of eggs currently taken is far below historical levels, with only one egg taken from each nest (T. Glass, pers. comm.).

As rockhoppers rear only a single chick, taking one egg from the clutch, particularly if it is the smaller, first-laid egg, is unlikely to cause any significant impact to the population. Little information is available about the historical harvest of northern rockhoppers at Amsterdam and Saint Paul Islands. Given the settlement of these islands and habitat modifications made to accommodate the introduction of livestock and other animals, it seems likely that penguins would have been exploited in a way similar to their past exploitation at Tristan da Cunha.

FISHERIES. There is little evidence that fishery operations currently pose a major threat to northern rockhoppers, with no information reported for bycatch of this species. Drift-net fishing in the 1980s was likely to have posed a significant threat to penguins off Tristan da Cunha (Ryan and Cooper 1991); however, this threat ended with the banning of drift nets. The use of penguins as bait for commercial rock lobster (*Jasus tristani*) fishing is likely to have posed a major threat to northern rockhoppers in the past, and heavy exploitation of penguins at Saint Paul Island in the 1930s may have caused a major reduction in numbers (Guinard et al. 1995). Penguins were also used as bait for rock lobster fishing at Tristan da Cunha in the 1970s (Cuthbert et al. 2009). This prac-

tice was banned, and the rock lobster fishery now uses imported bait and thus probably has no, or very little, effect on penguin numbers. Entanglement in marine debris, including fishing nets, is occasionally observed at Tristan da Cunha and Gough Island but is considered to be infrequent and to have little effect on the population (Cuthbert et al. 2009).

HABITAT DEGRADATION. Habitat destruction, particularly the burning of lowland tussock areas to create agricultural land, is likely to have been a major factor in the past, especially with the settlement of Amsterdam Island and Tristan da Cunha. Small landslides following heavy rains are observed at Gough and Tristan da Cunha (Cuthbert et al. 2009) and have been responsible for killing some breeding penguins (T. Glass, pers. comm.) and disrupting breeding efforts. Such events occur relatively infrequently and on a small spatial scale and are not likely to be a major issue. Increasing numbers of sub-Antarctic fur seals in both the Atlantic and Indian Ocean islands have increased the potential for prey competition and predation (Guinard et al. 1998; Cuthbert et al. 2009). Additionally, increasing numbers of fur seals have led to the displacement of penguins from breeding colonies on Nightingale Island (T. Glass, pers. comm.).

POLLUTION. The risk of pollution, particularly oil pollution, is an obvious threat to penguin populations, even in some of the remotest islands where this species breeds (fig. 12). Individual birds are occasionally seen with oiled plumage at Tristan da Cunha and Gough Island, but there are very few recorded observations (Cuthbert et al. 2009). The wrecking of an oil-rig platform on Tristan da Cunha in 2006 (Wanless et al. 2009) highlighted the potential risk that such an event could pose. On 16 March 2011, the MS *Oliva*, a bulk carrier en route from Brazil to Singapore, ran aground at Nightingale Island, spilling 1,400 metric tons of heavy fuel oil. Petroleum came ashore at Inaccessible Island and Tristan da Cunha, more than 30 kilometers from where the ship went aground. Approximately 3,700 oiled birds were collected for rehabilitation, but only 10% survived and could be released (Tristan da Cunha Website 2012). Tristan is a remote island accessible only by boat, and it proved impossible to get the equipment and drugs needed for cleaning the birds fast enough. The total number of oiled birds is unknown.

The oil spill from the *Oliva* occurred toward the end of the penguins' molting period, and a key action of the Tristan da Cunha Conservation Department was to "corral" birds onshore in order to prevent them from departing the colony and encountering oil. This temporary measure prevented many thousands of penguins from becoming oiled, and the fences were removed after heavy seas had broken up and dispersed much of the heavy oil on the beaches and in the surrounding waters. While these actions were successful at reducing the immediate impact of the spill, many birds were affected, and the long-term consequences of this oil spill on the population have yet to be assessed. Population counts have been undertaken on the two main affected islands since the spill. In October 2011, numbers of incubating birds on Nightingale and Middle Islands were estimated to number around 90,000 and 25,000 pairs, respectively. These counts indicated no major reduction in breeding numbers in comparison to previous counts from before the spill, with counts from 2009 and 2010 estimated at 25,000 and 20,000 pairs for Nightingale Island and 80,000 and 95,000 pairs for Middle Island (Robson et al. 2011). Determining the total damage from the spill will take three to five years of counts, but the initial counts suggest no major short-term impact.

FIG. 12 Petroleum discharge is a major problem for many species of penguins, including this northern rockhopper penguin, which has sought refuge on the beach, where it will starve to death. (P. Ryan)

FIG. 13 Northern rockhopper penguins incubating eggs in their breeding colony. (P. Ryan)

DISEASE. No disease issues have been observed or published with relation to northern rockhopper penguins, although at Amsterdam Island, infectious diseases have been linked to declines in albatross numbers (Weimerskirch 2004).

TOURISM. There is very limited tourism at the islands where northern rockhoppers occur. The small numbers of tourists that do visit land mostly at the main island of Tristan da Cunha and at Nightingale Island. Such visits are strictly controlled, with a local guide always present, and are unlikely to be detrimental to breeding birds. Visits to Amsterdam and Saint Paul are strictly controlled, and all visitors must comply with Terres Australes et Antarctiques Françaises (TAAF) regulations.

10. RECOMMENDED PRIORITY RESEARCH ACTIONS FOR CONSERVATION

The following recommendations for research are based on the output of the International Rockhopper Penguin Workshop, held in June 2008 in Edinburgh, Scotland, which formulated a plan for research and conservation actions for rockhopper penguins (BirdLife International 2010). The following were identified as priority areas for research:

1. Assess population trends by conducting regular censuses of breeding colonies at all of the main breeding islands.
2. Undertake tracking and foraging studies at breeding colonies.
3. Establish a transponder-tagged study population and quantify survival rates.
4. Evaluate the long-term impact of the MS *Oliva* oil spill at Tristan da Cunha.

Areas of research considered to be of medium priority include gathering data on the diet of the species, continuing to collect feather samples for stable isotope analysis, estimate breeding success for different colonies, and measure body mass and condition and growth rate of chicks at different sites annually (BirdLife International 2010).

11. CURRENT CONSERVATION EFFORTS

Due to the lack of knowledge on the cause of population declines and the remoteness of the breeding islands, there are currently no systematic conservation efforts in place for this species. Historical exploitation of penguins, losses due to their use as rock lobster fishing bait, and mortality from entanglement in drift nets have now ceased and been banned, and breeding colonies are protected as nature reserves at the majority of sites. Conservation actions at Nightingale Island, Tristan da Cunha, have recently included erecting a fence to prevent fur seals from encroaching on one of the main breeding sites (while still allowing penguins to enter this area) and covering a deep rock crevice that was trapping and killing breeding birds (T. Glass, pers. com).

12. RECOMMENDED PRIORITY CONSERVATION ACTIONS FOR INCREASING POPULATION RESILIENCE AND MINIMIZING THREATS AND IMPACTS

Priority conservation actions for northern rockhoppers depend on understanding the causal mechanism or mechanisms of their decline. It may then be possible to implement specific actions that will minimize the threats and impacts, although if the declines are a consequence of large-scale changes in oceanic systems, it is difficult to know what measures can be undertaken in the short-term. In the interim, priority actions are the following:

1. Protect the breeding colonies of this species at all sites.
2. Minimize disturbance.
3. Ensure no introduction of mammalian predators.
4. Protect the species at sea through minimizing any fisheries-related mortality.

ACKNOWLEDGMENTS

I am grateful to a large number of colleagues and field assistants who have led on aspects of penguin research or who have been pecked and flipper-bashed in the field by the always feisty northern rockhopper. They include Geoff Hilton, Peter Ryan, John Cooper, Trevor Glass, Norman Ratcliffe, Erica Sommer, Marie-Helene Burle, Johnny Wilson, Henk Louw, Paul Visser, Kalinka Rexer-Huber, and Graham Parker. Thanks to the Conservation Department and the Island Council of Tristan da Cunha and the administrators Mike Hentley and David Morley for permission to undertake research on Gough Island. The UK Overseas Territories Environment Programme and the Royal Society for the Protection of Birds provided financial support, and the South African National Antarctic Programme and Percy FitzPatrick Institute University of Cape Town supplied essential logistic support. Lastly, thanks to Helen Otley and Anne Brown for organizing the International Rockhopper Penguin Workshop and to the Royal Zoological Society of Scotland for hosting this event.

TABLE 8.1 Morphometric data for adult northern rockhoppers

	MALE	FEMALE	UNSEXED	REFERENCE
Body mass* (kg)	2.96 ± 0.33 (49)	3.12 ± 0.37 (45)		R. J. Cuthbert unpubl. data
Bill length (mm)	49.2 ± 7.0 (34)	44.3 ± 6.7 (31)		R. J. Cuthbert unpubl. data
Total head length (mm)**	123 ± 4.5 (15)	119 ± 5.3 (15)		R. J. Cuthbert unpubl. data
Tarsus-toe length (mm)	112.9 ± 5.5 (15)	109.5 ± 3.3 (15)		R. J. Cuthbert unpubl. data
Flipper length (mm)	185.3 ± 7.9 (15)	184.5 ± 5.3 (15)		R. J. Cuthbert unpubl. data
Egg length (mm)			A-egg 63.1 ± 3.1 (26) B-egg 70.4 ± 2.9 (26)	Wilson et al. 2010
Egg breadth (mm)			A-egg 50.2 ± 5.7 (26) B-egg 54.4 ± 3.7 (26)	Wilson et al. 2010
Egg mass (g)			A-egg 84.5± 15.0 (26) B-egg 113.2 ±10.9 (26)	Wilson et al. 2010

* Body mass is measured at the start of the breeding season on Gough Island
** Total head length only measured for immature birds at Gough Island

TABLE 8.2 Northern rockhopper breeding sites, geographical coordinates and number of breeding pairs

LOCATION	COORDINATES	NUMBER OF PAIRS
Gough	40° 19' 5" S, 9° 56' 7" W	32,000–65,000
Tristan da Cunha	37° 4' 0" S, 12° 19' 0" W	160,000–175,000 pairs
Nigthingale	37° 25' 10" S, 12° 28' 40" W	
Inaccessible	37° 18' 9" S, 12° 40' 28" W	
Alex (Middle)	37° 24' 40" S, 12° 28' 46" W	80,000
St. Paul	38° 43' 48" S, 77° 31' 20" E	9,000
Ile Amsterdam	37° 49' 33" S, 77° 33' 17" E	25,000

TABLE 8.3 Demographic rates

PARAMETER	MEAN ± SD	95% CI	PERIOD/SAMPLE	REFERENCE
Adult survival *	84.0 ± 1.1%	81.6–86.1%	5 years	Guinard et al. 1998
Immature survival **	39%	-	6 years	Guinard et al. 1998
Age of first breeding	4.7 ± 1.7 years	-	(n=13 birds)	Guinard et al. 1998
Maximum life span ***	25 years			BirdLife 2010
Breeding success				
Amsterdam (1993)	28%	-	1 year	Guinard et al. 1998
Amsterdam (1994)	35%	-	1 year	Guinard ct al. 1998
Amsterdam (1995)	52%	-	1 year	Guinard et al. 1998
Tristan da Cunha (1980)	51%	-	1 year	Williams & Stone 1981
Gough Island (2000)	36%	-	1 year	Cuthbert & Sommer 2004
Gough Island (2009)	53%	-	1 year	Cuthbert unpubl.. data

* Adult survival estimate may be low due to use of flipper bands (Guinard et al. 1998)

** Immature (first year) survival estimate adjusted due to higher estimated mortality from flipper bands (Guinard et al. 1998)

*** Maximum life span in captivity for Rockhopper Penguins *sensu lato* (BirdLife 2010)

REFERENCES

Banks, J., A. V. Buren, Y. Cherel, and J. B. Whitfield. 2006. Genetic evidence for three species of rockhopper penguins, *Eudyptes chrysocome. Polar Biology* 30(1):61–67.

BirdLife International. 2010. Rockhopper penguins: A plan for research and conservation action to investigate and address population changes. In *Proceedings of an International Workshop in Edinburgh, Scotland, 3–5 June 2008*. Cambridge: BirdLife International.

Cooper, J., C. R. Brown, M. A. Hindell, N. T. W. Klages, P. J. Moors, D. Pemberton, V. Ridoux, K. R. Thompson, and Y. M. van Heezik. 1990. Diets and dietary segregation of crested penguins (*Eudyptes*). In *Penguin Biology*, ed. L. S. Davis and J. T. Darby, 131–56. San Diego, CA: Academic Press.

Cooper, J., and P. G. Ryan. 1994. Management plan for the Gough Island Wildlife Reserve. Edinburgh, Tristan da Cunha, Government of Tristan da Cunha.

Cuthbert, R., J. Cooper, H. J. Burle, C. J. Glass, J. P. Glass, S. Glass, T. Glass, G. M. Hilton, E. S. Sommer, R. M. Wanless, and P. G. Ryan. 2009. Population trends and conservation status of the northern rockhopper penguin *Eudyptes moseleyi* at Tristan da Cunha and Gough Island. *Bird Conservation International* 19:109–120.

Cuthbert, R., and G. Hilton. 2004. Introduced house mice *Mus musculus*: A significant predator of endangered and endemic birds on Gough Island, South Atlantic Ocean? *Biological Conservation* 117:483–89.

Cuthbert, R., and E. Sommer. 2004. Population size and trends of four globally threatened seabirds at Gough Island, South Atlantic Ocean. *Marine Ornithology* 32:97–103.

Guinard, E., H. Weimerskirch, and P. Jouventin. 1998. Population changes and demography of the northern rockhopper penguin on Amsterdam and Saint Paul Islands. *Colonial Waterbirds* 21:222–28.

Hagen, Y. 1952. Birds of Tristan da Cunha. In *Results of the Norwegian Scientific Expedition to Tristan da Cunha, 1937–1938*, vol. 20, 1–248.

Hilton, G. M., D. R. Thompson, P. M. Sagar, R. J. Cuthbert, Y. Cherel, and S. J. Bury. 2006. A stable isotopic investigation into the causes of decline in a sub-Antarctic predator, the rockhopper penguin *Eudyptes chrysocome*. *Global Change Biology* 12:1–15.

IUCN (International Union for Conservation of Nature). 2011. IUCN Red List of Threatened Species. Version 2011.2. www.iucnredlist.org (accessed 16 March 2012).

Jackson, S., and R. P. Wilson. 2002. The potential costs of flipper-bands to penguins. *Functional Ecology* 16:141–48.

Jouventin, P. 1982. Visual and vocal signals in penguins, their evolution and adaptive characters. In *Advances in Ethology*, supplement to *Journal of Comparative Ethology*. Berlin and Hamburg: Verlag Paul Parey.

Jouventin, P., R. J. Cuthbert, and R. Ottvall. 2006. Genetic isolation and divergence in sexual traits: Evidence for the northern rockhopper penguin *Eudyptes moseleyi* being a sibling species. *Molecular Ecology* 15(11)3413–23.

Klages, N. T., M. De L. Brook, and B. P. Watkins. 1988. Prey of northern rockhopper penguins at Gough Island, South Atlantic Ocean. *Ostrich* 59:162–65.

Richardson, M. E. 1984. Aspects of the ornithology of the Tristan da Cunha group and Gough Island, 1972–1974. *Cormorant* 12:122–201.

Robson, B., T. Glass, N. Glass, J. Glass, C. Repetto, G. Rogers, R. A. Ronconi, P. G. Ryan, G. Swain, and R. J. Cuthbert. 2011. Revised population estimate and trends for the endangered northern rockhopper penguin *Eudyptes moseleyi* at Tristan da Cunha. *Bird Conservation International* 21:454–459.

Rogers, R. A. 1927. *The Lonely Isle*. London: George Allen & Unwin.

Ryan, P. G. 2007. *Field Guide to the Animals and Plants of Tristan da Cunha and Gough Island*. Newbury, United Kingdom: Pisces Publications.

Ryan, P. G., and J. Cooper. 1991. Rockhopper penguins and other marine life threatened by driftnet fisheries at Tristan da Cunha. *Oryx* 25:76–79.

Ryan, P. G., E. S. Sommer, and E. Breytenbach. 2008. Giant petrels *Macronectes* hunting northern rockhopper penguins *Eudyptes moseleyi* at sea. *Ardea* 96:191–94.

Tremblay, Y., E. Guinard, and Y. Cherel. 1997. Maximum diving depths of northern rockhopper penguins (*Eudyptes chrysocome moseleyi*) at Amsterdam Island. *Polar Biology* 17:119–22.

Tristan da Cunha Website. 2012. MS Oliva Disaster Home Page. http://www.tristandc.com/newsmsolivahome.php (accessed 20 July 2011).

Wace, N. M., and M. W. Holdgate. 1976. Man and nature in the Tristan da Cunha Islands. *IUCN Monographs* 6:1–114.

Wanless, R. M., A. Angel, R. J. Cuthbert, G. M. Hilton, and P. G. Ryan. 2007. Can predation by invasive mice drive seabird extinctions? *Biology Letters* 3:241–44.

Wanless, R. M., S. Scott, W. H. H. Sauer, T. G. Andrew, J. P. Glass, B. Godfrey, C. Griffiths, and Y. Yeld. 2009. Semi-submersible rigs: A vector transporting entire marine communities around the world. *Biological Invasions* 12(8):2573–83.

Weimerskirch, H. 2004. Diseases threaten Southern Ocean albatrosses. *Polar Biology* 27:374–79.

Williams, A. J., and C. Stone. 1981. Rockhopper penguins *Eudyptes chrysocome* at Tristan da Cunha. *Cormorant* 9:59–66.

Williams, T. D. 1995. *The Penguins*. Oxford: Oxford University Press.

Wilson, J. W., M. H. Burle, R. Cuthbert, R. L. Stirnemann, and P. G. Ryan. 2010. Breeding success in northern rockhopper penguins (*Eudyptes moseleyi*) at Gough Island, South Atlantic Ocean. *Emu* 110:1–5.

9

Erect-Crested Penguin

(Eudyptes sclateri)

Lloyd Spencer Davis

1. SPECIES (COMMON AND SCIENTIFIC NAMES)

Erect-crested penguin, *Eudyptes sclateri*

2. DESCRIPTION OF THE SPECIES

The erect-crested penguin is a medium-size penguin with a body length of 68 centimeters and a body mass of approximately 3.5–5.5 kilograms, depending on the time of the breeding season and the sex of the individual. They are similar in appearance to other crested penguins, especially the Fiordland and Snares penguins (table 9.1). When dry, erect-crested penguins are distinguished by the upright yellow plumes of their crests (fig. 1). When they are wet, identification is difficult because the crests droop, but the superciliary stripe starts closer toward the base of the bill than it does on Fiordland and Snares penguins. Feathers are black on the dorsal side of the body, the face, and the tail and white on the undersides. Flippers are black on their dorsal side, with a white trailing edge, and white on the ventral side. The bill is a dirty orange with a pronounced hook at the distal end of the maxilla and a white line of skin at the base of the mandible that forms a distinct gular pouch at the base of the gape. The eyes are brown. Legs and feet are pink. Immature erect-cresteds have a pale yellow superciliary stripe but lack the upright plumes and have a mottled gray throat.

3. TAXONOMIC STATUS

There are no subspecies. DNA analysis supports what

FIG. 1 (*FACING PAGE*) Erect-crested penguins nest mainly on the Bounty and Antipodes Islands near New Zealand. (L. S. Davis)

FIG. 2 Two erect-crested penguins are distinguished by the upright yellow feather plumes of their crests. (L. S. Davis)

was apparent from the penguin's external appearance: erect-crested penguins are most closely related to Fiordlands and Snares (Davis and Renner 2003). They have much in common in terms of their life-history patterns with other penguins of the genus *Eudyptes* (fig. 2).

4. RANGE AND DISTRIBUTION

Breeding range is now confined largely to the Bounty (Proclamation, Tunnel, Depot, Ruatara, Penguin, Lion, Spider, North Rock) and Antipodes (Antipodes, Bollons, Archway) Islands southeast of New Zealand (fig. 3) (Taylor 2000), but some birds apparently still breed at Disap-

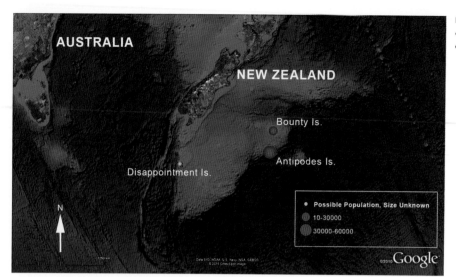

FIG. 3 Distribution and abundance of the erect-crested penguin, with counts based on pairs.

pointment Island in the Auckland Islands (Bartle and Paulin 1986). Several hundred erect-crested penguins used to breed on Campbell Island in the 1940s (Bailey and Sorensen 1962), but by 1986, no eggs or chicks were observed among the 20–30 pairs occupying nest sites (Marchant and Higgins 1990). Richdale (1941) followed a pair of erect-crested penguins that bred on the Otago Peninsula during the 1938 and 1939 breeding seasons. Subfossil remains suggest that the penguins were once abundant on the Chatham Islands.

5. SUMMARY OF POPULATION TRENDS

It seems beyond doubt that the population has suffered substantial declines since the late 1970s. There is an urgent need for more accurate, more frequently collected, and more consistent census data to truly ascertain the status of this species.

6. IUCN STATUS

The International Union for Conservation of Nature currently lists the erect-crested penguin as Endangered on its Red List of Threatened Species (IUCN 2011). This classification is based on both the restricted range of the species, which makes it vulnerable to local perturbations in the environment, and the evidence, such as it is, that suggests significant population declines in recent times. Hitchmough et al. (2007), in its review of New Zealand threat classification lists, describes the erect-crested penguin as "Nationally Endangered." In contrast, Miskelly et al. (2008) list the erect-crested penguin as "Naturally Uncommon" when assessing the conservation status of New Zealand

birds, presumably because the Bounty and Antipodes Islands fall outside New Zealand's Exclusive Economic Zone, which was the focus of the authors' compilation.

7. NATURAL HISTORY

BREEDING BIOLOGY. These penguins are offshore-foraging, migratory breeders (Croxall and Davis 1999). They arrive for breeding at their colonies on the Bounty and Antipodes Islands in September (fig. 4). The period between arrival and egg laying is quite long compared to penguins of similar size, with males spending about three weeks

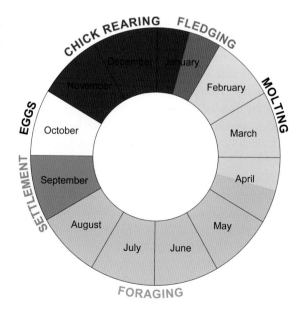

FIG. 4 Annual cycle of the erect-crested penguin.

FIG. 5 Erect-crested penguins lay their eggs directly on rock and have very rudimentary nests. (L. S. Davis)

FIG. 6 The crested penguins lay two eggs, and the first egg (A-egg) is much smaller than the second egg (B-egg). Erect-cresteds have the most extreme egg dimorphism of any bird. (L. S. Davis)

ashore before laying occurs. This time is, remarkably, not spent on nest building; erect-crested penguins lay their eggs directly onto rock platforms or, at best, the most rudimentary nests made of perhaps a few small stones and a little vegetation, such as grasses (fig. 5).

Typically, the penguin lays two eggs, with the first-laid egg (A-egg) being considerably smaller than the second-laid egg (B-egg) (table 9.2; fig. 6). The egg-size dimorphism is the most extreme for any bird, with B-eggs being 85% heavier in mass than A-eggs. The laying interval of 5.4 days is also the longest for any penguin (Davis et al., unpubl. data).

In most circumstances, the A-egg is lost before or on the day the B-egg is laid. Erect-cresteds, like crested penguins generally, are obligate brood reducers (Davis and Renner 2003) and rear only one chick. Unlike Snares and Fiordland penguins, for which brood reduction often takes place after hatching, erect-crested penguins exhibit a pattern similar to that of macaroni and royal penguins, which typically lose the A-egg well before hatching.

There is no information available on nest-site and mate fidelity.

PREY. The diet of erect-crested penguins has not been studied (Davis and Renner 2003), but taking into account their long foraging trips, they probably rely on krill, fish, and cephalopods, like other crested penguins.

PREDATORS. On the Antipodes Islands, brown skuas (*Catharacta skua*) prey on penguin eggs (Davis, unpubl. data). There are no mammalian predators on the Bounty

Islands and only mice on the Antipodes Islands (Taylor 2000). New Zealand fur seals (*Arctocephalus forsteri*) may kill adult penguins on land (Davis 2001). Leopard seals are seen occasionally in the waters around the Antipodes (Davis, unpubl. obs.) and, given their propensity for preying on penguins elsewhere, may well take erect-cresteds at sea. The reality is that very little is known of the threats these penguins face at sea.

MOLT. After their chicks fledge in late January, adults undertake a pre-molt foraging trip of about four weeks (Warham 1972). No exact data are available, except that almost all breeders had left for their pre-molt foraging trip by 4 February and seemingly many had returned by 2 March. Depending on whether or not they bred successfully, adults may start to molt anytime from early February to early March (Williams 1995). Molt takes 26–30 days, with birds completing molt and departing the colony by early to mid-April.

After their post-breeding molt, erect-crested penguins go to sea, where their winter distribution is unknown. Some birds molt regularly on other sub-Antarctic islands south of New Zealand and, less commonly, on New Zealand's South Island; vagrants have been recorded from New Zealand's North Island, southern Australia (including Tasmania) (Davis and Renner 2003), Heard Island (Speedie 1992), and the Falkland/Malvinas Islands (Napier 1968).

8. POPULATION SIZES AND TRENDS

The extremely isolated breeding locales for this species

FIG. 7 An erect-crested penguin colony on the isolated Antipodes Islands where they breed. (L. S. Davis)

have meant that there is very little census data available (fig. 7). Most publications refer to a figure of 115,000 pairs on the Bounty Islands (179°03′ E, 47°45′ S), derived from a count in 1978 (Robertson and van Tets 1982), and 110,000 pairs for the Antipodes Islands (178°45′ E, 49°41′ S), based on a 1978 survey cited in Marchant and Higgins (1990). However, partial censuses of the Antipodes in 1995 (Tennyson and Taylor 1997) and 1998 (Davis 1999) revealed that dramatically fewer erect-crested penguins were breeding there toward the end of the 20th century. The estimate in Tennyson and Taylor (1997) suggests that there were only 52,081 nests, a reduction of nearly half the breeding population from the Marchant and Higgins (1990) estimate. Three years later, counts at five erect-crested colonies, representing 25% of the Antipodes' population, according to Tennyson and Taylor, showed further reductions of between 8% to 41%, with an overall reduction of 26% (Davis 1999; Davis, unpubl. data).

There appears to be little if any subsequent data available. New Zealand's Department of Conservation suggests in its 2006 publication *Marine Protection for the New Zealand Subantarctic Islands* that the Bounty Islands are now an erect-crested stronghold, although this seemed to be based solely on the reduction in population at the Antipodes Islands noted by Tennyson and Taylor (1997). Strangely, the same report states that the population on the Bounty Islands, determined from a partial survey in 1997, was only 28,000 pairs (Department of Conservation 2006). If that were true, then the collapse of the erect-crested population on the Bounty Islands over a two-decade period from the late 1970s would mirror the one that occurred on the Antipodes Islands over the same period. However, Taylor (2000) points to substantial methodological differences between the 1978 and 1997 counts, although it then appears to accept the accuracy of the latter count when suggesting a worldwide total population of 81,000 ± 4,000 pairs. At the very least, there is enough evidence to suggest extreme concern for the welfare and long-term sustainability of the erect-crested population.

In sum, data on population sizes are very limited and tenuous in terms of their accuracy. Around 1978, it appears that there were about 225,000 pairs breeding on the Bounty and Antipodes Islands (Marchant and

Higgins 1990). By 1997, that figure may have dropped to an estimated 27,956 breeding pairs on the Bounty Islands (Taylor 2000) and 52,081 breeding pairs on the Antipodes Islands (Tennyson and Taylor 1997), giving a total population of just 80,037 pairs, or a 64% reduction in just two decades. Furthermore, the partial census in Davis (1999) reveals an average additional decline of 26% on the Antipodes between 1995 and 1998.

The problem is that the data are so few and the methodologies used in the different censuses are so inconsistent that the information is not wholly reliable. Perhaps the most consistent counts were those documented in Tennyson and Taylor (1997) and Davis (1999), which used the same methodology at the same time in the breeding cycle. While the 1998 counts sampled only a quarter of the population on the Antipodes Islands that were censused in 1995, they revealed reductions in breeding birds at all five colonies counted over that three-year interval. Photographs also reveal obvious contractions in colonies at some sites from 1978 to 1995 (Taylor 2000).

9. MAIN THREATS

Deducing from sketchy evidence that the population of erect-crested penguins has suffered a dramatic decline is one thing, but determining its cause is impossible given the dearth of research on either the penguins or their ecosystem. It is easier to rule out potential threats than it is to apportion blame.

Neither natural nor introduced predators would seem to be problematic for the penguins given the levels of observed predation and the fact that their breeding strongholds are free of the introduced mammals that have wreaked havoc on seabirds on New Zealand's South Island and other islands. (Mice found on the Antipodes Islands are not known to pose any threat to the penguins.) Competition with fisheries for food and the risk of becoming bycatch do not appear to threaten the penguins, in light of the isolation of the waterways around the Bounty and Antipodes Islands and the lack of fishing activity there. The same argument applies to risks from pollution and tourism: the isolation of the breeding areas is their protection. While some telltale signs remain on the Antipodes that penguins were harvested in the late 19th century for their skins, which were used to make ladies' muffs, that trade quickly died out, and apart from the few shipwrecked sailors who were forced to eat penguins until rescued, there has been no human harvesting of adult penguins or their eggs for more than 100 years. There are no resident human settlements on the Bounty or the Antipodes Islands, and, as a consequence, there has been no modification of the penguins' breeding habitat.

That really leaves three main suspects, which are not mutually exclusive: disease, food supply, and climate change. The limited studies conducted to date do not indicate that erect-crested penguins are dying because of disease, and certainly there is no evidence of anything on the scale of an epidemic. Unfortunately, nothing is known of the penguins' diet, where they forage, or whether there have been changes in their food supply. In the absence of any obvious land-based factors, it seems reasonable to suppose that food-related issues at sea may well be at the heart of the problem. Of course, even if that were true, food supply problems may be a symptom and not the ultimate cause. Climate change has an effect on sea surface temperatures, productivity, and ocean circulation.

Eastern rockhopper penguins (*Eudyptes chrysocome filholi*) breeding on Campbell Island declined by 94% from the 1940s to 1985 (Moors 1986; Cunningham and Moors 1994) and continued to decline (Taylor 2000).

FIG. 8 Female covered in mud after being "bullied" and driven off her nest by a nonbreeding bird. (L. S. Davis)

There was a similar 93% decline in the rockhopper population on the Antipodes Islands between 1978 and 1995 (Taylor 2000), coincident with the observed decline in numbers of erect-crested penguins breeding there. Increasing sea temperatures and consequent changes in the distribution of the rockhoppers' prey species have been blamed for their population crash on Campbell (Cunningham and Moors 1994). If so, a similar fate may well have befallen the rockhopper and erect-crested penguins breeding on the Antipodes and Bounty Islands, but research is needed to answer that.

10. RECOMMENDED PRIORITY RESEARCH ACTIONS FOR CONSERVATION

Almost any research on this, the least studied of penguins, might be useful, but the following actions are the most urgent:

1. Collect consistent and appropriate census data for monitoring the population status, which is arguably best done using aerial photography augmented by ground counts.
2. Conduct diet studies ascertaining the prey of these penguins.
3. Undertake foraging studies using the Global Positioning System (GPS) and satellite telemetry in combination with time-depth recorders to measure at-sea behavior and conditions.
4. Collect data on timing and causes of egg and chick mortality to perform survival analyses that can determine causes of breeding failure and critical risk factors and periods.

11. CURRENT CONSERVATION EFFORTS

There are essentially no conservation efforts under way at this time. New Zealand's Department of Conservation has jurisdiction over the Bounty and Antipodes Islands, and although its supposed research priorities mirror those suggested above (Taylor 2000), its approach to conservation seems to be to lock the places up and throw away the key. Landing on the islands is allowed only with a permit from the Department of Conservation, which places severe restrictions on access to the islands even for researchers, under the misguided assumption that keeping humans out and letting nature take its course minimizes risk. Unfortunately, that is patently not the case: the number of erect-crested penguins on Bounty and Antipodes Islands has declined despite an almost complete absence of humans.

12. RECOMMENDED PRIORITY CONSERVATION ACTIONS FOR INCREASING POPULATION RESILIENCE AND MINIMIZING THREATS AND IMPACTS

The top priority is to determine the scale and cause of this species' declining population.

1. Conduct more accurate and frequent censuses to determine population trends.
2. Monitor breeding success and identify causes of breeding failure.
3. Establish the penguins' forage areas and diet.

ACKNOWLEDGMENTS

I wish to thank Alexa Lynn Branesky for her assistance in preparing this chapter. Martin Renner and David Houston were very fine companions while we were conducting research on erect-crested penguins on the Antipodes, even if they did hog all the pillows. That research was supported by a New Zealand Lotteries Grant and a University of Otago Research Grant.

TABLE 9.1 Measurements of adult erect-crested penguins (showing mean, standard deviation and sample size)

ATTRIBUTE	MALES	FEMALES	REFERENCE
Body mass (kg)	5.24±0.3 (136)	5.11±0.3 (121)	Davis et al., unpubl data
Bill length (mm)	58.5±1.94 (44)	52.5±1.88 (44)	Warham (1972)
Bill depth (mm)	26.0±1.2 (44)	22.6±1.2 (44)	Warham (1972)
Flipper length (mm)	212±6.6 (44)	204±4.6 (44)	Warham (1972)

TABLE 9.2 Mean measurements of A-eggs and B-eggs of erect-crested penguins

ATTRIBUTE	A-EGG	B-EGG	REFERENCE
Length (mm)	69.2±3.8 (22)	83.8±3.3 (51)	Davis et al., unpubl data
Breadth (mm)	46.4±1.7 (22)	57.5±2.0 (51)	Davis et al., unpubl data
Mass (g)	81.6±7.6 (20)	150.9±17.9 (51)	Davis et al., unpubl data

REFERENCES

Bailey, A. M., and J. H. Sorensen. 1962. Subantarctic Campbell Island. *Proceedings of the Denver Museum of Natural History*, 2nd ser., 2(10).

Bartle, J. A., and C. D. Paulin. 1986. Bird observations, Auckland Islands, December 1976. In *Preliminary Reports of Expeditions to the Auckland Islands Nature Reserve, 1973–1984*, ed. A. Penniket, A. Garrick, and E. Breese, 51–61. Wellington, New Zealand: Reserves Series, Department of Lands and Survey.

Croxall, J., and L. S. Davis. 1999. Penguins: Paradoxes and patterns. *Marine Ornithology* 27:1–12.

Cunningham, D. M., and P. J. Moors. 1994. The decline of rockhopper penguins *Eudyptes chrysocome* at Campbell Island, Southern Ocean, and the influence of rising sea temperatures. *Emu* 94:27–36.

Davis, L. S. 1999. Report for the Department of Conservation: University of Otago Antipodes Island Expedition, September–November 1998. Report sent to Department of Conservation, Invercargill, 22 February.

———. 2001. A superlative penguin. *Natural History* 1(11):46–55.

Davis, L. S., and M. Renner. 2003. *Penguins*. London: T & A D Poyser.

Department of Conservation. 2006. Marine protection for the New Zealand Subantarctic Islands. Wellington, New Zealand: Department of Conservation. 48 pp.

Hitchmough, R., L. Bull, and P. Cromarty. 2007. New Zealand threat classification system lists, 2005. Science and Technical Publishing, Department of Conservation, Wellington, New Zealand.

IUCN (International Union for Conservation of Nature). 2011. IUCN Red List of Threatened Species. Version 2011.2. www.iucnredlist.org (accessed 22 March 2012).

Marchant, S., and P. J. Higgins. 1990. *Handbook of Australian, New Zealand and Antarctic Birds*. Vol. 1A. Melbourne: Oxford University Press.

Miskelly, C. M., J. E. Dowding, G. P. Elliot, R. A. Hitchmough, R. G. Powlesland, H. A. Robertson, P. M. Sagar, R. P. Scofield, and G. A. Taylor. 2008. Conservation status of New Zealand birds, 2008. *Notornis* 55:117–35.

Moors, P. J. 1986. Decline in numbers of rockhopper penguins at Campbell Island. *Polar Record* 23:69–73.

Napier, R. B. 1968. Erect-crested and rockhopper penguins interbreeding in the Falkland Islands. *British Antarctic Survey Bulletin* 16:71–72.

Richdale, L. E. 1941. The erect-crested penguin (*Eudyptes sclateri*) Buller. *Emu* 41:25–53.

Robertson, C. J. R., and G. F. van Tets. 1982. The status of birds at the Bounty Islands. *Notornis* 29:311–36.

Speedie, C. 1992. An erect-crested penguin in the southern Indian Ocean. *Notornis* 39:58–60.

Taylor, G. 2000. *Action Plan for Seabird Conservation in New Zealand*. Part A: *Threatened Seabirds*. Threatened Species Occasional Publication No. 16. Biodiversity Recovery Unit, Department of Conservation, Wellington.

Tennyson, A., and G. Taylor. 1997. Final report on penguins: Antipodes Island expedition, 30 October–26 November 1995. Manuscript, Department of Conservation, Wellington, 1 July.

Warham, J. 1972. Aspects of the biology of the erect-crested penguin, *Eudyptes sclateri*. *Ardea* 60:145–84.

Williams, T. D. 1995. *The Penguins:* Spheniscidae. Oxford: Oxford University Press.

10

Fiordland Penguin

(Eudyptes pachyrhynchus)

THOMAS MATTERN

1. SPECIES (COMMON AND SCIENTIFIC NAMES)

Fiordland penguin, *Eudyptes pachyrhynchus*

The Fiordland penguin is also known as Fiordland crested penguin, *tawaki* (Maori), New Zealand penguin, New Zealand crested penguin, and thick-billed penguin.

2. DESCRIPTION OF THE SPECIES

The Fiordland is a medium-size crested penguin with a body length of about 55 centimeters. The head, throat, and upperparts are blue black and turn dark brown near molt. The breast and abdomen are silky white. Feet and legs are pinkish white above and blackish brown behind the tarsi, soles, and at the front of webs (Marchant and Higgins 1990). The head features a sulfur-yellow crest that starts near the nostrils, extends back horizontally, and droops a little behind the eyes (Oliver 1953). The crest feathers are considerably shorter than those of the rockhopper (*E. chrysocome*) and, to a lesser extent, of the Snares (*E. robustus*) (Marchant and Higgins 1990). The eyes are brownish red, with some variance from brownish gray to orange red (Warham 1974). The bill, moderately large and heavy and of orange-brown color, is separated from the feathers by a thin strip of black skin (Marchant and Higgins 1990). Feathers in the cheek region feature whitish bases that are often displayed as three to six horizontal white stripes (Warham 1974).

Fiordland penguins are sexually dimorphic, with males larger than females (fig. 1). Sex can be determined by bill depth, which measures less than 24 millimeters in females and more than 24 millimeters in males. Morphometric data are provided in table 10.1.

The Fiordland is similar in appearance to the Snares, but some distinct features aid in identifying the species. The Snares's bill is fringed by an obvious pinkish-white band of skin, similar to that of the erect-crested penguin and which is absent in the Fiordland (Oliver 1953). The bill itself is more robust in Snares penguins. The bulbous culminicorn (i.e., the top-central plate of the upper mandible) is less pronounced in the Fiordland compared to that of the Snares penguin (Oliver 1953; Warham 1974; Marchant and Higgins 1990; Williams 1995). The Fiordland's lateral crest plumes are shorter (less than 5 cm) than those of the Snares (more than 5 cm) (Marchant and Higgins 1990). Furthermore, the Fiordland's crest is broader and narrows abruptly at the nostril, whereas the Snares's crest is thinner and narrows gradually toward the nostril (Warham 1974).

3. TAXONOMIC STATUS

The Fiordland penguin was described by Gray in 1845 (Buller 1905) and was considered a subspecies related to erect-crested and Snares penguins (Falla 1935). This classification was disputed in Oliver (1953), which dis-

FIG. 1 Male and female Fiordland penguin resting onshore at Stewart Island. (T. Mattern)

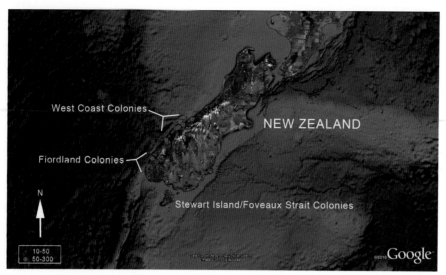

FIG. 2 Distribution and abundance of the Fiordland penguin, with counts based on pairs.

tinguished all three as separate species by pointing out morphological differences with a special emphasis on crest and beak shapes. While the separation from erect-crested penguins has been widely accepted, there is ongoing dispute as to whether Fiordland and Snares penguins should be considered a single species (Kinsky 1970; Christidis and Boles 2008) or remain classified as separate species (Stonehouse 1971; Falla et al. 1974; Warham 1974). Considering substantial differences in the life-history traits of Fiordlands and Snares, most notably breeding chronology that makes interbreeding unlikely, combining both into a single species is probably unjustified (see chapter 11 for a more detailed discussion of the subject).

4. RANGE AND DISTRIBUTION

The Fiordland breeding range is confined to the southwest coast of New Zealand's South Island as well as Solander, Codfish, and Stewart Islands (fig. 2; table 10.2). Historically, the distribution of the species might have been more widespread. Fossil records show that Fiordland penguins were common on northern South Island and probably even bred along the southern coastlines of New Zealand's North Island (Holdaway et al. 2001). A reported breeding attempt by a single pair of Fiordlands on the North Island (Falla 1954) may provide some support to this hypothesis.

Judging from nest densities determined during extensive surveys in the 1990s, it would seem that Fiordlands prefer to breed in areas where the marine habitat features a wider continental shelf, that is, the southern West Coast down to Yates Point on the South Island as well as the Foveaux Strait area (McLean and Russ 1991; Russ et al. 1992; McLean et al. 1993, 1997; Studholme et al. 1994). However, the question is whether low nest densities reported for central Fiordland might result from the vast expanse and inaccessible coastline within the fjord systems, making this area a logistical challenge for terrestrial surveys. Accordingly, published accounts suggest that the penguins prefer to breed close to the open ocean; however, some colonies have been found on islands well within the fjord systems. Rolla Island in Doubtful Sound (45.44° S, 167.13° E), for instance, is some 30 kilometers away from the open sea. Again, it is difficult to assess whether such "inland colonies" are exceptions or more common occurrences, but it is possible that there are more Fiordland penguins in Fiordland than the results of the surveys suggest. It is certain that the numbers reported for Stewart Island (Studholme et al. 1994) are an underestimate. The survey was limited principally to the southern ranges of the Stewart Island coast. Yet considerable numbers are known to breed along the northwestern and northern coasts, mostly in inaccessible sea caves (Mattern, pers. obs.; Phred Dobbins, Department of Conservation, Stewart Island, pers. comm.).

The distribution of Fiordland penguins outside the breeding period is largely unknown. While most breeders molt at or near their nest sites, juveniles and nonbreeders are known to molt on the Snares Islands (Warham 1974), along the Southland and Otago coastlines of New Zealand's South Island, as well as Tasmania and Victoria in Australia (Marchant and Higgins 1990). Generally it seems that Fiordlands tend to remain in waters north of the Subtropical Convergence within the Tasman Sea and southeastern Indian Ocean. Yet, there also are a few recorded sightings on sub-Antarctic islands such as Campbell Island and the Auckland Islands (Marchant and Higgins 1990; Reid et al. 1999).

5. SUMMARY OF POPULATION TRENDS

The current population is estimated at about 2,500–3,000 breeding pairs (McLean et al. 1997), although this likely represents an underestimate of the true population size (see section 8). The population is believed to have declined considerably in the 20th century (Taylor 2000), with ongoing declines being detected at some surveyed locations, but stable or seemingly increasing numbers at other sites (e.g., Eason 1994; Taylor 2000; Newton and Tansell 2005). Overall, population size, status, and trends are unclear.

6. IUCN STATUS

The International Union for Conservation of Nature (IUCN) currently ranks the Fiordland penguin as Vulnerable on its Red List of Threatened Species (IUCN 2011). This ranking is based on the suspected ongoing population decline (assumed to be ≥ 30% over the course of 10 years. No subpopulation is believed to contain more than 1,000 mature individuals (BirdLife International 2010). The Fiordland penguin was listed as Threatened under the U.S. Endangered Species Act (USFWS 2010).

Within New Zealand, the Fiordland is ranked Threatened—Nationally Vulnerable owing to low population size (1,000–5,000 mature individuals) and suspected ongoing population decline (50–70%) (Miskelly et al. 2008). Fiordland penguins share this assessment with white-flippered (*Eudyptula minor albosignata*) and yellow-eyed penguins (*Megadyptes antipodes*) and are ranked on the second-highest threat level of the New Zealand penguin species, being superseded only by the eastern rockhopper (*Eudyptes chrysocome filholi*), which is ranked Threatened—Nationally Critical. The remaining New Zealand species have been assigned lower threat levels ranging from At Risk—Declining (little penguin [*Eudyptula minor*]) to At Risk—Naturally Uncommon (Snares penguin [*E. Robustus*], erect-crested penguin [*E. sclateri*], and the Chatham Island subspecies of the little penguin [*Eudyptula minor chathamensis*]).

Fiordlands are an indigenous *taonga* species, one that is of significant cultural and spiritual value to New Zealand Maori, as established in the Ngāi Tahu Claims Settlement Act 1998. The species is fully protected under the Wildlife Act 1953, and any interactions (e.g., research) require permission from New Zealand's Department of Conservation (DOC); valid ethics approval is mandatory. While most of these penguins breed at fairly inaccessible locations on the South Island, some colonies in South Westland are exposed to increasing nature tourism (Taylor 2000). Commercial tourism is heavily regulated under the Wildlife Act 1953 through the DOC. However, New Zealand legislation grants public access to many areas of protected land used by Fiordland penguins (e.g., Munro Beach, Jackson Head), facilitating uncontrolled access that might detrimentally affect penguins. The Fiordlands' marine habitat is largely unprotected, so that commercial as well as recreational fisheries might have an impact on the species. So far no information on accidental bycatch has emerged. There is a ban on setnet fisheries offshore to four nautical miles, covering a small portion of the species breeding range—that is, the south-west coast to Awarua Point in Fiordland and from Sand Hill Point (East Fiordland) eastward along the South Island's southern coast (Ministry of Fisheries 2008—yet most of the Fiordland breeding range is exempt from the setnet ban. There are also plans to establish a 45-hectare mussel farm in the Jackson Bay area (Ministry of Fisheries 2006), which may have detrimental effects on the local Fiordland population (Lloyd 2003). In early 2009, local Maori groups applied to the Minister of Fisheries to establish eleven *mātaitai* reserves (protected areas of traditional importance to Maori for customary food gathering) in areas along the west coast that also encompass some important breeding sites, including Jackson Head (Ministry of Fisheries 2009).

7. NATURAL HISTORY

BREEDING BIOLOGY. Fiordlands are endemic to the southern west coast of New Zealand's South Island, Fiordland, and the Foveaux Strait area (i.e., Solander, Codfish, and Stewart Islands). Their nesting habitat is quite diverse, ranging from temperate rain forest (e.g., Codfish Island) and dense shrub (e.g., Breaksea Island) (fig. 3) to sea caves (e.g., Stewart Island) and rocky shorelines (e.g., Martin's Bay). Nests are often situated in hollows at the bases or underneath roots of trees, beneath boulders,

FIG. 3 Fiordland penguin on its nest (lower left corner) in typical breeding habitat on Breaksea Island, New Zealand. (T. Mattern)

or in rock crevices. Warham (1974) also describes a nest located on the horizontal trunk of a tree about 10 meters above the ground. Nests are sparsely lined with fern fronds, grass, twigs, and stones (Warham 1974; Davis and Renner 2003). Fiordlands breed in loose colonies with nests usually established several meters from one another (Davis and Renner 2003). Because of this, colonies are poorly delineated, and at some locations, nests are dispersed over considerable stretches of coastline (e.g., McLean et al. 1997).

The annual reproductive cycle is well defined (table 10.4). Birds arrive at their breeding colonies in the austral winter from mid-June onward, and the majority of nests are established by mid-July (Warham 1974) (table 10.5). About three-fourths of the penguins return to their previous nest sites, and about two-thirds eventually reunite with mates from previous seasons. Late-returning females that find their previous year's partners paired with other birds are known to expel the "intrusive" females in fights that are severe. Overall, the species is believed to be socially as well as genetically monogamous, that is, neither extra-pair fertilization nor intraspecific brood parasitism (in which eggs or chicks from another pair are raised) is common (McLean 2000). Reunited pairs tend to exhibit higher reproductive success than new pairs. Pairs that fail to raise a chick are more likely to separate the following season. The initiative for separation is more likely to come from the female (St. Clair et al. 1999).

Egg laying extends over a 10-day period around late July and early August. Therefore, breeding commences in austral midwinter, about six to eight weeks earlier

than the nearest relative species, the Snares penguin, resulting in the reproductive isolation of both species. The midwinter onset of breeding is facilitated by the fact that Fiordlands breed in a reasonably warm, continental region, at least compared to other crested penguin species. This geographic setting also provides Fiordlands with a coastal marine habitat with increased oceanic productivity and an early onset of the phytoplankton spring bloom (Murphy et al. 2001).

Eggs are laid 3–6 days apart (mean 4.3 ± 0.71 days, n = 46 [St. Clair 1992]) with the first-laid egg (A-egg) being smaller than the second-laid egg (B-egg) (table 10.1). Both Warham (1974) and St. Clair (1992) note the occasional reversal of this pattern, that is, the A-egg being larger than the B-egg. Incubation does not normally commence until the B-egg has been laid, a crucial behavioral trait that facilitates brood reduction (St. Clair 1992). Once incubation is under way, pairs tend to stay together at the nest, sharing incubation duties for 5–10 days until the female leaves for a two-week foraging trip (unlike all other crested penguin species, in which the male leaves first [Warham 1975]). By the time the males are relieved by their partners and depart to forage, they have fasted for 40–45 days and will have lost about one-fourth of their initial body mass (Warham 1974). The males tend to stay at sea for up to two weeks and usually return before the eggs hatch, providing their partners with the opportunity to take short-term foraging trips until the eggs have hatched.

Fiordlands are often described as obligate brood reducers, raising only one chick despite laying two eggs. However, there are records of Fiordland penguins raising both chicks, and it has been suggested that in years of good food supply, up to 12% of Fiordland pairs may raise both chicks to fledging (McLean 2000). In about 25% of nests, at least one egg is lost during the incubation period (Marchant and Higgins 1990). The smaller A-eggs compose the majority of eggs lost (St. Clair 1992), presumably as a result of accidental expulsion from the nest (Warham 1974). Egg loss commonly occurs before or on the day the B-egg is being laid but may occur throughout the incubation period. Pairs that lose both eggs do not re-lay (Warham 1974). The reasons for the frequent loss of the first egg remain unclear. St. Clair (1992) states that increased risk of predation by the weka (*Gallirallus australis*), the native New Zealand rail, in the early stages of incubation is the most likely explanation. However, this explanation applies only at sites where egg predators are present. The fact that Fiordland penguins tend to stand on or off the nest after the A-egg has been laid suggests that movement over the incomplete clutch facilitates accidental egg expulsion (Davis and Renner 2003).

Hatching occurs throughout September, some 31–36 days after the B-egg is laid (Warham 1974; St. Clair 1992). In the majority of nests (50–60%), only one egg hatches (Marchant and Higgins 1990; St. Clair 1992). In the rest of the nests, the larger B-eggs tend to hatch slightly earlier than the A-eggs (St. Clair 1992). The smaller chick hatched from the A-egg usually dies within a few days from starvation (Warham 1974) or because of accidental expulsion (McLean 1990).

Following incubation, the males guard the chicks for up to three weeks. During this period, females perform short foraging trips and return to their nest sites almost every night to feed the chicks; the fasting males do not contribute food during the guard stage (Warham 1974). When chicks are about three weeks of age, they are no longer guarded by the males and move away from the nest to form crèches with chicks from nearby nests. Crèche formation coincides with the departure of the males to forage. As the chicks get larger, in mid-October, crèches start to break up. While chicks are still fed primarily by females throughout the crèching period, males increase their food contribution in the later stages of post-guard. Chicks fledge around mid- to late November, about 75 days after hatching. Table 10.4 summarizes the breeding parameters reported for the Fiordland. Adults depart for their pre-molt foraging trips shortly after chicks have fledged, and by early December, the majority of adults have left their breeding colonies (Warham 1974).

MOLT. After completing their breeding season, adult penguins stay at sea for 60–80 days, fattening up in preparation for the annual molt. Successful breeders return to their colonies in late January and early February (Warham 1974). At this stage, the birds are nearly two kilograms heavier than at the end of breeding. During the first week of the molt, the plumage takes on a brownish cast and crests fade. Shedding of the old feathers takes about two weeks, after which another week or more is required for new body feathers to lengthen. Throughout the molt, the penguins' body mass is reduced by just about half. The birds start to leave their colonies around late February, and the number of birds in colonies dwindles rapidly throughout March (Warham 1974). Yearlings and nonbreeders tend to molt about a month earlier than breeding adults.

FORAGING ECOLOGY AND PREY. Very little is known about the Fiordland penguin's marine ecology. It seems that the prey composition of Fiordlands varies considerably between locations, presumably reflecting differences in their marine habitat. A diet study of their breeding on the west coast of the South Island found that the birds took primarily cephalopods (85%, chiefly arrow squid [*Nototodarus sloani*]), followed by crustaceans (13%, mainly krill [*Nyctiphanes australis*]), and fish (2%). The preference for arrow squid seems to suggest that they forage over the continental shelf, which, at the locations sampled (Martins Bay), extends no farther than 10 kilometers from the coast (van Heezik 1989). A study on Codfish Island showed a considerably different diet composition. Here, Fiordlands took primarily larval or juvenile stages of fish (85%, mainly ahuru [*Auchenoceros punctuates*] and red cod [*Pseudophycis bacchus*]), while the rest of the diet was made up of cephalopods (van Heezik 1990). Van Heezik (1990) also notes a considerable overlap of prey species taken by little penguins that are known to forage pelagically in inshore waters (Mattern et al. 2001; Chiaradia et al. 2007). Around northern Stewart Island, Fiordland penguins have been observed foraging very close to the coast (less than 2 km) during the later stages of the breeding season (November). The birds were found in large numbers (at least 30 individuals) in areas where tidal currents aggregate macroplanktonic biomass (Gull Rock, northeast Stewart Island) or in pairs, patrolling along the coastline (Mattern, unpubl. data). Along the Fiordland coast, anecdotal reports from tourism operators support the notion that breeding penguins forage close inshore or, in that case, within the fjord systems rather than farther offshore. So while Fiordlands are generally considered "offshore foragers" (Croxall and Davis 1999), it appears that the birds might forage inshore at least during the breeding period (fig. 4). Unless their foraging ecology is studied in greater detail, however, an important aspect of the species' biology remains a matter of conjecture.

PREDATORS. There is little information about marine predators but it is likely that Fiordlands are taken by

FIG. 4 Fiordland penguin at sea near Stewart Island, New Zealand. (T. Mattern)

predatory fishes such as sharks that are known to prey on other New Zealand seabird species (Hocken 2000, 2005). New Zealand fur seals (*Arctocephalus forsteri*) also inhabit the coastlines used for penguin breeding and occasionally prey on the birds (Spellerberg 1975). However, Warham (1974) states that pinnipeds are unlikely to play an important role as predators; thus, it seems that the main predatory threats are in the terrestrial environment. Introduced mammals most likely present the greatest threat to the species. Stoats (*Mustela ermine*) are known to prey on chicks and eggs on New Zealand's South Island (Alterio and Moller 1997). Stoats have even been reported to have attacked adult penguins (Warham 1974; Morrison 1980). Rats can prey on chicks and eggs, although there is no direct evidence for this (Taylor 2000). Apart from introduced mammalian predators, the weka is known to prey on Fiordland chicks and eggs (Warham 1974; St. Clair and St. Clair 1992). Historically, Fiordlands were taken for food by humans, as evidenced by remains in middens of early Polynesian settlers (Boessenkool et al. 2009).

ANNUAL CYCLE. See table 10.3 and figure 5.

8. POPULATION SIZES AND TRENDS

Information on the Fiordland's population size and trends is sketchy at best. Between 1990 and 1995, researchers attempted to conduct a series of comprehensive surveys across the entire Fiordland range (McLean and Russ 1991; Russ et al. 1992; McLean et al. 1993, 1997; Studholme et al. 1994) by means of ship-based searches of the coastlines for signs of penguin occupation (birds on shore, scat marks, etc.). Potential sites were subsequently examined on foot for short periods (30–90 minutes) to locate and count nests (McLean and Russ 1991). A total of 2,260 nests were found, but it was assumed that this figure represented an undercount by at least 20% (how the authors determined this proportion is unclear). Accordingly, the total population was estimated to range between 2,500 and 3,000 nests annually. McLean et al. (1997) stress that numbers can only be considered a minimum, as it was logistically impossible to search every stretch of Fiordland coastline. Seeing the penguins once they reach dense vegetation is dif-

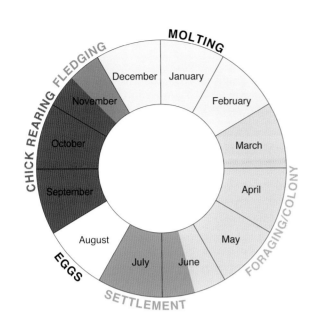

FIG. 5 Annual cycle of the Fiordland penguin.

ficult (fig. 6). The accuracy of the counts could not be validated, and it is quite possible that the inexperience of survey participants further contributed to undercounts. Nest searches performed in subsequent years support this notion. A nest count conducted by New Zealand's Department of Conservation on Shelter Island found considerably more nests, a total of 120 (Eason 1994), than the total of 50 nests reported in McLean and Russ (1991). Stewart Island numbers presented in Studholme et al. (1994) are underestimates, as mentioned. Considering this, the final figures provided in McLean et al. (1997) and reproduced elsewhere (e.g., Taylor 2000; Hitchmough et al. 2007; BirdLife International 2008) represent a significant underestimate of the Fiordland's population size.

FIG. 6 Fiordland penguin crossing rocks to reach dense vegetation on Breaksea Island, New Zealand. (T. Mattern)

The species' population trends and status are also unclear. There are reports of large colonies in Fiordland in the late 19th century (Hill and Hill 1987), suggesting that some substantial declines may have occurred throughout the 20th century (Taylor 2000). An often cited reference (e.g., Taylor 2000; BirdLife International 2008) is St. Clair (1998), which states that the penguin population on Open Bay Islands declined by about 33% between 1988 and 1995, but no absolute figures or supporting data are given. Data from other areas also found substantial declines, but nest numbers elsewhere remained reasonably stable or even increased (see table 10.5). Therefore, it seems unlikely that variations in nest numbers at some sites are unequivocally indicative of changes in the total Fiordland population. Search method and effort, observer experience and determination, and the timing of the surveys (e.g., Eason 1994) all likely affected the outcome. Monitoring programs initiated in the early 1990s by the DOC detected declining numbers at some of the regularly checked sites, but these declines could result from penguins moving out of the marked monitoring plots (McClellan 2009). The timidity of this species (Warham 1974) makes it particularly susceptible to disturbance associated with monitoring efforts, potentially contributing to the displacement of penguins. The extent to which observed declines might be an artifact of monitoring practices remains to be ascertained (McClellan 2009).

9. MAIN THREATS

CLIMATE VARIATION. Crested penguin populations in New Zealand's sub-Antarctic region have declined significantly in the past century, and rising ocean temperatures and subsequent changes in marine productivity are a suspected cause (Cunningham and Moors 1994; Hilton et al. 2006). However, it is unlikely that Fiordlands share this fate. They forage in a similar if not the same marine habitat as the Snares penguin, whose population has remained stable throughout the 20th century. If changes in the marine environment have affected Fiordland penguins, they would have also affected Snares. This indicates that ocean warming so far has had no significant effect on Fiordlands (Mattern et al. 2009). As is suspected with Snares, El Niño conditions might have a negative impact on Fiordlands' breeding success, but without more detailed knowledge of their marine ecology, it is difficult to assess how the climate pattern affects their foraging success and efforts.

FISHERIES. Fishing activities for species eaten by Fiordlands (i.e., arrow squid, red cod) are low along the Fiordland coastline (Ministry of Fisheries 2010), suggesting that the risk of significant interactions between trawl or jig fisheries and penguins may be relatively low during the breeding season. Setnetting is still permitted in most areas of Fiordland and poses a significant risk to penguins (Darby and Dawson 2000). Furthermore, the migratory habit of Fiordlands outside the breeding period might bring the species into areas of higher fishing efforts, for example, in the productive waters of the Southland and Otago shelf regions (Ministry of Fisheries 2010), increasing the bycatch risk and the potential for competition with commercial fisheries.

INTRODUCED PREDATORS. Introduced terrestrial predators, especially mustelids, are probably an important, if not the most important, factor determining population change in Fiordland penguins (Mattern et al. 2009). In the 19th and 20th centuries, stoats, known to prey on Fiordlands (Taylor 2000), had a significant impact on the populations of ground-breeding birds such as kakapo (*Strigops habroptila*), kiwi (*Apteryx* spp.), and weka throughout New Zealand (Wilson 2004). Rats have also been an important player in the demise of native bird species in Fiordland (Wilson 2004), but so far there is no evidence of rat predation on Fiordlands (Taylor 2000). Dogs and cats may prey on Fiordlands near human habitations, although this threat is relevant only for birds molting outside the main breeding range, which is largely uninhabited by humans (Marchant and Higgins 1990).

The introduction of the weka to some islands has been suggested to be detrimental to penguins living there (St. Clair and St. Clair 1992; Taylor 2000; McClellan 2009; USFWS 2010). However, it seems as if the concern expressed in some of these sources is somewhat overrated. Weka were released by sealers at some locations, such as Open Bay and Solander Islands, as a food source (Taylor 2000; Wilson 2004) and would have had the greatest impact on the local penguin populations in the decades following their release, that is, in the late 19th and early 20th centuries (King 1984). Furthermore, the natural ranges of Fiordland penguins and weka overlap, and both species have coexisted for millennia (Heather and Robertson 2000). Fiordlands are also known to aggressively defend their offspring against weka (McLean 1990). The impact of weka could be severe only if mammalian predators or human disturbance greatly facilitate their predation on penguin offspring.

Fiordlands are known to be very sensitive to human presence (Warham 1974; St. Clair and St. Clair 1992), which might be partly a result of early Polynesian settlers taking penguins for food (Ellenberg et al. 2006). While predatory activity on penguins may have ceased, humans can still have a negative impact through disturbance at breeding sites (e.g., unregulated tourism, research or monitoring) (Taylor 2000; McClellan 2009).

ENVIRONMENTAL DISASTERS. The Fiordland landscape is a complex jumble of peaks, ridges, and valleys with steep slopes. Earthquakes are common and sometimes cause catastrophic landslides (Craw and Norris 2003), which could wipe out an entire Fiordland colony in one blow. Such events, however, are unlikely to impact the population as a whole. Human activities within the breeding range may become more of a problem. Opening conservation land, including parts of the West Coast, to mining operations, as recently considered by the New Zealand government (Duthie 2010), could increase the risk for marine pollution through contamination of waterways and coastal waters (Osborn and Datta 2006). There are also plans to permit oil exploitation of the marine Waitutu and Hautere sub-basins, which are located in the western Foveaux Strait just south of Fiordland. This area encloses Solander Island and most likely covers significant Fiordland foraging grounds. If plans were to proceed, the penguins would be exposed to greater risks of oil pollution and spills.

10. RECOMMENDED PRIORITY RESEARCH ACTIONS FOR CONSERVATION

In light of prevailing uncertainty about the Fiordland's population status despite almost 20 years of monitoring (McClellan 2009), the following are the most important research actions for conservation:

1. Develop a reliable population assessment methodology. The unreliability of survey outcomes is compounded by the inaccessibility of nesting locations (e.g., narrow caves, impenetrable vegetation) (Warham 1974), differences in observer experience and search effort, and monitoring conducted

only within marked boundaries (McClellan 2009). Alternative methods suggested by McClellan (2009) include molting surveys and beach counts. However, surveys face problems similar to those associated with other terrestrial survey methods used in the past. Beach counts resolve the problem of the inaccessibility of nesting and molting sites, but displacement of breeding pairs in monitored areas is a concern. Another method would be to combine terrestrial surveys with sea-based surveys, that is, penguin counts at sea during the breeding period, as has been done for Humboldt penguins (Luna-Jorquera et al. 2000).

2. Examine basic life-history parameters, including adult survival rates, longevity, juvenile survival and recruitment, age of first breeding, nest-site fidelity, and natal philopatry to develop a better understanding of the population dynamics.

3. Study predator and researcher impacts on the population. For example, how do mustelids affect existing populations? Do native weka have significant impact on some local populations? Are there any environmental factors that facilitate the spread of disease or parasites (e.g., the blood parasite *Leucocytozoon tawaki* spread by native sandflies *Austrosimulium spp.* [Fallis et al. 1976])? To what extent have observers and researchers caused short- and long-term detrimental effects on the monitored populations?

4. Examine the potential for interactions between Fiordland penguins and fisheries (especially with setnetting in breeding regions). Detailed studies about the Fiordlands' dispersal and foraging ecology outside the breeding season could provide insights into potential conflicts with fisheries.

5. Study the foraging ecology of penguins breeding along the west coast of the South Island and within Fiordland and the Foveaux region. Our understanding of Fiordland penguin marine ecology is very limited and the marine habitats they use are diverse.

6. To clarify the species taxonomy, it may be worthwhile to determine whether Snares and Fiordland interbreed.

11. CURRENT CONSERVATION EFFORTS

In the early 1990s, New Zealand's Department of Conservation started monitoring some sites within the Fiordland's breeding range. Although logistical difficulties have undermined the primary goal of reliably assessing population changes, the monitoring provided information that will help determine best conservation measures.

The Fiordland penguin benefited from conservation efforts that did not target the species. On Codfish Island, all terrestrial predators (i.e., possums, rats, and weka) were eradicated, and strict quarantine measures were implemented in an effort to create a safe haven for the critically endangered kakapo (McClelland 2002), and the local population of Fiordland penguins benefited from the removal of potential predators. More recently, stoats were eradicated from Resolution Island, the third-largest island on the Fiordland coast (McMurtrie et al. 2008). While it is unclear if and how many Fiordlands breed on this island, the pest eradication program might aid Fiordland penguins in (re-)populating this location.

Along the west coast, some breeding sites are exposed to largely unregulated tourism. The DOC issued permits to two tourism companies for viewing Fiordland penguins on Murphy Beach. Unfortunately, the permit conditions were breached, and as the department lacks the resources to adequately police concessionaires, additional concessions are not recommended (Bull 2004).

12. RECOMMENDED PRIORITY CONSERVATION ACTIONS FOR INCREASING POPULATION RESILIENCE AND MINIMIZING THREATS AND IMPACTS

The most important four conservation actions are the following:

1. Examine the population status and trends.
2. Collect detailed information on the species biology, including its marine ecology, breeding biology, and susceptibility to human disturbance (e.g., research and tourism). This information is essential for effective conservation plans. Such a research effort would require 5–10 years.
3. Eradicate or at least control introduced predators within the species' breeding range.
4. Provide the relevant DOC offices with adequate resources for monitoring and effectively regulating tourism operations and tourist activities at publicly accessible Fiordland breeding sites (e.g., Murphy Beach, Jackson Head).

ACKNOWLEDGMENTS

The published data on Fiordland penguins is sparse. It was my good fortune that Dave Houston, Department of Conservation, provided me with access to many unpublished DOC reports on Fiordland penguins posted in a restricted area of his informative wildlife management web forum (http://www.wildlifemanagement.net.nz). Without Dave's help, this chapter would have been half as long as it is and probably not more than a repetition of information originally published by John Warham in the mid-1970s. Furthermore, I would like to thank Helen Otley (DOC, West Coast), Jo Hiscock (DOC, Southland), and an anonymous reviewer for valuable comments on early drafts of this chapter. Other helpful hints and suggestions came from Jane Tansell (DOC, Te Aanau). Further thanks are due to the Global Penguin Society for inviting me to attend the 7th International Penguin Conference and present a summary of this chapter.

TABLE 10.1 Fiordland penguin measurements (showing mean, standard deviation, range and sample size)

VARIABLE	MALES	FEMALES	JUVENILES	FLEDGINGS	EGGS	REFERENCES
Mass (g)						
pre-breeding	4530±373 (20)	4026±397 (17)				Warham 1974
	4160±360 (18)	3680±400 (12)				Murie et al. 1991
incubation (Aug)	3556±275 (9)	3208±226 (6)				Warham 1974
chick-guard (Sep)	2882±367 (11)	2475 (2)				Warham 1974
post-guard (Oct)	3096±354 (8)	2762 (4)				Warham 1974
end of brooding (Dec)	3285 (5)	3199 (4)				Warham 1974
pre-molt	4936±350 (11)	4820±356 (5)	3766±460 (5)			Warham 1974
post-molt	3004±289 (13)	2521±227 (14)	2177±534 (6)			Warham 1974
fledgling				2307±340 (17)		Warham 1974
Flipper length (mm)	185.7±3.7 (87)	178.5±5.8 (55)				Warham 1974
Bill length (mm)	51.1±2.0 (94)	45.0±1.7 (61)				Warham 1974
	51.7±1.2 (18)	46.0±2.2 (12)				Murie et al. 1991
Bill depth (mm)	26.1±1.7 (94)	21.8±1.3 (61)				Warham 1974
	27.6±1.2 (18)	22.9±1.0 (12)				Murie et al. 1991
Tarsus-toe length (mm)	124.0±5.3 (32)	116.5±4.5 (19)				Warham 1974
	116.9±4.5 (18)	109.7±2.9 (12)				Murie et al. 1991
Egg length					A-egg	
					68.0±2.6 (134)	Warham 1974
					68.2±1.3 (8)	Grau 1982
					67.3±2.8 (54)	St. Clair 1992
					B-egg	
					71.2±2.3 (121)	Warham 1974
					73.0±3.2 (7)	Grau 1982
					70.9±2.5 (54)	St. Clair 1992
Egg breadth					A-egg	
					51.9±1.9 (134)	Warham 1974
					50.8±1.5 (8)	Grau 1982
					51.1±2.1 (54)	St. Clair 1992
					B-egg	
					55.0±1.8 (121)	Warham 1974
					54.5±1.5 (7)	Grau 1982
					54.2±1.6 (54)	St. Clair 1992
Egg mass (g)					A-egg	
					99.9±7.8 (66)	Warham 1974
					96.8±7.7 (8)	Grau 1982
					98.9±9.1 (54)	St. Clair 1992
					B-egg	
					120.3±8.6 (52)	Warham 1974
					120.3±9.4 (7)	Grau 1982
					116.6±9.9 (54)	St. Clair 1992

TABLE 10.2 Range and distribution of Fiordland penguins

	LATITUDE	LONGITUDE	DATE	# NESTS	REFERENCE
West Coast					
Haretaniwha Point	-43.586381	169.556354	2003	47	Newton 2003
Paringa River	-43.632528	169.433575	1993	133	McLean et al. 1997
Monro Beach	-43.704918	169.249931	2008	10	McClellan 2009
Murphy Beach	-43.712315	169.232582	2008	19	McClellan 2009
Knights Point	-43.719399	169.210084	1993	11	McLean et al. 1997
Open Bay Island	-43.861936	168.879518	1990	120 150	McLean and Russ 1991
Jacksons Head	-43.968484	168.609450	2005	167	Newton 2005
Stafford River-Teer Creek	-44.001575	168.516178	2002	116	McLean et al. 1993
Teer Creek-Cascade Pt	-44.008693	168.434935	1994	260	McLean et al. 1997
Hope River-Gorge River	-44.129444	168.261822	1995	73	McLean et al. 1997
Gorge River	-44.185977	168.190358	1995	42	McLean et al. 1997
Fiordland Coast					
Martins Bay	-44.340804	168.001100	2008	47	McClellan 2009
Yates Point	-44.497042	167.817430	1994	177	Eason 1994
Harrison Cove	-44.624970	167.907776	1994	4	Eason 1994
Anita Bay	-44.586614	167.785914	1990	5	McLean and Russ 1991
Unnamed Island / George Sound	-44.986751	167.384259	1990	4	McLean and Russ 1991
Eleanor Island	-45.098330	167.141576	1990	2	McLean and Russ 1991
Fanny Island	-45.128946	167.137042	1990	7	McLean and Russ 1991
Catherine Island	-45.132681	167.138988	1990	1+	McLean and Russ 1991
Nancy Sound	-45.102999	167.031219	1990	3+	McLean and Russ 1991
Shelter Islands	-45.270974	166.893550	1994	36–120	McClellan 2009; Eason 1994
Seymour Island	-45.307213	167.006392	1992	9	McLean et al. 1993
Rolla Island	-45.440234	167.130738	1992	8	McLean et al. 1993
Breaksea Island	-45.580071	166.638562	1992	185	McLean et al. 1993
Johns Island	-45.569966	166.792163	1992	9	McLean et al. 1993
Hawea Island	-45.590874	166.644245	1992	8	McLean et al. 1993
Entry Island	-45.595500	166.701450	1992	22	McLean et al. 1993
Oke Island	-45.637846	166.853788	1992	22	McLean et al. 1993
Pigeon Island	-45.709124	166.559784	1992	8	McLean et al. 1993
Indian Island	-45.780308	166.585700	1992	2	McLean et al. 1993
Small Craft Harbour Island	-45.968436	166.647763	1992	8	McLean et al. 1993
Passage Islands	-46.023685	166.537547	1992	18	McLean et al. 1993
Chalky Island	-46.050565	166.522566	1992	3	McLean et al. 1993
Weka Island	-46.094675	166.691829	1992	11	McLean et al. 1993
Coal Island	-46.115032	166.627395	1992	17	McLean et al. 1993
Southern Islands					
Solander Island / Hautere	-46.578557	166.894583	1993	115	Studholme et al. 1994
Codfish Island / Whenua Hou	-46.770615	167.625734	1993	144	Studholme et al. 1994
Stewart Island / Rakiura					
Yankee River	-46.710720	167.936761	1993	?	Mattern pers. obs
Rollers Beach	-46.765041	167.987438	2005	2	Mattern pers. obs
Golden Beach	-46.802533	168.022363	2005	?	Mattern pers. obs
Gull Rock Point	-46.803677	168.033237	2005	?	Mattern pers. obs
Port Pegasus	-47.199361	167.689553	2000	?	Houston pers. comm

TABLE 10.3 Annual cycle of Fiordlands breeding at Jacksons Head, West Coast, New Zealand

PARAMETER	VALUE	REFERENCE
Arrival/ settlement	from 12 June	Warham 1974
Eggs laid	5 Aug ±5d	
Eggs hatch	9 Sep±9d	
Crèching	ca. 20 Sep	
Fledging	ca. 23 Nov	
Adults leave for pre-molt foraging	until 15 Dec	
Return / begin molt	early Feb	
Molt end / leave	early Mar	

TABLE 10.4 Breeding parameters reported for Fiordland penguins

PARAMETER	VALUE	REFERENCE
Incubation period	33.6 days (13)	Warham 1974
Chick rearing period	~75 days	Marchant and Higgins 1990
Hatching success	58.4% (83)	Marchant and Higgins 1990
Crèching success	~67% (83)	Marchant and Higgins 1990
Fledging success	~50% 63.7%±19.1 (12 sites)	Warham 1974 McClellan 2009*
Age at first breeding	5.5 years	Warham 1974

* Mean calculated from data ("median number of fledglings per nest") collected at 12 different sites over the course of 18 years.
Note: Hatching and fledging success represent proportion of chicks hatched/fledged per nest.

TABLE 10.5 Unpublished nest survey data from some sites within the Fiordlands' distributional range

LOCATION	DATE	# NESTS	REFERENCE
Jacksons Head	mid-Aug 1990 mid-Aug 2004	29 16	J. Tansell, DOC West Coast, unpubl. data J. Tansell, DOC West Coast, unpubl. data
Monro Beach	mid-Aug 1990 mid-Aug 2004	23 15	J. Tansell, DOC West Coast, unpubl. data J. Tansell, DOC West Coast, unpubl. data
Murphy Beach	mid-Aug 1990 mid-Aug 2004	27 15	J. Tansell, DOC West Coast, unpubl. data J. Tansell, DOC West Coast, unpubl. data
West Shelter Island	Aug 1994 Aug 1998	44 41	Willans 2000 Willans 2000
East Shelter Island	Aug 1994 Aug 1998	25 24	Willans 2000 Willans 2000
Martins Bay	Aug 1996 Aug 1998	55 63	Willans 2000 Willans 2000
Yates Point*	17 Sep 1986 22 Aug 1994	329 177	Eason 1994 Eason 1994
Teer Creek to Stafford River	14 Sep 1993 21/22 Aug 2002	45 116	Newton 2002 Newton 2002
Haretaniwha Point	12 & 28 Aug 1992 28 Aug 2003	54 47	Newton 2003 Newton 2003

* Eason suggests that the timing of the survey might have contributed to some of the differences in nest numbers found during both surveys. The first count was conducted during the early incubation stage (August), while the second count occurred in late September, i.e. late incubation/early chick rearing period, when nests that failed during incubation were likely missed.

REFERENCES

Alterio, N., and H. Moller. 1997. Diet of feral house cats (*Felis catus*), ferrets (*Mustela furo*) and stoats (*M. erminea*) in grassland surrounding yellow-eyed penguin (*Megadyptes antipodes*) breeding areas, South Island, New Zealand. *Journal of Zoology* 243:869–77.

BirdLife International. 2010. *Eudyptes pachyrhynchus.* IUCN (International Union for Conservation of Nature) 2011. IUCN Red List of Threatened Species. Version 2011.2. www.iucnredlist.org (accessed 22 March 2011).

Boessenkool, S., J. Austin, T. H. Worthy, P. Scofield, A. Cooper, P. Seddon, and J. Waters. 2009. Relict or colonizer? Extinction and range expansion of penguins in southern New Zealand. *Proceedings of the Royal Society Biological Sciences Series B* 276:815–21.

Bull, L. 2004. Recommendations for future research and concessions for West Coast Fiordland crested penguin (*tawaki*). Manuscript. Biodiversity Recovery Unit, Department of Conservation, Wellington, New Zealand.

Buller, W. 1905. A supplement to the birds of New Zealand. Vol 1. London: W. L. Buller.

Chiaradia, A., Y. Ropert-Coudert, A. Kato, T. Mattern, and J. Yorke. 2007. Diving behaviour of little penguins from four colonies across their whole distribution range: Bathymetry affecting diving effort and fledging success. *Marine Biology* 2007:1535–42.

Christidis, L., and W. E. Boles. 2008. Order Sphenisciformes. In *Systematics and Taxonomy of Australian Birds*, ed. L. Christidis, and W. E. Boles, 97–100. Collingwood, Vic., Australia: CSIRO Publishing.

Craw, D., and R. J. Norris. 2003. Landforms. In *The Natural History of Southern New Zealand*, ed. J. T. Darby, R. E. Fordyce, A. Mark, K. Probert, and C. Townsend, 17–35. Dunedin, New Zealand: University of Otago Press.

Croxall, J. P., and L. S. Davis. 1999. Penguins: Paradoxes and patterns. *Marine Ornithology* 27:1–12.

Cunningham, D., and P. Moors. 1994. The decline of rockhopper penguins *Eudyptes chrysocome* at Campbell Island, Southern Ocean, and the influence of rising sea temperatures. *Emu* 94:27–36.

Darby, J. T., and S. M. Dawson. 2000. Bycatch of yellow-eyed penguins (*Megadyptes antipodes*) in gillnets in New Zealand waters, 1979–1997. *Biological Conservation* 93:327–32.

Davis, L. S., and M. Renner. 2003. *Penguins*. London: T & A D Poyser.

Duthie, Q. 2010. *Forest & Bird* has discovered government plans to mine even more national park areas than ministers announced to the public in March. *Forest & Bird.* http://www.forestandbird.org.nz/what-we-do/publications/media-releases/forest-bird-finds-more-mining-plans-on-government-website.

Eason, D. 1994. Fiordland crested penguin survey season, 1994. Manuscript. Department of Conservation, Wellington, New Zealand.

Ellenberg, U., T. Mattern, P. Seddon, and G. Luna-Jorquera. 2006. Physiological and reproductive consequences of human disturbance in Humboldt penguins: The need for species-specific visitor management. *Biological Conservation* 133:95–106.

Falla, R. A. 1935. Notes on penguins of the genera Megadyptes and Eudyptes in Southern New Zealand. *Records of the Auckland Institute and Museum* 1:319–26.

———. 1954. Crested penguin (*Eudyptes pachyrhynchus pachyrhynchus*). *Notornis* 5:212.

Falla, R. A., J. Warham, and C. A. Fleming. 1974. Comments on the proposed preservation of *Eudyptes sclateri* (Buller 1888) and *Eudyptes robustus* (Oliver 1953). *Bulletin of Zoological Nomenclature* 30:136.

Fallis, A. M., S. A. Bisset, and F. R. Allison. 1976. Leucocytozoon tawaki n.sp. (Eucoccida: Leucocytozoidae) from the penguin *Eudyptes pachyrhynchus*, and preliminary observations on its development in *Austrosimulium* spp. (Diptera: Simuliidae). *New Zealand Journal of Zoology* 3:11–16.

Grau, C. R. 1982. Egg formation in Fiordland crested penguins (*Eudyptes pachyrhynchus*). *The Condor* 84:172–77.

Heather, B., and H. A. Robertson. 2000. *The Field Guide to the Birds of New Zealand.* Auckland, New Zealand: Viking.

Hill, S., and J. Hill. 1987. *Richard Henry of Resolution Island.* Dunedin, New Zealand: John McIndoe.

Hilton, G., D. Thompson, P. Sagar, C. RJ, Y. Cherel, and S. Brury. 2006. An isotopic investigation into the causes of decline in a sub-Antarctic predator, the rockhopper penguin *Eudyptes chrysocome. Global Change Biology* 12:611–25.

Hitchmough, R., L. Bull, and P. Cromarty. 2007. New Zealand threat classification system lists, 2005. Science and Technical Publishing. Department of Conservation, Wellington, New Zealand.

Hocken, A. G. 2000. Cause of death in blue penguins (*Eudyptula m. minor*) in North Otago, New Zealand. *New Zealand Journal of Zoology* 27:305–9.

———. 2005. Necropsy findings in yellow-eyed penguins (*Megadyptes antipodes*) from Otago, New Zealand. *New Zealand Journal of Zoology* 32:1–8.

Holdaway, R. N., T. H. Worthy, and A. J. D. Tennyson. 2001. A working list of breeding bird species of the New Zealand region at first human contact. *New Zealand Journal of Zoology* 28:119–87.

IUCN (International Union for Conservation of Nature). 2011. IUCN Red List of Threatened Species. Version 2011.2. www.iucnredlist.org/apps/redlist/search (accessed 18 April 2012).

King, C. 1984. *Immigrant Killers: Introduced Predators and the Conservation of Birds in New Zealand.* Auckland, New Zealand: Oxford University Press.

Kinsky, F. C. 1970. Annotated checklist of the birds of New Zealand including the birds of the Ross Dependency. Reed, Wellington, New Zealand.

Lloyd, B. D. 2003. Potential effects of mussel farming on New Zealand's marine mammals and seabirds: A discussion paper. Department of Conservation, Wellington, New Zealand.

Luna-Jorquera, G., S. Garthe, F. G. Sepulveda, T. Weichler, and J. A. Vásquez. 2000. Population size of Humboldt penguins assessed by combined terrestrial and at-sea counts. *Waterbirds* 23:506–10.

Marchant, S., and P. J. Higgins. 1990. *Eudyptes pachyrhynchus* Fiordland penguin. In *Handbook of Australian, New Zealand and Antarctic Birds*, vol. 1A, ed. S. Marchant and P. J. Higgins, 195–205. Melbourne, Australia: Oxford University Press.

Mattern, T., L. S. Davis, B. M. Culik, and D. M. Houston. 2001. Foraging ranges and breeding success of blue penguins (*Eudyptula minor*) at two different locations in New Zealand. Symposium 2001. *Proceedings of the Oamaru Penguin Symposium*, ed. A. G. Hocken. *New Zealand Journal of Zoology* 28:431–40.

Mattern, T., D. M. Houston, C. Lalas, A. N. Setiawan, and L. S. Davis. 2009. Diet composition, continuity in prey availability and marine habitat: Keystones to population stability in the Snares penguin (*Eudyptes robustus*). *Emu* 109:204–13.

McClellan, R. 2009. Tawaki monitoring review and analysis. Draft report. Department of Conservation, Hokitika, New Zealand.

McClelland, P. J. 2002. Eradication of Pacific rats (*Rattus exulans*) from Whenua Hou Nature Reserve (Codfish Island), Putauhinu and Rarotoka Islands, New Zealand. In *Turning the Tide: The Eradication of Invasive Species; Proceedings of the International Conference on Eradication of Island Invasives*, ed. C. R. Veitch and M. N. Clout, 173–81. Gland, Switzerland: IUCN, in collaboration with the IUCN/SSC Invasive Species Specialist Group, Auckland, New Zealand.

McLean, I. G. 1990. Chick expulsion by a Fiordland crested penguin. *Notornis* 37:181–82.

———. 2000. Breeding success, brood reduction and the timing of breeding in the Fiordland crested penguin (*Eudyptes pachyrhynchus*). *Notornis* 47:55–58.

McLean, I., M. Abel, C. Challies, S. Heppelthwaite, J. Lyall, and R. Russ. 1997. The Fiordland crested penguin (*Eudyptes pachyrhynchus*) survey, stage V: Mainland coastline, Bruce Bay to Yates Point. *Notornis* 44:37–47.

McLean, I., and R. Russ. 1991. The Fiordland crested penguin survey, stage I: Doubtful to Milford sounds. *Notornis* 38:183–90.

McLean, I., B. Studholme, and R. Russ. 1993. The Fiordland crested penguin survey, stage III: Breaksea Island, Chalky and Preservation inlets. *Notornis* 40:85–94.

McMurtrie, P., K.-A. Edge, D. Crouchley, and M. Willans. 2008. Resolution Island operational plan—stoat eradication. Internal report. Department of Conservation, Te Anau, New Zealand.

Ministry of Fisheries. 2006. Government gives go-ahead for marine farm in Jackson Bay. http://www.fish.govt.nz/en-nz/Press/Press+Releases+2006/June+2006/Government+gives+go-ahead+for+marine+farm+in+Jackson+Bay.htm.

———. 2008. Recreational set net prohibitions for Hector's dolphins. http://www.fish.govt.nz/NR/rdonlyres/CE9FF375–999E-4B72–9DE3–94C584D17B0D/0/MFishSetNetSI_ESBroWeb.pdf.

———. 2009. Proposed mātaitai reserves on the South Island west coast. http://www.fish.govt.nz/mi-nz/Press/Press+Releases+2009/March09/Proposed+mataitai+reserves+on+the+South+Island+West+Coast+-+Public+Meetings.htm.

———. 2010. NABIS—Internet mapping of New Zealand's marine environment, species and fisheries management. Available at http://www.nabis.govt.nz.

Miskelly, C. M., J. E. Dowding, G. P. Elliott, R. A. Hitchmough, R. G. Powlesland, H. A. Robertson, P. M. Sagar, R. P. Scofield, and G. A. Taylor. 2008. Conservation status of New Zealand birds, 2008. *Notornis* 55:117–35.

Morrison, K. 1980. Bird and stoat encounters in Fiordland. *Notornis* 27:324.

Murie, J., L. S. Davis, and I. McLean. 1991. Identifying the sex of Fiordland crested penguins by morphometric characters. *Notornis* 38:233–38.

Murphy, R. J., M. H. Pinkerton, K. M. Richardson, J. M. Bradford-Grieve, and P. W. Boyd. 2001. Phytoplankton distributions around New Zealand derived from SeaWiFS remotely-sensed ocean colour data. *New Zealand Journal of Marine and Freshwater Research* 35:343–62.

Newton, G. 2002. Tawaki (Fiordland crested penguin) survey. Teer Creek–Stafford River 2002. Manuscript. Department of Conservation, Haast, New Zealand.

———. 2003. Tawaki (Fiordland crested penguin) survey. Heretaniwha Point, Bruce Bay, 2003. Manuscript. Department of Conservation, Haast, New Zealand.

———. 2005. Tawaki (Fiordland crested penguin) survey, Jacksons Head West, 15/08/2005. Internal report. Department of Conservation, Hokitika, New Zealand.

Newton, G., and J. Tansell. 2005. Breeding success and recruitment in Fiordland crested penguin in South Westland from 1990 to 2004. In *Proceedings of the Oamaru Penguin Symposium 2005*, ed. A. G. Hocken. *New Zealand Journal of Zoology* 32:263–271.

Oliver, W. R. B. 1953. The crested penguins of New Zealand. *Emu* 53:185–87.

Osborn, D., and A. Datta. 2006. Institutional and policy cocktails for protecting coastal and marine environments from land-based sources of pollution. *Ocean and Coastal Management* 49:576–96.

Reid, T. A., C. L. Hull, D. W. Eades, R. P. Scofield, and E. J. Whoeler. 1999. Shipboard observations of penguins at sea in the Australian sector of the Southern Ocean, 1991–1951. *Marine Ornithology* 27:101–10.

Russ, R., I. McLean, and B. Studholme. 1992. The Fiordland crested penguin survey, stage II: Dusky and Breaksea sounds. *Notornis* 39:113–18.

Spellerberg, I. F. 1975. The predators of penguins. In *The Biology of Penguins*, ed. B. Stonehouse, 413–33. London: Macmillan.

St. Clair, C. 1992. Incubation behavior, brood patch formation and obligate brood reduction in Fiordland crested penguins. *Behavioral Ecology and Sociobiology* 31:409–16.

———. 1998. *Eudyptes pachyrhynchus*: Penguin conservation assessment and management plan. Report from a workshop held 8–9 September 1996, Cape Town, South Africa, ed. S. Ellis, J. Croxall and J. Cooper, 69–72. Apple Valley, MN, IUCN/SSC Conservation Breeding Specialist Group.

St. Clair, C., I. McLean, J. Murie, S. Phillipson, and B. Studholme. 1999. Fidelity to nest site and mate in Fiordland crested penguins *Eudyptes pachyrhynchus*. *Marine Ornithology* 27:37–41.

St. Clair, C., and R. St. Clair. 1992. Weka predation on eggs and chicks in Fiordland crested penguins. *Notornis* 39:60–63.

Stonehouse, B. 1971. The Snares Islands penguin *Eudyptes robustus*. *Ibis* 113:1–7.

Studholme, B., R. Russ, and I. McLean. 1994. The Fiordland crested penguin survey, stage IV: Stewart and offshore islands and Solander Island. *Notornis* 41:133–43.

Taylor, G. A. 2000. Action plan for seabird conservation in New Zealand. Part A: Threatened seabirds. Threatened Species Occasional Publication No. 16. Department of Conservation, Wellington, New Zealand.

USFWS (U.S. Fish and Wildlife Service). 2010. Endangered and threatened wildlife and plants; 12-month finding on a petition to list five penguin species under the Endangered Species Act, and proposed rule to list the five penguin species. *Federal Register* 73:77303–332.

van Heezik, Y. 1989. Diet of Fiordland crested penguins during the post-guard phase of chick growth. *Notornis* 36:151–56.

———. 1990. Diets of yellow-eyed, Fiordland crested, and little blue penguins breeding sympatrically on Codfish Island, New Zealand. *New Zealand Journal of Zoology* 17:543–48.

Warham, J. 1974. The Fiordland crested penguin. *Ibis* 116:1–27.

———. 1975. The crested penguins. In *The Biology of Penguins*, ed. B. Stonehouse, 189–269. London: Macmillan.

Willans, M. 2000. Fiordland crested penguin monitoring, Fiordland coast, 1994–2000. Manuscript. Department of Conservation, Wellington, New Zealand.

Williams, T. D. 1995. *The Penguins*. Oxford: Oxford University Press.

Wilson, K.-J. 2004. *Flight of the Huia*. Christchurch, New Zealand: Canterbury University Press.

11

Snares Penguin

(Eudyptes robustus)

Thomas Mattern

1. SPECIES (COMMON AND SCIENTIFIC NAMES)

Snares penguin, *Eudyptes robustus*

Other names are Snares crested penguin, Snares Islands penguin, and *pokotiwha* (Maori).

2. DESCRIPTION OF THE SPECIES

The Snares is a medium-size crested penguin with a body length of 51–61 centimeters; body mass ranges from 2.8 kilograms in females to 3.4 kilograms in males (Marchant and Higgins 1990). This species closely resembles the Fiordland penguin (*E. pachyrhynchus*), with which it is sometimes treated as conspecific. As with other crested penguins, the head, throat, and upperparts of the Snares are dark blue or black, turning to a dark brown near molt (fig. 1). The underparts are white and occasionally feature distinct melanism, with black dots spread randomly across the chest near the stomach (Warham 1974a, pers. obs.). Feet and legs are pinkish white above and blackish brown behind the tarsi and metatarsi, on the soles, and at the front of webs (Marchant and Higgins 1990). The bright yellow crest starts near the nostril, passes above the eyes, and spreads out horizontally or droops behind the eyes (Oliver 1953). The crest feathers are considerably shorter than in rockhopper penguins (*E. chrysocome*) but longer than in Fiordland penguins (Marchant and Higgins 1990). The eyes vary in color from reddish brown to a brighter pinkish red (Warham 1974a). A band of pinkish-white skin borders the bill. The bill itself is large and heavy, orange brown, and, especially in males, features a noticeable bulbous culminicorn (the top-central plate of the upper mandible) (Marchant and Higgins 1990).

FIG. 1 Young male Snares penguin. (T. Mattern)

MORPHOMETRIC DATA. See table 11.1.

Snares penguins are distinguished from Fiordlands by several morphological features. The bare skin at the base of the bill prominent in Snares is absent in Fiordlands (Oliver 1953). Snares have a longer, more robust bill (owing to the bulbous culminicorn) (Oliver 1953; Marchant and Higgins 1990; Williams 1995), and their lateral crest plumes are longer (more than 5 cm) than those of the Fiordlands (less than 5 cm) (Marchant and Higgins 1990). Fiordlands usually show a series of white stripes in their cheek region because the pale feather bases are exposed, but Snares rarely exhibit white stripes, although faint, narrow cheek stripes may be visible during threat displays (Warham 1974a, pers. obs.).

3. TAXONOMIC STATUS

Originally described as *Eudyptes atrata* by Hutton (Finsch 1875)—Buller (1888) corrected the name to *E. atratus*—the Snares penguin was considered only subspecifically distinct from the erect-crested (*E. sclateri*) and Fiordland penguins (Falla 1935). Based on close

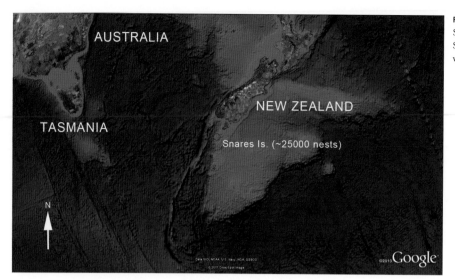

FIG. 2 The distribution of the Snares penguin is restricted to the Snares archipelago, New Zealand, with counts based on nests.

examination of crest appearances and other morphological features, Oliver (1953) concludes that all three should be classified as distinct species, naming the Snares *E. robustus*. Although this three-species distinction was widely accepted, the close relationship of Fiordland and Snares crested penguins, as noted by Oliver, has sparked debate as to whether Snares are a subspecies of Fiordlands. The 1970 edition of the *Checklist of New Zealand Birds* (Kinsky 1970) treats Snares penguins as conspecifics of erect-crested and Fiordland penguins. Stonehouse (1971), Warham (1974a), and Falla et al. (1974) consider them a separate species based on morphological distinction, lack of recorded hybrids, and reproductive isolation resulting from significant differences in timing of breeding.

In recent years, the taxonomy of Sphenisciformes has repeatedly been revisited incorporating molecular analysis of species relations using protein data and mitochondrial or nuclear DNA. Davis and Renner (2003) point out the close relationship between Snares and Fiordland penguins but keep them as distinct species. Baker et al. (2006) note the similarity of Snares and Fiordland penguins and observe that they diverged fairly recently (less than 2 million years ago) from macaroni (*E. chrysolophus*) and royal (*E. schlegeli*) penguins. A similar conclusion in Ksepka et al. (2006) is based on a combined analysis of morphological characters and sequence fragments from mitochondrial and nuclear genes. Interestingly, when only morphological characters are considered, the Snares penguin appears most closely related to the erect-crested penguin, which is the conclusion presented in Clarke et

al. (2007). Christidis and Boles (2008) review the taxonomic status of penguins and conclude, based on DNA sequence data from Ritchie (2001), that the genetic divergence of Snares and Fiordlands was insufficient to make the Snares a separate species. Besides classifying Snares as subspecies *E. pachyrhynchus*, the authors also dropped the common name in favor of "Fiordland penguin." However, whether a reclassification and upsetting of two long-established names is justified must be questioned and seems to oversimplify the situation. Warham (1974a), based on years of close study of both species, argues that there is an obvious lack of hybridization between Snares and Fiordlands. Although Christidis and Boles (2008) remark that hybridization would be difficult to detect in the field, they fail to acknowledge the different life-history traits of both species. Most notably, the temporal isolation of both species' reproductive cycles—Fiordlands start breeding in the austral midwinter, some two months earlier than Snares (see chapter 10)—makes interbreeding unlikely.

4. RANGE AND DISTRIBUTION

The Snares penguin breeds only on the Snares archipelago (fig. 2; table 11.2). Main breeding colonies are on North East Island (the main island), Broughton Island, and the islets Toru and Rima of the Western Chain (fig. 3). During the breeding season, Snares stay within 200–300 kilometers of their island, as indicated by GPS data loggers (Mattern 2007). Once the chicks have hatched, foraging activities seem to focus on the passage between the Snares Islands and Stewart Island to the north. Out-

FIG. 3 The steep terrain of the Snares Islands allows penguins to climb high above the surf. (T. Mattern)

side the breeding season, Snares disperse much farther, and landings have been recorded on the South Island of New Zealand, the Chatham Islands (Miskelly and Bell 2004), Tasmania (Marchant and Higgins 1990), Macquarie Island, and even the Falkland/Malvinas Islands (Lamey 1988; Demongin et al. 2010).

5. SUMMARY OF POPULATION TRENDS

The Snares population is considered stable, with estimates ranging between 20,000 in the 1960s and 30,000 in recent years (Warham et al. 1986; D. M. Houston, unpubl. data). Most estimates published in the 1970s and 1980s are based on indirect estimates of the breeding population via counts of fledglings. Some variance in inter-annual figures might be a result of inconsistent correction factors (e.g., compare Warham 1974b and Marchant and Higgins 1990). In the last decade, three comprehensive, more accurate counts found 30,979 (season 2000–2001 [Amey et al. 2001]), 26,053 (2008–9 [D. M. Houston, unpubl. data]), and 30,187 breeding pairs (2010–11 [D. M. Houston, unpubl. data]) (table 11.3).

6. IUCN STATUS

The International Union for Conservation of Nature ranks the Snares penguin as Vulnerable on its Red List of Threatened Species (IUCN 2011). The very limited breeding range, confined to the small Snares archipelago (total area about 3.5 square kilometers), exposes the species to anthropogenic influences or stochastic events (e.g., oil spills or introduction of terrestrial predators) that may render it Critically Endangered or even Extinct within a short time period (BirdLife International 2010).

The New Zealand Threat Classification System ranks the Snares penguin At Risk—Naturally Uncommon due to its limited distribution (Miskelly et al. 2008). The erect-crested penguin and the Chatham Island subspecies of the little penguin (*Eudyptula minor chathamensis*) are also classified At Risk.

The Snares is an indigenous *taonga* species, one that is of significant cultural and spiritual value to New Zealand Maori, as established in the Ngāi Tahu Claims Settlement Act 1998. The species is fully protected under the Wildlife Act 1953. Any interactions with the penguins at their colonies (e.g., research) require permission from the Department of Conservation and ethics approval. The Snares Islands were designated a nature reserve by the Reserves Act 1977 and became part of a UNESCO World Heritage site, the New Zealand Sub-Antarctic Islands, in 1998.

No protection plans are in place for the Snares's marine habitat, although the IUCN recommends the creation of a marine reserve and restrictions on fishing within a radius of 22 kilometers around the island (BirdLife International 2010).

7. NATURAL HISTORY

BREEDING BIOLOGY. Snares breed on North East Island, the smaller Broughton Island, and the unvegetated Western Chain, composed of rocky stacks located approximately four kilometers southwest of the main islands (fig. 4). On North East Island, most colonies are located under a forest canopy of tree daisy (*Olearia lyalli*) or Stewart Island tree groundsel (*Brachyglottis stewartiae*) or in *Hebe elliptica* evergreen shrubs (fig. 5). Some colonies are on rises dominated by tussock grass or bare rock (Warham 1974a; Amey et al. 2001). On Broughton Island, colonies predominantly occur on open ground (Warham 1974a). On the Western Chain, Snares breed principally under boulders and in crevasses, probably owing to competition with Salvin's mollymawks (*Thalassarche salvini*)

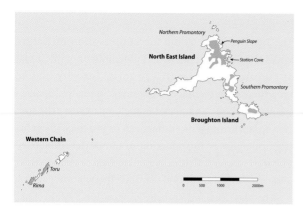

FIG. 4 Detailed distribution of Snares penguin colonies at Snares Islands, New Zealand. Green polygons delineate regions with aggregations of colonies. Sources: Northern and Southern Promontory, Amey et al 2001; Broughton Island, Warham 1974a; Western Chain, Miskelly 1984.

(Miskelly 1984). Colony sizes range from a handful of birds (less than 10 nests) to some 1,300 nests (Marchant and Higgins 1990; Amey et al. 2001). Nests generally are scooped-out hollows lined with mud and peat mixed with stones, twigs, or bones (fig. 7).

The Snares breeding period is well defined (table 11.4). Adults arrive in the first three weeks of September, the males about a week before the females (Warham 1974a; Amey et al. 2001). There is little information about mate and nest-site fidelity, but Warham (1974a) suggests that birds tend to be faithful to mates and nest sites for more than one season. Two eggs are laid an average of 4.5 days apart toward the end of September and beginning of October. As with other crested penguins, Snares exhibit a marked egg-size variation, with the first egg (A-egg) being about 78% the size of the second egg (B-egg) (Massaro and Davis 2005).

For the first two weeks after egg laying, both partners remain at their nest with the female incubating almost continuously (fig. 6). Toward the end of September, males start to leave the island (Warham 1974a). Their exodus peaks during the first week of October and is stable among years, so the most likely trigger is day length (Mattern 2007). Males remain at sea for 10–14 days and return to their colonies again in a fairly synchronized manner (within 4 days). The females then leave to forage for about a week and time their return to coincide with the hatching of the eggs (Mattern 2007). About 60% of the nests hatch both eggs, but chicks from the A-egg usually die within a week after hatching (table 11.5) (Warham 1974a). On rare occasions, both chicks

FIG. 5 Snares penguins standing under the vegetation near their breeding colony. (T. Mattern)

FIG. 6 Snares penguin pair in the nesting colony. (T. Mattern)

live until crèching (Mattern et al., unpubl. data). On one occasion, an abandoned chick adopted by a male already guarding two chicks was observed, and at least two of those chicks survived until crèching, but whether the adopted individual was among the surviving chicks could not be ascertained (pers. obs.).

For the first three weeks after hatching, the male guards and the female forages, returning on average every 30 hours to feed the chicks (fig. 7) (Mattern 2007). During the post-guard stage, both parents forage, but it seems that chicks are fed primarily by females. The young start to form crèches of up to 30 chicks around mid-November and fledge in mid- to late January. By early February, when adults leave for pre-molt foraging trips, chicks have left the island (Warham 1974a).

These patterns may not apply to Snares penguins breeding on the Western Chain. Reports indicate that the annual onset of breeding is delayed between 15 and 26 days in comparison to North East and Broughton Islands (Fleming and Baker 1973; Sagar 1977). Fledging chicks were estimated to be at least 44 days later in their development than fledglings on the main islands (Miskelly 1984). Closer examination of the Western Chain population of Snares is warranted.

MOLT. Chicks start to shed down from 24 December onward. The molt into juvenile plumage is complete in mid-January (Warham 1974a), and all chicks have gone to sea by the end of January (Horning and Horning 1974).

Yearlings appear on the island beginning in November and start to molt into adult plumage from around mid-January (fig. 8). The majority complete their molt and leave the island in early February (Warham 1974a).

Adult Snares return to the breeding colonies from their eight-week foraging trips around mid- to late March to molt. Breeding pairs usually reunite at their nest sites and engage in display, nest building, and occasional copulation. The duration of the molt is believed to range from 24 to 30 days. The colonies start to empty from late April onward, and by late May, no penguins remain on the island (Warham 1974a).

FORAGING ECOLOGY AND PREY. The foraging ecology of breeding Snares is known from GPS data loggers and dive recorders (Mattern 2007). When females are incubating the eggs, the males take long foraging trips (about two weeks); many forage at the Subtropical Front, an oceanographic feature some 200 kilometers east of the Snares Islands, where warm subtropical waters converge with cooler sub-Antarctic water. The males' foraging trips coincide with the onset of the spring phytoplankton bloom, which commences along the front. Foraging males dive to depths of up to 120 meters. When the males return, the females take shorter foraging trips (about one week), feeding in subtropical waters, closer to the island, where they dive to depths of 30–80 meters. After the chicks hatch, the females take short trips (mean trip duration 34.3 ± 8.2 hours, travel distance 121.1 ±

FIG. 7 Male Snares penguin guards his well-fed chick. (T. Mattern)

FIG. 8 A muddy yearling Snares penguin stands in the breeding colony. (T. Mattern)

31.6 kilometers, n = 19) in warmer, nutrient-rich waters between 50 and 90 kilometers north of the Snares. Dive depths there are considerably shallower than during the incubation (mean dive depth 23.5 ± 6.3 meters).

The prey composition of Snares outside the chick-guard stage is little known. Cooper et al. (1990) list 10 different species of crustaceans, two species of fish, and three species of squid found in stomachs of dissected chicks of unknown age. Marchant and Higgins (1990) note that the dominant crustacean was krill (*Nyctiphanes australis*). Mattern et al. (2009) confirm in a quantitative analysis of stomach contents that the food fed to chicks was primarily large quantities of *N. australis* (60% of stomach content wet weight), while fish (30%, principally Hector's lanternfish [*Lampanyctodes hectoris*] and long-snouted pipefish [*Leptonotus morae*]) and cephalopods (10%, chiefly arrow squid [*Nototodarus sloanii*]) made up the rest. Judging from otoliths and squid beaks also found in the stomachs, it seems that fish and squid are often digested before the penguins land and thus are prey for the adults. Species of importance for adults are the pelagic fish species redbait (*Emmelichthys nitidus*) and juvenile red cod (*Pseudophycis bacchus*), while arrow squid (*Nototodarus sloani*), warty squid (*Moroteuthis ingens*), and violet squid (*Histiotheuthis atlantica*) are all important cephalopod prey.

PREDATORS. Main predators of penguins around the Snares are both the Hooker's sea lion (*Phocarctos hookeri*) and the New Zealand fur seal (*Arctocephalus forsteri*). Both pinnipeds are abundant and kill penguins at sea (pers. obs.). Although not as numerous, leopard seals (*Hydrurga leptonyx*) may also prey heavily on Snares penguins, as evidenced by feathers in their scat (pers. obs.). Other marine predators probably include sharks and killer whales (*Orcinus orca*) (Davis and Renner 2003).

Chicks sometimes fall prey to brown skuas (*Catharacta lonnbergi*), although their effect on chick mortality seems to be slight (Warham 1974a). Brown skuas also prey on injured or sick adult penguins. Red-billed gulls (*Larus novaehollandiae*) scour the penguin colonies during the late incubation and early chick-rearing phases, primarily to feed on expelled eggs and dead or dying chicks (pers. obs.). Fledglings are preyed upon by southern giant petrels (*Macronectes giganteus*), which

FIG. 9 Annual cycle of the Snares penguin.

congregate near penguin launching places in January (Horning and Horning 1974; Warham 1974a).

ANNUAL CYCLE. See table 11.4 and figure 9.

8. POPULATION SIZES AND TRENDS

In the breeding season 1968–69, the population size was thought to be between 30,000 and 50,000 birds (Warham 1974a). The estimate was derived from fledgling counts (5,000 fledglings on the mainland, 500 on Broughton Island) and based on the assumption of 50% chick mortality during créching, suggesting that the breeding population was a minimum of 11,000 breeding pairs with an uncertain number of nonbreeders and penguins breeding on the Western Chain.

More population estimates were published in the 1980s. Warham et al. (1986) summarize assessments from further fledgling counts as about 20,000 pairs in the season 1982–83 and 37,600 pairs in 1984–85. Considering the preceding as well as the following population estimates, the 1984–85 figure appears high. Unfortunately, the authors referenced this figure as "Johns pers. comm.," which makes it difficult to verify, but a transcription error is possible. For the season 1985–86, Marchant and Higgins (1990) give an estimate of 23,250 pairs (19,000 pairs on North East Island, 3,500 pairs on Broughton Island, and 750 pairs on the Western Chain). Furthermore in 1986–87, the number of immature birds

FIG. 10 Snares penguins in a heavy rainstorm on North East Island, Snares Islands, New Zealand. (T. Mattern)

was estimated to be about 2,500 one-year-olds, 2,000 two-year-olds and 1,000 three-year-olds. From these figures, Marchant and Higgins (1990) conclude that in the season 1986–87, the world Snares population was around 54,000 birds.

In the past decade, three comprehensive censuses of the breeding population have been conducted (Amey et al. 2001; D. M. Houston, unpubl. data). During these surveys, all colonies on North East Island (fig. 10) were visited and breeding pairs were counted; Broughton Island's colonies were surveyed in 2008 and 2010. "Breeding pairs" were defined as pairs of adult birds guarding active nests (i.e., those that contained eggs) or were empty (i.e., failed nests, nests formed but eggs not, or not yet, laid). For the 2000 season, Amey at al. (2001) determine a total of 25,861 breeding pairs on North East Island. The other islands could not be visited, but numbers on Broughton Island were estimated to be 20% of the North East Island numbers (4,737 pairs), while an estimate of 381 pairs for the Western Chain was derived from other counts conducted in 1984 and 1996. In October 2008, considerably fewer pairs were encountered (North East Island, 21,891 pairs;

Broughton Island, 4,234 pairs; Western Chain, not visited [D. M. Houston, unpubl. data]). This apparent decline triggered a third survey, which was carried out in October 2010. This time, numbers were again comparable to the results from 2000, with 25,905 pairs found on North Island and 4,273 pairs counted on Broughton Island (D. M. Houston, unpubl. data). It is believed that low numbers encountered in 2008 only reflect a "bad year," when a portion of the penguin population did not breed. As such, there is no evidence for a population decline.

Historically, the Snares population seems to have remained stable, unlike other crested penguin populations breeding in the New Zealand sub-Antarctic region. Analysis of stable isotope ratios ($^{13}C/^{12}C$ and $^{15}N/^{14}N$) in feathers collected from living Snares and museum skins dating back to the 1880s indicated that the species did not experience any apparent changes in its prey composition and abundance. Carbon isotope ratios gave no evidence of gross disparities in marine productivity in the foraging areas in the course of the 20th century, while nitrogen isotope ratios showed that the penguins foraged at comparable trophic levels (Mattern et al. 2009).

9. MAIN THREATS

CLIMATE VARIATION. Unlike other species in New Zealand's sub-Antarctic region (Cunningham and Moors 1994; Hilton et al. 2006), Snares have not been affected by rising ocean temperatures in the past century (Mattern et al. 2009). However, El Niño conditions may result in a later onset of the spring bloom and reduced oceanic productivity, as reflected in increased foraging efforts by Snares penguins (Mattern, unpubl. data; Mattern 2007). The frequency of years with El Niño conditions has increased in the past decades (Mullan 1996) and could have a negative impact on the Snares population. If climate change also negatively affects the breeding habitat, the lack of an alternative habitat could present a predicament for the species.

FISHERIES. Commercial fishing activities in New Zealand's sub-Antarctic region have a negative impact on marine species (Chilvers 2009), but their potential effect on Snares penguins is unclear. The penguins could be caught in trawl nets (Davis and Renner 2003). So far, there are no records of Snares being taken as bycatch (Taylor 2000), but whether such incidents would be reported is unknown (Darby and Dawson 2000). Beyond the bycatch problem, there is also the potential for competition with fisheries for prey species (Taylor 2000). Considering the commercial fishing effort around the Snares Islands, interactions between fisheries and penguins seem possible. For example, in the 2008–9 season, commercial fisheries extracted up to 23.2 tons of arrow squid, 1.1 tons of redbait, and 0.6 tons of red cod (Ministry of Fisheries 2010)—besides krill, all three species are main prey items for the penguins (Mattern et al. 2009). It appears that a small number of individual fishing operators might be using setnets around the Snares, a technique that poses a significant threat to other penguin species in New Zealand (Rowe 2010; Ramm 2010). If setnetting were indeed used in the vicinity of the Snares Islands, its effect could potentially be devastating considering the sheer numbers of birds commuting to and from the island throughout the breeding season.

INTRODUCED PREDATORS. There are no mammalian predators at the Snares Islands (Taylor 2000); however, the introduction of exotic terrestrial species could have devastating effects for the entire island ecosystem. The greatest risk is the introduction of rats as a result of illegal landings by fishing vessels, as documented by Dingwall (1995), but the presence of millions of other ground-breeding seabird species such as the sooty shearwater (*Puffinus griseus*) (see Miskelly et al. 2001) probably render penguins only secondary targets for rodents. The accidental introduction of mustelids that would be an imminent risk to the penguins (King 1985) seems less likely but nevertheless is a worry.

ENVIRONMENTAL DISASTERS. Greater threats are sea-borne. The small size of the island group makes it particularly vulnerable to disastrous events, such as oil spills (Taylor 2000; BirdLife International 2010). On North East Island, Snares penguins use only a handful of landing sites along the east coast (fig. 11), with Penguin Slope and Station Cove being the most frequented (Warham 1974a). Oil spills within the vicinity of these sites could have severe consequences for landing and departing penguins, especially during the height of the breeding season from November to February, which also coincides with the peak season for cruise ship tourism in New Zealand's sub-Antarctic region. Of particular concern are plans for oil fields off southern Fiordland. A disastrous event similar to the British Petroleum Deepwater Horizon oil spill in the Gulf of Mexico in 2010 would have catastrophic effects on the Snares Island ecosystem, which is located less than 200 kilometers down current from the proposed extraction site (Waitutu and Hautere sub-basins) (see also chapter 10).

10. RECOMMENDED PRIORITY RESEARCH ACTIONS FOR CONSERVATION

The population dynamics of the Snares penguins are unclear. Before the nest counts initiated by the Department of Conservation (DOC) in 2000 and repeated in 2008 and 2010, most population estimates gave an indication of size and trends but lacked standardized assessment methodologies and detailed documentation thereof, making conclusions drawn from these data questionable. The following actions address the research needs:

1. Continue 10-year population surveys (as recommended in Taylor 2000; BirdLife International 2010), following methods established in the DOC's censuses.
2. Establish basic life-history parameters such as adult survival rates, longevity, juvenile survival and

FIG. 11 Snares penguins coming ashore in white water. (T. Mattern)

recruitment, and age at first breeding as well as nest-site fidelity and natal philopatry.

3. Investigate the particular influence of El Niño events on reproductive outcome and juvenile survival, even though the penguins may be less affected by ongoing environmental changes such as ocean warming. Similarly, developing an understanding of the influence of temporary climatic effects on foraging behavior and success during breeding will be valuable.

4. Clarify potential and actual interactions with commercial fisheries by collecting information on migration patterns outside the breeding season. Emphases should be fledgling dispersal at sea and investigation of mortality factors in studies.

5. Develop more detailed knowledge of the Snares's diet composition. Facilitate ongoing monitoring of diet composition through detection of intra- and inter-annual variations in diet composition, variation in chick diet during rearing, and meal sizes and feeding frequencies as well as study of stable isotopes from penguin tissues (e.g., blood, feathers) and isotopic signatures in their main prey species.

6. Clarify the species taxonomy by researching whether and to what degree Snares and Fiordland penguins are interbreeding. Similarly, the differences in timing of breeding warrant an investigation of the taxonomic status of the penguins from the Western Chain.

7. Collect information about chick growth, crèching behavior, or patterns of molt. The causes of chick

mortality (i.e., brood reduction, predation, environmental properties) also require clarification.

11. CURRENT CONSERVATION EFFORTS

To date, the main conservation effort specific to Snares penguins is a population assessment conducted every 10 years (Taylor 2000). Habitat-specific conservation efforts consist chiefly of enforcing strict quarantine measures to prevent introduction of exotic plant and animal species to the archipelago. Landings are prohibited (except by special permit), although offshore cruising is permitted (Dingwall 1995).

Various projects examining aspects of Snares biology have the potential to aid conservation. These projects address reproductive and social biology (Horning and Horning 1974; Warham 1974a; Lamey 1992; Massaro and Davis 2004, 2005; McGraw et al. 2009), vocalizations (Warham 1974a, 1975; Proffitt and McLean 1990), and foraging ecology and diet (Mattern 2007; Mattern et al. 2009). Overall, very little is known about the Snares Island ecosystem and particularly the endemic species that live there. Despite this, New Zealand's Department of Conservation restricts and even discourages research projects on the island quite rigorously—apparently in the hope that minimum human presence will suffice to ensure the well-being of the island's primarily marine avifauna. Thus, it seems unlikely that more information on Snares biology will emerge anytime soon.

12. RECOMMENDED PRIORITY CONSERVATION ACTIONS FOR INCREASING POPULATION RESILIENCE AND MINIMIZING THREATS AND IMPACTS

1. Conduct more research projects essential for developing an understanding of the Snares's ecology and population dynamics.

2. Monitor population developments with an emphasis on the effects of El Niño conditions on reproductive outcome and juvenile survival and recruitment.

3. Investigate the impact of commercial fisheries on the Snares population and, if required, work toward the application and enforcement of protective measures. A setnet fisheries ban needs to be established in the vicinity of the Snares archipelago.

4. Maintain stringent quarantine procedures and pest contingency plans.

ACKNOWLEDGMENTS

Substantial portions of this chapter could not have been written without the three-year research project on Snares penguins that I conducted between 2002 and 2004. I want to thank Lloyd S. Davis (University of Otago) for his endurance and stamina in getting our project organized despite seemingly insurmountable bureaucratic obstacles and opposition. I also extend my gratitude to Pete McClelland (Department of Conservation) for eventually issuing the permits for our project. Special thanks to Ursula Ellenberg, Katta Ludynia, Alvin Setiawan, and Dave Houston, who contributed significantly to the project and data compilation. Alvin Setiawan and David Houston furthermore contributed valuable comments on early drafts of this chapter. Special thanks go to Martin Renner for further helpful remarks on a later draft of this chapter.

TABLE 11.1 Snares penguin morphometric data (showing mean, standard deviation and sample size)

	ADULT MALE	ADULT FEMALE	YEARLING MALE	YEARLING FEMALE	FLEDGING (COMBINED SEXES)	REFERENCE
Mass (g)						
Incubation (pre-foraging)	2707±127 (7)	2450±257 (7)				Mattern, Houston and Ludynia (unpubl. data)
Incubation (post-foraging)	3685±246 (13)					Mattern, Houston and Ludynia (unpubl. data)
Early chick rearing (Nov)	2630±222 (12)	2484±137 (12)	2873±216 (24)	2647±206 (23)		Warham (1974a)
Late chick rearing (Jan/Feb)	3361±282 (12)	2700±155 (12) 2780±300 (32)				Warham (1974a) Stonehouse (1971)
Pre-molt			4340±311 (30)	3876±411 (39)		Warham (1974a)
Sheltered, inland colony					3082±316 (30)	
Exposed, near-sea colony					2893±384 (23)	
Flipper Length (mm)	183.0±4.2 (61)	177.3±3.9 (47)				Stonehouse (1971)
	183.1±4.6 (47)	177.6±3.2 (17)	180.0±3.9 (47)	174.6±4.2		Warham (1974a)
	189.0±3.8 (20)	180.0±3.7 (20)				Marchant and Higgins (1990)
Western Chain	189.0±5.0 (17)	179.0±5.5 (17)				Marchant and Higgins (1990)
Bill Length (mm)	59.2±2.2 (64)	52.5±2.1 (58)				Stonehouse (1971)
	58.9±2.7 (47)	52.2±1.7 (17)	55.0±2.2 (54)	49.9±2.3		Warham (1974a)
	58.6±2.2 (20)	52.5±1.9 (20)				Marchant and Higgins (1990)
Western Chain	56.8±1.8 (17)	50.6±1.8 (17)				Marchant and Higgins (1990)
	58.7±2.6 (16)	53.0±2.2 (59)				Mattern, Houston and Ludynia (unpubl. data)
Bill depth (mm)	28.0±1.4 (47)	24.4±1.6 (17)	23.0±1.1 (54)			Warham (1974a)
	26.8±1.1 (16)	22.9±1.3 (59)				Mattern, Houston and Ludynia (unpubl. data)
Tarsus-toe length (mm)	115.2±3.7 (58)	108.4±4.4 (41)				Stonehouse (1971)
	114.0±3.8 (12)	109.0±2.6 (41)				Warham (1974a)
Total head length (mm)	131.11±3.84 (9)	121.6±3.5 (16)				Mattern, Houston and Ludynia (unpubl. data)
Egg length					A-egg 67.1±2.6 (23) 68.6±2.4 (50) B-egg 72.1±2.6 (23) 73.8±2.3 (50)	Warham (1974a) Massaro and Davis (2005) Warham (1974a) Massaro and Davis (2005)
Egg breadth					A-egg 51.0±1.7 (23) 52.0±1.7 (50) B-egg 56.0±1.3 (23) 56.7±1.8 (50)	Warham (1974a) Massaro and Davis (2005) Warham (1974a) Massaro and Davis (2005)
Egg mass					A-egg 90.0±6.3 (6) 99.0±7.9 (50) B-egg 117.0±6.0 (6) 128.3±8.7 (50)	Marchant and Higgins (1990) Massaro and Davis (2005) Marchant and Higgins (1990) Massaro and Davis (2005)

TABLE 11.2 Range and distribution of Snares penguins

	LAT	LON	DATE	NO OF COLONIES	NO OF PAIRS	NO OF PAIRS PER COLONY		
North East Island								
Northern Promontory	48°1.2'	166°36.3'	4–23 April 2000	89	21,148	238	250	1–1351
Southern Promontory	48°1.5'	166°36.4'	16–17 Oct 2000	22	3,318	150	114	19–434
Broughton Island	48°2.5'	166°37.1'	Oct 2000	19[1]	4,737[2]	-	-	-
Western Chain								
Toru	48°3.5'	166°30.3'	ca. 1986	-	231	-	-	-
Rima	48°3.7'	166°29.8'	11 Feb. 1984	-	100–150	-	-	-

1   Warham (1974a)

2   Amey et al. (2001)

TABLE 11.3 Summary of breeding population estimates for Snares penguins

YEAR	METHOD	TOTAL	NORTH EAST IS.	BROUGHTON IS.	WESTERN CHAIN	REFERENCE
1968/1969	Chick counts	11,000[1]				Warham (1974a)
1982/1983	Chick counts	20,000				Warham et al. (1986)
1984/1985	Chick counts	37,600[2]				Warham at al. (1986)
1985/1986	Chick counts	23,250[3]	19,000	3,500	750	Marchant and Higgins (1990)
2000/2001	Nest counts	28,800	23,638	4,737	381	Amey et al (2001)
2008/2009	Nest counts	24,120	20,090	4,030	-	D.M. Houston, pers. com.

Note: All figures refer to number of pairs, excluding number of juvenile or non-breeding adult penguins.

1   Warham (1974a) provides a total range of 30,000–50,000 penguins as an estimate for the world population of Snares penguins.

2   Referenced as "Johns pers. comm.," potentially a transcription error

3   Marchant and Higgins 1990 also cite a world population estimate of 54,000 birds for the 1986–87 season.

TABLE 11.4 Annual cycle for penguins breeding on North East Island

ARRIVAL / SETTLEMENT	EGG LAID	EGGS HATCH	CRÈCHING	FLEDGING	ADULTS LEAVE FOR PRE-MOLT FORAGING	RETURN / BEGIN MOLT	MOLT END / LEAVE
1 Sep±12.5d	30 Sep±2d	3 Nov±5d	25 Nov±5.5d	24 Jan±7d	7 Feb±7d	21 Mar±4.5d	30 April±10d

Note: All dates are based on Warham 1974a). The annual cycle of birds breeding on the Western Chain is likely to be delayed by 15–26 days (Miskelly 1984).

TABLE 11.5 Breeding parameters determined for penguins on North East Island

	PARAMETER	REFERENCES
Incubation period (d)	33.3±1.4 (29)	Warham 1974a
Chick rearing period (d)	~75	Marchant and Higgins 1990
Hatching success (%)	81.0 (43) 75.1±6.6 (154*)	Marchant and Higgins 1990 Mattern, Houston and Ellenberg, unpubl. data
Crèching success (%)	46.6±1.7 (154*)	Mattern, Houston and Ellenberg, unpubl. data
Fledging success (%)	41.6 (43)	Marchant and Higgins 1990
First year survival (%)	15	Marchant and Higgins 1990
Juvenile survival (%)	57	Marchant and Higgins 1990
Age at first breeding (years)	5.8±1.1 (5)	Warham 1974a
Life span (years)	18+	Marchant and Higgins 1990

* Data of breeding season 2003/2004 from three different areas of colony A3 on North East Island, encompassing a total of 154 nests (45, 55 and 54 nests).
Note: Hatching, crèching, and fledging success relate to number of eggs laid. First year survival is the percentage of fledglings that survive the first year at sea. Juvenile survival is the percentage of yearlings that survive until three years (Marchant and Higgins 1990).

REFERENCES

Amey, J. M., D. M. Houston, A. J. D. Tennyson, and A. K. McAllister. 2001. Census of the Snares crested penguin (*Eudyptes robustus*) breeding population. Internal report. Department of Conservation, Wellington, New Zealand.

Baker, A. J., S. L. Pereira, O. P. Haddrath, and K.-A. Edge. 2006. Multiple gene evidence for expansion of extant penguins out of Antarctica due to global cooling. *Proceedings of the Royal Society Biological Sciences B* 273:11–17.

BirdLife International. 2010. *Eudyptes robustus*. In IUCN (International Union for Conservation of Nature) 2011. IUCN Red List of Threatened Species. www.iucnredlist.org (accessed 23 March 2012).

Buller, W. L. 1888. *A History of the Birds of New Zealand*. 2nd ed. London: W. L. Buller.

Chilvers, B. L. 2009. Foraging locations of female New Zealand sea lions (*Phocarctos hookeri*) from a declining colony. *New Zealand Journal of Ecology* 33:106–13.

Christidis, L., and W. E. Boles. 2008. Order Sphenisciformes. In *Systematics and Taxonomy of Australian Birds*, ed. L. Christidis and W. E. Boles, 97–100. Collingwood, Vic., Australia: CSIRO Publishing.

Clarke, J. A., D. T. Ksepka, M. Stucchi, M. Urbina, N. Giannini, S. Bertelli, Y. Narváez, and C. A. Boyda. 2007. Paleogene equatorial penguins challenge the proposed relationship between biogeography, diversity, and Cenozoic climate change. *Proceedings of the National Academy of Science* 104:11545–50.

Cooper, J., C. R. Brown, R. P. Gales, M. A. Hindell, N. T. W. Klages, P. J. Moors, D. Pemberton, V. Ridoux, K. R. Thompson, and Y. M. van Heezik. 1990. Diets and dietary segregation of crested penguins (*Eudyptes*). In *Penguin Biology*, ed. L. S. Davis and J. T. Darby, 131–56. San Diego, CA: Academic Press.

Cunningham, D., and P. Moors. 1994. The decline of rockhopper penguins *Eudyptes chrysocome* at Campbell Island, Southern Ocean, and the influence of rising sea temperatures. *Emu* 94:27–36.

Darby, J. T., and S. M. Dawson. 2000. Bycatch of yellow-eyed penguins (*Megadyptes antipodes*) in gillnets in New Zealand waters 1979–1997. *Biological Conservation* 93:327–32.

Davis, L. S., and M. Renner. 2003. *Penguins*. London: T & A D Poyser.

Demongin, L., M. Poisbleau, G. Strange, and I. Strange. 2010. Second and third records of Snares penguins (*Eudyptes robustus*) in the Falkland Islands. *The Wilson Journal of Ornithology* 122:190–93.

Dingwall, P. R. 1995. Progress in conservation of the subantarctic islands. International Union for Conservation of Nature, Gland, Switzerland, and Cambridge.

Falla, R. A. 1935. Notes on penguins of the genera Megadyptes and Eudyptes in Southern New Zealand. *Records of the Auckland Institute and Museum* 1:319–26.

Falla, R. A., J. Warham, and C. A. Fleming. 1974. Comments on the proposed preservation of *Eudyptes sclateri* (Buller, 1888) and *Eudyptes robustus* (Oliver, 1953). *Bulletin of Zoological Nomenclature* 30:136.

Finsch, O. 1875. On two apparently new species of penguin from New Zealand. *Ibis* 5:112–14.

Fleming, C. A., and A. N. Baker. 1973. The Snares Western Chain. *Notornis* 20:37–45.

Hilton, G., D. Thompson, P. Sagar, C. RJ, Y. Cherel, and S. Brury. 2006. A stable isotopic investigation into the causes of decline in a sub-Antarctic predator, the rockhopper penguin *Eudyptes chrysocome*. *Global Change Biology* 12:611–25.

Horning, D. S., and C. J. Horning. 1974. Bird records of the 1971–1973 Snares Islands, New Zealand, expedition. *Notornis* 21:13–24.

IUCN (International Union for Conservation of Nature) 2011. IUCN Red List of Threatened Species. Version 2011.2. www.iucnredlist.org (accessed 20 April 2012).

King, C. 1985. *Immigrant Killers: Introduced Predators and the Conservation of Birds in New Zealand*. Oxford: Oxford University Press.

Kinsky, F. C. 1970. Annotated checklist of the birds of New Zealand, including the birds of the Ross Dependency. Reed, Wellington, New Zealand.

Ksepka, D. T., S. Bertelli, and N. P. Giannini. 2006. The phylogeny of the living and fossil Sphenisciformes (penguins). *Cladistics* 22:412–41.

Lamey, T. C. 1988. Snares crested penguin in the Falkland Islands. *Notornis* 37:78.

———. 1992. Egg-size differences, hatch asynchrony and obligate brood reduction in crested penguins. PhD diss., University of Oklahoma.

Marchant, S., and P. J. Higgins. 1990. *Eudyptes robustus* Snares penguin. In *Handbook of Australian, New Zealand and Antarctic Birds*, vol. 1A, ed. S. Marchant and P. J. Higgins, 205–14. Melbourne: Oxford University Press.

Massaro, M., and L. S. Davis. 2004. Preferential incubation positions for different sized eggs and their influence on incubation period and hatching asynchrony in Snares crested (*Eudyptes robustus*) and yellow-eyed penguins (*Megadyptes antipodes*). *Behavioral Ecology and Sociobiology* 56:426–34.

———. 2005. Differences in egg size, shell thickness, pore density, pore diameter and water vapour conductance between first and second eggs of Snares penguins *Eudyptes robustus* and their influence on hatching asynchrony. *Ibis* 147:251–58.

Mattern, T. 2007. Marine ecology of offshore and inshore foraging penguins: The Snares penguin *Eudyptes robustus* and yellow-eyed penguin *Megadyptes antipodes*. PhD diss., University of Otago, Dunedin, New Zealand.

Mattern, T., D. M. Houston, C. Lalas, A. N. Setiawan, and L. S. Davis. 2009. Diet composition, continuity in prey availability and marine habitat: Keystones to population stability in the Snares penguin (*Eudyptes robustus*). *Emu* 109:204–13.

McGraw, K. J., M. Massaro, T. J. Rivers, and T. Mattern. 2009. Annual, sexual, size- and condition-related variation in the colour and fluorescent pigment content of yellow crest-feathers in Snares penguins (*Eudyptes robustus*). *Emu* 109:93–99.

Ministry of Fisheries. 2010. NABIS—Internet mapping of New Zealand's marine environment, species and fisheries management. Ministry of Fisheries, New Zealand. http://www.nabis.govt.nz.

Miskelly, C. M. 1984. Birds of the Western Chain, Snares Islands. *Notornis* 31:209–23.

Miskelly, C. M., and M. Bell. 2004. An unusual influx of Snares crested penguins (*Eudyptes robustus*) on the Chatham Islands, with a review of other crested penguin records from the islands. *Notornis* 51:235–37.

Miskelly, C. M., J. E. Dowding, G. P. Elliott, R. A. Hitchmough, R. G. Powlesland, H. A. Robertson, P. M. Sagar, R. P. Scofield, and G. A. Taylor. 2008. Conservation status of New Zealand birds, 2008. *Notornis* 55:117–35.

Miskelly, C. M., P. M. Sagar, A. J. D. Tennyson, and R. P. Scofield. 2001. Birds of the Snares Islands, New Zealand. *Notornis* 48:1–40.

Mullan, B. 1996. Effects of ENSO on New Zealand and the South Pacific. In *Prospects and Needs for Climate Forecasting*, ed. D. Braddock, 23–27. Wellington: Royal Society of New Zealand.

Oliver, W. R. B. 1953. The crested penguins of New Zealand. *Emu* 53:185–87.

Proffitt, F. M., and I. G. McLean, 1990. Recognition of parents' calls by chicks of the Snares crested penguin. *Bird Behavior* 9:130–113.

Ramm, K., 2010. Conservation Services Programme observer report: 1 July 2008 to 30 June 2009. Final draft report. Department of Conservation, Wellington. 126 pp. http://www.doc.govt.nz/mcs (accessed 1 May 2012).

Ritchie, P. A. 2001. The evolution of the mitochondrial DNA control region in the Adélie penguins of Antarctica. PhD diss., Massey University, New Zealand.

Rowe, S. 2010. Level 1 risk assessment for incidental seabird mortality associated with New Zealand fisheries in the NZ-EEZ. Marine Conservation Services, Department of Conservation, Wellington. 75 pp. http://www.

doc.govt.nz/documents/conservation/marine-and-coastal/marine-conservation-services/level1-seabird-risk-assessment.pdf (accessed 29 October 2012).

Sagar, P. M. 1977. Birds of the Western Chain, Snares Islands, New Zealand. *Notornis* 24:178–83.

Stonehouse, B. 1971. The Snares Islands penguin *Eudyptes robustus*. *Ibis* 113:1–7.

Taylor, G. A. 2000. Action plan for seabird conservation in New Zealand. Part A: Threatened seabirds. Threatened Species Occasional Publication No. 16. Department of Conservation, Wellington, New Zealand.

Warham, J. 1974a. The breeding biology of behaviour of the Snares crested penguin. *Journal of the Royal Society of New Zealand* 4:63–108.

———. 1974b. The Fiordland crested penguin. *Ibis* 116:1–27.

———. 1975. The crested penguins. In *The Biology of Penguins*, ed. B. Stonehouse, 555. London: Macmillan.

Warham, J., E. B. Spurr, and W. C. Clark. 1986. Research on penguins in New Zealand. Report prepared by the Penguin Research Review Subcommittee, 14–15. Wildlife Research Liaison Group, Wellington, New Zealand.

Williams, T. D. 1995. *The Penguins*. Oxford: Oxford University Press.

12

Macaroni Penguin
(Eudyptes chrysolophus)

and

Royal Penguin
(Eudyptes schlegeli)

GLENN T. CROSSIN, PHIL N. TRATHAN, AND ROBERT J. M. CRAWFORD

1. SPECIES (COMMON AND SCIENTIFIC NAMES)

Macaroni penguin, *Eudyptes chrysolophus* (J. F. von Brandt, 1837); Royal penguin, *Eudyptes schlegeli* (O. Finsch, 1876)

2. DESCRIPTION OF THE SPECIES

Adult macaroni and royal penguins, like most penguins, have black bodies and tails and white underparts (little penguins [*Eudyptula minor*] and white-flippered penguins [*Eudyptula minor albosignata*] are the exceptions). Both species have the conspicuous golden-yellow and orange crest feathers characteristic of all crested penguins, but unlike the others, the plumes of the macaronis and royals rise from a common patch on the forehead approximately one centimeter back from the bill and extend backward over the eyes. For both species, the bill is stout and varies in color from red to orange to brown, the eyes are dark red, and there is a patch of bare pink skin from the base of the bill to the eye. Legs and feet are pink (fig. 1). The principal morphological distinction between the species lies in the highly variable facial coloration. Royals have pure white to pale gray

FIG. 1 (*FACING PAGE*) Macaroni and royal penguin at Marion Island, South Africa, showing their pink feet. (P. Ryan)

FIG. 2 A young adult macaroni penguin has yet to develop a full crest and holds a feather in its bill. Note the down attached at the base of each feather. (P. Ryan)

faces, whereas macaronis have black faces, though light-faced macaronis are reported at some sites (see section 3 below). Both species are similar in dimensions and morphology, but royals are generally larger by 10–20% depending on the trait in question (table 12.1). Within species, the sexes are similar, but males are larger than

FIG. 3 Distribution and abundance of the macaroni penguin, with counts based on pairs.

females in all respects by 15–20% (table 12.1); female royals are similar in size to male macaronis (Woehler 1995). Juvenile macaronis and royals (i.e., immature fledglings and one-year-olds) are smaller than mature adults and do not have well-developed crest feathers, which may be absent altogether or present as an irregular patch of small, stubbly, yellow feathers on the forehead (fig. 2). Juveniles also have an ashy-gray coloring on the throat (compared to the black throat of the adults) and have smaller, less robust bills that are a darker brown-black color. Diagnostic characteristics for adult macaronis and royals are compared in table 12.1.

Occasionally, there are sightings of Isabelline macaroni and royal penguins, a term used to describe a partial form of albinism in which a uniform lightening of the plumage results in a grayish-yellow coloration instead of black. This is sometimes termed "leucistic" coloration, but the more accepted term is "Isabellinism." Isabellinism has been observed in 12 species and in all genera except *Eudyptula* (Everitt and Miskelly 2003).

3. TAXONOMIC STATUS

The macaroni and royal lineage diverged from other crested penguins, specifically the Snares (*E. robustus*) and Fiordland (*E. pachyrhynchus*) penguins, circa 2 million years ago, coinciding with the onset of Pleistocene glaciations (Baker et al. 2006). Similarities in morphology, behavior, and life history once suggested that royal penguins were a distinct subspecies of macaroni penguins (i.e., *Eudyptes chrysolophus schlegeli*) rather than a separate species (Falla and Mougin 1979; Marchant

and Higgins 1990; Christidis and Boles 1994), and early arguments for single-species classification noted that the polymorphisms exhibited by royals and some Indian Ocean populations of macaronis produce phenotypes that cannot be reliably differentiated by morphological or plumage characteristics alone (Shaughnessy 1975; Barre et al. 1976; Jouventin 1982). Christidis and Boles (1994) speculate that this polymorphism could be due to hybridization, but this hypothesis has not been tested. Nevertheless, the weight of evidence points toward the current two-species classification (Sibley and Monroe 1993; Baker et al. 2006). The reproductive isolation of royals (which breed principally at Macquarie Island and two adjacent islets) and some morphological distinctions (Williams 1995; Woehler 1995; Hull 1996) suggest that royals may have diverged from macaronis, though this, too, has not been tested.

4. RANGE AND DISTRIBUTION

Macaronis are circumpolar in distribution (fig. 3). Breeding colonies are found on a number of sub-Antarctic islands throughout the South Atlantic and Indian Oceans between latitudes 46° and 65° south, and these are summarized in table 12.2. At present, there is only minimal breeding on the Antarctic continent, situated along the Western Antarctic Peninsula (Naveen et al. 2000; Gorman et al. 2010). Ranges during the nonbreeding season in winter are not well known (Woehler 1992; Reid et al. 1995), but recent telemetry studies show that macaronis originating from Kerguelen can cover an area greater than 3 million square kilometers during their six-month-

FIG. 4 Distribution and abundance of the royal penguin.

long migrations throughout the Indian Ocean basin (Bost et al. 2009). Royals are restricted in their breeding distribution to Macquarie Island and its adjacent Clerk and Bishop islets (table 12.2; fig. 4). Nonbreeding distributions are also not well known, but, like macaronis, royal penguins are absent from breeding colonies and local waters in winter, so they are probably migratory. Indeed, sightings have been reported at Australia, New Zealand, and parts of Antarctica in winter (Woehler 1992), and some "white-faced" birds were seen at Heard, Marion, Crozet, and Kerguelen Islands, though whether these were royals, polymorphic macaronis, or royal-macaroni hybrids is not known (see section 3 above). A summary of breeding locales and population estimates is provided in table 12.2.

5. SUMMARY OF POPULATION TRENDS

Macaroni penguins are decreasing throughout their range (BirdLife International 2010a), particularly at South Georgia (Trathan et al. 2012), and a population estimated at 1 million breeding pairs was believed to have been eliminated recently from McDonald Island due to volcanic activity in the Kerguelen plateau, southern Indian Ocean (Woehler 2006). However, recent satellite imagery of McDonald Island shows unidentified colonies on the fresh lava flows, which may be recolonizing macaroni penguins. Royal penguins are restricted to a breeding area smaller than 100 square kilometers that contains fewer than five discrete breeding locations, all of which are decreasing (BirdLife International 2010b).

6. IUCN STATUS

The International Union for Conservation of Nature has listed macaroni penguins as Vulnerable on its Red List of Threatened Species (IUCN 2011; BirdLife International 2010a), based on its criteria of global population reduction of 30% over 10 years or three generations within the past 30 years, and on the persistence of anthropogenic stressors (e.g., fisheries activities and changes in the marine environment). This classification relies heavily on global population estimates extrapolated from small-scale data sets, which at times are incomplete or data deficient for the most recent years. The lack of recent population assessment data at a time when many other penguin species are decreasing highlights the importance of long-term monitoring programs to confirm population trends.

Royal penguins are listed as Vulnerable, a designation stemming from the species' extremely restricted breeding range, past degradations of breeding habitat, and its comparatively low and variable population numbers (BirdLife International 2010b). However, habitat quality is likely to improve due to ongoing conservation measures, particularly rodent eradication programs (see Tasmanian Parks and Wildlife Service 2003), which may lead to an improved conservation status on the island and an IUCN downlisting in the future (fig. 5).

LEGAL STATUS. The Antarctic Treaty was signed by 12 nations in 1959, and it was further strengthened in 1991 with the agreement on the Protocol on Environmental Protection to the Antarctic Treaty. The Antarctic Treaty

FIG. 5 Royal penguins squabble on the beach at Macquarie Island, Australia. (P. D. Boersma)

established a committee for environmental protection to formulate advice and recommendations to the signatory parties and includes the Convention for the Conservation of Antarctic Seals (1972) and the Convention on the Conservation of Antarctic Marine Living Resources (1980). A 1964 Antarctic Treaty Consultative Meeting adopted the Agreed Measures for the Conservation of Antarctic Fauna and Flora, which makes it illegal to harm or in any way interfere with penguins and their eggs. The agreement requires researchers seeking penguin specimens for scientific purposes to apply for a special collection permit, and all subsequent research activities are to be reported to the Scientific Committee for Antarctic Research. The treaty applies only to treaty areas below 60° south, so penguins breeding on sub-Antarctic islands outside the treaty area must depend on other national and international agreements for protection.

The Convention on International Trade in Endangered Species of Wild Fauna and Flora (CITES) was ratified in 1973 and regulates trade in certain wildlife species, including penguins. Macaroni and royal penguins are not traded and therefore do not qualify for protection. At this time, only African penguins (*Spheniscus demersus*) (Linnaeus 1758) and Humboldt penguins (*S. humboldti*) (Meyen 1834) are listed with the convention.

The Penguin Conservation Assessment and Management Plan (Ellis et al. 1998) details existing threats and outlines priority research needs for macaroni and royal penguins, but it is not binding on any state. UNESCO has inscribed some locales as World Heritage sites, which puts in place internationally binding habitat protections, thereby indirectly protecting penguins. At this time,

many macaroni and royal breeding sites are included on the World Heritage List and the Ramsar List of Wetlands of International Importance, among other conservation measures.

At the national level, many governments with jurisdiction over penguin breeding localities have enacted species protection or habitat conservation plans with the aim of protecting all important wildlife. Breeding populations of macaronis at South Georgia Island and the South Sandwich Islands benefit from a restrictive visitation policy, and some areas, such as Cooper, Annenkov, and Bird Islands, are specially protected and reserved for permitted scientific activities only. In the marine waters surrounding South Georgia and the South Sandwich Islands, there is a complete prohibition on fishing (through licensing policy) within approximately 22 kilometers of the South Georgia coast, and 5 kilometers of the South Sandwich Islands.

South Africa's Prince Edward Islands, which includes both Marion Island and Prince Edward Island, were declared a special nature reserve under the Environmental Conservation Act of 1989 and maintain this status under the National Environmental Management Protected Areas Act of 2003. From the mid-1990s onward, activities in these islands have been regulated by a management plan.

In the French sub-Antarctic territories, the French government established a 7,000-square-kilometer nature reserve that includes all of the terrestrial and adjacent marine areas of the Crozet archipelago, Kerguelen Islands, Saint Paul Island, and Amsterdam Island (Cook and Davaine 2009). There is also a concerted effort under way to protect and restore all indigenous floral and faunal communities that have been impacted by introduced species (mice, rats, rabbits, cats, sheep, bighorn sheep, cows, reindeer, and even salmonids) (Johnstone 1985; Duchêne 1989; Chapuis et al. 1994).

In Australia, macaroni populations breeding at Heard Island and McDonald Island are afforded protection via the Heard Island and McDonald Island Marine Reserve Management Plan (Australian Antarctic Division 2005). These islands are also protected internationally as a World Heritage site and are covered by the Ramsar convention. There is presently a stringent no-visitation policy in place at McDonald Island, implemented to facilitate the recovery of macaroni penguins after recent volcanic activities, and even scientific visits are prohibited. Commercial fishing that occurs around Heard and McDonald Islands is restricted to long-lining.

The Australian government lists royal penguins as a marine species under the Environment Protection and Biodiversity Conservation Act 1999. The birds' principal breeding locale at Macquarie Island is a UNESCO World Heritage site and biosphere reserve. It is also designated a Tasmanian Nature Reserve (Holmes 2007; Tasmanian Parks and Wildlife Service 2003) and is on the Ramsar List of Wetlands of International Importance (fig. 6).

7. NATURAL HISTORY

BREEDING BIOLOGY. Macaronis are circumpolar in distribution (46–65° S) and breed on sub-Antarctic islands close to the Antarctic Convergence in the South Atlantic and Indian Oceans and in limited areas of the Antarctic Peninsula that are free from pack ice. Royal breeding is restricted to Macquarie Island, Australia, and adjacent Bishop Island and Clerk Island. The breeding colonies for both macaronis and royals generally are found on steep, rocky scree slopes or on more level open areas with little vegetation (fig. 7). Entrances to colonies are usually along rocky ledges, which may be exposed to heavy surf, and pathways or "highways" from the coast lead inland to nesting areas (fig. 8). In many places, creeks provide access routes into and out of colonies. Colonies can range in size from a few hundred individuals to several hundreds of thousands. At Bird Island, South Georgia, for example, one colony contains approximately 500 pairs, and less than one kilometer away, another colony contains approximately 40,000 pairs. Nests typically consist of little more than a shallow scrape or depression lined with small stones or sometimes with bits of vegetation (fig. 9). Some macaronis nest directly in tussock grass (*Poa* spp.), which tends to grow along colony margins. Royals sometimes nest in vegetated and unvegetated sandy areas.

After the breeding season, both species are dispersive and highly pelagic, unseen on land during the six-month over-winter period (April–October). Recent telemetry studies show that macaronis can range over 3 million square kilometers throughout the southern latitudes in winter (Bost et al. 2009). Pairs are generally monogamous (fig. 10). In a four-year period, 71–79% of pairs

FIG. 6 Royal penguins show variation in their head and neck coloration, Macquarie Island, Australia. (P. D. Boersma)

remained together for two consecutive years, and more than 50% for three consecutive years (Williams 1995). Pairs also tend to reoccupy nest sites from previous years. At South Georgia, site fidelity in a subsequent year was 69–87%. When pairs do split, 69% of females change nest sites compared to 31% of males (Williams 1995).

Typical of all *Eudyptes*, macaroni and royal penguins exhibit an unusual reproductive strategy characterized by reversed hatching asynchrony and obligate brood reduction (Williams 1995; Crossin et al. 2010). In this system, females produce two eggs but fledge only one chick (fig. 11). Unusual, too, is that the first-laid egg (A-egg) is 55–75% smaller in mass than the second-laid egg (B-egg). Such extreme egg-size dimorphism is unparalleled in birds (Slagsvold et al. 1984; Christians 2002) and, despite many long-standing hypotheses, remains unexplained (e.g., Lack 1968; Johnson et al. 1987; St. Clair 1998), though a recent study links this dimorphism to a direct physiological conflict of migration, that is, a migratory carry-over effect (Crossin et al. 2010). A forthcoming phylogenetic analysis also suggests that constraint on

A-egg size has occurred over evolutionary time scales and that the females' inability to rid themselves of the first ovulation leads to the evolution of reversed hatching asynchrony, egg ejection behaviors, and maternal infanticide (Crossin et al. 2010 and references therein).

PREY. Throughout their biogeographic range, macaronis forage principally on euphausid crustaceans and myctophid fishes along with some amphipods and squid. The relative importance and diversity of euphausid and fish components vary geographically. Antarctic krill (*Euphausia superba*) dominates the diet at South Georgia and in the southwestern Atlantic in most seasons but is largely absent from diets in more northerly and easterly locales, where other species of krill (*E. vallentini* and *Thysanoessa gregaria*) tend to be the dominant crustaceans and myctophid fish compose a greater proportion of the diet (Crawford et al. 2003; Ratcliffe and Trathan, in press). Prey is caught by pursuit-diving, normally at depths of 15–60 meters, though the penguins may dive 90 meters or deeper. Some night foraging occurs,

FIG. 7 Macaroni penguins breeding on a steep cliff at South Georgia Island. (P. D. Boersma)

FIG. 8 Adult macaroni penguin balances on a rock before moving inland. (P. Ryan)

FIG. 9 Macaroni penguins on their nests trying to guard their chicks. (P. Ryan)

FIG. 10 Macaroni penguin pair engage in mutual preening at their nest site. (P. Ryan)

FIG. 11 Royal penguin on its nest brooding a young chick. (P. Ryan)

but night dives are much shallower, only 3–6 meters in depth. Dives rarely exceed two minutes. At Marion Island, macaroni penguins foraged closer to the island (up to 50 km) during early chick rearing than during late chick rearing (50–100 km) (Brown and Klages 1987). The diet and foraging behavior of royals is less studied, but unlike macaronis, royals rely more heavily on myctophid fishes (59%) than on euphausiid crustaceans (37%) (Hindell 1988; Hull et al. 1997; Hull 1999; Goldsworthy et al. 2001).

PREDATORS. At sea, principal predators are killer whales (*Orcinus orca*) (Condy et al. 1978), leopard seals (*Hydrurga leptonyx*) (Walker et al. 1998), and, less frequently, Antarctic fur seals (*Arctocephalus gazella*) (Bonner and Hunter 1981). On land, principal predators are giant petrels (*Macronectes* spp.), skuas (*Catharacta* spp.), sheathbills (*Chionis* spp.), and gulls (*Larus* spp.), which take eggs, chicks, and vulnerable juveniles and adults and scavenge carcasses. Introduced animals, such as rats (*Rattus* spp.), mice (*Mus* spp.), and cats (*Felis*

catus), can have significant impacts on ground- and burrow-nesting seabirds. Their effect on macaroni penguins and their eggs is unquantified but is likely to be directed at small chicks.

For many years, royal penguins were massively exploited for their rich oil, which was a valued commodity, but by the early 20th century, the Australian government had legislated their protection. On Macquarie Island, introduced species caused enormous damage, which was subsequently exacerbated by attempts to eradicate those species with additional introduced species (Bergstrom et al. 2009; Raymond et al. 2010). Rats, mice, and rabbits were introduced at various times and have had noticeable though largely unquantified effects on royals. Cats were subsequently introduced to control the rodent populations, but cats also preyed on seabirds. Whether they killed penguins is not well documented, but penguin predation has been inferred from studies of feral cat diet at Marion Island (van Aarde 1980). The Australian government is attempting to eradicate all introduced species from Macquarie Island by 2015. To

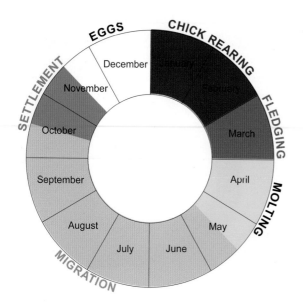

FIG. 12 Annual cycle of the macaroni penguin.

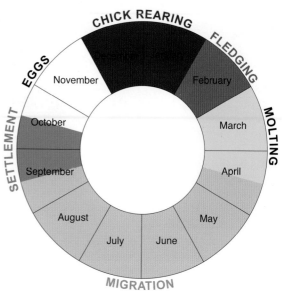

FIG. 13 Annual cycle of the royal penguin.

date, feral cats have been fully eradicated, and efforts to remove rabbits and other rodents are ongoing.

MOLT. Adult macaronis leave their breeding colonies as soon as their chicks fledge (generally by early March) and undertake pre-molt foraging trips lasting two to three weeks. Body mass increases by as much as 70% in preparation for the molting fast, and penguins return to their nest sites for a complete body molt lasting 25–35 days at South Georgia and Marion Island. Endogenous fat reserves are essential for fueling molt and new feather production. Upon completion of molt, penguins return to sea and depart for over-wintering areas. The last individuals are seen at South Georgia and Marion Island in late April, and birds are not seen again locally until the following October.

The pre-molt foraging trips of royals last considerably longer than those of macaronis (30–35 days vs. 14–16 days), which may reflect the better foraging conditions found around South Georgia and Marion Island (Williams and Croxall 1991) or other unknown factors. Like macaronis, most royals leave their colonies in mid- to late April, with the last birds seen as late as mid-May.

ANNUAL CYCLE. See figures 12 and 13 and table 12.3.

REPRODUCTIVE AND FITNESS PARAMETERS. Some estimates of reproductive and fitness parameters are indicated in table 12.4. Croxall and Davis (1999) note that male maca-

ronis bred for the first time at the age of 7.5 years; birds may breed at considerably younger ages, but breeding success is generally lower with younger breeders. Breeding success for macaroni penguins may vary considerably between years and localities (Crawford et al. 2003; Crawford et al. 2006a). At Marion Island, a correlation of breeding numbers over a 26-year period suggests that over-wintering conditions may influence the relative abundances of breeding birds (Crawford et al. 2006a). At the same locality, the pre-breeding body mass of males for breeding was significantly correlated to pre-breeding female mass over an 11-year period, and the mass of chicks at fledging was significantly related to reproductive success (Crawford et al. 2006a). In general, over-winter feeding conditions can influence pre-breeding body mass in many seabird species, which can influence not only whether or not birds breed but patterns of reproductive investment (e.g., egg size). However, links between body mass at arrival, reproductive investment, and reproductive success are not strong in pre-breeding female macaronis (Crossin et al. 2010).

8. POPULATION SIZES AND TRENDS

The global macaroni population includes at least 258 colonies at approximately 55 breeding sites on many islands throughout sub-Antarctic and Antarctic waters north of the pack ice, and the current global population is estimated at 6.3 million breeding pairs (see table 12.2). This represents a 30% decrease from the most recent global population estimate of approximately 9 million breeding pairs (Woehler 1993; Reilly 1994; Woehler and Croxall 1997; Bingham and Majias 1999; Ellis et al. 1998). The decrease in population abundance is driven mostly by recent decreases at South Georgia (Trathan et al. 2012), and several long-term monitoring programs at other locales indicate rapid decreases in abundance over the past 50 years (Woehler and Croxall 1997; Trathan et al. 1998; Trathan 2004; Barlow et al. 2002; Crawford et al. 2003; Crawford et al. 2006a; Crawford et al. 2009; Pütz et al. 2003). The most striking rate of decrease is at South Georgia, where nearly half of the global population once bred. The number breeding was estimated at roughly 5 million pairs in the 1980s (Croxall and Prince 1979, 1980) but fell to approximately 2.7 million in the mid-1990s (Trathan et al. 1998) and to less than 1 million pairs in 2002 (Trathan et al. 2012).

Although the South Georgia population as a whole has decreased very significantly, data from individual monitoring colonies show that local increases have occurred since the early part of this decade (British Antarctic Survey, unpubl. data). Elsewhere, other examples of recent decreases include the populations at Heard Island and McDonald Island. Each island supported a population of about 1 million macaroni pairs in 1987 (Woehler 1991), but volcanic activity recently eliminated the McDonald Island population (Woehler 2006), though unidentified penguins are visible in satellite images. Surveys at Heard Island suggest a population decrease due to the loss of some smaller colonies (typically less than 1,000 pairs) (Woehler 2006), but comprehensive survey data are lacking. The overall population at Marion Island decreased by more than 30% over 14 years between 1994–2005 and 2008–9, from 434,000 pairs to 290,000 pairs (Crawford et al. 2003, 2009). One small colony that has been monitored at Marion Island since the late 1970s decreased after the early 1980s (Cooper et al. 1997). Other macaroni colonies have shown decreases since the early 1980s (e.g., Woehler et al. 2001; Ryan et al. 2003), but some colonies at various locations may be stable or perhaps increasing. For example, at Seal Island, South Shetland Islands, there was little evidence of change in population size between 1988 and 1995 (Woehler and Croxall 1997), and at the Kerguelen Islands, several of the more populous colonies (2000–40,000 pairs) show a slight increase at times (Weimerskirch et al. 1989; Woehler et al. 2001). The sizes of some

large macaroni populations have not been assessed since 1989 (e.g., the Crozet and Kerguelen Islands), limiting the ability to assess regional and global population trends accurately (Woehler et al. 2001). Given that some other main breeding sites have shown marked decreases over the past 36 years (BirdLife International 2000), a reassessment of the global population is an urgent priority following the collection of contemporary population data. Despite the lack of new assessment data from the Crozet and Kerguelen Islands, we nevertheless estimate a 30% decrease in global population since 1993, which is likely to be an underestimate.

The estimated breeding population of royals on Macquarie Island in the mid-1980s was 850,000 pairs, with an additional 1,000 pairs on Bishop and Clerk Islands (Garnett and Crowley 2000). Recent estimates at Macquarie Island suggest a decrease of approximately 40% to 500,000 pairs (E. J. Woehler, pers. comm.).

TRENDS. See table 12.2.

9. MAIN THREATS

Different species might experience a common set of factors that cause reductions in their populations. Introduced mammals are a substantial problem for breeding seabirds on a number of sub-Antarctic islands, including Gough, Kerguelen, Crozet, Auckland, and Macquarie Islands (Grant et al. 2011). For example, cats, mice, and rabbits have posed problems for nesting seabirds at the Kerguelen and Crozet Islands (Jouventin and Micol 1995), and rats are an ongoing problem at Macquarie Island, though quantifying the effects of these introduced species on penguin population dynamics is difficult and sometimes speculative (but see Raymond et al. 2010) (also see "Habitat degradation" below). Despite this, recent documentation of mouse predation on live wandering albatross (*Diomedia exulans*) chicks at Gough Island (Cuthbert and Hilton 2004) and Tristan albatrosses (*D. dabbenena*) at Marion Island (Jones and Ryan 2010) underscores the potential for introduced rodents to have significant and unpredictable impacts on seabird populations.

Eradication programs at various sub-Antarctic islands are under way in an effort to restore and conserve native faunal and floral communities. Overfishing may also have serious effects on penguin demographics. In particular, the commercial harvesting of Antarctic krill is now increasing and is of concern, as krill is the main food source for macaronis in the southwestern Atlantic, where most krill harvesting takes place. Elsewhere, fisheries for sardines and anchovies have already been linked with reductions in populations of other penguin species (e.g., Crawford et al. 2006b, 2007). Further pressures include potential oil spills and increasing tourism, as well as predicted climate change. The latter is particularly important to the long-term maintenance of penguin numbers, as penguins are very sensitive to changes in regional and broad-scale climate, sea surface temperature, and ocean currents and the consequent effects on prey availability (Ellis et al. 1998; Reid and Croxall 2001; Trathan et al. 2007; Forcada and Trathan 2009; Scheffer et al. 2010). The following factors affect (or have the potential to affect) penguin demography.

CLIMATE VARIATION. By its influence on prey productivity and distribution, climate change is currently the factor most likely to affect macaroni and royal demographies, as has already been documented for other penguin species (see Forcada and Trathan 2009). Throughout the Southern Ocean, evidence for increases in air and sea surface temperature can be found in the Atlantic, Indian, and Pacific sectors (Moors 1986; Cunningham and Moors 1994; Jacka and Budd 1998; Gille 2002; Jenouvrier et al. 2005a, 2005b, 2009; Forcada and Trathan 2009; Ainley et al. 2010), thus underlying a warming trend and major change in key properties of the Southern Ocean's physical environment. Global, atmospheric-oceanographic teleconnections have direct and indirect effects on the Southern Ocean ecosystem, and phenomena such as El Niño appear to affect sea ice processes in the Southern Ocean (Yuan 2004). Indeed, some of the clearest signals of regional climate warming are found on the Antarctic Peninsula (Smith et al. 1999; Vaughan et al. 2001; Gille 2002; Cook et al. 2005), including significant increases in air temperature, the retreat of glaciers, and the collapse of large ice shelves (Oppenheimer and Alley 2004; Clarke et al. 2007; Ducklow et al. 2007; Stammerjohn et al. 2008).

The recent reduction in the number of cold years in which there is a heavy volume of winter sea ice has had significant consequences for resident penguin populations, especially the Adélie, chinstrap, and gentoo (Fraser et al. 1992; Fraser and Trivelpiece 1996; Croxall et al. 2002; Smith et al. 2003). The western Antarctic Penin-

sula is one of only three high-latitude locales that are warming faster than the global average, and this warming has been associated with changes in the abundance and phenology of sea ice, which is in turn associated with krill productivity (Atkinson et al. 2004). At South Georgia, where about half of the world's macaroni penguins once bred, the ecosystem depends on Antarctic krill that are advected from the Antarctic Peninsula (Murphy et al 1998; Atkinson et al 2004). Because food supplies are known to affect animal population dynamics (Sinclair and Krebs 2002), major population decreases are likely to result from reduced prey availability caused by the effects of warming on sea ice. Recent decreases in macaroni penguin numbers at South Georgia may be related to the effect of warming on krill advection from the Antarctic Peninsula (Atkinson et al. 2004; Trivelpiece et al. 2011), though some authors have suggested that the decreases are more likely to be linked to resource competition following the Antarctic fur seal's recovery from historical harvesting (Barlow et al. 2002). Climate warming affects not just a population's demography but also its distribution, and one consequence of recent warming events on macaroni populations is the sighting of macaronis at Avian Island on the Antarctic Peninsula, which lies just one degree south of the Antarctic Circle, making this the most southerly observation of macaronis on record (Gorman et al. 2010). Though not breeding at this location, breeding macaronis have been observed in recent years at other locales along the Antarctic Peninsula near Palmer Station (Gorman et al. 2010). Whether these observations are due to vagrancy by macaroni penguins or to complex relationships among ocean-climate and population-demographic processes as well as biogeographic range dynamics is presently unknown. The warming trend along the Antarctic Peninsula shows no sign of abating, and species that require ice-free areas for breeding (e.g., gentoo and chinstrap and possibly macaroni penguins) are expected to expand along the Antarctic Peninsula, while those that are ice obligate (e.g., Adélie penguins) are predicted to contract poleward and further decrease in number (Fraser and Patterson 1997; Ainley et al. 2003; Fraser and Hoffman 2003; Forcada et al. 2006; Forcada and Trathan 2009).

HISTORICAL OR CURRENT HARVEST. Internationally, all 18 species of penguins, including macaronis and royals, are protected from hunting and egg collection through the

FIG. 14 Royal penguins resting on the beach at Macquarie Island. Individuals show variation in their coloring. Note the gray feathers under the chin of the penguin in the foreground. (P. D. Boersma)

Convention on the Illegal Trade in Endangered Species of Wild Flora and Fauna (CITES) and by the IUCN. Similar protections are afforded to penguins by countries that have breeding colonies within their jurisdictions. Although macaronis and their eggs were once harvested, most information is anecdotal, and the scale of harvest is unknown, though it is likely to have been small and localized. Past harvest of royals in the 19th century was more important due to the species' small population size and restricted geography. For many years, royals were harvested and rendered for their oil. It was estimated that 2,700 penguins were slaughtered each day from early January to late March during the years of peak production in the early part of the 20th century. This went on for more than 25 years, but the practice was eventually terminated when the Australian government declared Macquarie Island a wildlife sanctuary in 1920 (fig. 14).

FISHERIES. Krill fisheries in the Southern Ocean began more than 30 years ago and have been held at a relatively low level of harvest (approximately 200,000 tons in 2010), well below the maximum total allowable catch (5.6 million tons in 2010) established by the Commission for the Conservation of Antarctic Marine Living Resources (CCAMLR) which regulates the krill fishery. The greatest threat to macaroni and royal penguins is not necessarily the abundance of krill harvested but the

timing and location of harvest and the spatial concentration (i.e., spatial and temporal overlap between commercial fishing activities and penguin foraging during the breeding season) (Trathan et al. 1998; Croxall and Nicol 2004; Cresswell et al. 2008). Though krill harvests tend to occur in summer when penguins are breeding, the fishery around South Georgia operates only in the winter and so minimizes any direct conflict with the high number of breeding penguins (as well as other breeding krill-dependent predators). Additionally, licensing policy completely prohibits fishing within approximately 20 kilometers of the South Georgia coast, with a restriction of 5 kilometers around the South Sandwich Islands. The French sub-Antarctic territories have similar limits on fisheries within the exclusive economic zone. A similar zone restricts fishing around Australia's Heard, McDonald, and Macquarie Islands. In general, fisheries pose significant threats to penguins, not through incidental mortality, but by their effect on prey abundance. These effects are likely to be local near breeding areas, but as the over-winter distributions for some penguins, including royals, are not well known, fisheries may have a greater effect than estimated during the breeding season alone (though low densities of penguins at sea and low fishing catch mass in winter suggest that impacts may be low).

Effective fisheries management regimes that minimize competition with penguins are therefore critical to ensuring the sustainability of penguin populations. This is especially important in the Southern Ocean, where illegal, unreported, and unregulated (IUU) fisheries are pervasive and undermine existing management schemes (Agnew 2000; Croxall and Nicol 2004). Though climate change and marine pollution have clear and profound influences on the Antarctic ecosystem, we can do relatively little to mitigate their effects, as these processes operate in complex ways over long time scales and large spatial scales, generally originating from outside the Southern Ocean. In contrast, fishing is the single major activity in the Antarctic region that actively sets out to exploit, and thus alter, the existing ecosystems, and it must be managed effectively (Croxall and Nicol 2004).

HABITAT DEGRADATION. In the French sub-Antarctic territories, introduced species have had wide-ranging, negative effects on habitat quality and demography via predation (Johnstone 1985; Chapuis et al. 1994), and the French government is committed to restoring indigenous floral and faunal communities (fig. 15). Heard and McDonald Islands are free of introduced predators, but habitat loss due to volcanic activity is responsible for the extinction of macaroni penguins on McDonald Island (Woehler 2006). At Macquarie Island, helicopter pilots and sharpshooters use aerial baiting for eradication of cats and rabbits that have affected many burrow-nesting seabird species. Springer spaniels and Labrador retrievers detect and flush out any surviving rabbits (Tasmanian Parks and Wildlife Service 2003). Feral cats, which were long ago introduced to the island to control introduced rodent populations, were eradicated. Rats remain, though their effect on penguin population dynamics is not known. Ongoing eradication efforts are not without complications (Tasmanian Parks and Wildlife Service 2010). There are no introduced mammals at Prince Edward Island. At Marion Island, cats and mice were both introduced, but cats have been eradicated (Bester et al. 2002).

Introduced plant species, documented at many locales, do not appear to have any influence on habitat choice or use by macaroni or royal penguins.

POLLUTION. Oil pollution from tanker discharge and spills is an important cause of mortality for temperate-zone penguins (Williams 1995; Boersma 2008), but spills also pose a threat to penguins farther south. In 1989, for example, an Argentine resupply vessel ran aground on

FIG. 15 Pair of young macaroni penguins are likely to benefit from habitat restoration on their breeding islands. (P. Ryan)

the Antarctic Peninsula near Palmer Station and spilled an estimated 600,000 liters of oil (an event overshadowed by the *Exxon Valdez* spill, which occurred that same year), resulting in the loss of about 16% of Adélie penguins breeding in the region (Fraser and Patterson 1997; for other impacts, see Woehler et al. 2001). In 1995, the *Iron Baron* oil spill near northern Tasmania affected little penguins (*Eudyptula minor*), and subsequent monitoring showed that oiled and rehabilitated females had significantly lower breeding success compared to non-oiled females (Giese et al. 2000). In March 2011, the Greek vessel MS *Olivia* ran aground on Nightingale Island at Tristan da Cunha en route from Rio de Janeiro to Singapore. The ship foundered and spilled more than 1,400 metric tons of heavy crude oil around the island, which is home to nearly half the world's population of endangered northern rockhopper penguins, and many rockhoppers were heavily oiled. Exploration for oil has recently begun at the Falkland/Malvinas Islands (led, ironically, by a company named Rockhopper Petroleum PLC), putting at risk many local and adjacent penguin populations, including the low number of macaroni penguins breeding on the Falkland/Malvinas Islands.

Oil pollution has already had an impact on royal populations, and remediation techniques are currently being used to clean up small spills at three sites on Macquarie Island (Pople et al. 1990; Smith and Simpson 1998). The Australian government is developing risk assessment guidelines and target hydrocarbon levels for remediation (Pyper 2009). Pollutants other than oil are a growing problem for some penguins, but documentation is lacking for macaronis and royals. Threats include ingestion of and entanglement in a variety of plastic debris and stray fishing gear.

DISEASE. The number of human visitors to Antarctica is rapidly increasing, and with this comes the risk of introducing infectious microorganisms, viruses, and parasites (Alexander et al. 1989; Grimaldi et al. 2011). Potential vectors include humans as well as introduced species. Precautionary measures have been instituted at some breeding areas in order to prevent pathogen introduction and spread. At Bird Island, South Georgia, at the Prince Edward Islands, and at Heard and Macquarie Islands, all visitors must disinfect their boots before setting foot on land, and no landings are permitted on McDonald Island. Despite this, bacterial (*Campylobacter jejuni,* the most common agent of gastroenteritis in humans), viral

(paramyxovirus), and parasitic (*Ixodes uriae, Tegophorus macronectes, Contracaecum heardi, Tetrabothrius* spp.) organisms have been detected in macaroni penguins at South Georgia (Broman et al. 2000; Barbosa and Palacios 2009 and references therein; Grimaldi et al. 2011). Similar pathogens have been found in royal penguins, including *Chlamydia* spp. (Barboas and Palacios 2009), which is a common sexually transmitted disease in humans. In 1993, an unknown disease killed an estimated 5,000–10,000 macaronis at Marion Island, and in 2004, avian cholera (*Pasteurella multocida*) killed about 2,000 more (Cooper et al. 2009). The hematozoa responsible for avian malaria, the most common agent of mortality of penguins in captivity, was not found in wild macaronis and royals (Schultz and Petersen 2003). Nevertheless, introduction of pathogens and susceptibility to infection are both predicted to increase as the climate continues to change and insect vectors expand (Chown et al. 1998). On Macquarie Island, the *Myxoma* virus, which was spread by introduced rabbits, was first detected in 1978, but rabbits have now been almost fully eradicated. Myxomatosis does not appear to affect the island's royal penguins.

HUMAN VISITATION. Tourism is increasing throughout the Antarctic and sub-Antarctic regions (see Naveen et al. 2001). Visitation of breeding colonies can be both a

FIG. 16 Royal penguin on the beach at Macquarie Island, Australia. (P. D. Boersma)

chronic and an acute stressor that can lead to increased predation or desertion, and as such poses the most serious risk to penguins. Getting people to remote colonies, usually by ship, poses the additional risk of oil pollution. On the other hand, tourism can be proactively managed so as to minimize these impacts, as with the development of minimum-distance approach guidelines for all wildlife, including penguins (e.g., Naveen et al. 2001; Holmes et al. 2005; IAATO 2012). Visitors arrive at macaroni penguin colonies at South Georgia each year and are expected to conform to guidelines that minimize disturbance, but there are no data on whether visitation has caused changes in demography.

10. RECOMMENDED PRIORITY RESEARCH ACTIONS FOR CONSERVATION

Ellis (1999) summarizes a number of priority research and management goals identified in the IUCN's Penguin Conservation Assessment and Management Plan (see also Ellis et al. 1998). It deems the following actions to be critical for macaroni penguins:

1. Assess the status and trends of populations breeding at colonies in the Indian Ocean, especially at Crozet and Kerguelen Islands.
2. Assess and further document the decrease in population at South Georgia Island.
3. Conduct ongoing studies of demography and foraging ecology.

It identifies similar goals for royal penguins:

1. Conduct studies of demography and foraging ecology.
2. Determine the status of and trends in the Macquarie Island population (fig. 16). Because royals breed only at Macquarie Island, the removal of introduced predators and the establishment of a marine reserve are also necessary for long-term conservation.

11. CURRENT CONSERVATION EFFORTS

It is often difficult to quantify the effects of current conservation programs, such as regulation of fisheries around South Georgia, Kerguelen, Crozet, Heard, and Macquarie Islands, preservation of nearly all breeding habitats, restrictions on human visitation of breeding colonies, and eradication of introduced species; nevertheless, these activities can be regarded as proactive measures in the effort to conserve macaroni and royal populations. The global macaroni population is decreasing, and this is likely the combined effect of many different processes. Changes in the ocean environment and the effects of fisheries are two of the greatest threats to long-term conservation. Climate change is largely beyond our ability to control, but many regulations and acts are in place to control or minimize the direct and indirect effects of fishing pressure on penguins. Although many national habitat preservation acts and local fishing regulations do much to protect penguins during the breeding season, development and ratification of international agreements and management plans that further minimize the fisheries' competition with and the incidental mortality of penguins are of particular importance. Penguins are highly pelagic during the nonbreeding period and thus at particular risk from the pervasive illegal, unreported, and unregulated fisheries operating throughout the Southern Ocean (Agnew 2000; Croxall and Nicol 2004).

12. RECOMMENDED PRIORITY CONSERVATION ACTIONS FOR INCREASING POPULATION RESILIENCE AND MINIMIZING THREATS AND IMPACTS

For both species, the recommendations put forth by Ellis (1999) are still relevant. In order of priority, macaroni and royal penguins would benefit from the following:

1. Contemporary population assessments.
2. Programs that monitor ingestion of and entanglement in marine debris, including the frequency of marine oiling.
3. Programs that monitor and mitigate conflicts with fisheries.
4. Continued eradication of introduced species.
5. Limitations on tourism in and disturbance of breeding colonies.
6. Development of demographic models with robust estimates of survival at different sex and age classes.
7. Ongoing studies of the potential impacts of climate change.

ACKNOWLEDGMENTS

Thanks are extended to Eric Woehler and Norman Ratcliffe, for comments regarding penguin demography and survival, and to Tony D. Williams, for editorial suggestions. Thanks, too, to an anonymous reviewer.

TABLE 12.1A Representative diagnostic characteristics of macaronis (*Eudyptes chrysolophus*) from Bird Island, South Georgia

	MEAN±SD (SAMPLE SIZE)	RANGE	REFERENCES †
ADULT FEMALES			
Mass-colony arrival (g)	5050±250 (106)	4800–5310	[1], 2, 3
Mass-pre-laying (g)	4890±140 (9)	4700–5100	[2], 3
Mass–start of incubation (g)	3851±267 (44)	3300–4400	2, [3]
Mass–brood/guard (g)	3432±285 (17)	3000–3900	2, [4]
Mass–crèche (g)	3787±323 (27)	3200–4300	2, [5]
Mass–pre-molt (g)	5700±350 (6)	5400–6300	[2]
Mass–post-molt (g)	3180±340 (6)	2800–3700	[2]
Wing length (mm)	204±6.15 (30)	185–214	2, [13]
Bill length (mm)	53.7±1.6 (15)	50.9–56.1	[6]
Total head length (mm)	127±4.06 (30)	119–135	[13]
Tarsus length (mm)	49±3.10 (30)	42–53	[13]
Tarsus-toe length (mm)	113±4.79 (30)	104–127	[13]
ADULT MALES			
Mass (pre-breeding, colony arrival) (g)	4982±280 (71)	4685–5240	1, 2
Mass-pre-laying (g)	4650±450 (11)	3900–5500	[2]
Mass–start of incubation (g)	3380±250 (6)	3100–3700	[2]
Mass–brood/guard (g)	4220±200 (6)	4000–4500	[2]
Mass–crèche (g)	4560±280 (16)	4200–5000	[2]
Mass–pre-molt (g)	6410±200 (7)	6100–6600	[2]
Mass–post-molt (g)	3720±340 (6)	3100–4200	[2]
Wing length (mm)	211±7.61 (30)	192–226	[13]
Bill length (mm)	61.3±2.3 (15)	56.5–65.1	[6]
Total head length (mm)	135±2.51 (30)	131–138	[13]
Tarsus length (mm)	51±2.91 (30)	47–58	[13]
Tarsus-toe length (mm)	117±4.41 (30)	107–125	[13]
EGGS			
A-egg:			
Mass (g)	94.4±9.7 (94)	74–121	3, [7]
Length (mm)	68.9±2.0 (42)	62.4–75.0	2, [14]
Breadth (mm)	50.3±3.0 (42)	44.9–53.5	2, [14]
B-egg:			
Mass (g)	149.2±12.6 (94)	108–172	3, [7]
Length (mm)	78.4±2.8 (46)	72.5–84.0	2, [14]
Breadth (mm)	58.3±1.7 (46)	54.0–62.1	2, [14]

TABLE 12.1B Representative diagnostic characteristics of royals (*E. schlegeli*) from Macquarie Island, Australia

	MEAN±SD (SAMPLE SIZE)	RANGE	REFERENCES†
ADULT FEMALES			
Mass-colony arrival (g)	—	4200–6300	[2], 11
Mass-pre-laying (g)			
Mass–start of incubation (g)			
Mass–brood/guard (g)	—	3000–3500 3200–5000	[2], 11, 12
Mass–crèche (g)			
Mass–pre-molt (g)	—	5200–8100	[2], 11
Mass–post-molt (g)			
Wing length (mm)	185.1±8.8 (7)	176–203	2, [8]
Bill length (mm)	61.1±2.63 (71)	58.0–65.5	[9], 10
Total head length (mm)	133.8±4.34 (71)	—	[9], 10
Tarsus length (mm)			
Tarsus-toe length (mm)	76.0±1.7 (7)	73.0–79.2	2, [8]
ADULT MALES			
Mass (pre-breeding, colony arrival) (g)	—	4300–7000 3800–5500	[2], 11
Mass-pre-laying (g)			
Mass–start of incubation (g)			
Mass–brood/guard (g)	—	3800–5500	[2], 11
Mass–crèche (g)			
Mass–pre-molt (g)	—	5700–8100	[2], 11
Mass–post-molt (g)			
Wing length (mm)	198.6±4.5 (5)	183–195	2, [8]
Bill length (mm)	68.7±2.85 (67)	64.2–68.6	[9], 10
Total head length (mm)	143.5±4.72 (67)	—	[9], 10
Tarsus length (mm)			
Tarsus-toe length (mm)	79.7±1.1 (4)	78.8–81.5	2, [8]
EGGS			
A-egg:			
Mass (g)	100.3 (31)	79.7–126.9	[2], 8
Length (mm)	69.7 (31)	62.5–75.2	[2], 8
Breadth (mm)	50.8 (31)	47.5–56.2	[2], 8
B-egg:			
Mass (g)	159.3 (28)	125.6–181.0	[2], 8
Length (mm)	80.7 (28)	72.0–88.0	[2], 8
Breadth (mm)	59.2 (28)	53.5–62.8	[2], 8

† Reference numbers in brackets provide the principal data for means and standard deviations presented in this table while numbers without brackets are secondary sources. (1) Williams and Croxall 1991, (2) Williams 1995, (3) Crossin et al. 2010, (4) Green et al. 2002, (5) Green et al. 2003, (6) Croxall and Prince 1980, (7) Williams 1990, (8) Marchant and Higgins 1990, (9) Hull 1996, (10) Woehler 1995, (11) Carrick 1972, (12) Warham 1971, (13) P.N. Trathan, unpublished data, (14) G.T. Crossin, unpublished data.

TABLE 12.2 Data summarizing the number of known macaroni and royal breeding locales and colonies within each locale (including geographic coordinates, approximate sizes, census date, population trend, and references)

BREEDING LOCALE	LOCALE NO.	COLONY NAME	NO. OF COLONIES	LAT	LON	COLONY SIZE (PAIRS)	MOST RECENT CENSUS DATE	% DECREASE SINCE WOEHLER 1993	REFERENCES†
MACARONI PENGUINS									
Bouvetoya	1	Nyrøysa	7	54°25' S	03°22' E	1100	2007/08	63.3	1
Prince Edward Islands	2	Marion Island		45°54' S	37°45' E	291907		27.9	2
		Macaroni Bay south	1			135	2008/09		3
		Archway Bay	1			107	2008/09		3
		East Cape	1			153	2008/09		3
		Bullard Beach north	1			173	2008/09		3
		Bullard Bay main	1			129622	2008/09		3
		Bullard Beach south	1			3 224	2008/09		3
		Waterfall	1			261	2008/09		3
		Whale Bird Promontory	1			310	2008/09		3
		Kildalkey Rocks	1			789	2008/09		3
		Kildalkey Bay	1			127976	2008/09		3
		Puisie Beach	1			288	2008/09		3
		Goodhope Bay central	1			240	2008/09		3
		Goodhope Bay west	1			1523	2008/09		3
		Rook's Peninsula	1			5563	2008/09		3
		Rook's Peninsula west	1			730	2008/09		3
		Rook's Bay	1			460	2008/09		3
		Rook's Cave	1			2380	2008/09		3
		Rook's Palin east	1			151	2008/09		3
		Rook's Plain west	1			269	2008/09		3
		Swartkop Point south	1			224	2008/09		3
		Swartkop Point main	1			13525	2008/09		3
		Kaalkoppie south	1			750	2008/09		3
		Kaalkoppie main	1			365	2008/09		3
		Kaalkoppie Crater	1			1023	2008/09		3
		Triegaardt Bay	1			4492	2008/09		3
		Sea Elephant Bay	1			260	2008/09		3
		Ship's Cove	1			380	2008/09		3
	3	Prince Edward Island		45°38' S	37°56' E	12037		29.4	
		Vaalkop west	1			480	2008/09		2,4,5
		Vaalkop east	1			1223	2008/09		2,4,5
		Albatross Valley west	1			194	2008/09		2,4,5
		Albatross Valley east	1			331	2008/09		2,4,5
		McCall Kop	1			4522	2008/09		2,4,5
		RSA Point north	1			4573	2008/09		2,4,5
		RSA Point south	1			714	2008/09		2,4,5
Isle Crozet		TOTAL				2172000	1985	unknown	6,7
	4	Ile Possession	12	46°24' S	51°45' E	400000	1981–82		6,7
	5	Ile de l'Est	12	46°24' S	52°12' E	400000	1981–82		6,7
	6	Ile aux Cochons	20	46°05' S	50°14' E	272000	1982		6,7
	7	Iles des Apotres	1	45°57' S	50°25' E	100000	1981–82		6,7

TABLE 12.2 (cont.)

BREEDING LOCALE	LOCALE NO.	COLONY NAME	NO. OF COLONIES	LAT	LON	COLONY SIZE (PAIRS)	MOST RECENT CENSUS DATE	% DECREASE SINCE WOEHLER 1993	REFERENCES[†]
	8	Ilots des Pingouins	1	45°25' S	50°24' E	1000000	1981–82		6,7
Isle Kerguelen	9	TOTAL				1812000	1985	unknown	8
		Courbet Peninsula	5	49°08' S	70°13' E	336000	1989		8
		Ronarch Peninsula	3	49°36' S	70°18' E	30000	1989		8
		Jeanne d'Arc Peninsula	9	49°42' S	70°00' F	130000	1989		8
		Coast south of Mt Ross	2	49°38' S	69°31' E	6000	1989		8
		Rallier du Baty Peninsula	6	49°42' S	69°46' E	560000	1989		8
		Nuageuses Islands	7	48°38' S	68°39' E	500000	1989		8
		Loranchet Peninsula	9	48°41' S	68°56' E	250000	1989		8
Heard Island	10	Many	Many	53°03' S	73°30' E	<1000000	2003–04	unknown	9,22
McDonald Island	11	Indeterminate		53°03' S	73°31' E		2006	100	9,22
Antarctic Peninsula	12	Humble Island	1	60°46' S	64°06' W	1	1985	unknown	
South Shetland Islands & Elephant and Clarence Island Group		TOTAL				13199	1979	unknown	10,11
	13	Deception Island	3	62°57' S	60°38' W	133	1957–58		10
	14	Livingston Island: Hannah Point	1	62°39' S	60°37' W	8	1987		6
	15	Livingston Island: King George & Nelson Island	1	62°15' S	59°15' W	2	1976		6
	16	Ridley Island	1	61°51' S	58°00' W	1	1980		6
	17	Aspland Island	1	61°30' S	50°55' W	21	1977		6
	18	Gibb Island	2	61°27' S	55°29' W	1750	1977		10,11
	19	Clarence Island	5	61°14' S	54°08' W	4142	1977		10,11
	20	Seal Islands	2	60°50' S	55°20' W	62	1987		12
	21	Elephant Island	19	61°05' S	55°00' W	7080	1987		10,6
South Orkney Islands		TOTAL				287–412		unknown	10
	22	Michelsen Island	1	60°43' S	45°00' W	6	1983		10,6
	23	Matthews Island	1	60°44' S	45°09' W	27	1979		10
	24	Signy Island	4	60°40' S	45°38' W	11	1979		6
	25	Acuna Island (Laurie Is.)	1	60°46' S	44°36' W	2	1994		13
	26	Graptolite Island (Laurie Is.)	1	60°43' S	44°27' W	3	1994		13
South Sandwich Islands		TOTAL				56128	1997	unknown	
	27	Cook	0	59°26' S	27°11' W	0	1997		10,14
	28	Thule	0	59°27' S	27°20' W	0	1997		10,14
	29	Bellingshausen	0	59°26' S	27°05' W	0	1997		10,14
	30	Bristol	0	59°02' S	26°34' W	0	1997		10,14
	31	Montagu	0	58°26' S	26°18' W	0	1997		10,14
	32	Saunders	0	57°46' S	26°28' W	0	1997		10,14
	33	Vindication	1	56°06' S	26°47' W	1000	1997		10,14
	34	Candlemas	5	57°05' S	26°42' W	1128	1997		10,14
	35	Visokoi	2	56°42' S	27°09' W	2000	1997		10,14
	36	Zavodovski	3	56°17' S	27°34' W	52000	1997		10,14
South Georgia		TOTAL				>1000000	2002	65.3	15

TABLE 12.2 (cont.)

BREEDING LOCALE	LOCALE NO.	COLONY NAME	NO. OF COLONIES	LAT	LON	COLONY SIZE (PAIRS)	MOST RECENT CENSUS DATE	% DECREASE SINCE WOEHLER 1993	REFERENCES†
		Counted colonies	99			938,017			15
		Uncounted colonies	(+13)						P.N. Trathan pers. comm..
	37	Willis Islands	20	54°00' S	38°11' W				15
	38	Bird Island	3	54°00' S	38°03' W				15
	39	Mainland and SW Islands				938017			15
		Elsehul	1	54°01' S	37°58' W				15
		Welcome Island	3	53°38' S	37°29' W				15
		Hercules Bay	10	54°07' S	36°40' W				15
		Barff Peninsula	4	54°16' S	36°21' W				15
		Calf Head	1	54°30' S	36°02' W				15
		Royal Bay	7	54°33' S	35°55' W				15
		Cooper Bay	3	54°46' S	35°48' W				15
		Cooper Island	8	54°48' S	35°47' W				15
		Natriss Head	14	54°52' S	36°01' W				15
		Cape Paryadin	1	54°04' S	38°00' W				15
		Cape North	1	53°58' S	37°44' W				15
		Sheathbill Bay	1	54°00' S	37°27' W				15
		Clerke Rocks	8	55°01' S	34°41' W				15
		Pickersgill Island	2	54°37' S	36°45' W				15
		Annekov Island	8	54°29' S	37°03' W				15
Falkland/Malvinas Islands	40	Falkland/Malvinas Islands total				25	2005–06	unknown	16,17
	41	Beauchene Island	>9	52°45' S	59°09' W	50	1980	unknown	18
South America		TOTAL				14010		81.3	6
		Chile							
	42	Isla Diego Ramirez	1	51°50' S	75°02' W	10000		unknown	19,20
	43	Isla Hornos				3			19,20
	44	Barnevelt Island				54			19,20
	45	Ildefonso Group				>3000			19,20
	46	Isla Noir	17	45°25' S	50°24' W	3470	2005		21
	47	Isla Recalada	few	45°25' S	50°24' W	300	1989/90		13
	48	Isla Desolacion	few	52°60' S	74°10' W	few			20,21
	49	Isla Bueneventure	1			50			21
	50	Islote Leonard	1	53°23' S	74°04' W	132	2005		21
	51	Cabo Pilar	1	52°43' S	74°42' W				19,20
	52	Isla Deceit	1	55°51' S	67°06' W				19,20
	53	Isla Terhalten	1	55°26' S	67°03' W				19,20
		Argentina							
	54	Isla de los Estados (Staten Is.)	1	54°46' S	64°15' W	1	2010		
		TOTAL GLOBAL ESTIMATE (breeding pairs)				6,310,798			
		TOTAL ESTIMATED DECREASE SINCE WOEHLER 1993				30%			

TABLE 12.2 (cont.)

BREEDING LOCALE	LOCALE NO.	COLONY NAME	NO. OF COLONIES	LAT	LON	COLONY SIZE (PAIRS)	MOST RECENT CENSUS DATE	% DECREASE SINCE WOEHLER 1993	REFERENCES†
ROYAL PENGUINS									
Macquerie Island	1		Many	54°S	158°E	500000	Since 1985	Decreasing	9,23,24, E.J. Woehler pers. comm.
Bishop Island & Clerk Island	2			54°S	158°E	1000	1984–85		9,24 E.J. Woehler pers. comm.

† References: (1) Biuw et al. 2010; (2) Crawford et al. 2006b; (3) Crawford et al. 2003; (4) Ryan et al. 2003; (5) Crawford et al. 2009; (6) Woehler 1993; (7) Guinet et al. 1996; (8) Weimerskirch et al. 1989; (9) Woehler 2006; (10) Croxall & Kirkwood 1979; (11) Croxall & Furse 1980; (12) Shuford & Spear 1988; (13) Woehler & Croxall 1997; (14) Convey et al. 1999; (15) Trathan et al., in review; (16) Croxall et al. 1984; (17) Huin 2007; (18) Smith & Prince 1985; (19) Araya & Millie 1986; (20) Wallace 1991; (21) Oehler et al. 2008; (22) Woehler and Green 1992; (23) Goldsworthy et al. 2001; (24) Copson & Rounsevell 1987.

TABLE 12.3 Timing of events in the annual cycles of macaronis and royals

EVENT	MEAN DATES	REFERENCES†
Macaroni penguins		
Colony arrival:		
Bird Island	14–23 Oct	1, 2
Marion Island	16–21 Oct	3
Egg laying:		
Bird Island	10–30 Nov	1, 2
Marion Island	1–5 Nov	3
Incubation:		
Bird Island	15 Nov–25 Dec	1, 4
Marion Island	1 Nov-20 Dec	3
Hatching:		
Bird Island	25 Dec–10 Jan	1, 4
Marion Island	7–20 Dec	3
Chick rearing:		
Bird Island	10 Jan–20 Feb	1, 4
Marion Island	7 Dec–14 Feb	3
Fledging:		
Bird Island	20 Feb–10 Mar	1, 4
Marion Island	29 Jan–14 Feb	3
Molting:		
Bird Island	10 Mar–1 Apr	1, 4
Marion Island	15 Mar–15 Apr	5
Overwinter migration:		
Bird Island	Apr-Oct	1, 4, 6
Marion Island	Apr–Oct	5
Royal penguins		
Colony arrival	20 Sep–15 Oct	1
Egg laying	12–13 Oct	1
Incubation	14 Nov–4 Dec	1, 6
Hatching		1
Chick rearing	1 Dec -15 Jan	1
Fledging	15 Jan -1 Feb	1
Molting	Feb–Mar	1
Overwinter migration	Mar–Sept	1

† (1) Williams 1995; (2) Crossin et al. 2010; (3) Crawford et al. 2003; (4) P. N. Trathan, unpublished data; (5) R. J. M. Crawford and B. M. Dyer, unpublished observations; (6) Bost et al. 2009.

TABLE 12.4 Reproductive and fitness parameters of macaronis and royals

ATTRIBUTE	MACARONI	ROYAL
Incubation period (days)	30–36	30–35
Chick-rearing period (days)	20–40	20–30
Age of 1st breeding (years)	3#–6	7–8
Adult survival estimate	0.765^–0.892*	≥ 0.80
Maximum life-span (years)	9*†	

Note: Approximate trait values obtained from Williams (1995) and references therein, except for #Crawford et al. (2003), ^Croxall and Davis (1999) and *N. Ratcliffe & P. Trathan, unpublished data. †This value was derived from TIRIS marked birds and represents the average lifespan for those that have already survived to fledging, not the maximum age. Approximately 50% of individuals live longer than 9 years.

REFERENCES

Agnew, D. J. 2000. The illegal and unregulated fishery for toothfish in the Southern Ocean, and the CCAMLR Catch Documentation Scheme. *Marine Policy* 24:361–74.

Ainley, D. G., G. Ballard, S. D. Emslie, W. R. Fraser, P. R. Wilson, and E. J. Woehler. 2003. Adélie penguins and environmental change. *Science* 300:429.

Ainley, D., J. Russell, S. Jenouvrier, E. Woehler, P. Lyver, W. R. Fraser, and G. L. Kooyman. 2010. Antarctic penguin response to habitat change as Earth's troposphere reaches 2°C above preindustrial levels. *Ecological Monographs* 80:49–66.

Alexander, D. J., R. J. Manvell, M. S. Collins, S. J. Brockman, H. A. Westbury, I. Morgan, and F. J. Austin. 1989. Characterization of paramyxoviruses isolated from penguins in Antarctica and Sub-Antarctica during 1976–1979. *Archives of Virology* 109:135–43.

Araya, B., and G. Millie. 1986. Guia de campo de las aves de Chile. Editorial Universitaria Santiago de Chile.

Atkinson, A., V. Siegel, E. Pakhomov, and P. Rothery. 2004. Long-term decline in krill stock and increase in salps within the Southern Ocean. *Nature* 432:100–103.

Baker, A. J., S. L Pereira, O. P. Haddrath, and K. A. Edge. 2006. Multiple gene evidence for expansion of extant penguins out of Antarctica due to global cooling. *Proceedings of the Royal Society of London B* 217:11–17.

Barbosa, A., and M. J. Palacios. 2009. Health of Antarctic birds: A review of their parasites, pathogens and diseases. *Polar Biology* 32:1095–1115.

Barlow, K. E., I. L. Boyd, J. P. Croxall, K. Reid, I. J. Staniland, and A. S. Brierley. 2002. Are penguins and seals in competition for Antarctic krill at South Georgia? *Marine Biology* 140:205–13.

Barré, H., J. L. Mougin, J. Prevost, and M. Van Beveren. 1976. Bird ringing in the Crozet archipelago, Kerguelen, New Amsterdam and St. Paul Islands. *The Ring* 86–87:1–16.

Bergstrom, D. M., A. Lucieer, K. Kiefer, J. Wasley, L. Belbin, T. K. Pedersen, and S. L. Chown. 2009. Indirect effects of invasive species removal devastate World Heritage Island. *Journal of Applied Ecology* 46:73–81.

Bester, M. N., J. P. Bloomer, R. J. van Aarde, B. H. Erasmus, P. J. J. van Rensburg, J. D. Skinner, P. G. Howell, and T. W. Naude. 2002. A review of the successful eradication of feral cats from sub-Antarctic Marion Island, southern Indian Ocean. *South African Journal of Wildlife Research* 31:65–73.

Boersma, P. D. 2008. Penguins as marine sentinels. *BioScience* 58:597–607.

Bingham, M., and E. Mejias. 1999. Penguins of the Magellan region. *Scientia Marina* 63:485–93.

BirdLife International. 2000. *Threatened Birds of the World*. Barcelona and Cambridge: Lynx Edicions and BirdLife International.

———. 2010a. Species factsheet: *Eudyptes chrysolophus*. www.birdlife.org (accessed 18 October 2010).

———. 2010b. Species factsheet: *Eudyptes schlegli*. www.birdlife.org (accessed 18 October 2010).

Biuw, M., C. Lydersen, P. J. Nico De Brun, A. Arriola, G. G. J. Hofmeyr, P. Kritzinger, and K. M. Kovacs. 2010. Long-range migration of a chinstrap penguin from Bouvetøya to Montagu Island, South Sandwich Islands. *Antarctic Science* 22:157–62.

Bonner, W. N., and S. Hunter. 1981. Predatory interactions between Antarctic fur seals, macaroni penguins and giant petrels. *British Antarctic Survey Bulletin* 56:75–79.

Bost, C. A., J. B. Thiebot, D. Pinaud, Y. Cherel, and P. N. Trathan. 2009. Where do penguins go during the inter-breeding period? Using geolocation to track the winter dispersion of the macaroni penguin. *Biology Letters* 5:473–76.

Broman, T., S. Bergstrom, S. L. W. On, H. Palmgren, D. J. McCafferty, M. Sellin, and B. Olsen. 2000. Isolation and characterization of *Campylobacter jejuni jejuni* from macaroni penguins (*Eudyptes chrysolophus*) in the subantarctic region. *Applied Environmental Microbiology* 66:449–52.

Brown, C. R., and N. T. W. Klages. 1987. Travelling speed and foraging range of macaroni and rockhopper penguins at Marion Island. *Journal of Field Ornithology* 58:118–25.

Carrick, R. 1972. Population ecology of the Australian black-backed magpie, royal penguin and silver gull. In *Population Ecology of Migratory Birds: A Symposium*, 441–99. U.S. Department of the Interior Wildlife Research Report No. 2.

Chapuis, J. L., P. Boussèr, and G. Barnaud. 1994. Alien mammals, impact and management in the French subantarctic islands. *Biological Conservation* 67:97–104.

Chown, S. L., N. J. M. Gremmen, and K. J. Gaston. 1998. Ecological biogeography of southern ocean islands: Species-area relationships, human impacts, and conservation. *The American Naturalist* 152:562–75.

Christians, J. K. 2002. Avian egg size: Variation within species and inflexibility within individuals. *Biological Reviews* 77:1–26.

Christidis, L., and W. E. Boles. 1994. The taxonomy and species of birds of Australia and its territories. *R Australasian Ornithological Union Monographs* 2:1–112.

Clarke, A., E. J. Murphy, M. P. Meredith, J. C. King, L. S. Peck, D. K. A. Barnes, and R. C. Smith. 2007. Climate change and the marine ecosystem of the western Antarctic Peninsula. *Philosophical Transactions of the Royal Society B* 362:149–66.

Condy, P. R., R. J. Van Aarde, and M. N. Bester. 1978. The seasonal occurrence and behaviour of killer whales *Orcinus orca* at Marion Island. *Journal of Zoology* (London) 184:449–64.

Convey, P., A. Morton, and J. Poncet. 1999. Survey of marine birds and mammals of the South Sandwich Islands. *Polar Record* 35:107–24.

Cook, A. J., A. J. Fox, D. G. Vaughan, and J. G. Ferrigno. 2005. Retreating glacier fronts on the Antarctic Peninsula over the past half-century. *Science* 308:541–44.

Cook, T. R., and P. Davaine. 2009. Freshwater fishing in seabirds from the sub-Antarctic Kerguelen Islands. *Marine Ornithology* 37:245–47.

Cooper, J., R. J. M. Crawford, M. S. de Villiers, B. M. Dyer, G. J. G Hofmeyr, and A. Jonker. 2009. Disease outbreaks among penguins at sub-Antarctic Marion Island: A conservation concern. *Marine Ornithology* 37:193–96.

Cooper, J., A. C. Wolfaardt, and R. J. M. Crawford. 1997. Trends in population size and breeding success at colonies of macaroni and rockhopper penguins, Marion Island, 1979/80–1995/96. *CCAMLR Science* 4:89–103.

Copson, G. R., and D. E. Rounsevell. 1987. The abundance of royal penguins (*Eudyptes schlegeli*) breeding at Macquarie Island. *ANARE Research Notes* 41:1–11.

Crawford, R. J. M., J. Cooper, and B. M. Dyer. 2003. Population of the macaroni penguin *Eudyptes chrysolophus* at Marion Island, 1994/95–2002/03, with information on breeding and diet. *African Journal of Marine Science* 25:475–86.

Crawford, R. J. M., B. M. Dyer, J. Cooper, and L. G. Underhill. 2006a. Breeding numbers and success of *Eudyptes* penguins at Marion Island, and the influence of mass and time of arrival of adults. *CCAMLR Science* 13:175–90.

Crawford, R. J. M., E. Goya, J-P. Roux, and C. B. Zavalaga. 2006b. Comparison of assemblages and some life-history traits of seabirds in the Humboldt and Benguela systems. *African Journal of Marine Science* 28:553–60.

Crawford, R. J. M., L. G. Underhill, L. Upfold, and B. M. Dyer. 2007. An altered carrying capacity of the Benguela upwelling ecosystem for African penguins (*Spheniscus demersus*). *ICES Journal of Marine Science* 64:570–76.

Crawford, R. J. M., P. A. Whittington, L. Upfold, P. G. Ryan, S. L. Petersen, B. M. Dyer, and J. Cooper. 2009. Recent trends in numbers of four species of penguins at the Prince Edward Islands. *African Journal of Marine Science* 31:419–26.

Crossin, G. T., P. N. Trathan, R. A. Phillips, A. Dawson, F. Le Bouard, and T. D. Williams. 2010. A carry-over effect of migration underlies individual variation in reproductive readiness and extreme egg size dimorphism in macaroni penguins (*Eudyptes chrysolophus*). *The American Naturalist* 176:357–66.

Croxall, J. P., and L. S. Davis. 1999. Penguins: Paradoxes and patterns. *Marine Ornithology* 27:1–12.

Croxall, J. P., and J. R. Furse. 1980. Food of chinstrap penguins *Pygoscelis antarctica* and macaroni penguins *Eudyptes chrysolophus* at Elephant Island Group, South Shetland Islands. *Ibis* 122:237–45.

Croxall, J. P., and E. D. Kirkwood. 1979. The distribution of penguins on the Antarctic Peninsula and islands of the Scotia Sea. British Antarctic Survey, Cambridge.

Croxall, J. P., S. J. McInnes, and P. A. Prince. 1984. The status and conservation of seabirds at the Falkland Islands. In *Status and Conservation of the World's Seabirds*, ed. J. P. Croxall, P. G. H. Evans, and R. W. Schreiber, 271–91. ICBP Technical Publication 2. Cambridge: ICBP.

Croxall, J. P., and S. Nicol. 2004. Management of Southern Ocean resources: Global forces and future sustainability. *Antarctic Science* 16:569–84.

Croxall, J. P., and P. A. Prince. 1979. Antarctic seabird and seal monitoring studies. *Polar Record* 19:573–95.

———. 1980. The food of gentoo penguins *Pygoscelis papua* and macaroni penguins *Eudyptes chrysolophus* at South Georgia. *Ibis* 122:245–53.

Croxall, J. P., P. N. Trathan, and E. J. Murphy. 2002. Environmental change and Antarctic seabird populations. *Science* 297:1510–14.

Cunningham, D. M., and P. J. Moors. 1994. The decline of rockhopper penguins *Eudyptes chrysocome* at Campbell Island, Southern Ocean, and the influence of rising sea temperatures. *Emu* 94:27–36.

Cuthbert, R., and G. Hilton. 2004. Introduced house mice *Mus musculus*: A significant predator of threatened and endemic birds on Gough Island, South Atlantic Ocean? *Biological Conservation* 117:483–89.

Duchêne, J. C. 1989. *Kerguelen: Recherches au bout monde.* Tours, France: Terres Australes et Antarctiques Françaises.

Ducklow, H. W., K. Baker, W. R. Fraser, D. G. Martinson, L. B. Quetin, R. M. Ross, R. C. Smith, S. Stammerjohn, and M. Vernet. 2007. Marine ecosystems: The West Antarctic Peninsula. *Philosophical Transactions of the Royal Society B* 362:67–94.

Ellis, S. 1999. The Penguin Conservation Assessment and Management Plan: A description of the process. *Marine Ornithology* 27:163–69.

Ellis, S., J. P. Croxall, and J. Cooper. 1998. Penguin Conservation and Assessment Plan. Apple Valley, MN: IUCN/SSC Conservation Breeding Specialist Group.

Everitt, D. A., and C. M. Miskelly. 2003. A review of isabellinism in penguins. *Notornis* 50:43–51.

Falla, R. A., and J. L. Mougin. 1979. Order Sphenisciformes. In *Checklist of Birds of the World*, ed. E. Mayr and G. W. Cottrell, 121–34. Cambridge, MA: Museum of Comparative Zoology.

Forcada, J., and P. N. Trathan. 2009. Penguin responses to climate change in the Southern Ocean. *Global Change Biology* 15:1618–30.

Forcada, J., P. N. Trathan, K. Reid, E. J. Murphy, and J. P. Croxall. 2006. Contrasting population changes in sympatric penguin species in association with climate warming. *Global Change Biology* 12:411–23.

Fraser, W. R., and E. E. Hofmann. 2003. A predator's perspective on causal links between climate change, physical forcing and ecosystem response. *Marine Ecology Progress Series* 265:1–15.

Fraser, W. R., and D. L. Patterson. 1997. Human disturbance and long-term changes in Adélie penguin populations: A natural experiment at Palmer Station, Antarctic Peninsula. In *Antarctic Communities: Species, Structure and Survival*, ed. B. Battaglia, J. Valencia, and D. W. H. Walton, 445–52. Cambridge: Cambridge University Press.

Fraser, W. R., and W. Z. Trivelpiece. 1996. Factors controlling the distribution of seabirds: Winter-summer heterogeneity in the distribution of Adélie penguin populations. In *Foundations for Ecological Research West of the Antarctic Peninsula*, ed. R. M. Ross, E. E. Hofmann, and L. B. Quetin, 257–72. Antarctic Research Series 70. Washington, DC: American Geophysical Union.

Fraser, W. R., W. Z. Trivelpiece, D. G. Ainley, and S. G. Trivelpiece. 1992. Increases in Antarctic penguin populations: Reduced competition with whales or loss of sea ice due to environmental warming? *Polar Biology* 11:525–31.

Garnett, S. T., and G. M. Crowley. 2000. Action Plan for Australian Birds 2000. Environment Australia, Canberra.

Giese, M., S. D. Goldsworthy, R. Gales, N. Brothers, and J. Hamill. 2000. Effects of the *Iron Baron* oil spill on little penguins (*Eudyptula minor*). 3: Breeding success of rehabilitated oiled birds. *Wildlife Research* 27:583–91.

Gille, S. T. 2002. Warming of the Southern Ocean since the 1950s. *Science* 295:1275–77.

Goldsworthy, S. D., X. He, G. N. Tuck, M. Lewis, and R. Williams. 2001. Trophic interactions between the Patagonian toothfish, its fishery, and seals and seabirds around Macquarie Island. *Marine Ecology Progress Series* 218:283–302.

Gorman, K. B., E. S. Erdmann, B. C. Pickering, P. J. Horne, J. R. Blum, H. M. Lucas, D. L. Patterson-Fraser, and W. R. Fraser. 2010. A new high-latitude record for the macaroni penguin (*Eudyptes chrysolophus*) at Avian Island, Antarctica. *Polar Biology* 8:1155–58.

Grant, S., P. Convey, K. Hughes, R. A. Phillips, and P. N. Trathan. 2011. Conservation and management of Antarctic ecosystems. In *Antarctica: An Extreme Environment in a Changing World*, ed. A. Rogers. Oxford: Wiley-Blackwell.

Green, J. A., P. J. Butler, A. J. Woakes, and I. L. Boyd. 2002. Energy requirements of female macaroni penguins breeding at South Georgia. *Functional Ecology* 16:671–81.

———. 2003. Energetics of diving in macaroni penguins. *Journal of Experimental Biology* 206:43–57.

Grimaldi, W., J. Jabour, and E. J. Woehler. 2011. Considerations for minimizing the spread of infectious diseases in Antarctic seabirds and seals. *Polar Record* 47:56–66.

Guinet, C., Y. Cherel, V. Ridoux, and P. Jouventin. 1996. Consumption of marine resources by seabirds and seals in Crozet and Kerguelen waters: Changes in relation to consumer biomass, 1962–85. *Antarctic Science* 8:23–30.

Hindell, M. A. 1988. The diet of royal penguins *Eudyptes schlegeli* at Macquarie Island. *Emu* 88:219–26.

Holmes, N. 2007. Comparing king, gentoo, and royal penguin responses to pedestrian visitation. *Journal of Wildlife Management* 71:2575–82.

Holmes, N., M. Giese, and L. K. Kriwoken. 2005. Testing the minimum approach distance guidelines for incubating royal penguins *Eudyptes schlegeli*. *Biological Conservation* 126:339–50.

Huin, N. 2007. Falkland Islands penguin census 2005/06. Falkland Islands Conservation Report.

Hull, C. L. 1996. Morphometric indices for sexing adult royal, *Eudyptes schlegeli*, and rockhopper penguins, *E. chrysocome*, at Macquarie Island. *Marine Ornithology* 24:23–27.

———. 1999. Comparison of the diets of breeding royal (*Eudyptes schlegeli*) and rockhopper (*Eudyptes chrysocome*) penguins on Macquarie Island over three years. *Journal of Zoology* (London) 247:507–29.

Hull, C. L., M. A. Hindell, and K. J. Michael. 1997. The foraging zones of royal penguins during the breeding season, and their association with oceanographic features. *Marine Ecology Progress Series* 153:217–28.

IAATO (International Association of Antarctic Tour Operators). 2012. Guidelines for visitors to the Antarctic. guidelines- http://www.iaato.org/guidelines.html (accessed 3 April 2012).

IUCN (International Union for Conservation of Nature). 2011. IUCN Red List of Threatened Species. Version 2011.2. www.iucnredlist.org (accessed 3 April 2012).

Jacka, T. H., and W. F. Budd. 1998. Detection of temperature and sea-ice-extent changes in the Antarctic and Southern Ocean, 1949–96. *Annals of Glaciology* 27:553–59.

Jenouvrier, S., C. Barbraud, B. Cazelles, and H. Weimerskirch. 2005a. Modeling population dynamics of seabirds: Importance of the effects of climate fluctuations on breeding proportions. *Oikos* 108:511–22.

Jenouvrier, S., H. Caswell, C. Barbraud, M. Holland, J. Stroeve, and H. Weimerskirch. 2009. Demographic models and IPCC climate projections predict the decline of an emperor penguin population. *Proceedings of the National Academy of Sciences* 106:1844–47.

Jenouvrier, S., H. Weimerskirch, C. Barbraud, Y. H. Park, and B. Cazelles. 2005b. Evidence of a shift in the cyclicity of Antarctic seabird dynamics linked to climate. *Proceedings of the Royal Society of London* 272:887–95.

Johnson, K., J. C. Bednarz, and S. Zack. 1987. Crested penguins: Why are first eggs smaller? *Oikos* 49:347–49.

Johnstone, G. W. 1985. Threats to birds on subantarctic islands. In *Conservation of Island Birds*, ed. P. J. Moors, 101–21. ICBP Technical Publication No. 3.

Jones, M. G. W., and P. G. Ryan. 2010. Evidence of mouse attacks on albatross chicks on sub-Antarctic Marion Island. *Antarctic Science* 22:39–42.

Jouventin, P. 1982. *Visual and Vocal Signals in Penguins: Their Evolution and Adaptive Characters*. Berlin: Paul Parey.

Jouventin, P., and T. Micol. 1995. Conservation status of the French sub-Antarctic islands. In *Progress in Conservation of the Subantarctic Islands*, ed. P. R. Dingwall, 31–52. Gland, Switzerland, and Cambridge: IUCN.

Lack, D. 1968. *Ecological Adaptations for Breeding in Birds*. London: Methuen.

Marchant, S., and P. J. Higgins. 1990. *Handbook of Australian, New Zealand and Antarctic Birds*. Vol. 1, part A. Melbourne: Oxford University Press.

Moors, P. J. 1986. Decline in numbers of rockhopper penguins at Campbell Island. *Polar Record* 23:69–73.

Murphy, E. J., J. L. Watkins, K. Reid, P. N. Trathan, I. Everson, J. P. Croxall, J. Priddle, M. A. Brandon, A. S. Brierley, and E. H. Hofmann. 1998. Interannual variability of the South Georgia marine ecosystem: Biological and physical sources of variation in the abundance of krill. *Fisheries Oceanography* 7:381–90.

Naveen, R., S. C. Forrest, R. G. Dagit, L. K. Blight, W. Z. Trivelpiece, and S. G. Trivelpiece. 2000. Census of penguin, blue-eyed shag, and southern giant petrel populations in the Antarctic Peninsula region, 1994–2000. *Polar Record* 36:323–34.

———. 2001. Zodiac landings by tourist ships in the Antarctic Peninsula region, 1989–99. *Polar Record* 37:121–32.

Oehler, D. A., S. Pelikan, W. Roger Fry, L. Weakley, A. Kush, and M. Marin. 2008. Status of crested penguins *Eudyptes* spp. on three islands in southern Chile. *Wilson Journal of Ornithology* 120:575–81.

Oppenheimer, M., and R. Alley. 2004. The West Antarctic ice sheet and long term climate policy. *Climatic Change* 64:1–10.

Pople, A., R. D. Simpson, and S. C. Cairns. 1990. An incident of Southern Ocean oil pollution: Effects of a spillage of diesel fuel on the rocky shore of Macquarie Island (Sub-Antarctic). *Australian Journal of Marine and Freshwater Research* 41:603–20.

Pütz, K., A. P. Clausen, N. Huin, and J. P. Croxall. 2003. Re-evaluation of historical rockhopper penguin population data in the Falkland Islands. *Waterbirds* 26:169–75.

Pyper, W. 2009. Cleaning up fuel spills on Macquarie Island. *Australian Antarctic Magazine* 16:21. http://www.antarctica.gov.au/about-antarctica/australian-antarctic-magazine/issue-16–2009/cleaning-up-fuel-spills-on-macquarie-island.

Ratcliffe, N., and P. N. Trathan. In press. A review of the diet and foraging movements of penguins breeding within the CCAMLR area. *CCAMLR Science*.

Raymond, B., J. McInnes, J. M. Dambacher, S. Way, and D. M. Bergstrom. 2010. Qualitative modeling of invasive species eradication on subantarctic Macquarie Island. *Journal of Applied Ecology* 48:181–91.

Reid, K., and J. P. Croxall. 2001. Environmental response of upper trophic-level predators reveals a system change in an Antarctic marine ecosystem. *Proceedings of the Royal Society of London B* 268:377–84.

Reid, T. A., C. A. Hull, D. W. Eades, R. P. Scofield, and E. J. Woehler. 1995. Shipboard observations of penguins at sea in the Australian Sector of the Southern Ocean, 1991–1995. *Marine Ornithology* 27:101–10.

Reilly, P. 1994. *Penguins of the World*. Oxford: Oxford University Press.

Ryan, P. G., J. Cooper, B. M. Dyer, L. G. Underhill, R. J. M. Crawford, and M. N. Bester. 2003. Counts of surface-nesting seabirds breeding at Prince Edward Island, summer 2001/02. *African Journal of Marine Science* 25:441–51.

Scheffer, A., P. N. Trathan, and M. Collins. 2010. Foraging behaviour of king penguins (*Aptenodytes patagonicus*) in relation to predictable mesoscale oceanographic features in the Polar Front Zone to the north of South Georgia. *Progress in Oceanography* 86:SI 232–45.

Schultz, A., and S. L. Petersen. 2003. Absence of haematozoa in breeding macaroni *Eudyptes chrysolophus* and rockhopper *E. chrysocome* penguins at Marion Island. *African Journal of Marine Science* 25:499–502.

Shaughnessy, P. D. 1975. Variation in facial colour of the royal penguin. *Emu* 75:147–52.

Shuford, W. D., and L. B. Spear. 1988. Surveys of breeding penguins and other seabirds in the South Shetland Islands, Antarctica, January–February 1987. NOAA Technical Memorandum NMFS/NEC-59, National Marine Fisheries Service, Seattle.

Sibley, C. G., and B. L. Monroe. 1993. *Distribution and Taxonomy of Birds of the World.* New Haven, CT: Yale University Press.

Sinclair, A. R. E., and C. J. Krebs. 2002. Complex numerical responses to top-down and bottom-up processes in vertebrate populations. *Philosophical Transactions of the Royal Society B* 357:1221–31.

Slagsvold, T., J. Sandvik, G. Rofstad, O. Lorentsen, and M. Husby. 1984. On the adaptive value of intraclutch egg-size variation in birds. *The Auk* 101:685–97.

Smith, R. C., D. Ainley, K. Baker, E. Domack, S. Emslie, B. Fraser, J. Kennett, A. Leventer, E. Mosley-Thompson, S. Stammerjohn, and M. Vernet. 1999. Marine ecosystem sensitivity to climate change. *BioScience* 49:393–404.

Smith, R. C., W. R. Fraser, and S. E. Stammerjohn. 2003. Climate variability and ecological response of the marine ecosystem in the Western Antarctic Peninsula (WAP) region. In *Climate Variability and Ecosystem Response at Long-Term Ecological Research Sites*, ed. D. Greenland, D. G. Goodin, and R. C. Smith, 158–73. Oxford: Oxford University Press.

Smith, R. I. L., and P. Prince. 1985. The natural history of Beauchêne Island. *Biological Journal of the Linnean Society* 24:233–83.

Smith, S. D. A., and R. D. Simpson. 1998. Recovery of benthic communities at Macquarie Island (sub-Antarctic) following a small oil spill. *Marine Biology* 131:567–81.

St. Clair, C. C. 1998. What is the function of first eggs in crested penguins? *The Auk* 115:478–82.

Stammerjohn, S. E., D. G. Martinson, R. C. Smith, and R. A. Iannuzzi. 2008. Sea ice in the western Antarctic Peninsula region: Spatio-temporal variability from ecological and climate change perspectives. *Deep Sea Research, Part II* 55:2041–58.

Tasmanian Parks and Wildlife Service. 2003. Macquarie Island Nature Reserve and World Heritage Area draft management plan 2003. Tasmanian Parks and Wildlife Service, Hobart, Australia.

———. 2010. Winter weather frustrates eradication team. *Macquarie Island Pest Eradication Project Newsletter* 6:1–2. http://www.parks.tas.gov.au/file.aspx?id=19827.

Trathan, P. N. 2004. Image analysis of color aerial photography to estimate penguin population size. *Wildlife Society Bulletin* 32:332–43.

Trathan, P. N., J. Forcada, and E. J. Murphy. 2007. Environmental forcing and Southern Ocean marine predator populations: Effects of climate change and variability. *Philosophical Transactions of the Royal Society B* 362:2351–65.

Trathan, P. N., E. J. Murphy, J. P. Croxall, and I. Everson. 1998. Use of at-sea distribution data to derive potential foraging ranges of macaroni penguins during the breeding season. *Marine Ecology Progress Series* 169:263–75.

Trathan, P. N., N. Ratcliffe, and E. A. Masden. 2012. Ecological drivers of change at South Georgia: The krill surplus, or climate variability. *Ecography* 34. doi: 10.1111/j.1600–0587.2012.07330.x.

Trivelpiece, W. Z., J. T. Hinke, A. K. Miller, C. S. Reiss, S. G. Trivelpiece, and G. M. Watters. 2011. Variability in krill biomass links harvesting and climate warming to penguin population changes in Antarctica. *Proceedings of the National Academy of Sciences* 108:7625–28.

van Aarde, R. J. 1980. The diet and feeding behaviour of feral cats *Felis catus* at Marion Island. *South African Journal of Wildlife Research* 10:123–28.

Vaughn, D. G., G. J. Marshall, W. M. Connelley, J. C. King, and R. Mulvaney. 2001. Devil in the detail. *Science* 293:1777–79.

Walker, T. R., I. L. Boyd, D. J. McCafferty, N. Huin, R. I. Taylor, and K. Reid. 1998. Seasonal occurrence and diet of leopard seals (*Hydrurga leptonyx*) at Bird Island, South Georgia. *Antarctic Science* 10:75–81.

Wallace, G. E. 1991. Noteworthy bird records from southernmost Chile. *Condor* 93:175–76.

Warham, J. 1971. Aspects of breeding behaviour in the royal penguin (*Eudyptes chrysolophus schlegeli*). *Notornis* 18:91–115.

Weimerskirch, H., R. Zotier, and P. Jouventin. 1989. The avifauna of the Kerguelen Islands. *Emu* 89:15–29.

Woehler, E. J. 1991. Status and conservation of the seabirds of Heard Island and the McDonald Islands. In *Seabird Status and Conservation: A Supplement*, ed. J. P. Croxall, 263–77. Cambridge: International Council for Bird Preservation.

———. 1992. Records of vagrant penguins from Tasmania. *Marine Ornithology* 20:61–74.

———. 1993. The distribution and abundance of Antarctic and Subantarctic penguins. Cambridge: Scientific Committee on Antarctic Research.

———. 1995. Bill morphology of royal and macaroni penguins, and geographic variation within eudyptid penguins. In *The Penguins: Ecology and Management*, ed. P. Dann, I. Norman, and P. Reilly, 319–30. Chipping Norton, NSW, Australia: Surrey Beatty & Sons.

———. 2006. Status and conservation of the seabirds of Heard Island. In *In Heard Island: Southern Ocean Sentinel*, ed. K. Green and E. Woehler, 128–65. Chipping Norton, NSW, Australia: Surrey Beatty and Sons.

Woehler, E. J., J. Cooper, J. P. Croxall, W. R. Fraser, G. L. Kooyman, G. D. Miller, D. C. Ner, D. L. Patterson, H. U. Peter, C. A. Ribic, K. Salwick, W. R. Trivelpiece, and H. Weirmirskirch. 2001. A statistical assessment of the status and trends of Antarctic and Subantarctic seabirds. Scientific Committee for Antarctic Research, Cambridge.

Woehler, E. J., and J. P. Croxall. 1997. The status and trends of Antarctic and sub-Antarctic seabirds. *Marine Ornithology* 25:43–66.

Woehler, E. J., and J. A. Green. 1992. Consumption of marine resources by seabirds and seals at Heard Island and the Macdonald Islands. *Polar Biology* 12:659–65.

Williams, T. D. 1990. Growth and survival in macaroni penguin, *Eudyptes chrysolophus*, A- and B-chicks: Do females maximize investment in the large B-egg? *Oikos* 59:349–54.

———. 1995. *The Penguins.* Oxford: Oxford University Press.

Williams, T. D., and J. P. Croxall. 1991. Annual variation in breeding biology of macaroni penguins, *Eudyptes chrysolophus*, at Bird Island, South Georgia. *Journal of Zoology* 223:189–202.

Yuan, X. 2004. ENSO-related impacts on Antarctic sea ice: Synthesis of phenomenon and mechanisms. *Antarctic Science* 16:415–25.

V

Banded Penguins

Genus *Spheniscus*

13

African Penguin

(Spheniscus demersus)

Robert J. M. Crawford, Jessica Kemper, and Les G. Underhill

1. SPECIES (COMMON AND SCIENTIFIC NAMES)

African penguin, *Spheniscus demersus*

The African penguin is also known as the black-footed penguin and the jackass penguin.

2. DESCRIPTION OF THE SPECIES

ADULT. Face, throat, crown stripe, nape, and upperparts are black, including the upper flipper. The underparts are mostly white. A broad white supercilium runs from in front of the eye, around the back of the face, and to the neck and joins the white breast (fig. 1). A black stripe runs up the side of the body and across the upper breast. Some birds have a second black band on the neck. The chest and abdomen have a variable number of irregular black spots. The black plumage becomes browner with wear. A patch of naked pink (rarely black) skin above the eye is white and feathered immediately after molt. The bill is black with a vertical pale horn band at the gonys. Eyes are brown; legs and feet are black with pink blotches.

IMMATURE. Plumage is the same as the juvenile's but fades to gray and then brownish before molt. Some individuals undergo partial head molt at sea before complete molt into adult plumage; the extent varies from the supercilium to the entire head and upper body to the level of the breast-band.

FIG. 1 African penguin pair showing the broad white neck-band characteristic of the species. (J. Kemper)

JUVENILE. Plumage is blue gray on the upperparts, shading to white below. The chin and lower throat are gray. There are few darker spots on the breast and flanks. Legs and feet are pale flesh (Ryan et al. 1987; Randall 1989; Hockey et al. 2005).

MEASUREMENTS. In general, males are larger than females, but measurements overlap. See table 13.1.

3. TAXONOMIC STATUS

No subspecies are recognized. The species is one of four in the genus *Spheniscus*. The Magellanic penguin (*S. magellanicus*) occurs as a rare vagrant off southern Africa: it is larger, always with two breast-bands (like some African penguins), a broader upper band, narrower white supercilium, less pink on the face, and a diagnostic thin white line at the base of the bill that is absent in the African penguin.

4. RANGE AND DISTRIBUTION

The African penguin is endemic as a breeding species to southern Africa, where it breeds at 28 localities between Hollamsbird Island, Namibia, and Bird Island, South Africa (fig. 2) (Kemper et al. 2007c). Of these localities, 11 are in Namibia, 11 in the Western Cape, and

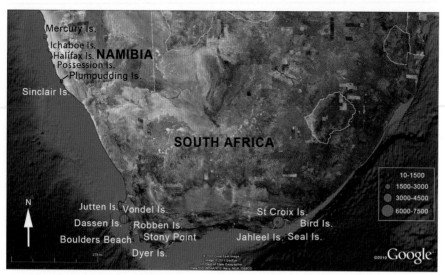

FIG. 2 Breeding distribution and abundance of the African penguin, with counts based on pairs. Extinct colonies are not shown, nor are all extant colonies shown.

6 in the Eastern Cape. The usual nonbreeding range extends along some 3,200 kilometers of coast between about 18° south on the Namibian coast and 29° south on the coast of KwaZulu-Natal. Vagrant birds have been recorded north to Sette Cama (2°32′ S), Gabon, on the West African coast (Malbrant and Maclatchy 1958), to the Limpopo River mouth (25° S), and to Mozambique on the east coast (Shelton et al. 1984). This species has have been recorded up to 100 kilometers offshore (Rand 1960), but most occur within 20 kilometers of the coast (Wilson et al. 1988), except on the Agulhas Bank, where the distribution of their prey extends farther offshore (Shelton et al. 1984).

5. SUMMARY OF POPULATION TRENDS

Breeding colonies of African penguins are grouped in three regions: Namibia in the north and South Africa's Western Cape and Eastern Cape provinces in the south. Colonies in the Western Cape are separated from those in the other two regions by distances of about 600 kilometers.

According to Namibia's Ministry of Fisheries and Marine Resources (MFMR), African penguins breeding in Namibia decreased from 49,000 pairs in 1956–57 to 12,000 pairs in 1978–79 and 4,500 pairs in 2009–10 (MFMR, unpubl. data). In the Western Cape, there may have been close to 1 million pairs at Dassen Island in the 1920s (Crawford et al. 2007). Some 92,000 pairs bred in this region in 1956, decreasing to 41,000 pairs in 1978, 34,000 pairs in 2001, and 11,000 pairs in 2009 (Crawford et al. 2011). In the Eastern Cape, numbers increased from about 6,000 pairs in 1956 to an average of 22,000 pairs from 1985 to 2001 and then decreased to an average of 10,000 pairs from 2003 to 2009 (Crawford et al. 2009). The population trends for Namibia, the Western Cape, and the Eastern Cape are illustrated by the number of nests reported for Possession, Dassen, and St. Croix Islands (fig. 3). The overall population may have been

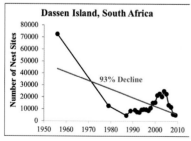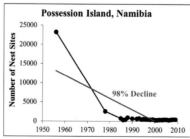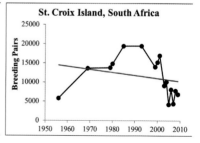

FIG. 3 Population trends of nests in African penguin colonies: (a) Dassen Island, South Africa; (b) Possession Island, Namibia; (c) St. Croix Island, South Africa.

FIG. 4 African penguin colony at Boulders Beach, South Africa, showing the boardwalk for visitors and some nesting penguins. (P. Ryan)

on the order of 1 million pairs in the 1920s but decreased to about 147,000 pairs in 1956–57. It fell to about 75,000 pairs in 1978, 63,000 pairs in 2001, and 25,000 pairs in 2009 (Crawford et al. 2011).

6. IUCN STATUS

The International Union for Conservation of Nature upgraded the species to Endangered on its Red List of Threatened Species in 2010 because of a large, sustained decrease in the 20th century, at a rate of 13% per generation since 1976 (IUCN 2011; Ellis et al. 1998; Kemper et al. 2007c). The breeding population has decreased by more than 50% in the three most recent generations and continues to decline. A large decrease in South Africa, from 57,000 breeding pairs in 2001 to 21,000 pairs in 2009, halved the global population in the 21st century.

The species is a specially protected bird in Namibia and is included in Appendix 2 of the Convention on International Trade in Endangered Species of Wild Fauna and Flora (CITES) and Appendix 2 of the Convention for the Conservation of Migratory Species of Wild Animals (CMS). The U.S. Fish and Wildlife Service lists it under the Endangered Species Act of 1973.

7. NATURAL HISTORY

BREEDING BIOLOGY. Coastal islands represent 24 out of 28 extant breeding localities. Caves are used at 2 localities on the Namibian mainland (Loutit and Boyer 1985; Bartlett et al. 2003; Simmons and Kemper 2003), and there are 2 mainland localities in South Africa's Western Cape, although occasional breeding has been attempted at other coastal sites. Breeding is usually colonial (fig. 4), but solitary nests occur. Both sexes build nests in burrows in guano or sand, in clefts between rocks, in disused buildings, and on the surface, preferably under shade (Shelton et al. 1984; Crawford et al. 1995a; Kemper et al. 2007d). African penguins make scrapes and dig burrows by pushing sand backward with their feet (Eggleton and Siegfried 1979). Burrows have a more constant microclimate than surface nests. In burrows, relative humidity is higher, air temperatures fluctuate less, wind effect is negligible, and birds are not exposed to direct sunlight (Frost et al. 1976a) (fig. 5). Nesting material includes

FIG. 5 Adult African penguin panting to keep cool while incubating a clutch of two eggs. (P. Ryan)

seaweed, pieces of vegetation, rocks, shells, bones, and feathers, but some nests are not lined.

Breeding is monogamous. At St. Croix Island, 80–92% of birds whose mates did not disappear bred with the same mate in the following breeding season; 89% of males and 78% of females returned to breed at the same site (Randall 1983). At Robben Island, over a four-year period, 80 of 85 adults retained the same mate; for the other 5, at least one partner had died. In the same period, 53% of 85 adults nested at the same site, 44% used two sites, 2% used three sites, and 1% used four sites (Crawford et al. 1995a). Breeding locality fidelity is strong, with no confirmed records of adults breeding at more than one locality. However, first-time breeders may choose to breed at localities or colonies other than their natal site, having the flexibility to emigrate and hence to take advantage of long-term changes in the distribution of food (Shelton et al. 1984; Crawford 1998; Crawford et al. 1999).

The clutch is usually two eggs, sometimes one, and rarely three (Crawford et al. 1999, 2000b). The egg is a

rounded oval and white, becoming stained as incubation proceeds. Egg size and dimensions are shown in table 13.2. "Runt" eggs are laid occasionally. The laying interval is 3–3.2 days (Williams 1981; Williams and Cooper 1984). Lost clutches may be replaced, and successful breeders may re-lay (Randall and Randall 1981; La Cock and Cooper 1988). Incubation starts with the first-laid egg, lasts 38–41 days (about 37–38 days per egg), and is shared equally by both sexes (Rand 1960; Williams and Cooper 1984; Randall 1989). The incubation shift for eggs hatched successfully is 1–2 days; for eggs eventually abandoned, it is 1–14 days. Mean hatching interval is 2.1 days (Williams and Cooper 1984).

The newly hatched chick is blind, helpless, and covered in a sooty gray protoptile down (Randall 1989). Mass at hatching is about 72 grams (Williams and Cooper 1984). Chicks are brooded by adults until about 10 days after hatching, at which time secondary down starts to develop. They are partly covered by adults until about 15 days (fig. 6). After 16 days, they are able to sit beside their parents (Seddon and van Heezik 1993). Chicks reach full thermoregulatory capacity at a mass of about 400 grams, which they attain at 21–25 days (Erasmus and Smith 1974). From 26–30 days, they are often left unguarded and may form crèches of up to 25 chicks. At 31–35 days, contour feathers start to grow beneath the down. Chicks have full juvenile plumage from 61–65 days and fledge when between 55–130 days old (Seddon and van Heezik 1993; Kemper 2006). Mortality of chicks aged 0–34 days is due primarily to burrow collapse, exposure, drowning, accidental death in the nest, and predation by kelp gulls (*Larus dominicanus*); from 42–90 days, it is mostly from starvation or heat stress (Seddon and van Heezik 1991;

FIG. 6 Adult African penguin guarding its two chicks. (J. Kemper)

Kemper 2006). The daily food intake of chicks increases from about 250 grams at 20–30 days to about 500 grams at 40–50 days (Cooper 1977). Most feeding of chicks takes place in the afternoon. Growth of two chicks in the same brood is similar for the first 25 days, after which the growth of the B-chick lags behind that of the older sibling and returns to the level of the A-chick only after the A-chick fledges (Williams and Cooper 1984). Last-hatched chicks are more likely than their siblings to die (Seddon and van Heezik 1991). Full growth of foot and flipper is achieved before fledging, but the bill matures during the first year at sea (Williams and Cooper 1984).

FORAGING AND PREY. African penguins feed solitarily and in small groups (Rand 1960; Wilson and Wilson 1990), sometimes in conjunction with other seabirds. Foraging behavior is similar to that of other *Spheniscus* penguins (Ryan et al. 2007). They are visual hunters, but may use scent cues such as dimethyl sulfide (DMS) to locate prey patches at sea (Wright et al. 2011). Adults hunt cooperatively, swimming rapidly around a school of fish and compressing it (Wilson 1985b; Wilson and Wilson 1990); fish behavior may be manipulated by the adult penguins' striking plumage (Wilson et al. 1987). Adults feed mainly on active, free-swimming prey, usually schooling pelagic fish. Especially important are anchovy (*Engraulis encrasicolus*), sardine (*Sardinops sagax*) (e.g., Crawford 2007), and, in Namibia, bearded goby (*Sufflogobius bibarbatus*) (Ludynia et al. 2010). Other prey includes cephalopods, horse mackerel (*Trachurus capensis*), juvenile hake (*Merluccius* sp.), and redeye (*Etrumeus whiteheadi*) (Rand 1960; Randall and Randall 1986; Ludynia et al. 2010). Juveniles are thought to forage singly on slow-moving prey (Rand 1960); they are underrepresented in adult feeding flocks (Ryan et al. 1987). The few diet samples from juveniles contained squid, clinid fish, stomatopods, and even polychaete worms (Rand 1960).

Most dives are less than 30 meters deep, although some reach up to 85 meters (MFMR, unpubl. data) and last up to 2.5 minutes (Ryan et al. 2007). Initial reports of dives up to 130 meters were based on measurements made with capillary depth gauges (Wilson 1985b) but have not been repeated with more accurate digital depth loggers (Ryan et al. 2007; L. Pichegru; MFMR, unpubl. data). Almost all dives take place during daylight (Ryan et al. 2007); foraging is assumed to be largely visual. At longer distances, African penguins may direct foraging movements by scent (Cunningham et al. 2008). Most food is caught between 10:00 A.M. and 6:00 P.M., with a lull in feeding activity around midday (Wilson and Wilson 1995). Birds generally do not feed at night (Wilson 1985a). Adults provisioning downy chicks generally forage within 40 kilometers of colonies (Heath and Randall 1989; Petersen et al. 2006; Pichegru et al. 2010; MFMR, unpubl. data) but may travel up to 120 kilometers (Ludynia 2007). Although penguins can swim at speeds of up to 5 meters per second in short bursts (Wilson 1985b), commuting birds typically travel at 1.4–1.9 meters per second (Petersen et al. 2006).

PREDATORS. At St. Croix Island, injuries inflicted by great white sharks (*Carcharodon carcharias*) were second only to oil pollution as a cause of mortality of penguins (Randall et al. 1988). At Robben Island, mole snakes (*Pseudaspis cana*) eat penguin eggs (Crawford et al. 1995a). Kelp gulls scavenge deserted clutches and dying chicks and prey on poorly guarded eggs and small chicks, especially at surface colonies (Cooper 1974; Kemper et al. 2007d, 2007e). At Dassen Island, great white pelicans (*Pelecanus onocrotalus*) occasionally eat penguin chicks (de Ponte Machado 2007).

There are many accounts of Cape fur seals (*Arctocephalus pusillus pusillus*) killing penguins (e.g., du Toit et al. 2004). This phenomenon has been recorded at Mercury, Ichaboe, Halifax, Possession, Lambert's Bay, Malgas, Dassen, Robben, and Dyer Islands. Few observations have been conducted at other islands. Mortality may be high—at least 25 penguins were killed on one day at Dassen Island (Cooper 1974)—and unsustainable (e.g., Marks et al. 1997; Crawford et al. 2001; MFMR, unpubl. data). Young male seals seem to inflict the most mortality (e.g., David et al. 2003). There are isolated records of killer whales (*Orcinus orca*) feeding on African penguins (Williams et al. 1990).

Feral cats (*Felis catus*) prey on penguin eggs or chicks at Robben Island (Crawford et al. 1995a) and caused mortality at Dassen Island (Berruti 1986), where they have been exterminated. Penguins are vulnerable to various predators at mainland localities and at the two islands (Lambert's Bay and Marcus) that are joined to the mainland. Leopards (*Panthera pardus*) ate adult penguins at Stony Point (Crawford et al. 1995c). Other potential predators recorded near the breeding colony at Stony Point include caracals (*Felis caracal*), water mongooses

(*Atilax paludinosus*), small gray mongooses (*Galerella pulverulenta*), large-spotted genets (*Genetta tigrina*), and Cape clawless otters (*Aonyx capensis*) (Whittington et al. 1996), as well as domestic dogs. Rats (*Rattus* spp.) are potential predators of eggs at mainland and some island localities.

MOLT. Birds usually molt at landing beaches or near their breeding sites and do not enter the sea except to preen and drink (Cooper 1978). Some birds molt at localities other than their natal or breeding sites (Kemper and Roux 2005). Duration of molt, from the time birds come ashore until they depart for sea again, is approximately 21 days (Randall and Randall 1981). All feathers are replaced, those on the head and flippers being among the last to be shed (fig. 7) (Cooper 1978). The feather-shedding phase of molt (from the time the first feathers stand out until the last loose feathers fall away) has a mean duration of 12.7 days (n = 45) (Randall 1983). This value is critical for the calculation of population sizes using counts of molting individuals; the duration of molt therefore requires more thorough investigation (Kemper 2007; Kemper et

FIG. 7 Juvenile African penguin is ready to start the molt. (P. Ryan)

al. 2007b). The mean intermolt interval is almost exactly one year (Randall and Randall 1981; Kemper et al. 2008), although the timing of the first molt varies in relation to the fledging period and may be as long as 23 months (Kemper and Roux 2005). Flippers become enlarged during molt (Jarvis 1970). The pre-molt fattening period lasts about 35 days, and the post-molt refattening period is 42 days (Randall 1983, 1989; Adams and Brown 1990). Pre-molt adults are on average 31% heavier than breeding adults. Birds lose 47% of their mass, 45% of water, 56% of fat, and 43% of protein during molt. Post-molt birds have sunken breasts and atrophied pectoral muscles (Cooper 1978).

ANNUAL CYCLE. At most localities, molt tends to be synchronized, but the timing of adult molt varies around the southern African coast, as does the timing of breeding (fig. 8). In Namibia, peaks in molt of adult birds occur in December–January and April–May (Kemper et al. 2008). At Dassen Island, the peak for adults molting in burrows is August–November, and along the shore, the peak is November–January (Wolfaardt et al. 2009a). At Robben Island and Boulders, most adults molt from November to January (Underhill and Crawford 1999; Hemming 2001; Crawford et al. 2006b). At Dyer and St. Croix Islands, the peak for adults is October–December, with a substantial proportion at Bird Island commencing molt in September (Randall et al. 1986a; Crawford et al. 2006b). At all localities, most immature birds molt in October–March (Kemper and Roux 2005; Crawford et al. 2006b).

In Namibia, breeding occurs throughout the year but mostly in July–February; egg laying peaks during October–November and, to a lesser degree, during June–July (Kemper and Roux 2005; Kemper et al. 2007e, 2008). At Dassen Island, eggs are laid throughout the year but mostly in December–June, and most chicks are encountered in January–August (Wolfaardt et al. 2009b). At Robben Island, first clutches are laid in January–August, most (38%) in February, and 94% by the end of May; chicks with down are abundant in April–September, and chicks with feathers in May–October (Crawford et al. 1995a, 1999). At St. Croix Island, egg laying peaks in January, with a second peak in March–April attributable to replacement laying after clutch failure and a third peak in June that results from second clutches or further replacement laying (Randall and Randall 1981; Randall et al. 1986a).

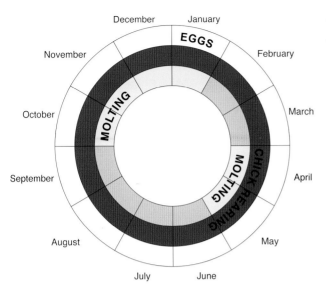

December   January
November   EGGS   February
October   MOLTING   March
September   CHICK REARING / MOLTING   April
August   May
July   June

FIG. 8 Annual cycle of the African penguin.

DEMOGRAPHIC PARAMETERS. Annual adult survival was 0.60–0.86 in Namibia in 1994–2003 (mean 0.78) (Kemper 2006). During 1987–98, it was 0.81 for both Dassen and Robben Islands (Whittington 2002). At Robben Island, it ranged between 0.82 and 0.89 in 1993–94 (Crawford et al. 1999). At Dyer Island, it was 0.69 in 1979–85 (La Cock and Hänel 1987). At St. Croix Island, annual adult survival averaged 0.91 from 1976–77 to 1981–82 (Randall 1983). Up to June 2000, there were 26 records of birds living for more than 20 years, with a maximum recorded longevity of more than 27 years (Whittington et al. 2000).

First-year survival in Namibia was in the range of 0.31–0.89 (mean 0.71) between 1993 and 2004 (Kemper 2006). For cohorts banded between 1987 and 1998, the range was 0.10–0.53 (mean 0.38) at Dassen Island and 0.17–0.44 (mean 0.31) at Robben Island (Whittington 2002). At Dyer Island, it was 0.69 (La Cock and Hänel 1987), and at St. Croix Island, the first-year survival averaged 0.32 (Randall 1983).

Mean age of first breeding for cohorts banded between 1990 and 1995 was 6 years at Mercury Island and 5 years at Ichaboe, Dassen, Robben, and Bird (Algoa Bay) Islands (Whittington et al. 2005). At Robben Island, during 1988–95, of 53 birds of known age, 1 bred unsuccessfully for the first time at the age of around 2 years, 4 (outcome unknown) when age 2–3 years, 12 when 3–4 years, 24 when 4–5 years, and 12 when 5–6 years (Crawford et al. 1999). At St. Croix Island, of 9 birds whose ages at first breeding were known, 3 were age 3–4 years, 3 were 4–5 years, and 3 were 5–6 years (Randall 1983). The proportion of sexually mature birds that nested in any year at Robben Island was 0.7–1.0 and was related to the spawner biomass of sardine and anchovy (Crawford et al. 1999). At Stony Point in 1982–86 and 1989–96, 20% of possible breeding attempts were skipped (Whittington et al. 1996).

In Namibia in 1996–2004, an average of 0.61 chicks fledged per breeding attempt (Kemper et al. 2007e). In the same period, averages of 0.70 chicks were fledged per breeding attempt at Mercury Island (SD = 0.80, n = 674 monitored attempts) and 0.61 chicks at Ichaboe Island (SD = 0.80, n = 1240). Averages of 0.44 chicks were fledged at Halifax Island (SD = 0.74, n = 571) in 2000–2004 and 0.71 chicks at Possession Island (SD = 0.87, n = 295) in 1999–2004 (Kemper et al. 2007e). At Marcus and Jutten Islands, the number of chicks fledged per nest increased from about 0.15 in 1984 to 0.55 in 1989, breeding success being significantly related to the biomass of sardine (Adams et al. 1992). At Dassen Island, pairs fledged 0.37–0.67 chicks per year (Frost et al. 1976b). In 1994–2000, an average of 0.89 chicks were fledged per breeding attempt, with breeding success significantly related to the combined biomass of anchovy and sardine (Wolfaardt et al. 2008b). At Robben Island, pairs fledged 0.32–0.59 chicks per year during 1989–95 (Crawford et al. 1999); 0.65 chicks in 1996, and 0.60 chicks in 1998 (Crawford et al. 2000b). During 1989–2004, when biomass of anchovy and sardine was less than 2 million metric tons, 0.46 chicks per year were fledged; when biomass was more than 2 million metric tons, 0.73 chicks per year were fledged. Breeding success was significantly related to the combined biomass of these two prey species (Crawford et al. 2006a). Pairs fledged 0.61 chicks per year at Boulders (in 1998) and 0.38 chicks at St. Croix Island (Randall 1983; Crawford et al. 2000b).

Of 428 first clutches monitored in a season at Robben Island, 48% failed during incubation, 28% failed at the brood stage, and 24% produced at least one fledged chick. When the first clutch failed at incubation stage, 32% of birds re-layed. When failure occurred at brood stage, 23% re-layed. When a chick was successfully fledged, 21% re-layed. Overall, 27% of breeding pairs produced a second clutch in the season. Of 112 second clutches, 30% failed at incubation stage, 40% failed at brood stage, and 30% produced at least one fledged chick. One unsuccessful third clutch was produced (Crawford et al. 1999).

FIG. 9 (a) During the 1930s, the Halifax Island colony in Namibia was covered with breeding penguins; (b) by 2004, only small, isolated patches of breeding penguins remained. (Eberlanz Museum, Lüderitz)

8. POPULATION SIZES AND TRENDS

NAMIBIA. Between 1956 and 2008, breeding occurred at the following 13 localities: from north to south, Hollamsbird Island, Sylvia Hill, Oyster Cliffs, Mercury Island, Neglectus Island, Ichaboe Island, Penguin Island, Halifax Island, North Reef, Possession Island, Pomona Island, Plumpudding Island, and Sinclair Island. One pair was present at Hollamsbird Island, the northernmost presumed extant colony, in 1988 (Williams and Dyer 1990); there have been no subsequent observations. At Sylvia Hill and Oyster Cliffs, both mainland colonies, up to 45 and 54 nests, respectively, have been recorded (Loutit and Boyer 1985; Bartlett et al. 2003; Simmons and Kemper 2003). Neglectus Island, where penguins were present in the 19th century (Crawford et al. 1995b), was recolonized in 1996 (Roux et al. 2003), and since that time, as many as 12 pairs have been counted at the island. The colony at Penguin Island was probably extinct before 1900 (Shelton et al. 1984) and was recolonized with one pair in the 2005–6 breeding season (Kemper et al. 2007c).

In 1956, at each of the eight other Namibian breeding localities, the estimated number of breeding pairs exceeded 500 (table 13.3). Overall, numbers of penguins in Namibia decreased by more than 90% after 1956, especially south of Lüderitz, where large decreases occurred and two colonies became extinct (e.g., see figs. 9a and 9b). The colony at North Reef became extinct in 2001 as did the colony at Pomona Island in 2004. In the vicinity of Lüderitz and farther north, numbers at two islands have shown recent stability, and two former island breeding localities have been recolonized. Throughout the region, the decreases probably were influenced by the collapse of the Namibian stock of sardine (Crawford 2007), the main food of African penguins in Namibia in the 1950s (Matthews 1961).

For more discussion of trends in numbers at localities in Namibia, see Berry et al. (1974), Shelton et al. (1984), Crawford et al. (1985), Crawford et al. (1990), Crawford et al. (1995c), Crawford et al. (2001), Cordes et al. (1999), Kemper et al. (2001), Kemper et al. (2007a), and Kemper et al. (2007b).

WESTERN CAPE. Between 1956 and 2008, breeding occurred at 13 localities: from north to south and east, Lambert's Bay, Malgas Island, Marcus Island, Jutten Island, Von-

deling Island, Dassen Island, Robben Island, Boulders, Seal Island (False Bay), Stony Point, Dyer Island, Geyser Island, and De Hoop. After 1956, four new colonies were established: Stony Point in 1982 (Whittington et al. 1996); Robben Island, a former colony that was extinct by about 1800 and was recolonized in 1983 (Crawford et al. 1995a, 1999); Boulders in 1985 (Crawford et al. 2000b); and De Hoop in 2003 (Underhill et al. 2006). Three of the new localities (Boulders, Stony Point, De Hoop) are on the mainland. The other colonies in the Western Cape are all at islands, although Lambert's Bay and Marcus Island are joined to the mainland by causeways. The colony at Lambert's Bay became extinct in 2006 and that at De Hoop in 2008. The latter attained a maximum of 18 pairs in 2006 but was accessible to terrestrial predators.

Estimates of numbers of penguins at 10 colonies in the Western Cape from 1956 to 2009 are shown in table 13.4, excluding De Hoop and Seal and Geyser Islands. There were about 250 pairs at Seal Island in 1956, but since 1979, this colony has fluctuated between 48 and 95 pairs and is stable. The colony at Geyser Island decreased from 318 pairs in 1979 to 4 pairs in 2009.

Trends in numbers of penguins at colonies in the Western Cape have reflected changes in the abundance and distribution of sardines and anchovies and have probably been influenced by local patterns of fishing. Large decreases at the more northern colonies in the Western Cape (Lambert's Bay, Malgas, Marcus, Jutten, and Dassen Islands) followed a collapse of South Africa's stock of sardine in the 1960s. Three new colonies (Robben Island, Boulders, Stony Point) formed between Dassen and Dyer Islands in the early 1980s as South Africa's sardine stock recovered (Crawford et al. 2001). Substantial numbers of birds (thought to be first breeders) emigrated to the new colonies from Dyer Island, where numbers decreased (Crawford 1998; Crawford et al. 2001). In the late 1990s and early 2000s, sardines and anchovies were both abundant off South Africa, and there was rapid growth of the Robben Island colony and of colonies farther north, especially at Dassen Island. As the distribution of anchovies and sardines began shifting to the south and east, from the mid-2000s, penguins at all colonies from Malgas Island to Robben Island underwent large decreases, with a smaller decrease at Boulders. The northernmost colony at Lambert's Bay became extinct. To the east of Boulders, penguins increased at Stony Point and attempted to breed at De Hoop. Numbers at Dyer Island, between Stony Point and De Hoop, did not increase (Crawford et al. 2011).

See Shelton et al. (1984), Crawford et al. (1990), Crawford et al. (1995c), Crawford et al. (2001), Crawford et al. (2011), Underhill et al. (2006), and Kemper et al. (2007b) for more discussion of trends in penguins in South Africa's Western Cape localities.

EASTERN CAPE. Breeding occurs at six localities: from west to east, Jahleel, Brenton, St. Croix, Seal, Stag and Bird Islands (table 13.5). A decrease in the number of penguins breeding in the region after 2001 took place at the same time as a large increase in the catch of sardine in the region and the construction of Coega harbor near Jahleel, Brenton, and St. Croix Islands (Crawford et al. 2009).

Further discussion of trends in numbers at localities in South Africa's Eastern Cape is in Shelton et al. (1984), Crawford et al. (1990), Crawford et al. (1995c), Crawford et al. (2001), Crawford et al. (2009), Crawford et al. (2011), and Kemper et al. (2007b).

9. MAIN THREATS

FOOD ABUNDANCE, DISTRIBUTION, AND QUALITY AND COMPETITION WITH FISHERIES. A combination of competition with the commercial fishing industry and environmental variability has led to a lack of food for all regional penguin populations. In Namibia, relatively abundant but energy-poor pelagic goby replaced the energy-rich sardine as the main prey following the collapse of the sardine in the late 1960s (Crawford 2007; Ludynia et al. 2010). The range of sardine in Namibia contracted to the north, out of the reach of penguins breeding in Namibia (MFMR, unpubl. data). Although commercial purse-seine fisheries operate far north of the breeding range, they may be preventing the stock from expanding south.

In South Africa, commercial purse-seine fisheries compete with African penguins for the birds' two main prey items, anchovy and sardine. Sardine off South Africa collapsed in the 1960s (Crawford and Shelton 1978), and its range contracted to the south. This was accompanied by an increase in anchovy, which largely replaced sardine off South Africa (Crawford 1998). In the early 1980s, South Africa's sardine stock recovered, and in the late 1990s and early 2000s, sardine and anchovy were both abundant off South Africa. In the early 2000s, a shift of South Africa's anchovy and sardine stocks to the south and east caused a mismatch in the distribution of penguins and their prey in breeding localities in the Western Cape (Crawford et al. 2008; Grémillet et al. 2008). It also caused a discrepancy in the distribution of fish-processing plants and fish, which resulted in intensified fishing around penguin colonies. Both the mismatch and the intensified fishing around colonies probably influenced the collapse of the penguin population in the Western Cape from 39,000 pairs in 2004 to 11,000 pairs in 2009, equivalent to a loss of more than half the species' population. The fishery made large catches of sardine in the immediate vicinity of Dyer Island in the early 2000s, heavily outcompeting the penguins there for available food.

There are unconfirmed reports of penguins being used as bait in rock lobster traps (Cooper 1974; Hockey and Hallinan 1981; MFMR, unpubl. data).

CLIMATE VARIATION. Climate variation is likely to have a major impact on African penguins. The recent shift in distribution of anchovy and sardine stocks in South Africa is not yet fully understood but may be due, at least in part, to climate variation (Coetzee et al. 2008; Grémillet et al. 2008). Significant changes in sea surface temperature and upwelling have been recorded in the highly variable northern Benguela upwelling system in Namibia over the past three decades. Observed and expected effects of climate change on the penguin population in Namibia include the following: decreased juvenile recruitment caused by a rise in sea surface temperature and a decrease in upwelling intensity (Roux 2003; Kemper 2006); decreased breeding success caused by an increase in sea level and coastal rainfall (Kemper 2006; Kemper et al. 2007e) and a loss of low-lying breeding localities; and decreased breeding success and survival due to an increase in the frequency and intensity of Benguela Niño events (Roux 2003). The entire food web is affected (Kemper 2006). A reduction in food has been shown to impact seabird colonies, in particular those of the African penguin (Cury et al. 2011).

HARVESTS. Historically, penguins were killed for food, for fuel for ship boilers, and for their fat (Randall 1989). Egg collections may have taken up to 48% of the total number of eggs produced and caused population decreases (Shannon and Crawford 1999). In 1897, 762,400 eggs were collected; in 1899, 801,500 eggs; in 1905, 745,250 eggs. The last authorized egg collections were in 1967 (Shelton et al. 1984).

HABITAT DEGRADATION. The removal of accumulated deposits of guano forced birds to nest on the surface of islands (Frost et al. 1976b; Cooper 1980), where nests are sometimes vulnerable to being flooded (Randall et al. 1986b; Kemper 2006), eggs and chicks are more accessible to aerial predators (Kemper et al. 2007d, 2007e), and adults and chicks are subject to heat stress, which sometimes causes abandonment of breeding attempts and the death of eggs and chicks (Randall 1983; Kemper et al. 2007d). Pipes and other forms of artificial nests have been placed at some islands (Wilson and Wilson 1989), and the birds often use them (Crawford et al. 1994), which has improved breeding success in some cases (Kemper et al. 2007d). Surface nesting also may have rendered African penguins more susceptible to displacement from breeding sites by larger animals such as seals (e.g., Crawford et al. 1989).

Coastal and inshore mining operations along Namibia's southern coast threaten foraging habitats through the large-scale release of sediment into coastal waters. Water turbidity may reduce prey availability and is likely to affect foraging behavior. Sediment movement is also likely to contribute to the formation of temporary land bridges, offering access for land predators at Pomona and Sinclair Islands (Kemper 2006).

POLLUTION. Oiling (including crude, fuel, and fish oil) is a major threat to African penguins. Oil causes feathers to clump, which breaks down their insulating properties. As a result, birds become hypothermic and are forced to leave cold waters (Erasmus et al. 1981). They become dehydrated, mobilize stored energy reserves, and rapidly lose body mass (Morant et al. 1981; Erasmus and Wessels 1985). If not rescued, they eventually starve. Oil ingested by preening can cause ulceration of the mouth, esophagus, and stomach and, in severe cases, leads to substantial blood loss. Oil absorbed into the system causes red blood cells to rupture, leading to anemia (Birrel 1994). Further, an immunosuppressant effect makes birds more susceptible to diseases (Morant et al. 1981) such as pneumonia and aspergillosis. Ingested oil may promote a greater diversity of pathogenic bacteria in affected penguins (Kerley and Erasmus 1987). Untreated oiling can lead to ulceration of the cornea and blindness (Crawford et al. 2000a). When an oiled bird molts, oil is transferred from pre-molt to post-molt feathers (Kerley et al. 1985).

Oiled penguins in poor condition suffer higher mortality than those with a body mass greater than two kilograms at capture (Kerley and Erasmus 1987). After removal of oil, a bird's plumage must regain its water-repelling properties before it is released (Kerley et al. 1987). There is no difference in the immune status of rehabilitated oiled birds and non-oiled birds during molt (Hemming 2001) or in time elapsed between banding and recovery of cleaned and non-oiled birds (Whittington 1999). Mean annual survival rate of rehabilitated adult African penguins is 79% per year (Whittington 2003). Many rehabilitated birds make a successful transition back to the wild (Underhill et al. 1999, 2000; Wolfaardt et al. 2001, 2008a, 2009b; Barham et al. 2006) and breed again (Randall et al. 1980; Wolfaardt and Nel 2003; Wolfaardt et al. 2008a), but they may not do so for up to six years (Wolfaardt et al. 2001; Barham et al. 2007).

Namibia has escaped large-scale oil spills to date, but persistent chronic oiling of penguins occurs regularly from sunken vessels or vessels illegally cleaning bilges at sea. In South Africa, in addition to numerous chronic oiling incidents, there were at least 14 major oiling incidents between 1968 and 2000, and more than 47,000 oiled African penguins were caught and admitted to the South African National Foundation for the Conservation of Coastal Birds (SANCCOB) for rehabilitation during that period (Nel et al. 2003). In addition, a further 19,500 non-oiled penguins were caught at Dassen and Robben Islands during the *Treasure* spill in 2000 and were relocated to Port Elizabeth, 800 kilometers east, to prevent their coming into contact with the oil; during this incident, more than 3,000 orphaned/abandoned chicks were collected for captive rearing (Crawford et al. 2000a).

Low levels (fresh weight) of DDE (dichlorodiphenyldichloroethylene) (0.06 μg/g), an organochlorine metabolite of DDT (dichlorodiphenyltrichloroethane), and PCB (polychlorinated biphenyls) residues (0.24 μg/g) were found in eggs collected at St. Croix Island, Algoa Bay, Eastern Cape, during 1981–83 but are insufficient to cause reproductive impairment or eggshell thinning (De Kock and Randall 1984).

The entanglement of penguins in fishing nets, discarded line, and other materials (Hockey et al. 2005) and the ingestion of other marine debris such as plastics are causing limited mortality. A growing mariculture industry in Namibia, particularly in the vicinity of Lüderitz, is increasing the risk of direct entanglement in fixed structures and use of associated debris by penguins as nesting material.

DISEASE. Heavy infestation of the first 0.5 meter of the small intestine by trematodes (*Cardiocephaloides physalis*) caused mortality of chicks and recently fledged juveniles at St. Croix Island (Randall and Bray 1983). Hemoparasites, including avian malaria (*Plasmodium relictum*), were found in 22% of penguins at a mainland SANCCOB rehabilitation center in summer months (often with fatal outcomes) but only in 0.7% of penguins from Saldanha Bay (Brossy 1992; Brossy et al. 1999). If diagnosed, malaria can be treated (Ellis et al. 1998). African penguins are also infected by *Leucocytozoon tawaki* (Brossy 1993); an avian piroplasm (*Babesia peircei*), for which the vector is probably the tick (*Ornithodoros capensis*) (Earlé et al. 1993; Brossy et al. 1999); and avian cholera (*Pasteurella multocida*) (Crawford et al. 1992). Other diseases include aspergillosis, pneumonia (viral or coccal), Newcastle disease, and borreliosis (Yabsley et al. 2012). In captivity, bumblefoot may be caused by *Staphylococcus* bacteria associated with damp floors (Ellis et al. 1998). Captive birds may die of infections of *Salmonella typhimurium*, *Escherichia coli*, and *Staphylococcus aureus* (Westphal and Rowan 1971).

HUMAN DISTURBANCE. Starting in the 1840s, guano collection caused considerable disturbance at breeding sites (Berry et al. 1974). Guano is no longer collected at South African islands but continues to be harvested in Namibia. Humans approaching colonies too closely may decrease breeding success (Hockey and Hallinan 1981). They also may cause burrows to collapse. However, tourism to colonies is generally well controlled, with defined paths or walkways at the most visited colonies. Access to the Namibian breeding localities is strictly controlled, although illegal trespassing by tourists and fishermen occurs occasionally. Fire is a potential hazard at some colonies (Ellis et al. 1998).

OTHER THREATS. This species is also threatened by predation (detailed above) as well as competition with Cape fur seals and other seabirds for food and breeding areas (e.g., Shaughnessy 1980, 1984; Crawford et al. 1989).

10. RECOMMENDED PRIORITY RESEARCH ACTIONS FOR CONSERVATION

One of the greatest threats facing African penguins is lack of good-quality, reliably available food. Detailed information on foraging ranges, habitats, and behavior as well as identification of key foraging areas of breeding and nonbreeding birds are essential for conservation management. While several studies on the foraging ecology of breeding African penguins have been completed or are under way (e.g., Petersen et al. 2006; Ludynia 2007; Pichegru et al. 2010; MFMR, unpubl. data), data are lacking for juveniles and nonbreeding adults, as well as for penguins fattening up before and after molt.

Understanding the factors that influence the selection of breeding localities by first-time breeders will aid in establishing new colonies near food resources or bolstering decreasing colonies. Such insights may be gained through releases of captive-reared chicks held at release localities over different periods.

The following recommendations are presented in order of priority.

1. Apply mitigating measures, such as placing restrictions on fisheries when appropriate.
2. Collect information on the diet and how it varies between localities and over time.
3. Collect information on the influence of fishing on the local availability of prey.
4. Study the effects of oceanographic anomalies, environmental (possibly driven by climate) change, and ecosystem shifts and prey availability.
5. Extend monitoring programs for commercially important prey species to other main prey species such as pelagic goby in Namibia.
6. Continue efforts to monitor the status of African penguins and to gauge the success of conservation management measures. A thorough mark-recapture program will yield valuable information on demographic parameters, such as survival, movement, and breeding productivity, provided that the methods used do not compromise marked individuals, a high-level re-sighting effort is maintained, and data are curated properly, although much information already exists (e.g., Whittington 2002; Whittington et al. 2000, 2005; Kemper 2006).
7. Complete comprehensive health assessments throughout this penguin's range for disease management. Screen populations for emerging diseases and the presence of pathogens and vectors, including external, internal, and blood parasites. The identification of disease agents and information on their presence at a local, regional, or global scale, their

variability over time, and their potential effects on survival and production rates are essential for the formulation of contingency plans and protocols for preventing and managing disease.

8. Investigate the risk of transmitting disease from rehabilitated penguins to wild populations and implement measures to prevent transmission. Similar assessments should precede the use of captive-bred individuals to bolster the wild population.

9. Quantify the extent of kelp gull predation on eggs and chicks at breeding colonies and develop control programs where needed.

11. CURRENT CONSERVATION EFFORTS

Temporary bans on purse-seine fishing in areas around Dassen and St. Croix Islands were made to examine the feasibility of imposing the bans as a means of securing food availability for penguins within their foraging range during breeding. Preliminary results suggest that the foraging effort of breeding African penguins decreased by 30% within three months of the closure at St. Croix Island and that most penguins shifted their feeding effort to the closed area. Birds breeding at a colony that remained open to fishing 50 kilometers away increased their foraging effort during the same period (Pichegru et al. 2010). Additional monitoring is needed to validate these results.

Namibia declared its first marine protected area in February 2009. One of the key objectives of the Namibian Islands' Marine Protected Area (NIMPA) is to improve the conservation status of species of concern by protecting their breeding colonies and foraging habitats. The Namibian government and other organizations developed strategies, action plans, and comprehensive monitoring and research programs to address this objective in 2010. NIMPA covers an area of roughly 1 million hectares and embraces all Namibian African penguin breeding localities as well as some known key foraging areas in Namibia and provides legal protection status to all breeding localities. The declaration follows a zoned approach, with strictest conservation measures afforded to seabird breeding localities, and includes measures to limit fishing activities and reduce threats posed by other activities, such as mining and tourism (Currie et al. 2009).

South Africa is developing its Biodiversity Management Plan for the African Penguin in terms of section 43 of its National Environmental Management: Biodiversity Act (2004). A number of stakeholders participated in a workshop in 2010, which initiated the development of the management plan (Shaw et al. 2011).

Interventions during oil spills have included the rescue and rehabilitation of oiled birds, the translocation of non-oiled birds so that they will not become oiled, and captive rearing and release of orphaned chicks (e.g., Crawford et al. 2000a). Each of these interventions proved successful, although 27% of rehabilitated oiled birds did not breed and those that did had reduced breeding success and increased cost of reproduction (Barham et al. 2006, 2007, 2008; Wolfaardt et al. 2008a, 2008b, 2009a, 2009b).

Artificial nests have been placed at Ichaboe, Halifax, Possession, Marcus, Robben, Seal (False Bay), Dyer, and Bird Islands and at Boulders. Kemper et al. (2007d) recorded improved breeding success at artificial nests at Halifax Island. Cape fur seals that had displaced penguins from breeding sites were removed from Mercury Island, allowing recovery in the numbers of penguins breeding there (Crawford et al. 1989; Kemper et al. 2001).

Feral cats have been eradicated from Dassen Island and reduced in number at Robben Island. Control of individual Cape fur seals known to kill African penguins is carried out around Mercury, Ichaboe, Possession, and Dyer Islands. At Possession Island, where kelp gulls prey on eggs and chicks, an egg-sterilization program helps control the large kelp gull population. Culling of kelp gulls preying on African penguins is undertaken at Dyer and Bird (Algoa Bay) Islands.

In Namibia, three of the four main breeding localities (Mercury, Ichaboe, and Possession Islands) are permanently staffed; a fourth locality, Halifax Island, is visited regularly. The four islands support roughly 96% of the penguin population in Namibia. Bimonthly counts of molting penguins and monthly counts of nests containing eggs or chicks provide reliable measures of population numbers and allow accurate monitoring of population trends.

12. RECOMMENDED PRIORITY CONSERVATION ACTIONS FOR INCREASING POPULATION RESILIENCE AND MINIMIZING THREATS AND IMPACTS

The governments of Namibia and South Africa are committed to implementing an ecosystem approach to fisheries to improve the management of fish stocks, taking into account ecosystem health and functioning as well as the needs of African penguins. In South Africa, several national and provincial governmental institutions, as well as nongovernmental organizations (e.g., SANCCOB), share responsibility for managing African penguins and their habitats. To ensure concerted management efforts, all relevant stakeholders are developing a national biodiversity management plan in terms of South Africa's National Environmental Management: Biodiversity Act, 2004. Conservation actions should include the following:

1. Promote the recovery and expansion of depleted commercially important fish stocks through conservative total allowable catches, limits on the number of fishing permits, and the closure of relevant areas during key periods. Fisheries management plans and total allowable catch allocations should take into account food requirements (Cury et al. 2011, Pikitch et al. 2012).

2. Implement a representative system of marine protected areas by 2012 as recommended in the international legal requirements stemming from the World Summit on Sustainable Development held in 2002 (FAO 2003) and the Convention on Biodiversity.

3. Protect key foraging areas used by breeding African penguins and by penguins fattening before and after molt.

4. Improve legislation and its enforcement in Namibia, particularly with respect to the illegal dumping of oil and other pollutants. Draft regulations specific to NIMPA. Revise the Namibian National Oil Spill Contingency Plan so that it guarantees immediate and comprehensive action in case of a spill.

5. Implement strict operational guidelines ensuring minimal negative impacts of guano harvesting on penguins. Although Currie et al. (2009) and other studies recommend that no guano harvesting take place within NIMPA, the right to harvest guano from Ichaboe Island, and possibly from Mercury Island, was renewed in 2012.

6. All Namibian islands need to draft and implement individual management plans, with priority given to Penguin Island, which has been earmarked for tourism activities. In Namibia, the management of African penguins (including breeding localities and foraging areas) is the responsibility of the Ministry of Fisheries and Marine Resources. Comprehensive management plans for the species and associated habitats are lacking; however, NIMPA management and action plans take African penguins into account.

7. The Namibian government should formulate contingency plans for the rescue and rehabilitation of African penguins. The plans should then be incorporated into individual island management plans and the Namibian National Oil Spill Contingency Plan.

8. Namibia and South Africa should sign a memorandum of understanding on concerted conservation management of African penguins across their breeding range.

9. Use artificial nests to prevent fragmentation and the decreased breeding success associated with edge effects, as needed. Assess the suitability of recent nest designs.

10. Consider reestablishing breeding colonies in abandoned localities or establishing them at new localities, as long as they provide suitable breeding habitat with a good-quality, reliable food source nearby and are not threatened by human activities, predators, or disease.

11. Train rehabilitation volunteers, field technicians, reserve managers, and scientists to ensure continuity of research and conservation efforts for the species. Standardize monitoring techniques across the species' range; a manual of basic monitoring techniques is available (Kemper et al. 2007a) and widely distributed.

12. Namibia and South Africa should develop a comprehensive educational and awareness program designed to sensitize the public (including children, the fishing community, the mining and tourism industries, and government institutions) about African penguins, their status and importance, the threats they face, and conservation measures to improve their status.

ACKNOWLEDGMENTS
We thank N. Parsons, P. G. Ryan, and S. L. Petersen for reviewing earlier drafts of this chapter.

TABLE 13.1 Measurements of the adult African penguins (showing mean, standard deviation or range, and sample size)

	MALE	FEMALE	REFERENCE
Flipper (mm)	191±5.6 (127)	180±2.9 (180)	Cooper 1972, Duffy 1987
Culmen (mm)	60.5, 57–65 (41)	55.5, 51–59 (25)	Cooper 1972, Duffy 1987
Mass (kg)	3.31±0.26 (127)	2.96±0.31 (127)	Cooper 1972, Duffy 1987

TABLE 13.2 Measurements of African penguin eggs (showing mean, range and sample size)

	A-EGG	B-EGG	REFERENCE
Mass (g)	106.8, 75–132 (70)	104.8 (83–129)	Williams & Cooper 1984
Dimensions (mm)	69.6x52.1, 62.1–76.0x47.2–55.9 (70)	67.7x52.0, 54.1–72.8x48.2–55.8 (70)	Williams & Cooper 1984

TABLE 13.3 Estimates of numbers of African penguins (pairs) at eight Namibian islands, 1956–2009

YEAR	MERCURY	ICHABOE	HALIFAX	NORTH REEF	POSSESSION	POMONA	PLUMPUDDING	SINCLAIR
1956	4750	4200	5500	850	23245	6000	3000	1500
1978	3218	3598	1750	151	2568	123	438	246
1979		4200		58		20	100	124
1981			1007					
1985	2092	2070	334	2	735	13	92	51
1986	1126	739	389	1	327	15		
1987	2181	1372	678	2	454	13	228	29
1988				0	946		90	44
1989	3394	2427						
1990	3552	1937	1050	0	702	2	1	27
1991						8	2	38
1992	3576	2418	690		669			
1993	3666	2655	709	0	751		27	63
1994	2122	2736	681		506			
1995	2261	3343	473		608	1	38	123
1996	2387	2640	384		498			
1997	2662	1903	431		512			
1998	2884	1562	444		465			
1999	3041	1731	497	1	477		54	55
2000	2822	1345	370		359		67	75
2001	2615	2260	520	0	362	1	54	86
2002	3462	2182	508	0	406	2		
2003	2783	2158	534	0	372			
2004	2037	864	669	0	445	0	36	77
2005	2455	1077	612	0	470	0		
2006	1813	715	305	0	356	0	57	88
2007	2026	531	530	0	459	0	171	70
2008	3188	520	635	0	474	0	86	68
2009	2398	615	843	0	481	0		

Note: Estimates are from Rand (1963b), Shelton et al. (1984), Crawford et al. (1995c, 2001), Cordes et al. (1999), Kemper (2006), Kemper et al. (2001 2007a, 2007b), and MFMR (unpubl. data). Estimates for 1956, except for North Reef and Possession Island, were obtained by halving the estimated population. Estimates since 1996 are of numbers of nests containing eggs or chicks. Estimates in other years are of nest sites. The maximum count obtained in any year was used. At Mercury, Ichaboe, Halifax, and Possession islands, counts were conducted more regularly after the early 1990s.

TABLE 13.4 Estimates of African penguin pairs at 10 breeding localities in the Western Cape, 1956–2009

YEAR	LAMBERT'S	MALGAS	MARCUS	JUTTEN	VONDELING	DASSEN	ROBBEN	BOULDERS	STONY PT	DYER
1956	250	2500	4750	7500	300	72500				4000
1976			717		1028					
1978	37									
1979	50	1022	1243	2878	495	12646				22655
1982									1	
1983							9		1	
1984							24		5	
1985							103	2	11	
1986							227		27	18481
1987	21	142	214		196	4588	476	7		
1988	23						849	34		
1989		101			109	8428	829	38	69	
1990	30	118					1278	54	89	8349
1991	25	80	207	806	229	9012	1879	131	77	6115
1992	18	99		991	133	7563	2027	158	57	7579
1993	22	70		526	141	7199	2176	241	40	2374
1994	16	80	204	1349	169	9389	2799	359	44	4649
1995	26	74	160	891	205	9792	2279	366		4260
1996	20	71	99	779	258	9502	3097	416		3279
1997	23	43	73	947	361	8651	3336	726		2745
1998	27	61	122	962	157	10918	3467	555	72	1963
1999	11	67	116	759	333	15155	4399	906		2363
2000	10	48	96	898	528	15598	5705	949	104	2220
2001	9	55	114	1338	649	21409	6723	1054	111	2088
2002	9	58	57	1042	544	22883	7252	1083		2145
2003	18	43	65	719	622	20319	6433	1033	123	1929
2004	22	28	78	848	612	24901	8524	1196	98	2216
2005	10	26	53	801	564	22687	7152	1227	186	2053
2006	0	11	33	435	396	13283	3697	1075	265	2057
2007	0	8	59	329	345	11785	5935	824	260	1513
2008	0	12	64	669	507	5719	2234	913	310	1605
2009	0	20	41	408	361	5138	2415	704	487	1260

Note: Estimates are from Rand (1963a), Shelton et al. (1984), Crawford et al. (1995c, 2001), and Underhill et al. (2006), which have been updated. Estimates for 1956 were obtained by halving the estimated population. Other estimates are of numbers of nest sites, the maximum count obtained in any year being used.

TABLE 13.5 Estimates of African penguin pairs at the 6 breeding localities in the Eastern Cape, 1956–2009 (updated from Crawford et al. 2009)

YEAR	JAHLEEL	BRENTON	ST CROIX	SEAL	STAG	BIRD
1956			5764	150	40	60
1969			13654			
1975				607		
1977					45	345
1979			13821			
1980			14850			
1981	578				23	572
1985			19448			
1986		38				
1990				510	50	3703
1991					15	3194
1992				372	24	3784
1993	549	31	19478	375	21	2293
1995						3651
1997	454	35				1621
1999	243	20	14005	316	23	3883
2000	538	32	15211	433	24	4093
2001			16950	345	24	5376
2003	141		9116			936
2004	479	32	10088	295	20	2951
2005	316	36	4153	319	23	3203
2006	301	17	8077	237	13	2822
2007	141	3	4363	87	6	1403
2008	65	3	7739	95	8	2765
2009	181	8	6824	137	4	2624

REFERENCES

Adams, N. J., and C. R. Brown. 1990. Energetics of molt in penguins. In *Penguin Biology*, ed. L. S. Davis and J. T. Darby, 297–315. San Diego, CA: Academic Press.

Adams, N. J., P. J. Seddon, and Y. M. van Heezik. 1992. Monitoring of seabirds in the Benguela upwelling system: Can seabirds be used as indicators and predictors of change in the marine environment? *South African Journal of Marine Science* 12:959–74.

Barham, P. J., B. J. Barham, L. G. Underhill, R. J. M. Crawford, and T. M. Leshoro. 2007. Differences in breeding success between African penguins (*Spheniscus demersus*) that were and were not oiled in the *Treasure* oil spill in 2000. *Emu* 107:1–7.

Barham, P. J., R. J. M. Crawford, L. G. Underhill, A. C. Wolfaardt, B. J. Barham, B. M. Dyer, T. M. Leshoro, M. A. Meÿer, R. A. Navarro, D. Oschadleus, L. Upfold, P. A. Whittington, and A. J. Williams. 2006. Return to Robben Island of African penguins that were rehabilitated, relocated, or reared in captivity following the *Treasure* oil spill of 2000. *Ostrich* 77:202–9.

Barham, P. J., L. G. Underhill, R. J. M. Crawford, R. Altwegg, T. M. Leshoro, D. Bolton, B. M. Dyer, and L. Upfold. 2008. Hand-reared African penguin chicks orphaned in the *Treasure* oil spill in 2000: Survival, age at first breeding and breeding productivity. *Bird Conservation International* 18:1–9.

Bartlett, P. A., J-P. Roux, R. Jones, and J. Kemper. 2003. A new mainland breeding locality for African penguins, bank and crowned cormorants on the Namib desert coast. *Ostrich* 74:222–25.

Berruti, A. 1986. The predatory impact of feral cats *Felis catus* and their control on Dassen Island. *South African Journal of Antarctic Research* 16:123–27.

Berry, H. H., M. K. Seely, and R. E. Fryer. 1974. The status of the jackass penguin *Spheniscus demersus* on Halifax Island off South West Africa. *Madoqua*, 2nd ser. (3):27–29. In *Scientific Papers of the Namib Desert Research Station* 71.

Birrel, J. 1994. General principles of disease control. In *Proceedings: Coastal Oil Spills: Effect on Penguin Communities and Rehabilitation Procedures*, ed. J. Barrett, Z. Erasmus, and A. J. Williams, 34–37. Cape Town, South Africa: Cape Nature Conservation.

Brossy, J. J. 1992. Malaria in wild and captive jackass penguins *Spheniscus demersus* along the southern African coast. *Ostrich* 63:10–12.

———. 1993. Haemoparasites in the African (jackass) penguin (*Spheniscus demersus*). *Penguin Conservation* 6:20–21.

Brossy, J. J., A. L. Plös, J. M. Blackbeard, and A. Kline. 1999. Diseases acquired by captive penguins: What happens when they are released into the wild? *Marine Ornithology* 27:185–86.

Coetzee, J. C., C. D. van der Lingen, L. Hutchings, and T. P Fairweather. 2008. Has the fishery contributed to a major shift in the distribution of South African sardine? *ICES Journal of Marine Science* 65:1676–88.

Cooper, J. 1972. Sexing the jackass penguin. *Safring News* 1:23–25.

———. 1974. The predators of the jackass penguin *Spheniscus demersus*. *Bulletin of the British Ornithologists Club* 94:21–24.

———. 1977. Energetic requirements for growth of the jackass penguin. *Zoologica Africana* 12:201–13.

———. 1978. Moult of the black-footed penguin. *International Zoo Yearbook* 18:22–27.

———. 1980. Breeding biology of the jackass penguin with special reference to its conservation. In *Proceedings of the 4th Pan-African Ornithological Congress*, ed. D. N. Johnson, 227–31. Johannesburg, South Africa: Southern African Ornithological Society.

Cordes, I., R. J. M. Crawford, A. J. Williams, and B. M. Dyer. 1999. Decrease of African penguins at the Possession Island Group, 1956–1995: Contrasting trends for colonial and solitary breeders. *Marine Ornithology* 27:129–38.

Crawford, R. J. M. 1998. Responses of African penguins to regime changes of sardine and anchovy in the Benguela system. *South African Journal of Marine Science* 19:355–64.

———. 2007. Food, fishing and seabirds in the Benguela upwelling system. *Journal of Ornithology* 148 (suppl. 2):S253–60.

Crawford, R. J. M., D. M. Allwright, and C. W. Heÿl. 1992. High mortality of Cape cormorants (*Phalacrocorax capensis*) off western South Africa in 1991 caused by *Pasteurella multocida*. *Colonial Waterbirds* 15:236–38.

Crawford, R. J. M., R. Altwegg, B. J. Barham, P. J. Barham, J. M. Durant, B. M. Dyer, D. Geldenhuys, A. B. Makhado, L. Pichegru, P. G. Ryan, L. G. Underhill, L. Upfold, J. Visagie, L. J. Waller, and P. A. Whittington. 2011. Collapse of South Africa's penguins in the early 21st century. *African Journal of Marine Science* 33(1):139–56.

Crawford, R. J. M., P. J. Barham, L. G. Underhill, L. J. Shannon, J. C. Coetzee, B. M. Dyer, T. M. Leshoro, and L. Upfold. 2006a. The influence of food availability on breeding success of African penguins *Spheniscus demersus* at Robben Island, South Africa. *Biological Conservation* 132:119–25.

Crawford, R. J. M., H. G. v. D. Boonstra, B. M. Dyer, and L. Upfold. 1995a. Recolonization of Robben Island by African penguins, 1983–1992. In *The Penguins: Ecology and Management*, ed. P. Dann, I. Norman, and P. Reilly, 333–63. Chipping Norton, NSW, Australia: Surrey Beatty & Sons.

Crawford, R. J. M., R. A. Cruickshank, P. A. Shelton, and I. Kruger. 1985. Partitioning of a goby resource amongst four avian predators and evidence for altered trophic flow in the pelagic community of an intense, perennial upwelling system. *South African Journal of Marine Science* 3:215–28.

Crawford, R. J. M., J. H. M. David, L. J. Shannon, J. Kemper, N. T. W. Klages, J-P. Roux, L. G. Underhill, V. L. Ward, A. J. Williams, and A. C. Wolfaardt. 2001. African penguins as predators and prey—coping (or not) with change. *South African Journal of Marine Science* 23:435–47.

Crawford, R. J. M., J. H. M. David, A. J. Williams, and B. M. Dyer. 1989. Competition for space: Recolonising seals displace endangered, endemic seabirds off Namibia. *Biological Conservation* 48:59–72.

Crawford, R. J. M., S. A. Davis, R. T. Harding, L. F. Jackson, T. M. Leshoro, M. A. Meÿer, R. M. Randall, L. G. Underhill, L. Upfold, A. P. van Dalsen, E. van der Merwe, P. A. Whittington, A. J. Williams, and A. C. Wolfaardt. 2000a. Initial impact of the *Treasure* oil spill on seabirds off western South Africa. *South African Journal of Marine Science* 22:157–76.

Crawford, R. J. M., B. M. Dyer, and R. K. Brooke. 1994. Breeding nomadism in southern African seabirds: Constraints, causes and conservation. *Ostrich* 65:231–46.

Crawford, R. J. M., B. M. Dyer, and P. C. Brown. 1995b. Absence of breeding by African penguins at four former colonies. *South African Journal of Marine Science* 15:269–72.

Crawford, R. J. M., M. Hemming, J. Kemper, N. T. W. Klages, R. M. Randall, L. G. Underhill, A. D. Venter, V. L. Ward, and A. C. Wolfaardt. 2006b. Molt of the African penguin, *Spheniscus demersus*, in relation to its breeding season and food availability. *Acta Zoologica Sinica* 52 (suppl.):444–47.

Crawford, R. J. M., L. J. Shannon, and P. A. Whittington. 1999. Population dynamics of the African penguin *Spheniscus demersus* at Robben Island, South Africa. *Marine Ornithology* 27:139–47.

Crawford, R. J. M., L. J. Shannon, P. A. Whittington, and G. Murison. 2000b. Factors influencing growth of the African penguin colony at Boulders, South Africa, 1985–1999. *South African Journal of Marine Science* 22:111–19.

Crawford, R. J. M., and P. A. Shelton. 1978. Pelagic fish and seabird interrelationships off the coasts of South West and South Africa. *Biological Conservation* 14:85–109.

Crawford, R. J. M., L. G. Underhill, J. C. Coetzee, T. Fairweather, L. J. Shannon, and A. C. Wolfaardt. 2008. Influences of the abundance and distribution of prey on African penguins *Spheniscus demersus* off western South Africa. *African Journal of Marine Science* 30:167–75.

Crawford, R. J. M., L. G. Underhill, L. Upfold, and B. M. Dyer. 2007. An altered carrying capacity of the Benguela upwelling ecosystem for African penguins (*Spheniscus demersus*). *ICES Journal of Marine Science* 64:570–76.

Crawford, R. J. M., P. A. Whittington, A. P. Martin, A. J. Tree, and A. B. Makhado. 2009. Population trends of seabirds breeding in South Africa's Eastern Cape, and the possible influence of anthropogenic and environmental change. *Marine Ornithology* 37:159–74.

Crawford, R. J. M., A. J. Williams, J. H. Hofmeyer, N. T. W. Klages, R. M. Randall, J. Cooper, B. M. Dyer, and Y. Chesselet. 1995c. Trends of African penguin *Spheniscus demersus* populations in the 20th century. *South African Journal of Marine Science* 16:101–18.

Crawford, R. J. M., A. J. Williams, R. M. Randall, B. M. Randall, A. Berruti, and G. J. B. Ross. 1990. Recent population trends of jackass penguins *Spheniscus demersus* off southern Africa. *Biological Conservation* 52:229–43.

Cunningham, G. B., V. Strauss, and P. G. Ryan. 2008. African penguins (*Spheniscus demersus*) can detect dimethyl sulphide, a prey-related odour. *Journal of Experimental Biology* 211:3123–27.

Currie, H., C. A. F. Grobler, and J. Kemper, eds. 2009. Namibian Islands' Marine Protected Area. Concept note, background document and management proposal for the declaration of marine protected areas on and around the Namibian offshore islands and adjacent coastal area. Ministry of Fisheries and Marine Resources, Namibia. http://www.nacoma.org.na/FindOutMore/ReportsPublications.htm.

Cury, P. M., I. L. Boyd, S. Bonhommeau, T. Anker-Nilssen, R. J. M. Crawford, R. W. Furness, J. A. Mills, E. J. Murphy, H. Österblom, M. Paleczny, J. F. Piatt, J-P. Roux, L. Shannon, and W. J. Sydeman. 2011. Global seabird response to forage fish depletion—one-third for the birds. *Science* 334:1703–6.

David, J. H. M., P. Cury, R. J. M. Crawford, R. M. Randall, L. G. Underhill, and M. A. Meÿer. 2003. Assessing conservation priorities in the Benguela ecosystem, South Africa: Analysing predation by seals on threatened seabirds. *Biological Conservation* 114:289–92.

de Kock, A. C., and R. M. Randall. 1984. Organochlorine insecticide and polychlorinated biphenyl residues in eggs of coastal birds from the Eastern Cape, South Africa. *Environmental Pollution Series A, Ecological and Biological* 35:193–201.

de Ponte Machado, M. 2007. Is predation on seabirds a new foraging behaviour for great white pelicans? History, foraging strategies and prey defensive strategies. In *Final Report of the BCLME (Benguela Current Large Marine Ecosystem) Project on Top Predators as Biological Indicators of Ecosystem Change in the BCLME*, ed. S. P. Kirkman, 131–42. Cape Town, South Africa: Avian Demography Unit.

du Toit, M., P. A. Bartlett, M. N. Bester, and J-P. Roux. 2004. Seabird predation by individual seals at Ichaboe Island, Namibia. *South African Journal of Wildlife Research* 34:45–54.

Duffy, D. C. 1987. Ecological implications of intercolony size-variation in jackass penguins. *Ostrich* 58:54–57.

Earlé, R. A., F. W. Huchzermeyer, G. F. Bennett, and J.-J. Brossy. 1993. *Babesia peircei* sp. nov. from the jackass penguin. *South African Journal of Zoology* 28:88–90.

Eggleton, P., and W. R. Siegfried. 1979. Displays of the jackass penguin. *Ostrich* 50:139–67.

Ellis, S., J-P. Croxall, and J. Cooper. 1998. Penguin Conservation Assessment and Management Plan. Apple Valley, MN: IUCN/SSC Conservation Breeding Specialist Group.

Erasmus, T., R. M. Randall, and B. M. Randall. 1981. Oil pollution, insulation and body temperatures in the jackass penguin *Spheniscus demersus*. *Comparative Biogeochemistry and Physiology: A Comparative Physiogy* 69:169–71.

Erasmus, T., and D. Smith. 1974. Temperature regulation of young jackass penguins, *Spheniscus demersus*. *Zoologica Africana* 9:195–203.

Erasmus, T., and E. D. Wessels. 1985. Heat production studies on normal and oil-covered jackass penguins (*Spheniscus demersus*) in air and water. *South African Journal of Zoology* 20:209–12.

FAO (Food and Agriculture Organization) Committee on Fisheries. 2003. World Summit on Sustainable Development 2002 and its implications for fisheries. Summary report. Fisheries and Aquaculture Department, FAO, Rome. http://www.fao.org/docrep/meeting/005/y8294E.htm.

Frost, P. G. H., W. R. Siegfried, and A. E. Burger. 1976a. Behavioural adaptations of the jackass penguin, *Spheniscus demersus*, to a hot, arid environment. *Journal of Zoology* 179:165–87.

Frost, P. G. H., W. R. Siegfried, and J. Cooper. 1976b. Conservation of the jackass penguin (*Spheniscus demersus* [L.]). *Biological Conservation* 9:79–99.

Grémillet, D., S. Lewis, L. Drapeau, C. van der Lingen, J. A. Huggett, J. C. Coetzee, H. M. Verheye, F. Daunt, and S. Wanless. 2008. Spatial match-mismatch in the Benguela upwelling zone: Should we expect chlorophyll and sea-surface temperature to predict marine predator distributions? *Journal of Applied Ecology* 45:610–21.

Heath, R. G. M., and R. M. Randall. 1989. Foraging ranges and movements of jackass penguins (*Spheniscus demersus*) established through radio telemetry. *Journal of Zoology* 217:367–79.

Hemming, M. 2001. The *Treasure* oil spill and its influence on moulting African penguins *Spheniscus demersus* at Robben Island. MSc thesis, University of Cape Town.

Hockey, P. A. R., W. R. J. Dean, and P. G. Ryan, eds. 2005. *Roberts Birds of Southern Africa*. 7th ed. Cape Town: John Voelcker Bird Book Fund.

Hockey, P. A. R., and J. Hallinan. 1981. Effect of human disturbance on the breeding behaviour of jackass penguins *Spheniscus demersus*. *South African Journal of Wildlife Research* 11(2):59–62.

IUCN (International Union for Conservation of Nature). 2011. IUCN Red List of Threatened Species. Version 2011.2. www.iucnredlist.org (accessed 20 April 2012).

Jarvis, M. J. F. 1970. A problem in banding penguins. *Ostrich* 41:120–21.

Kemper, J. 2006. Heading towards extinction? Demography of the African penguin in Namibia. PhD diss., University of Cape Town.

———. 2007. Estimating African penguin population size: A comparison of census techniques. In *Final Report of the BCLME (Benguela Current Large Marine Ecosystem) Project on Top Predators as Biological Indicators of Ecosystem Change in the BCLME*, ed. S. P. Kirkman, 77–80. Cape Town, South Africa: Avian Demography Unit.

Kemper, J., J. Braby, B. M. Dyer, J. James, R. Jones, K. Ludynia, R. Mullers, J-P. Roux, L. G. Underhill, and A. C. Wolfaardt. 2007a. Annex 3: Monitoring seabirds in the BCLME: Data collection manual. In *Final Report of the BCLME (Benguela Current Large Marine Ecosystem) Project on Top Predators as Biological Indicators of Ecosystem Change in the BCLME*, ed. S. P. Kirkman. Cape Town, South Africa: Avian Demography Unit.

Kemper, J., and J-P. Roux. 2005. Of squeezers and skippers: Factors determining the age at moult of immature African penguins *Spheniscus demersus* in Namibia. *Ibis* 147:346–52.

Kemper, J., J-P. Roux, P. A. Bartlett, Y. J. Chesselet, J. A. Delport, J. A. C. James, R. Jones, L. G. Underhill, N. N. Uhongora, and S. Wepener. 2007b. The African penguin *Spheniscus demersus*: Population estimates, trends, adult survival and age structure from molt and nest counts. In *Final Report of the BCLME (Benguela Current Large Marine Ecosystem) Project on Top Predators as Biological Indicators of Ecosystem Change in the BCLME*, ed. S. P. Kirkman, 69–76. Cape Town, South Africa: Avian Demography Unit.

Kemper, J., J-P. Roux, P. A. Bartlett, Y. J. Chesselet, J. A. C. James, R. Jones, S. Wepener, and F. J. Molloy. 2001. Recent population trends of African penguins *Spheniscus demersus* in Namibia. *South African Journal of Marine Science* 23:429–34.

Kemper, J., J-P. Roux, and L. G. Underhill. 2008. Effect of age and breeding status on molt phenology of adult African penguins (*Spheniscus demersus*) in Namibia. *The Auk* 125:808–19.

Kemper, J., L. G. Underhill, R. J. M. Crawford, and J-P. Roux. 2007c. Revision of the conservation status of seabirds and seals in the Benguela ecosystem. In *Final Report of the BCLME (Benguela Current Large Marine Ecosystem) Project on Top Predators as Biological Indicators of Ecosystem Change in the BCLME*, ed. S. P. Kirkman, 325–42. Cape Town, South Africa: Avian Demography Unit.

Kemper, J., L. G. Underhill, and J-P. Roux. 2007d. Artificial burrows for African penguins on Halifax Island, Namibia: Do they improve breeding success? In *Final Report of the BCLME (Benguela Current Large Marine Ecosystem) Project on Top Predators as Biological Indicators of Ecosystem Change in the BCLME*, ed. S. P. Kirkman, 101–6. Cape Town, South Africa: Avian Demography Unit.

Kemper, J., L. G. Underhill, J-P. Roux, P. A. Bartlett, Y. J. Chesselet, J. A. C. James, R. Jones, N.-N. Uhongora, and S. Wepener. 2007e. Breeding patterns and factors influencing breeding success of African penguins *Spheniscus demersus* in Namibia. In *Final Report of the BCLME (Benguela Current Large Marine Ecosystem) Project on Top Predators as Biological Indicators of Ecosystem Change in the BCLME*, ed. S. P. Kirkman, 89–99. Cape Town, South Africa: Avian Demography Unit.

Kerley, G. I. H., C. G. Crellin, and T. Erasmus. 1987. Gravimetric determination of water-repellancy in rehabilitated oiled seabirds. *Marine Pollution Bulletin* 18:609–11.

Kerley, G. I. H., and T. Erasmus. 1987. Cleaning and rehabilitation of oiled jackass penguins. *South African Journal of Wildlife Research* 17:64–70.

Kerley, G. I. H., T. Erasmus, and R. P. Mason. 1985. Effect of moult on crude oil load in a jackass penguin *Spheniscus demersus*. *Marine Pollution Bulletin* 16:474–76.

La Cock, G. D., and J. Cooper. 1988. The breeding frequency of jackass penguins on the west coast of South Africa. *Journal of Field Ornithology* 59:155–56.

La Cock, G. D., and C. Hänel. 1987. Survival of African penguins *Spheniscus demersus* at Dyer Island, southern Cape, South Africa. *Journal of Field Ornithology* 58:284–87.

Loutit, R., and D. Boyer. 1985. Mainland breeding by jackass penguins *Spheniscus demersus* in South West Africa/Namibia. *Cormorant* 13:27–30.

Ludynia, K. 2007. Identification and characterisation of foraging areas of seabirds in upwelling systems: Biological and hydrographic implications for foraging at sea. PhD diss., University of Kiel.

Ludynia, K., J-P. Roux, R. Jones, J. Kemper, and L. G. Underhill. 2010. Surviving off junk: Low-energy prey dominates the diet of African penguins *Spheniscus demersus* at Mercury Island, Namibia, between 1996 and 2009. *African Journal of Marine Science* 32: 563–72.

Malbrant, R., and A. Maclatchy. 1958. A propos de l'occurrence de deux oiseaux d'Afrique austral au Gabon: Le manchot du Cap, *Spheniscus demersus* Linné, et la grue couronnée, *Balearica regulorum* Bennett. *L'Oiseau et la Revue Française d'Ornithologie* 28:84–86.

Marks, M. A., R. K. Brooke, and A. M. Gildenhuys. 1997. Cape fur seal *Arctocephalus pusillus* predation on Cape cormorants *Phalacrocorax capensis* and other birds at Dyer Island, South Africa. *Marine Ornithology* 25:9–12.

Matthews, J-P. 1961. The pilchard of South West Africa (*Sardinops ocellata*) and the marsbanker (*Trachurus trachurus*): Bird predators, 1957–1958. *Investigational Report South West Africa Marine Research Laboratory* 3:1–35.

Morant, P. D., J. Cooper, and R. M. Randall. 1981. The rehabilitation of oiled jackass penguins *Spheniscus demersus*, 1970–1980. In *Proceedings of the Symposium on Birds of the Sea & Shore*, ed. J. Cooper, 267–301. Cape Town, South Africa: African Seabird Group.

Nel, D. C., R. J. M. Crawford, and N. Parsons. 2003. The conservation status and impact of oiling on the African penguin. In *The Rehabilitation of Oiled African Penguins: A Conservation Success Story*, ed. D. C. Nel and P. A. Whittington, 1–7. Cape Town, South Africa: BirdLife South Africa.

Petersen, S. L., P. G. Ryan, and D. Grémillet. 2006. Is food availability limiting African penguins *Spheniscus demersus* at Boulders? A comparison of foraging effort at mainland and island colonies. *Ibis* 148:14–26.

Pichegru, L., D. Grémillet, R. J. M. Crawford, and P. G. Ryan. 2010. Marine no-take zone rapidly benefits endangered penguin. *Biology Letters* 6:498–501.

Pikitch, E., P. D. Boersma, I. L. Boyd, D. O. Conover, P. Cury, T. Essington, S. S. Heppell, E. D. Houde, M. Mangel, D. Pauly, É. Plagányi, K. Sainsbury, and R. S. Steneck. 2012. *Little Fish, Big Impact: Managing a Crucial Link in Ocean Food Webs.* Washington, DC: Lenfest Ocean Program. 108 pp.

Rand, R. W. 1960. The biology of guano-producing seabirds. 2: The distribution, abundance and feeding habits of the Cape penguin, *Spheniscus demersus*, off the south-western coast of the Cape Province. *Investigational Report Sea Fisheries Research Institute South Africa* 41:1–28.

———. 1963a. The biology of guano-producing seabirds. 4: Composition of colonies on the Cape islands. *Investigational Report Sea Fisheries Research Institute South Africa* 43:1–32.

———. 1963b. The biology of guano-producing seabirds. 5: Composition of colonies on the South West African islands. *Investigational Report Sea Fisheries Research Institute South Africa* 46:1–26.

Randall, R. M. 1983. Biology of the jackass penguin *Spheniscus demersus* (L.) at St. Croix Island, South Africa. PhD diss., University of Port Elizabeth.

———. 1989. Jackass penguins. In *Oceans of Life Off Southern Africa*, ed. A. I. L. Payne and R. J. M. Crawford, 244–56. Cape Town, South Africa: Vlaeberg Publishers.

Randall, R. M., and R. A. Bray. 1983. Mortalities of jackass penguin *Spheniscus demersus* chicks caused by trematode worms *Cardiocephaloides physalis*. *South African Journal of Zoology* 18:45–46.

Randall, R. M., and B. M. Randall. 1981. The annual cycle of the jackass penguin *Spheniscus demersus* at St. Croix Island, South Africa. In *Proceedings of the Symposium of Birds of the Sea & Shore*, ed. J. Cooper, 427–50. Cape Town, South Africa: African Seabird Group.

———. 1986. The diet of jackass penguins *Spheniscus demersus* in Algoa Bay, South Africa, and its bearing on population declines elsewhere. *Biological Conservation* 37:119–34.

Randall, R. M., B. M. Randall, and J. Bevan. 1980. Oil pollution and penguins: Is cleaning justified? *Marine Pollution Bulletin* 11:234–37.

Randall, B. M., R. M. Randall, and L. J. V. Compagno. 1988. Injuries to jackass penguins (*Spheniscus demersus*): Evidence for shark involvement. *Journal of Zoology* 214:589–59.

Randall, R. M., B. M. Randall, J. Cooper, and P. G. H. Frost. 1986a. A new census method for penguins tested on jackass penguins *Spheniscus demersus*. *Ostrich* 57:211–15.

Randall, R. M., B. M. Randall, and T. Erasmus. 1986b. Rain-related breeding failures in jackass penguins. *Gerfaut* 76:281–88.

Roux, J-P. 2003. Risks. In *Namibia's Marine Environment*, ed. F. Molloy and T. Reinikainen, 137–52. Windhoek, Namibia: Directorate of Environmental Affairs of the Ministry of Environment and Tourism.

Roux, J-P., J. Kemper, P. A. Bartlett, B. M. Dyer, and B. L. Dundee. 2003. African penguins *Spheniscus demersus* recolonise a formerly abandoned nesting locality in Namibia. *Marine Ornithology* 31:203–5.

Ryan, P. G., S. L. Petersen, A. Simeone, and D. Grémillet. 2007. Diving behaviour of African penguins: Do they differ from other *Spheniscus* penguins? *African Journal of Marine Science* 29:153–60.

Ryan, P. G., R. P. Wilson, and J. Cooper. 1987. Intraspecific mimicry and status signals in juvenile African penguins. *Behavioural Ecology & Sociobiology* 20:69–76.

Seddon, P. J., and Y. M. van Heezik. 1991. Hatching asynchrony and brood reduction in the jackass penguin: An experimental study. *Animal Behaviour* 42:347–56.

———. 1993. Behaviour of the jackass penguin chick. *Ostrich* 64(1):8–12.

Shannon, L. J., and R. J. M. Crawford. 1999. Management of the African penguin *Spheniscus demersus*: Insights from modelling. *Marine Ornithology* 27:119–28.

Shaughnessy, P. D. 1980. Influence of Cape fur seals on jackass penguin numbers at Sinclair Island. *South African Journal of Wildlife Research* 10(1):18–21.

———. 1984. Historical population levels of seals and seabirds on islands off southern Africa, with special reference to Seal Island, False Bay. *Investigational Report Sea Fisheries Research Institute South Africa* 127:1–61.

Shaw, K. A., L. J. Waller, R. J. M. Crawford, and W. H. Oosthuizen, eds. 2011. *Proceedings of the African penguin BMP-S Stakeholder Workshop 26–28 October 2010, Die Herberg, Arniston, South Africa*. Stellenbosch, South Africa: CapeNature.

Shelton, P. A., R. J. M. Crawford, J. Cooper, and R. K. Brooke. 1984. Distribution, population size and conservation of the jackass penguin *Spheniscus demersus*. *South African Journal of Marine Science* 2:217–57.

Simmons, R. E., and J. Kemper. 2003. Cave breeding by African penguins near the northern extreme of their range: Sylvia Hill, Namibia. *Ostrich* 74:217–21.

Underhill, L. G., P. A. Bartlett, L. Baumann, R. J. M. Crawford, B. M. Dyer, A. Gildenhuys, D. C. Nel, T. B. Oatley, M. Thornton, L. Upfold, A. J. Williams, P. A. Whittington, and A. C. Wolfaardt. 1999. Mortality and survival of African penguins *Spheniscus demersus* involved in the *Apollo Sea* oil spill: An evaluation of rehabilitation efforts. *Ibis* 141:29–37.

Underhill, L. G., and R. J. M. Crawford. 1999. Season of moult of African penguins at Robben Island, South Africa, and its variation, 1988–1998. *South African Journal of Marine Science* 21:437–41.

Underhill, L. G., R. J. M. Crawford, A. C. Wolfaardt, P. A. Whittington, B. M. Dyer, T. M. Leshoro, M. Ruthenberg, L. Upfold, and J. Visagie. 2006. Regionally coherent trends in colonies of African penguins *Spheniscus demersus* in the Western Cape, South Africa, 1987–2005. *African Journal of Marine Science* 28:697–704.

Underhill, L. G., P. A. Whittington, R. J. M. Crawford, and A. C. Wolfaardt. 2000. Five years of monitoring African penguins (*Spheniscus demersus*) after the *Apollo Sea* oil spill: A success story identified by flipper bands. *Vogelwarte* 40:315–18.

Westphal, A., and M. K. Rowan. 1971. Some observations on the effects of oil pollution on the jackass penguin. *Ostrich Supplement* 8:521–26.

Whittington, P. A. 1999. The contribution made by cleaning oiled African penguins *Spheniscus demersus* to population dynamics and conservation of the species. *Marine Ornithology* 27:177–80.

———. 2002. Survival and movements of African penguins, especially after oiling. PhD diss., University of Cape Town.

———. 2003. Post-release survival of rehabilitated African penguins. In *The Rehabilitation of Oiled African Penguins: A Conservation Success Story*, ed. D. C. Nel and P. A. Whittington, 8–17. Cape Town: BirdLife South Africa.

Whittington, P. A., B. M. Dyer, and N. T. W. Klages. 2000. Maximum longevities of African penguins *Spheniscus demersus* based on banding records. *Marine Ornithology* 28:81–82.

Whittington, P. A., J. H. Hofmeyr, and J. Cooper. 1996. Establishment, growth and conservation of a mainland colony of jackass penguins *Spheniscus demersus* at Stony Point, Betty's Bay, South Africa. *Ostrich* 67:144–50.

Whittington, P. A., N. T. W. Klages, R. J. M. Crawford, A. C. Wolfaardt, and J. Kemper. 2005. Age at first breeding of the African penguin. *Ostrich* 76:14–20.

Williams, A. J. 1981. Why do penguins have long laying intervals? *Ibis* 123:202–4.

Williams, A. J., and J. Cooper. 1984. Aspects of the breeding biology of the jackass penguin *Spheniscus demersus*. In *Proceedings of the 5th Pan-African Ornithological Congress*, ed. J. A. Ledger, 841–53. Johannesburg, South Africa: Southern African Ornithological Society.

Williams, A. J., and B. M. Dyer. 1990. The birds of Hollamsbird Island, least known of the southern African guano islands. *Marine Ornithology* 18:13–18.

Williams, A. J., B. M. Dyer, R. M. Randall, and J. Komen. 1990. Killer whales *Orcinus orca* and seabirds: "Play," predation, and association. *Marine Ornithology* 18:37–41.

Wilson, R. P. 1985a. Diurnal foraging patterns of the jackass penguin. *Ostrich* 56:212–14.

———. 1985b. The jackass penguin (*Spheniscus demersus*) as a pelagic predator. *Marine Ecology Progress Series* 25:219–27.

Wilson, R. P., P. G. Ryan, A. James, and M. P. Wilson. 1987. Conspicuous coloration may enhance prey capture in some piscivores. *Animal Behaviour* 35:1558–60.

Wilson, R. P., and M. P. T. Wilson. 1989. Substitute burrows for penguins on guano-free islands. *Gerfaut* 79:125–31.

———. 1990. Foraging ecology of breeding *Spheniscus* penguins. In *Penguin Biology*, ed. L. S. Davis and J. T. Darby, 181–206. San Diego, CA: Academic Press.

———. 1995. The foraging behaviour of the African penguin *Spheniscus demersus*. In *The Penguins: Ecology and Management*, ed. P. Dann, I. Norman, and P. Reilly, 244–65. Chipping Norton, NSW, Australia: Surrey Beatty & Sons.

Wilson, R. P., M. P. T. Wilson, and D. C. Duffy. 1988. Contemporary and historical patterns of African penguin *Spheniscus demersus* distribution at sea. *Estuarine Coastal Shelf Science* 26:447–58.

Wolfaardt, A. C., and D. C. Nel. 2003. Breeding productivity and annual cycle of rehabilitated African penguins following oiling. In *The Rehabilitation of Oiled African Penguins: A Conservation Success Story*, ed. D. C. Nel and P. A. Whittington, 18–24. Cape Town: BirdLife South Africa.

Wolfaardt, A. C., L. G. Underhill, R. Altwegg, and J. Visagie. 2008a. Restoration of oiled African penguins *Spheniscus demersus* a decade after the *Apollo Sea* oil spill. *African Journal of Marine Science* 30:421–36.

Wolfaardt, A. C., L. G. Underhill, and R. J. M. Crawford. 2009a. Comparison of moult phenology of African penguins *Spheniscus demersus* at Robben and Dassen Islands. *African Journal of Marine Science* 31:19–29.

Wolfaardt, A. C., L. G. Underhill, R. J. M. Crawford, and N. T. W. Klages. 2001. Results of the 2001 census of African penguins *Spheniscus demersus*: First measures of the impact of the *Treasure* oil spill on the breeding population. *Transactions of the Royal Society of South Africa* 56:45–49.

Wolfaardt, A. C., L. G. Underhill, D. C. Nel, A. J. Williams, and J. Visagie. 2008b. Breeding success of African penguins *Spheniscus demersus* at Dassen Island, especially after oiling following the *Apollo Sea* spill. *African Journal of Marine Science* 30:565–80.

Wolfaardt, A. C., A. J. Williams, L. G. Underhill, R. J. M. Crawford, and P. A. Whittington. 2009b. Review of the rescue, rehabilitation and restoration of oiled seabirds in South Africa, especially African penguins *Spheniscus demersus* and Cape gannets *Morus capensis*, 1983–2005. *African Journal of Marine Science* 31:31–54.

Wright, K. L. B., L. Pichegru, and P. C. Ryan. 2011. Penguins are attracted to dimethyl sulphide at sea. *Journal of Experimental Biology* 214:2509–11.

Yabsley, M. J., N. J. Parsons, E. C. Horne, B. C. Shock, and M. Purdee. 2012. Novel relapsing fever *Borrelia* detected in African penguins (*Spheniscus demersus*) admitted to two rehabilitation centers in South Africa. *Parasitology Research* 110:1125–30.

14

Magellanic Penguin

(Spheniscus magellanicus)

P. Dee Boersma, Esteban Frere, Olivia Kane, Luciana M. Pozzi, Klemens Pütz, Andrea Raya Rey, Ginger A. Rebstock, Alejandro Simeone, Jeffrey Smith, Amy Van Buren, Pablo Yorio, and Pablo Garcia Borboroglu

1. SPECIES (COMMON AND SCIENTIFIC NAMES)

Magellanic penguin, *Spheniscus magellanicus* (J. R. Forster, 1781)

The Magellanic penguin is also known as the Patagonian penguin, *pingüino de Magallanes* (Spanish), and *pinguims-de-Magalhães* (Portuguese).

2. DESCRIPTION OF THE SPECIES

The Magellanic is a medium-size penguin. Body length from the bill tip to the middle toenail of the outstretched foot is 48 centimeters (± 0.5 cm, n = 59); body mass varies depending on season but ranges between 1.8 kilograms when near starvation to 6 kilograms at start of molt. The heaviest weighed 8.6 kilograms (n = 28,493) (Boersma, unpubl. data). Males are usually larger, with deeper bills and smaller cloacae than females (Scolaro et al. 1983; Boersma and Davies 1987; Gandini et al. 1992). See table 14.1 for details.

ADULT. Magellanic penguins are similar in appearance to other *Spheniscus* penguins, with a white crescent on each side of the head extending from the crown above each eye and joining at the throat to make a white band (fig. 1) (Williams 1995). The black throat band is bordered by white feathers, and a black inverted U-shape band on the chest extends down each flank. The bands on the flanks may confuse fish, scatter schools, and enhance prey capture (Wilson et al. 1987). Magellanics usually have two black throat bands, but individual variation appears higher than for other species in the genus (Boersma, pers. obs.). Shortly after coming ashore at the breeding colony, individuals shed the white feathers around the bill and eye, exposing bare skin that has black or no pigment. Unpigmented areas are bright pink on hot days and white on cold days (fig. 1). The dorsal side of the flipper is black, and the ventral side is white with black feathers. After the molt, the trailing edge of the flipper is edged with white feathers that may disappear with wear. The pattern of white and black feathers on the ventral side of the flipper, the bare facial skin, and the feet patterns are individually distinct. The black bill has a lighter band toward the tip and a pronounced hook at the distal end of the upper mandible that hooks over a groove in the lower mandible (fig. 2). The eyes are usually brown with a red ring around the pupil in adults, pink and gray in juveniles, and gray in chicks (Boersma, unpubl. data). Tarsi and feet are black with mottled white and pink blotches, depending on temperature and the extent of

FIG. 1 (*FACING PAGE*) A Magellanic male penguin in the setting sun shows the characteristic feather loss around the bill and eyes that happens during the spring and summer breeding season. (P. Garcia Borboroglu)

FIG. 2 The Magellanic penguin's hook upper mandible fits into the groove of the lower mandible. Note the denticles on the tongue and roof of the mouth that help hold the prey. The penguin is panting, and the bright pink coloration indicates that the bird is dissipating heat by shunting blood to the bare skin. (P. D. Boersma)

FIG. 3 The Magellanic penguin's white tail spot varies in size. The tail spot of this adult is quite small. About half of Magellanic penguins, but no Galápagos penguins, have a tail spot. (P. D. Boersma)

black-pigmented skin. Feet have black soles and claws with a black line up the back of the tarsus as on Adélie and many other species of penguins. At Punta Tombo, Argentina, 54% of adults have white tail spots (fig. 3) (n = 1,742 [Boersma, unpubl. data]). Isabellinism is rare, with two adults and two chicks seen at Punta Tombo between 1983 and 2010 out of more than 200,000 breeding pairs. Likewise, melanism is rare, and only 4 chicks with dark gray down chests occurred out of more than 50,000 chicks handled over 28 years (Boersma, pers. obs.).

JUVENILE. Immature Magellanics have grayer plumage than adults and lack the bands on the head, neck, and breast. Their cheeks vary from white to very dark gray (fig. 4). When cheeks are dark in juveniles, the lighter feathers around the cheeks make a less pronounced facial crescent than an adult's, but juveniles always lack a band of black feathers on the sides. During the winter, some juveniles grow patches of long black feathers on their backs or may develop black or white feathers on their heads that give them partial adult head markings (Boersma, pers. obs.).

CHICK. Chicks hatch with dark gray down on their backs and lighter gray on their bellies (fig. 5). The secondary

down sprouts all over the body by about two weeks of age. The hatching down rubs off over time as the secondary down grows. By about 30 days, the secondary down predominates and is gray on the back and white on the front (fig. 6). The juvenile plumage grows in under the secondary down starting at about 40 days and predominates by 70 days or before fledging (Boersma, unpub. data). Chicks of all four *Spheniscus* species are so similar

FIG. 4 Juvenile Magellanic penguins lack the two black bands of adults. This juvenile grew extra feathers on the sides of its head during the winter and is on the beach with other penguins, many of them molting, in January. (P. D. Boersma)

in appearance that they are more reliably distinguished by location.

3. TAXONOMIC STATUS

There are no subspecies recognized.

The status of African (*S. demersus*), Humboldt (*S. humboldti*), and Magellanic penguins as separate species is questioned as the three interbreed readily in captivity (references in Simeone et al. 2009) and the *Spheniscus* species share alleles suggesting recent divergence, rapid speciation, or hybridization (Kikkawa et al. 2005). In the wild, Humboldt and Magellanic penguins interbreed on occasion where their ranges overlap in Chile (Simeone et al. 2009). DNA analysis suggests that Magellanics are most closely related to African penguins, and Humboldt and Galápagos (*S. mendiculus*) penguins are more closely related to each other (Davis and Renner 2003). Magellanic penguins have a relatively high heterozygosity of 46% (Akst et al. 2002), and the population has limited genetic structuring, meaning that Magellanics from different colonies are not genetically well differentiated, consistent with a recent expansion of their population (Bouzat et al. 2009).

4. RANGE AND DISTRIBUTION

BREEDING. The breeding range is from about 55° south to approximately 40° south on islands and coasts of South America including the Falkland/Malvinas Islands (fig. 7). On the Atlantic side of South America, this species breeds from Cape Horn to Complejo Islote Lobos (41°33′

S). On the Pacific side, Isla Algarrobo, Chile (33° S), is the northernmost breeding site reported (Schlatter 1984), although Osorno Province, Chile (40°32′ S), is about the northern end of current significant breeding (Cursach et al. 2009). Johow (in Murphy 1936) mentions nesting Magellanic penguins at Santa Clara Island in the Juan Fernández Islands, Chile (33° S), in 1896, but they do not currently breed there (Hahn et al. 2009).

NONBREEDING. After their post-breeding molt, Magellanic penguins leave their breeding colonies to go to sea, rarely coming ashore until they return to land to breed. In the Atlantic, they move northward for the winter (Stokes et al. 1998; Pütz et al. 2000, 2007) and occasionally reach Ceara state, Brazil (2°52′ S), close to the equator (Garcia Borboroglu et al. 2010). They swim faster (2.3–3.7 km per hour) the first few days of the northbound migration than after the first week (0.9–1.9 km per hour) (Stokes et al. 1998; Pütz et al. 2000). Most winter off northern Argentina to southern Brazil, but some from northern and southern colonies remain off southern Argentina (Jehl 1980; Boersma et al. 1990; Frere et al. 1996a; Stokes et al. 1998; Pütz et al. 2000, 2007; Garcia Borboroglu et al. 2010). Out of 22 individuals from a colony in Tierra del Fuego, 21 migrated northeast in the Atlantic Ocean and one went northwest into the Pacific Ocean (Pütz et al. 2007). Preferred migration and foraging areas are in inshore waters (Stokes et al. 1998; Pütz et al. 2000, 2007), but they risk encountering marine debris (Brandão et al. 2011). Some penguins from the Falkland/

FIG. 5 Adult Magellanic penguin in its burrow guarding a well-fed two-week-old chick. (P. D. Boersma)

FIG. 6 Adult Magellanic penguin with two chicks begging for food. The chick on the right has the normal grayish down, and its sibling, yellowish in color, is called an Isabelline or leucistic penguin. The light plumage is caused by a recessive allele. (P. D. Boersma)

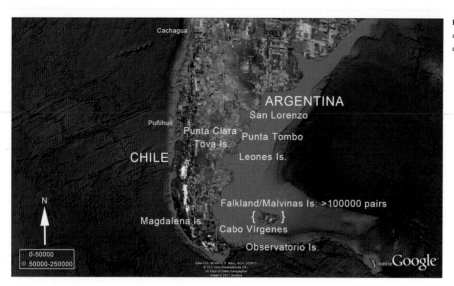

FIG. 7 Distribution and abundance of the Magellanic penguin, with counts based on pairs.

Malvinas Islands use the Patagonian Shelf slope during their northward migration (Pütz et al. 2000). Starting in August, breeding birds head south to their colonies (Boersma 2012).

No difference in foraging patterns was seen between males and females tracked from Tierra del Fuego (Pütz et al. 2007). In Chile, they disperse from their breeding colonies starting in March (austral autumn) to more productive waters north of 40° south and are present and abundant in waters as far as Coquimbo (30° S) (Skewgar et al., unpubl. data; A. Simeone, pers. obs.). Zavalaga and Paredes (2009) report four sightings of Magellanic penguins at Punta San Juan (15°45′ S, 75°42′ W), Peru, coinciding with the autumn-winter dispersal. In the Pacific Ocean, Magellanics tend to forage in coastal, productive waters, but the migration is not as directional as it is in the Atlantic Ocean, with birds sometimes reversing their swim direction (Skewgar et al., unpubl. data; Pütz et al., unpubl. data).

VAGRANTS. Vagrants are recorded from Africa, Australia, and New Zealand. The southernmost record is of a molting Magellanic at Avian Island (67°46′ S, 68°43′ W), Marguerite Bay, Antarctic Peninsula (Barbosa et al. 2007). The species is occasionally seen at South Georgia Island in the South Atlantic (Prince and Payne 1979). Magellanic penguins found in New Zealand were tame and may have been transported on ships (Robertson et al. 1972; Darby 1991; McGrouther, pers. comm.). A dead bird was found on Phillip Island, Australia, in 1979 (McEvey 1980), and a juvenile was photographed on Marion Island (sub-Antarctic Indian Ocean) in February 2006 (Norbert Klages, pers. comm.).

5. SUMMARY OF POPULATION TRENDS

The population of Magellanics is unknown but is likely between 1.2 and 1.6 million breeding pairs with a minimum of 138 colonies. The overall population trend is uncertain. Population sizes and trends for the Falkland/Malvinas Islands and the Pacific are unknown. In the Atlantic, recent data show mixed trends. Two of the largest known colonies—Punta Tombo (fig. 8) and Isla Leones, in central Patagonia, Argentina—declined substantially (Boersma 1987, 2008; Boersma, unpubl. data; Pozzi and Garcia Borboroglu, unpubl. data). Colonies in other sectors throughout Patagonia are increasing, while the Atlantic breeding range has continued to expand northward (Schiavini et al. 2005; Boersma 2008; Pozzi and Garcia Borboroglu, unpubl. data).

6. IUCN STATUS

The International Union for Conservation of Nature lists the Magellanic penguin as Near Threatened on its Red List of Endangered Species (BirdLife International 2009; IUCN 2009).

7. NATURAL HISTORY

Table 14.2 shows the life-history parameters.

BREEDING BIOLOGY. Most Magellanic pairs breed under bushes or dig burrows, which provide protection from the sun as well as inclement weather and aerial preda-

FIG. 8 Magellanic penguins nest in both bushes and burrows. This high-density burrow area at Punta Tombo has about 12,000 nests. (P. D. Boersma)

tors, although in desert environments at higher latitudes, penguins can nest successfully in the open (Scolaro 1978, 1984; Stokes and Boersma 1991, 1998, 2000; Frere et al. 1992; Gandini et al. 1997, 1999a; Garcia Borboroglu et al. 2002). Nests with more cover have higher breeding success (Frere et al. 1992; Gandini et al. 1997, 1999a; Stokes and Boersma 1998; Garcia Borboroglu et al. 2002). Magellanic penguins dig with their feet and often plant their bills in the soil while digging, especially when digging on their backs (Boersma, pers. obs.).

The nest cup is usually a hollow where adults have deposited a few stones, shrub branches, feathers, bones, grass, or even seaweed. Adults often bring nesting material from nearby to the nest when exchanging nesting duties (Boersma and Clark, pers. obs.). Pairs as well as single males and females fight over nest sites (Renison et al. 2002, 2003, 2006). Fights between females are longer but less intense than those between males (Renison et al. 2003). Individuals have distinct calls, allowing mates to recognize each other vocally and chicks to recognize their parents (Clark et al. 2006).

Extra-pair copulations are rare; none was observed in 509 copulations (Hood 1996). The cloacae of both female and male distend during courtship, increasing the chance for successful transfer of sperm to the female's cloaca (Boersma and Davies 1987). After copulation, a female will pulsate her cloaca and envelop the sperm sometimes for longer than a minute while remaining prone on the ground (Boersma, pers. obs.).

The sex hormones progesterone and testosterone are highest in both males and females when penguins arrive at the breeding colony or just before egg laying begins. Estradiol-17β is also highest on arrival in females. These hormones drop to low levels during incubation and chick rearing. After a nest failure, levels of testosterone and luteinizing hormone increase again in both males and females, and estradiol increases in females, even though these penguins do not lay replacement clutches (Fowler et al. 1994).

Males have lower cell-mediated immunocompetence than females around the time of chick hatching. Males that bred later were in better condition and had higher immunocompetence than males that bred earlier (Moreno et al. 2001). They also had lower white blood cell counts than females. Female mass, body condition, and hematocrit and egg volume decreased with laying date. Females in poor health, indicated by hematological indices, laid smaller eggs and fledged fewer chicks than females in better health (Moreno et al. 2002). See table 14.2 for life-history details.

Typically two eggs are laid (table 14.3), with the first-laid egg being longer and sometimes narrower than the second-laid egg but similar in volume and weight in 95% of clutches (table 14.4) (Boersma et al. 1990; Yorio et al. 2001b; Rafferty et al. 2005; Boersma and Rebstock 2009a, 2010a). Because first and second eggs differ in shape, separate equations should be used for calculating volume from length and breadth (Boersma and Rebstock 2010a). Females often ingest bivalve shells before egg laying, probably to supply calcium for making thick eggshells, which reduces breakage (Boersma et al. 2004). Runt eggs are rare; 24 of 27,973 eggs from 1983 to 2009 had a length of less than 5 centimeters (Boersma, unpubl. data). The laying interval is four days (range 0–6, n = 1148 clutches [Rebstock and Boersma 2011]). The number of egg pores is similar for first (48.2 ± 10.9 pores per cm²) and second (50.3 ± 10.2 pores per cm²) eggs (paired t = 1.2, p = 0.24, n = 33 clutches) and is not related to the difference in incubation period between first and second eggs (Boersma and Rebstock 2009b). The incubation temperature is around 34° C, body temperature 38° C, and brood patch temperature 37° C (Rebstock and Boersma 2011). Adults will sometimes incubate egglike objects, such as rocks (Wagner et al., in revision). Females normally take the first incubation bout unless they are in poor condition (fig. 9). A penguin in poor condition is likely to desert the nest, and the eggs are then lost to predators (Yorio and Boersma 1994a, 1994b). The length of the incubation period depends on the adult's attentiveness to the egg as well as on nest type and weather (Frere et al. 1992; Rebstock and Boersma 2011). The hatching interval is usually two days because the first egg is not well incubated until the second egg is laid, delaying development (Rebstock and Boersma 2011). Replacement eggs are rare. In an experiment in which first eggs were removed after they were laid at Punta Tombo, only 1 of 10 females laid two more eggs, and it was the heaviest female, weighing 5.1 kilograms after laying her first egg (Boersma, unpubl. data).

Chicks hatch fully covered in down, but with eyes closed and unable to thermoregulate. First and sec-ond chicks are similar in size at hatching (table 14.5) (Boersma and Stokes 1995). However, the first chick is usually fed before the second chick hatches, resulting in a size asymmetry that lasts 20–60 days (table 14.6) (Boersma and Stokes 1995; Yorio et al. 2001b). Stress response develops as chicks grow, similar to development in altricial birds, but well-fed chicks had lower baseline levels of corticosterone than their underfed siblings (Walker et al. 2005a). Chicks given antibiotics grew faster than control chicks, but only while the antibiotics were being administered (Potti et al. 2002).

Hatching success is variable among colonies and years (table 14.7). Magellanics can raise two chicks depending on food availability (table 14.8; fig. 10). Brood reduction takes place after hatching, and the larger of the two chicks, usually the first hatchling, receives more food (Boersma 1992; Blanco et al. 1996). Generally, parents exchange nest duties more frequently around the time of hatching. Lone chicks grow faster than chicks with siblings, first chicks are more likely to fledge than second chicks, and the size asymmetry of chicks can be reversed by parental feeding preferences (Boersma 1992).

The sex ratio is skewed at Punta Tombo, with approximately 1.5 adult males attending the colony for every adult female (Boersma, unpubl. data). The sex ratio at hatching is one to one, and the sex ratio of first and second eggs is similar (Akst and Boersma, unpubl. data).

Magellanic penguins are long-lived seabirds and breed for many years (table 14.9). Although the maximum age is unknown, 103 of 5,592 penguins banded as

FIG. 9 Adult Magellanic penguin in a thorny dead bush nest with two recently laid eggs. (P. D. Boersma)

adults (assumed to be at least 4 years of age) at Punta Tombo were seen alive when at least 26 years old as of the 2009–10 breeding season (Boersma, unpubl. data). A 15-year marking study showed that males and females with small web tags had similar survival and that the impact of wearing two stainless steel bands compared to small web tags was small (Boersma and Rebstock 2010b). Mate fidelity (table 14.10) depends on reproductive success, with divorces more likely to occur following a failed reproductive attempt (Fowler 1993). One pair at Punta Tombo stayed together for 17 consecutive years until the female failed to return (Boersma, unpubl. data). Both females and males have high nest fidelity between years (79% and 70%, respectively), and the female is slightly more likely to move following a nest failure (Boersma, unpubl. data). Fidelity to the natal colony is also high (Boersma 2008).

PREY AND FORAGING BEHAVIOR. Magellanic penguins are opportunistic foragers and flexible in their diet, which is composed of fish, cephalopods, and crustaceans, depending on the season and species availability in the region (Frere et al. 1996b; Scolaro et al. 1999; Thompson 1993; Clausen and Pütz 2002; Forero et al. 2002; Wilson et al. 2005). The penguins rely mainly on small, schooling, oily fish during the breeding season but will eat larger fish as well as squid and crustaceans. They can consume 1.86–4.87 grams of fish per minute at sea (Wilson et al. 2007). In the Atlantic Ocean north of about 46° south, Argentine anchovy (*Engraulis anchoita*) is

the main prey. South of that, Chilean sprat (*Sprattus fuegensis*) replace the anchovy in the ecosystem and in the penguins' diet. In the center of their range in the Atlantic Ocean (northern Santa Cruz Province), where both anchoita and sprat are in low densities, penguins feed on a variety of prey, such as silversides (*Austroatherina* spp.) and squid (*Illex* spp. and *Loligo* spp.). Crustaceans, especially lobster krill (*Munida gregaria*) and squid, are important in the Falkland/Malvinas Islands and to a lesser extent during the incubation period in the Beagle Channel (Thompson 1993; Pütz et al. 2001; Otley et al. 2004; Scioscia 2011). In Chile, prey includes Peruvian anchoveta (*E. ringens*), South American pilchard (*Sardinops sagax*), Chilean sprat, and squid (Wilson et al. 1995; Radl and Culik 1999). At Isla Magdalena, in the Strait of Magellan, prey includes the sprat (*Ramnogaster arcuata*) and lobster krill, while at Isla Noir, Chilean sprat is the major prey (Venegas 1999). Less is known about the winter diet, and much of that information comes from stranded penguins that were likely starving (fig. 11). Squid, various fish, isopods, and salps were in penguin stomachs in Brazil in winter (Fonseca et al. 2001).

Foraging-trip distance and duration differ among colonies (table 14.11), as does the time of day that penguins leave or return to the colony (Radl and Culik 1999; Pütz et al. 2002; Otley 2005; Boersma and Rebstock 2009d). Other foraging behaviors (e.g., dive depth and duration, dive rate) also differ among colonies (Radl and Culik 1999). At Punta Tombo, most foraging dives are less than

FIG. 10 Feeding time is often a chance for a neighboring chick or a kelp gull to try to get some food. (P. D. Boersma)

FIG. 11 The most common cause of death for Magellanic penguins at Punta Tombo is starvation. The collapsed chest and algae on the breast of this adult penguin shows that it has been at sea but did not find food. (P. D. Boersma)

three minutes and to depths of about 30 meters, where anchovy are found during the day, but dives can be as deep as 90 meters (Walker and Boersma 2003). At a colony on the Peninsula Valdés, Argentina, the modal depth was 55–60 meters and modal dive durations were 56–64 seconds and 120–28 seconds (Peters et al. 1998). At Isla Magdalena and Seno Otway, Chile, the penguins dive to 91.5 meters, but the means of the dives are 16.5 meters and 14.9 meters, respectively (Radl and Culik 1999). Magellanic penguins dive more often and deeper during the day than at night, during both the breeding season and migration (Stokes et al. 1998; Radl and Culik 1999; Walker and Boersma 2003; Wilson et al. 2005). Penguins captured prey from below 89% of the time, using buoyancy to aid in their ascent, at a mean speed of 1.94 meters per second (Wilson et al. 2010). Per undulation (short ascent and descent from the bottom of the dive), 1 prey item was captured 90% of the time, 2 prey items were captured 8% of the time (Simeone and Wilson 2003), and up to 20 sprat were captured in a single dive (Wilson et al. 2007). Penguins took more pre-dive breaths before long, deep dives and took more pre-dive breaths for a given depth if they had captured prey on the previous dive (Wilson 2003). When returning to the surface from a dive, they reach speeds of 3.5 meters per second (Wilson and Zimmer 2004; Wilson et al. 2010).

Primary and secondary marine productivity is higher at southern than at northern colonies in the Atlantic Ocean (Sánchez and de Ciechomsky 1995; Boersma et al. 2009), and breeding Magellanics generally forage closer to their nests at southern colonies (Wilson et al. 2005; Boersma et al. 2002, 2009; Raya Rey et al. 2010; Scioscia et al. 2010). Penguins at the new colonies in the northernmost part of their Atlantic range also forage close to colonies (Garcia Borboroglu et al., unpubl. data). Penguins forage farthest from the colony during incubation and stay closest when chicks are small and still being guarded (table 14.11). Distance increases again when chicks are older and left unguarded but is still much less than during incubation. There is no difference between males and females in foraging-trip distance at Punta Tombo, and reproductive success decreases as distance increases (Boersma and Rebstock 2009b). Magellanic penguins, like other seabirds, are central place foragers during the breeding season. They can cover more than 170 kilometers in a day while foraging and swim at a constant speed day and night when returning to the colony to feed their chicks (Boersma et al. 2009).

PREDATORS Predation by marine mammals at breeding colonies is usually rare. On a local scale, however, it can be substantial and important. Southern sea lion (*Otaria flavescens*) bulls may take a few birds every day (Pütz, pers. obs.; Godoy, unpubl. data). Southern sea lions can be a major cause of death for fledglings near Punta Norte, Peninsula Valdés, Argentina (Boersma and Frere, unpubl. data). One elephant seal (*Mirounga leonina*) killed about 90 breeding adults on land at Punta Tombo over three years (Clark and Boersma 2006). At Martillo Island, leopard seals (*Hydrurga leptonix*) attack adult penguins as they return to the colony, especially at the beginning of the breeding season (Raya Rey et al., unpubl. data).

Magellanic penguins are an important component in the diet of southern giant petrels (*Macronectes giganteus*) breeding in northern Patagonia (Copello et al. 2008), which attack penguins in the water and on land but are most successful at capturing fledglings when they first exit the breeding colony (Boersma, pers. obs.). Foxes and wild cats kill adult penguins, mostly in early spring when their usual prey is rare (Smith, pers. comm.). A few pumas (*Puma concolor*) have killed thousands of penguins at Monte Leon, Argentina (Frere et al. 2010; Martínez et al. 2012).

Many species prey on eggs and chicks in the colonies. Kelp gulls (*Larus dominicanus*), Antarctic skuas (*Stercorarius antarcticus*), chimango caracaras (*Milvago chimango*), armadillos (*Chaetophractus villosus*), skunks (*Conepatus humboldtii*), red foxes (*Lycalopex culpaeus*), gray foxes (*Lycalopex griseus*), and weasels (*Galictis cuja*) prey on eggs and chicks, particularly in nests that lack cover (Conway 1971; Boersma et al. 1990; Yorio and Boersma 1994b; Garcia Borboroglu 1998; Gandini et al. 1999a). On Isla Magdalena, the Chilean skua (*Stercorarius chilensis*) preys on chicks that are in poor condition or in poorly covered burrows (Godoy, pers. obs.).

MOLT. Molting is an annual event. Individuals molt on land, avoiding the water, and take about 19 days to grow their new feathers. Penguins stop using their preen glands and oiling their feathers, so that feathers become brownish before the molt (Boersma, per. obs.). They shed the old feathers on their heads, necks, and flippers last. Juveniles and young nonbreeding adults molt

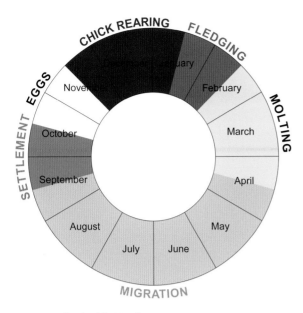

FIG. 12 Annual cycle of the Magellanic penguin.

on the beach or near the shore in the shade of bushes (Boersma et al. 1990; Pozzi and Boersma, unpubl. data). Older adults (more than 14 years old, n = 112) molted about 7 days later than young adults (4–5 years old, n = 26). Successful breeders sometimes molt in their nests with their mates, but females often start molting before males after a pre-molt foraging trip of about three weeks. Skipping a molt appears to be very rare, but one penguin banded as a chick at Punta Tombo retained juvenile plumage for an additional year, skipping its first molt. The feathers were very worn, and feathers on the lower back were missing (Boersma, unpubl. data).

ANNUAL CYCLE. In the Atlantic, the Magellanic's breeding phenology shows latitudinal and yearly variations (fig. 12). In Patagonia, penguins arrive for breeding at their colonies in September, with males arriving a week or so before the first females. The period between adult arrival and egg laying is variable, but males often may be on shore for three to four weeks before females lay their eggs and the males leave to forage (Scolaro 1984; Boersma et al. 1990; Yorio and Boersma 1994a). Farther south, eggs are laid later and chicks fledge earlier (Frere et al. 1996a; Garcia Borboroglu and Yorio, unpubl. data; Yorio et al 2001b). Penguins return in October on the north coast of East Falkland and a few days later in the western Falkland/Malvinas Islands (Strange 1992).

In the Atlantic colonies, first eggs are usually laid in October (table 14.12). At Punta Tombo, 98.7% of 2,602 first eggs were laid in October (Boersma and Rebstock, in prep.). Late laying is rare, although eggs have been laid in late November and early December, and one clutch was laid in January. Egg laying is synchronous, with most eggs laid over about a two-week period (Boersma et al. 1990; Frere et al. 1996a).

The incubation period is about 40 days, longer for first eggs than for second eggs (table 14.13). In most colonies, the majority of the chicks hatch between the first and third weeks of November (table 14.14). At Punta Tombo, 98.2% of first chicks in 5,298 clutches hatched in November (Boersma and Rebstock, in prep.).

The chick-guarding period lasts three to four weeks (table 14.15). After the guard stage, both parents forage and the chicks are often left unattended (fig. 13). Chicks fledge in January and February (table 14.16). Peak fledging ranged from 24 January to 17 February at Punta Tombo in 1989–2010 (n = 7,510 chicks [Boersma and Rebstock, in prep.]). Fledging chicks often follow adults to the water (Boersma, pers. obs.).

Juveniles return to beaches at breeding colonies after chicks have hatched, and their numbers increase until they leave for a pre-molt foraging trip of about three or four weeks in late December and early January (Boersma et al. 1990; Pozzi and Boersma, unpubl. data).

The number of penguins on the beach changes over the season (fig. 14). In September, penguins go to the beach early in the day. Later in the season, with the increase in juveniles and nonbreeding young adults and the failure of nests, more penguins are on the beach. Particularly on hot days, the number on the beach increases as many leave poorly covered nests to seek cooler tem-

FIG. 13 Magellanic penguin chicks do not really form crèches, but sometimes several chicks stay close together in burrow nests. (P. D. Boersma)

FIG. 14 From November to March, beaches at Punta Tombo often have a ring of penguins resting onshore and bathing in the shallows. (P. D. Boersma)

peratures (fig. 15) (Pozzi and Boersma, unpubl. data). On the beach, penguins drink, bathe, rest, preen, and socialize and are most active early and late in the day (Boersma, pers. obs.).

Molting begins in late January and early February with juveniles and young adults and continues until April (Boersma et al. 1990; Pozzi and Boersma, unpubl. data). In northern colonies, many penguins molt in their nests; at more southern Atlantic colonies, penguins often leave the colony, move northward, gain weight, and eventually molt on a beach away from their breeding colony; some may remain on the beach and in the colony until May (Boersma, unpubl. data).

8. POPULATION SIZES AND TRENDS

ARGENTINA. There are 66 colonies along the coast of Patagonia, Argentina, with a little more than a million breeding pairs (Schiavini et al. 2005; Pozzi and Garcia Borboroglu, unpubl. data). There are few reliable population estimates before the 1990s.

The population in Chubut Province, Argentina, rapidly expanded early in the 20th century, and Magellanic penguins colonized many locations on the mainland coast and expanded their breeding range northward (Daciuk 1976; Scolaro and Kovaks 1978; Perkins 1984; Boersma et al. 1990; Carribero et al. 1995; Schiavini et al. 2005; Boersma 2008; Pozzi and Garcia Borboroglu, unpubl. data). Punta Tombo was colonized by penguins in the early 1920s and grew to be the largest breeding colony of the species by the 1970s (Boswall and McIver 1975; Boersma et al 1990). Colonies on Peninsula Valdés were established in the late 1960s (Daciuk 1976; Perkins 1984; Scolaro and Kovaks 1978) and are increasing at annual growth rates of 9–21% (Pozzi and Garcia Borboroglu, unpubl. data). The penguins settled a new colony in northern Chubut in 2009 and colonized three islets in Rio Negro Province between 2002 and 2009 (Schiavini et al. 2005; Garcia Borboroglu 2010; Pozzi and Garcia Borboroglu, unpubl. data); however, some large colonies in Chubut, representing a large proportion of the population, have declined (table 14.17). The colony at Punta Tombo, with more than 200,000 pairs, declined over 20% between 1987 and 2008 (Boersma 2008, unpubl. data). Isla Leones had almost 100,000

FIG. 15 On hot days, Magellanic penguins at Punta Tombo, Argentina, often go to the beach or intertidal areas to stay cool. (P. D. Boersma)

pairs in 1995, and the population has declined more than 50% since then (Garcia Borboroglu et al. 2002; Pozzi and Garcia Borboroglu, unpubl. data). Colonies at Caleta Malaspina and Cabo Dos Bahías have been relatively stable from 1993 until 2008 (Garcia Borboroglu et al. 2002; Pozzi and Garcia Borboroglu, unpubl. data).

In Santa Cruz Province, breeding pairs increased 14% in 19 years (from 315,000 in 1994 to 360,000 in 2003) (table 14.17) (Frere and Gandini 1998; Gandini and Frere 1998). In Tierra del Fuego Province, colonies increased at Bahía Franklin during the 1990s and at Isla Martillo during the 2000s (Raya Rey and Schiavini, unpubl. data)

The range of the Magellanic has expanded in Argentina, and colonies are dynamic, as evidenced by the lack of genetic structuring among colonies (Bouzat et al. 2009). Given the high number of colonies and the sporadic and fragmented data on colony size, it is not possible to determine a population trend for the Atlantic population. (For more on colony size in Argentina, see, for example, Carrara 1952; Godoy 1963; Capurro et al. 1988; Carribero et al. 1995; Frere et al. 1996a; Gandini et al. 1996; Yorio et al. 1998a; Garcia Borboroglu et al. 2002; and Schiavini et al. 2005.)

FALKLAND/MALVINAS ISLANDS. Magellanics are widespread in the Falkland/Malvinas Islands. They nest in deep burrows and are not densely aggregated, making censuses difficult; hence, population size and trends are unknown. Croxall et al. (1984) list 41 known colonies, with a minimum population of 100,000 pairs, but this is likely an underestimate. A breeding-bird atlas project in the 1990s estimated the population at 76,000–142,000 pairs (Woods and Woods 1997). In 1972, Strange estimated that the total population in the Falklands/Malvinas must have approached some 2 or 2.5 million pairs (Boswall and McIver 1975) and believes it has declined substantially (pers. comm.). Bingham (2002) reports a 76% decline in breeding pairs from 1989 to 2001, but the methods and numbers are not well documented and this rate of decline has not been substantiated by other studies. Pütz et al. (2001) show declines at several sites, but monitoring stopped after 1995–96 out of concern that the methods caused penguins to abandon the study areas. Breeding success is higher in the Falkland/Malvinas Islands than in most studied colonies in Argentina, leading to the conclusion in Pütz et al. that the population may be stable but that any increase in adult mortality rates could result in population declines. More recently, in Pistorius (2009), the author reports mixed trends in breeding numbers and burrow occupancy rates at two locations between 1998 and 2008. He suggests that the population was doing reasonably well, with no apparent directional change in breeding density or nest occupancy, but reiterates that difficulties in the censuses may bias the results.

CHILE Although little is known about the Magellanic population size in Chile, 31 breeding sites are known between 32° and 56° south (table 14.18). The bulk of the Magellanic population in Chile breeds south of 42° south, where it is very difficult to conduct counts. Population size was estimated for 14 of the 31 colonies, with 4 colonies counted more than once, but due to inconsistencies in methods, no trends could be assessed. The 14 colonies are estimated to have 144,000 pairs, but Venegas (1999) suggests that a rough estimate of the Magellanic population in the Pacific could be as high as 500,000 pairs.

9. MAIN THREATS

The main threats to Magellanic penguins are at sea, but the birds are also at risk on land. Climate variation, fishing bycatch and entanglement, competition with fisheries for prey, petroleum pollution, toxic algae blooms, disease, penguin harvests, egg collection, and unregulated tourism all harm penguins.

CLIMATE VARIATION. Global climate variation may cause penguin population oscillations (Boersma 2008). Changes in ocean productivity have an impact on penguins. Phytoplankton respond to decadal-scale climate oscillations at a basin scale (Martinez et al. 2009), affecting penguin prey such as fish and squid. Small pelagic fish, the main prey of Magellanic penguins, undergo large population fluctuations with decadal-scale variability in climate (Lluch-Belda et al. 1992). Magellanics are traveling greater distances to find food at Punta Tombo, Argentina, than they did a decade ago (Boersma 2008), and reproductive success decreases as foraging distance increases (Boersma and Rebstock 2009b). Surface water temperature anomalies resulted in a lack of

food in 2008, killing thousands of individuals in winter from northern Argentina to near the equator in Brazil (Garcia Borboroglu et al. 2010).

FISHING BYCATCH. Entanglement in fishing gear in both the Atlantic and the Pacific is likely a more important cause of penguin mortality than is currently recognized (Boersma and Stokes 1995; Gandini et al. 1999b; Simeone et al. 1999; Yorio and Caille 1999; Schlatter et al. 2009; Pütz et al. 2011). Penguins are not only vulnerable to entanglement during foraging trips but also during their migration in autumn and winter (Petry and Da S. Fonseca 2002; Schlatter et al. 2009; Cardoso et al. 2011).

In the Atlantic, Magellanic foraging areas overlap with trawl fisheries in central Patagonia (Yorio et al. 2010). In San Jorge Gulf, trawl fisheries for Argentine hake (*Merluccius hubbsi*) and Argentine red shrimp (*Pleoticus muelleri*) kill hundreds of penguins (Gandini et al. 1999b; González-Zevallos and Yorio 2006; González-Zevallos et al. 2007, 2011). Magellanic penguins are less likely to be killed in Atlantic coastal fisheries than in more offshore fisheries (Yorio and Caille 1999; Tamini et al. 2002; Marinao and Yorio 2011). Penguins associating with fishing vessels become entangled while diving to take prey from the net during haulback and die (González-Zevallos and Yorio 2006). Incidental mortality of penguins at high-seas trawlers occurs mostly in fishing areas close to breeding colonies (fig. 16), indicating that spatial management has the potential to reduce the impact of fisheries on breeding populations (Boersma and Parrish 1999; Yorio et al. 2010). Magellanic penguins are also killed incidentally in artisanal fisheries operating in southern Santa Cruz and Tierra del Fuego (Gandini et al. 2000; Schiavini and Raya Rey 2001; Alegre et al. 2004).

During migration and winter foraging, penguins encounter fishing vessels in the Atlantic. The cornalito (*Sorgentinia incisa*) fishery in northern Argentina kills a few penguins in May and June (Tamini et al. 2002). In Brazil, hundreds of penguins are killed annually during winter in gill nets and trawls (Petry and Da S. Fonseca 2002; Cardoso et al. 2011).

In central Chile, a few Magellanic penguins are killed in gill nets, but mostly it is Humboldt penguins that are killed, as they are the more common species (Simeone et al. 1999). In southern Chile, Magellanics frequently become entangled in fishing nets; for example, 1,380

FIG. 16 Magellanic penguins get caught and often drown in fishing nets. Here, a dead adult penguin and a fishing box washed ashore on a beach near Punta Tombo, Argentina. (P. D. Boersma).

drowned in nets in one incident in March 2009 (Schlatter et al. 2009).

Although bycatch rates for penguins are low in many individual fisheries, the impact on the species is the sum of the rates in all the fisheries (Dillingham and Fletcher 2008). Most important, many or most of the penguins killed in fishing gear are adults (Gandini et al. 1999b; Tamini et al. 2002; Schlatter et al. 2009), and mortality of adults has a disproportionately negative impact on populations of long-lived seabirds (Sæther and Bakke 2000).

COMPETITION WITH FISHERIES. Little is known about the effects of trawl fisheries on penguin prey, although reduction of the prey by commercial fisheries is a potential threat (Boersma et al. 1990; Gandini et al. 1999b; Skewgar et al. 2007; Boersma 2008). Penguin diet composition overlaps with commercial fisheries in some areas, such as northern Patagonia and the Falkland/Malvinas Islands (Frere et al. 1996b; Clausen and Pütz 2002). Anchovies represent an attractive alternative target species for other partially overexploited fish in Argentina, and if anchovy harvests increase, that will likely harm Magellanic penguins (Skewgar et al. 2007). In other areas of the world, overharvest of fish that are important in seabird diets has led to decreases in penguin and other seabird populations (Jahncke et al. 2004; Crawford et al. 2007; Field et al. 2010; Pikitch 2012). New food webs emerge when fish are overharvested, resulting in less productive ecosystems (Pennisi 2010).

The introduction of salmon may also negatively impact penguins and alter the ecosystem in southern Argentina and Chile (Wagner and Boersma 2011). Introduced salmonids in southern Patagonia represent both a potential prey base and potential competitors for food resources (Ciancio et al. 2010). Although the extent of their impact is minor, at local scales, salmonid trophic interactions could become more significant (Pascual et al. 2009; Ciancio et al. 2010).

PETROLEUM POLLUTION. In the 1980s, petroleum pollution killed tens of thousands of penguins each year in Chubut Province, Argentina (Gandini et al. 1994). Even small amounts of petroleum on feathers suppress breeding (Fowler et al. 1995). Although tanker routes have been altered and less oiling occurs in Chubut, chronic petroleum pollution remains a problem from Brazil to northern Argentina (Boersma 2008; Garcia Borboroglu et al. 2006, 2008, 2010). Most of the Atlantic population spends the winter at sea in a restricted, narrow, offshore area from northern Argentina to southern Brazil, where they frequently encounter petroleum pollution (Boersma et al. 1990; Petry and Da S. Fonseca 2002; Garcia Borboroglu et al. 2006, 2008, 2010; Pütz et al. 2000, 2007). Large oil spills in Chile in the 1970s affected seabirds (Schlatter 1984), and oiling continues to affect penguins (Matus and Blank 2008).

TOXIC ALGAE BLOOMS AND DISEASE. Toxic algae blooms kill penguins and other seabirds (Shumway et al. 2003) and are occurring more frequently (Howarth 2008). A new disease found in 2007 in Magellanic penguins a year after it emerged in African penguins caused chicks and juveniles to lose feathers and grow more slowly (fig. 17) (Kane et al. 2010). In a warmer world, outbreaks of malaria and other diseases may increase and harm penguins. Avian pox can be fatal but so far has not been a large source of mortality on the breeding grounds (Kane et al. 2012). Aspergillosis occurs at Punta Tombo and at some colonies in Chile (Boersma, pers. obs.; Godoy, unpubl. data).

DIRECT HARVEST AND EGGING. Magellanic penguins were harvested for oil and leather from colonies in Chubut and Santa Cruz Provinces (Carrara 1952). Eggs were harvested in the past, but egg collecting in Argentina is rare, it is controlled in the Falkland/Malvinas Islands, and its impact in Chile is unknown (Croxall et al. 1984; Schlatter 1984; Schiavini and Yorio 1995). The use of penguins for fish bait and food was common in Chile (Schlatter

FIG. 17 In 2007, a new problem, loss of the second coat of down, was found in Magellanic penguins at colonies in Chubut. Although the chicks grew more slowly, some survived and grew their juvenile plumage. (P. D. Boersma)

1984), and penguins were frequently used in the king crab (*Paralomis granulosa*) fishery in the Beagle Channel, Argentina (Schiavini and Yorio 1995). In Chubut Province, a Japanese company proposed harvesting penguins for gloves, oil, and protein in the early 1980s, but this is not currently a threat (Boersma 2008).

TOURISM. Wildlife-based tourism has been growing rapidly along the Atlantic coast during the past two decades (Yorio et al. 2001a; Losano and Tagliorette 2009), and although it generates large revenues, it can harm penguins if it is inadequately managed (fig. 18). Punta Tombo had a few hundred visitors annually in the 1960s but received more than 100,000 people in 2008 (Punta Tombo Management Plan 2006; Boersma 2008). Numbers of visitors have similarly increased at most of the 24 Magellanic colonies visited in Patagonia. Magellanics appear to be tolerant of human visitation, and penguins breeding in areas where tourism is controlled become accustomed to people (Yorio and Boersma 1992; Fowler 1999; Cevasco et al. 2001; Walker et al 2006). Although they do habituate (Walker et al. 2006), penguins show physiological stress with human visitation (Walker et al. 2005a, 2005b; Villanueva et al 2012). Stress responses are higher when penguins are in poor body condition (Fowler et al. 1994; Hood et al. 1998). Unless tourism is carefully controlled and monitored, visitors may diminish nest quality, mainly by collapsing burrows (Skewgar et al. 2009). In some cases, adults were killed by irresponsible visitors or their pets, and habitat was degraded by visitors setting bush nests on fire or throwing garbage in the field (Garcia Borboroglu 2010). Human presence can delay the return of adults or induce temporary abandonment of nests, thereby reducing chick growth and increasing predation. Placing restrictions on where and when people visit penguin colonies and requiring visitors to have a guide reduce the impacts of tourism on penguins (Yorio et al. 2001a; Scioscia et al. 2009). Penguins are the main reason people visit breeding colonies, but money from tourism often does not return to fund management or conservation of the resources people come to see.

CHANGES IN LAND USE AND SPECIES INTERACTIONS. Land-use changes have affected predation and other species interactions at breeding colonies. Wild cats, foxes, and pumas were hunted intensively in Argentina in the interest of sheep ranching starting in the early 1900s and continued until the 1980s, when predator control decreased because of high labor costs (Bellati and Thungen 1990; Novaro et al. 2005; Boersma, pers. obs.). The rapid expansion of Magellanic penguins on the Argentine mainland was likely a response to the elimination of many terrestrial predators. Now, native predators like the culpeo (red) fox, puma, and other cats are returning and killing breeding Magellanics (Frere et al. 2010; Martínez et al. 2012). In the Falkland/ Malvinas Islands, large-scale decreases in tussock grass (*Poa flabellata*) habitat following European settlement affected many seabird species, and Magellanic penguins were killed because their burrowing contributed to ero-

FIG. 18 A throng of tourists flocks to see the penguins. At Punta Tombo, when cruise ship passengers arrive, the penguins have trouble crossing the blockade of humans to reach their nests or the ocean. (P. D. Boersma)

sion on denuded land (Croxall et al. 1984). Dogs, cats, rats, and other invasive species introduced by humans harm penguins.

10. RECOMMENDED PRIORITY RESEARCH ACTIONS FOR CONSERVATION

The following scientific data and long-term research will serve to guide conservation actions:

1. Assess overall population size and trends. Consistent population size estimations will enable measurement and comparison of trends within and among major regions. In Chile and Cape Horn, a complete survey to identify colonies is needed. Methods for estimating population size should be standardized to identify trends and make comparisons among colonies.

2. Measure demographic parameters, especially those that drive population trends, and their spatial and temporal variability. Understanding the population dynamics of the species requires a regional approach.

3. Learn where and when conflicts occur between penguins and human activities at sea, such as fisheries and petroleum exploration, extraction, and transport. Penguin distributions outside of the breeding season are not known in detail.

4. Improve and standardize studies that measure mortality caused by fishing gear to determine if and how it currently affects population trends. Develop and test mitigation methods for reducing this mortality. Determine the extent of overlap in resource use between fisheries and penguins. Estimate the amount of food required by penguins in different regions and seasons.

5. Implement rapid assessment methods of confirming penguin mortality from toxic algae bloom events and determine if they are increasing in frequency or intensity.

6. Monitor emerging diseases and their effects on penguin populations.

7. Assess and monitor the effects of changes in climate and ocean productivity on penguin breeding parameters and migration patterns.

8. Study species interactions at the breeding colonies, particularly with introduced species, and species that are increasing or declining because of land-use and management changes.

11. CURRENT CONSERVATION EFFORTS

Many nongovernmental organizations, academic institutions, and individuals work on the conservation of Magellanic penguins, and data from many studies have advanced conservation. Tanker lanes were moved 40 kilometers farther offshore along the coast of Chubut in 1997, which greatly reduced the number of oiled penguins (Boersma 2008). Trawling by large commercial fishing vessels is banned in the Isla Escondida area from October to April to protect the reproduction and juvenile survival of hake, one of the Magellanic's prey species. The management plan for Punta Tombo is in place but has not yet been effectively implemented by Chubut Province authorities. The Argentine government has created new marine protected parks along the coast that include some penguin breeding lands and fragments of foraging areas (San Jorge, Isla Pingüino, San Julian, and Monte Leon). Unfortunately, many of the parks lack effective planning and/or implementation. As marine protected areas are in general ineffective for the protection of highly mobile species such as penguins (Boersma and Parrish 1999), protection of penguin populations requires new conservation tools (Boersma et al. 2002, 2007; Garcia Borboroglu et al 2008; Yorio 2009; Reyes and Garcia Borboroglu, unpubl. data).

12. RECOMMENDED PRIORITY CONSERVATION ACTIONS FOR INCREASING POPULATION RESILIENCE AND MINIMIZING THREATS AND IMPACTS

Research is still needed, but enough is known now to take action to protect Magellanic penguins, although the political will is often lacking.

1. Implement ocean zoning and protected areas for penguins in their wintering areas, along their migration routes, and surrounding their breeding colonies.

2. Minimize chronic petroleum discharge into waters year-round.

3. Ensure that current fisheries and potential new fisheries do not and will not negatively affect Magellanic populations through bycatch or excessive removal of prey. Manage fisheries in an integrated way through ecosystem-based management or similar sustainable approaches, so that they do not harm penguins. Adopt a precautionary approach when information is lacking and implement mitigation measures for incidental mortality.

4. Limit areas and times for tourist visits to colonies, and monitor visitors so that they do not disturb penguins or damage breeding habitat.

5. Establish educational programs for tourists and local people explaining the penguins' importance to ecosystems and local economies, the local and international conservation actions that can help penguins, and the importance of scientific research for determining how to conserve penguins.

6. Support and facilitate scientific projects that conserve penguins. Projects should be funded, in part, by revenues generated by tourism at penguin colonies.

7. Promote coordination and integration of management among decision makers to avoid duplication, conflict and overlap among jurisdictions.

8. Prepare contingency plans for emerging diseases that start to affect penguin populations.

9. Design effective legal instruments for achieving rapid protection for new breeding colonies. Prevent the introduction of invasive species into breeding colonies.

ACKNOWLEDGMENTS

The Global Penguin Society, the Wadsworth Endowment for Conservation Science, the University of Washington, Consejo Nacional de Investigaciones Científicas y Técnicas (CONICET), Argentina, and Leiden Conservation Foundation made it possible to write this chapter. The Wildlife Conservation Society, Rufford Small Grants for Conservation, Whitley Fund for Nature, and many others supported the field research.

TABLE 14.1 Mean (± SE) measurements of adult Magellanics at colonies in Argentina

	MALES	FEMALES	REFERENCES
Mass (kg)1			
Punta Tombo	4.6 ± 0.01 (n = 1300)	3.8 ± 0.01 (n = 1208)	Boersma, unpubl.[2]
Cabo Virgenes	4.1 ± 0.04 (n = 144)	3.2 ± 0.04 (n = 144)	Gandini et al. 1992
Isla Martillo	4.7 ± 0.5 (n = 112)	3.9 ± 0.5 (n = 113)	Raya Rey et al., unpubl.[3]
Isla Vernacci Norte	4.0 ± 0.04 (n = 61)	3.5 ± 0.03 (n = 143)	Yorio et al. 2001b
Bill length (cm)4			
Punta Tombo	5.82 ± 0.004 (n = 1284)	5.34 ± 0.004 (n = 1189)	Boersma, unpubl.
Cabo Virgenes	5.77 ± 0.02 (n = 144)	5.36 ± 0.02 (n = 144)	Gandini et al. 1992
Isla Martillo	5.76 ± 2.2 (n = 228)	5.32 ± 2.2 (n = 238)	Raya Rey et al, unpubl.
Isla Vernacci Norte	5.77 ± 0.26 (n = 61)	5.33 ± 0.17 (n = 143)	Yorio et al. 2001b
Bill depth (cm)			
Punta Tombo	2.48 ± 0.002 (n = 1284)	2.11 ± 0.002 (n = 1189)	Boersma, unpubl.
Cabo Virgenes	2.40 ± 0.009 (n = 144)	2.06 ± 0.01 (n = 144)	Gandini et al. 1992
Isla Martillo	2.44 ± 1.1 (n = 228)	2.10 ± 1.1 (n = 238)	Raya Rey et al., unpubl.
Isla Vernacci Norte	2.39 ± 0.1 (n = 61)	2.07 ± 0.23 (n = 143)	Yorio et al. 2001b
Flipper length (cm)			
Punta Tombo[5]	15.6 ± 0.009 (n = 1280)	14.7 ± 0.01 (n = 1185)	Boersma, unpubl.
Cabo Virgenes[5]	15.74 ± 0.05 (n = 144)	14.99 ± 0.06 (n = 144)	Gandini et al. 1992
Isla Martillo [6]	18.6 ± 0.7 (n = 222)	17.8 ± 0.8 (n = 234)	Raya Rey et al., unpubl.
Isla Vernacci Norte[5]	15.6 ± 0.72 (n = 61)	14.8 ± 0.39 (n = 143)	Yorio et al. 2001b
Foot length (cm)			
Punta Tombo	12.2 ± 0.007 (n = 1280)	11.5 ± 0.008 (n = 1184)	Boersma, unpubl.
Cabo Virgenes	12.3 ± 0.035 (n = 144)	11.52 ± 0.046 (n = 144)	Gandini et al. 1992
Isla Vernacci Norte	12.2 ± 0.48 (n = 61)	11.5 ± 0.3 (n = 143)	Yorio et al. 2001b

1 Weights vary throughout each breeding season.

2 Data for 1982–2009, all stages of the breeding season.

3 Weight for chick-rearing period.

4 Bill length measured from the tip of the bill to the base of the "V," where feathering starts on the forehead.

5 Flipper length measured from the elbow joint to the tip of the flipper.

6 Flipper length measured from the shoulder joint to the tip of the flipper.

TABLE 14.2 Key life-history parameters of the Magellanic penguin

PARAMETER	VALUE	REFERENCES
Age at first breeding[1]	4 years	Table 14.9
Sex ratio[1]	1.5:1 (males:females)	Boersma unpubl. data
Eggs	2, similar volume	Tables 14.3 and 14.4
Laying interval	4 days	Rebstock and Boersma 2011
Incubation period	1st egg: 40–42 days, 2nd egg: 38–40 days	Table 14.13
Hatching interval	2 days	Rebstock and Boersma 2011
Guard period	21–29 days	Table 14.15
Crèche behavior	Some chicks crèche in nests, most remain in or near natal nests	Boersma unpubl.
Chick period	49–101 days, typically 65–75 days	Table 14.16
Breeding success	0.03–1.96 chicks per nest	Table 14.8
Molt	Annual, after breeding	Boersma et al. 1990
Molt duration[1]	19 days	Boersma unpubl. data
Nest density	Varies with habitat: <1 to >60 nests/100 m² Typical means: 6–9 nests/100 m²	Capurro et al. 1988, Garcia Borboroglu 1998, Stokes and Boersma 2000
Max dive depth	~97 m	Peters et al. 1998, Radl and Culik 1999, Walker and Boersma 2003
Adult survival rate[1]	0.87 (n = 200)	Boersma and Rebstock 2010b
Maximum lifespan[1]	unknown (≥ 30 years)	Boersma unpubl. Data

1 Long-term study at Punta Tombo, Argentina. May vary by colony or region, but no other data are available.

TABLE 14.3 Clutch size (eggs per nest)

COLONY	YEAR(S)	MEAN CLUTCH SIZE[1]	REFERENCES
Caleta Valdés	1978	1.91 (n = 431)	Perkins 1984
Punta Tombo	1985–1987	1.92 (n = 2531)	Boersma et al. 1990
Isla Vernacci Norte 2	1999	1.91 ± 0.29 (n = 156)	Yorio et al. 2001b
Cabo Vírgenes	1989–1991	1.90–1.97 (n=468)	Frere et al 1998

1 May underestimate 2-egg clutches because predators likely took some eggs that were not found.

TABLE 14.4 Egg measurements (± SE)

	1ST EGG	2ND EGG	REFERENCES
Mass (g)			
Punta Tombo	124.9 ± 0.26 (n = 1704)	124.7 ± 0.28 (n = 1599)	Boersma and Rebstock 2010a
Volume (cm3)			
Punta Tombo[1]	113 ± 1.1 (n = 135)	113 ± 1.1 (n = 114)	Boersma and Rebstock 2010a
Cabo Virgenes[2]	223.3 ± 0.9 (n = 432)	219.7 ± 1.0 (n = 358)	Frere et al. 1996a
Length (cm)			
Caleta Valdes[3]	7.44 (6.74–9.83)		Perkins 1984
Punta Tombo	7.46 ± 0.004 (n = 7866)	7.31 ± 0.004 (n = 7562)	Boersma and Rebstock 2010a
Isla Vernacci Norte	7.47 ± 0.03 (n = 141)	7.31 ± 0.03 (n = 141)	Yorio et al. 2001b
Breadth (cm)			
Caleta Valdes[3]	5.42 (4.11–5.93)		Perkins 1984
Punta Tombo	5.51 ± 0.002 (n = 7862)	5.54 ± 0.002 (n = 7559)	Boersma and Rebstock 2010a
Isla Vernacci Norte	5.54 ± 0.02 (n = 141)	5.54 ± 0.02 (n = 141)	Yorio et al. 2001b

1 Measured by water displacement.
2 Volume = LB², where L = length and B = breadth.
3 First and second eggs combined.

TABLE 14.5 Chick size at hatching (means ± SE)

	1ST CHICK	2ND CHICK	REFERENCES
Mass (g)			
Punta Tombo	90.0 ± 0.5 (n = 6284)	90.2 ± 0.6 (n = 5404)	Boersma unpubl. data
Isla Vernacci Norte	90.0 ± 1.87 (n = 40)	88.3 ± 1.68 (n = 40)	Yorio et al. 2001b
Bill length (cm)			
Punta Tombo	1.47 ± 0.004 (n = 6030)	1.51 ± 0.003 (n = 5303)	Boersma unpubl. data
Isla Vernacci Norte	1.54 ± 0.009 (n = 40)	1.55 ± 0.012 (n = 40)	Yorio et al. 2001b
Bill depth (cm)			
Punta Tombo	0.80 ± 0.002 (n = 6027)	0.82 ± 0.002 (n = 5301)	Boersma unpubl. data
Isla Vernacci Norte	0.80 ± 0.007 (n = 40)	0.81 ± 0.006 (n = 40)	Yorio et al. 2001b
Flipper length (cm)			
Punta Tombo	2.8 ± 0.007 (n = 6031)	2.9 ± 0.006 (n = 5303)	Boersma unpubl. data
Isla Vernacci Norte	2.95 ± 0.028 (n = 40)	2.98 ± 0.028 (n = 40)	Yorio et al. 2001b
Foot length (cm)			
Punta Tombo	3.3 ± 0.009 (n = 6031)	3.4 ± 0.007 (n = 5299)	Boersma unpubl. data
Isla Vernacci Norte	3.42 ± 0.031 (n = 40)	3.37 ± 0.027 (n = 40)	Yorio et al. 2001b

TABLE 14.6 Mass of chicks near fledging

COLONY	YEAR(S)	MASS (KG) ± SD	REFERENCES
Argentina:			
Punta Tombo[1]	2006–2010	2.9 ± 0.4 (n = 361)	Boersma unpubl.
Isla Vernacci Norte[2]	1999	1st chicks: 2.67 ± 0.49 (n = 30) 2nd chicks: 2.32 ± 0.63 (n = 26)	Yorio et al. 2001b
Cabo Virgenes[3]	1989–1991	2.9–3.0 (n = 327)	Frere et al. 1998
Chile:			
Isla Magdalena[4]	1996	3.3 ± 0.5 (n = 37)	Radl and Culik 1999
Otway Sound[4]	1996	3.4 ± 0.3 (n = 32)	Radl and Culik 1999

1 Chicks leaving the colony (fledging).
2 60 days of age.
3 Last day seen in the colony (nests checked daily).
4 Older than 50 days of age.

TABLE 14.7 Hatching success

COLONY	YEAR(S)	HATCHING SUCCESS	REFERENCES
Caleta Valdes	1978	76.7% (n = 290)	Perkins 1984
Punta Tombo	1983–2006	1.44 ± 0.8 eggs per nest (n = 7076 clutches)	Boersma, unpubl.
Isla Vernacci Norte	1999	1.2 ± 0.85 eggs per nest (n = 146 clutches)	Yorio et al. 2001b
Cabo Virgenes	1989–1991	32–79% (n = 908 eggs)	Frere et al. 1998

TABLE 14.8 Breeding success (chicks per nest)

COLONY	YEAR(S)	BREEDING SUCCESS (CHICKS PER PAIR)	REFERENCES
Argentina			
Punta Tombo	1983–2009	0.03–0.95 (n = 5287)	Boersma and Stokes 1995; Boersma 2008, Boersma et al unpubl. Data
Isla Vernacci Norte	1999	0.56 ± 0.67 (n = 142)	Yorio et al. 2001b
Isla de los Pàjaros, Ria Deseado	1996	1.7 (n = 17)	Cevasco et al. 2001
Isla Quiroga, Ria Deseado	1996	1.2 (n = 18)	Cevasco et al. 2001
Cabo Virgenes	1989–1991	0.19–0.83 (n = 468)	Frere et al. 1998
Isla Martillo	2003–2010	1.1–1.7	Scioscia et al. in press and Raya Rey, Schiavini and Scioscia unpubl. data
Chile			
Isla Magdalena	1996	1.79 ± 0.4 (n = 28)	Radl and Culik 1999
Otway Sound	1996	1.96 ± 0.3 (n = 24)	Radl and Culik 1999
Falkland/Malvinas Islands			
East Falkland Island	2001	0.16 (n = 15 study nests) 0.20 (n = 14 control nests)	Otley et al. 2001
Volunteer Point, Sea Lion Island, Steeple Jason	1989–2008	0.82 (sample size not given)	Pistorius 2009
Multiple sites	1989–1995	0.78 (sample size not given)	Pütz et al. 2001

TABLE 14.9 Demographic variables for Punta Tombo, 1983–2009

VARIABLE	VALUE
Age at first breeding (years)	Males: 1% breed at age 4, 50% breed by 7–8 (n=721)
	Females: 9% breed by age 4, 50% breed by 6 (n=296)
Age at first successful breeding (years)	Males: 3 of 7 males breeding at age 4 were successful
	Females: 7 of 29 females breeding at age 4 were successful
Reproductive life span (years)[1]	Males: mean = 6.6, median = 6, maximum = 20[2] (n=570)
	Females: mean = 5.6, median = 4, maximum = 20[2] (n=226)
Oldest known-age individual (years)	26 (as of 2009–10) (n=4133)
Oldest not known-age individual (years)	≥ 30[3] (as of 2009–10) (n=9097)

Note: Source: Boersma, unpublished data.

1 Years between last and first breeding records. Known-age birds that started breeding between 1990 and 2003 only.
   Did not necessarily breed in each year. This is an underestimate as many of these birds are still breeding.

2 20 years is the maximum detectable for birds that started breeding in 1990 (up to 2009).
   The true maximum is > 20 years, but unknown.

3 Banded as a breeding adult in 1983, seen alive in 2009. Assumes birds were at least 4 years old when banded.

TABLE 14.10 Nest and mate fidelity

	NEST FIDELITY	MATE FIDELITY
Males	79% (n = 8212 nest records for 3723 individuals)	86% (n = 7171 nest records for 3962 individuals)
Females	70% (n = 7341 nest records for 3120 individuals)	83% (n = 7483 nest records for 4197 individuals)

Note: Nest fidelity is the percentage of times that an individual used the same nest in two consecutive years.

Mate fidelity is the percentage of times that an individual had the same mate in two consecutive years.

Years that are not consecutive are not included (Boersma unpublished data).

TABLE 14.11 Foraging trip distance (km) or duration (h), mean ± SE

COLONY OR REGION	DISTANCE (KM) OR DURATION (H)	STAGE	REFERENCES
Argentina			
San Lorenzo (n = 21)	50% of time within 100 km	Incubation (long trip)	Wilson et al. 2005
La Ernestina (n = 6)	91 ± 17.7 km	Late chick	Boersma et al. 2009
Punta Tombo (n = 80)	411 ± 11.8 km (max 605 km)	Incubation (long trip)	Boersma and Rebstock 2009c
Punta Tombo (n = 35)	61 ± 3.9 km	Early chick	Boersma and Rebstock 2009c
Punta Tombo (n = 101)	111 ± 5.0 km	Late chick	Boersma and Rebstock 2009c
Punta Loberia (n = 6)	50% of time within 200 km	Incubation (long trip)	Wilson et al. 2005
Cabo Dos Bahias (n = 6)	63 ± 20.9 km	Late chick	Boersma et al. 2009
Golfo San Jorge (n = 15)	26–120 km	Early chick	Yorio et al. 2010
Puerto Deseado (n = 5)	76 ± 4.3 km	Late chick	Boersma et al. 2009
San Julian (n = 5)	50% of time within 150 km	Incubation (long trip)	Wilson et al. 2005
San Julian (n = 5)	79 ± 10.8 km	Late chick	Boersma et al. 2009
Monte Leon (n = 6)	43 ± 4.1 km	Late chick	Boersma et al. 2009
Cabo Virgenes (n = 4)	50% of time within 70 km	Incubation (long trip)	Wilson et al. 2005
Cabo Virgenes (n = 18)	13.8 ± 7.1 h	Early chick	Wilson et al. 2007
Isla Martillo (n=79)	34.2 ± 17 h	Incubation	Scioscia et al. in press
Isla Martillo (n=135)	12 ± 3 h	Early chick	Scioscia et al. in press
Isla Martillo (n=73)	19 ± 7 h	Late chick	Scioscia et al. in press
Isla Martillo (n=7)	24 ± 10 km, 15 ± 7 h	Early chick	Raya Rey et al. 2010
Chile			
Magdalena Island (n = 15)	17.7 ± 1.5 h	Chick	Radl and Culik 1999
Seno Otway (n = 14)	9.2 ± 0.8 h	Chick	Radl and Culik 1999
Falkland/Malvinas Islands			
New Island (n = 11)	50% of time within 70 km	Incubation (long trip)	Wilson et al. 2005
New Island (n = 9)	16 km	Incubation (late)	Boersma et al. 2002
New Island (n = 13)	6 km	Late chick	Boersma et al. 2002

TABLE 14.12 Egg dates (showing median date of 1st egg, range, and sample size)

COLONY	YEARS	MEDIAN DATE OF 1ST EGG	REFERENCES
Punta Tombo	1986–2009	8–21 Oct (26 Sep–15 Nov) n=2602	Rebstock and Boersma 2011
Isla Vernacci Norte 2	1999	14 Oct (6 Oct–1 Nov) n=156	Yorio et al. 2001b
Cabo Virgenes	1990 and 1991	16 Oct (2–26 Oct) n=224 (1990), n=280 (1991)	Frere et al. 1996a
East Falkland Island	2001	19 Oct* n=14	Otley et al. 2004

* Mean

TABLE 14.13 Incubation period, days ± SE

COLONY	YEAR(S)	1ST EGG	2ND EGG	REFERENCES
Punta Tombo	1983–2009	40.5 ± 0.04 (n = 1148)	38.8 ± 0.03 (n = 1148)	Rebstock and Boersma 2011
Isla Vernacci Norte	1999	40.9 ± 0.18 (n = 50)	38.6 ± 0.16 (n = 48)	Yorio et al. 2001b
Cabo Virgenes	1990–1991	42 ± 0.14 (n = 117)	40.3 ± 0.09 (n = 132)	Frere et al. 1996a

TABLE 14.14 Hatching dates (showing median date of 1st egg, range, and sample size)

COLONY	YEARS	MEDIAN DATE OF 1ST CHICK	REFERENCES
Punta Tombo	1987–2009	18–29 Nov (2 Nov–23 Dec) n=5298	Boersma and Rebstock in prep
Cabo Virgenes	1990–1991	25 Nov (13 Nov–6 Dec) n=117	Frere et al. 1996a
East Falkland / Malvinas Island	2001	(21 Nov–10 Dec) n=14	Otley et al. 2004

TABLE 14.15 Guard period—days from hatching until chicks are left unattended

COLONY	YEAR(S)	BROODING PERIOD	REFERENCES
Punta Tombo	1983–1987	29 ± 4 (n = 38)	Boersma et al. 1990
East Falkland/ Malvinas Island	2001	21 ± 1 (n = 8)	Otley et al. 2004

TABLE 14.16 Chick-rearing period—days from hatching to fledging) ± SE

COLONY	YEAR(S)	CHICK-REARING PERIOD	REFERENCES
Argentina			
Punta Tombo	1983–2009	73.9 ± 0.16 (49–101) (n = 2860)[1]	Boersma unpubl.
Cabo Virgenes	1989–1991	66.4–70.8 (n = 235)	Frere et al. 1998
Cabo Virgenes	1989–1994	69 (60–84)	Frere et al. 1996a
Falkland/Malvinas Islands			
East Falkland/ Malvinas Island	2001	64–73 (n = 2)	Otley et al. 2004

1 This is an underestimate for some clutches as chicks were still present on last check of season.

TABLE 14.17 Location, size, and growth trend of Magellanic colonies in Argentina

#	COLONY	LOCATION	PAIRS	YEAR	TREND	SOURCE
1–3	Islote Redondo, Pastosa and de los Pájaros	41°26'S,65°01'W	22	2002	Increasing	1, 2
4	Islote Notable	42°25'S,64°31'W	624	2003		1
5	Asentamiento Oeste	42°06'S,63°56'W	1,621	2003	Increasing	1, 2
6	Estancia San Lorenzo	42°05'S,63°51'W	56,737	2003	Increasing	1, 2
7	Caleta Externa	42°16'S,63°38'W	9,322	1998	Increasing	1, 2, 3
8	Isla Primera (Caleta Valdés)	42°21'S,63°37'W	12,539	2003		1
9	Isla Segunda (Caleta Valdés)	42°20'S,63°37'W	380	2003		1
10	Caleta Interna	42°27'S,63°36'W	461	2003		1
11	El Pedral	42°56'S, 64°20 W	13	2009	Increasing	2
12	Punta Clara	43°58'S,65°15'W	70,000	1989		3
13	Punta Tombo	44°02'S,65°11'W	~200,000	2005	Decreasing	4, 5
14	Punta Lobería	44°35'S,65°22'W	6,745	1995		5
15	Isla Cumbre	44°35'S,65°22'W	31	1994		5
16	Isla Blanca Mayor	44°46'S,65°38'W	510	1994		5
17	Cabo Dos Bahías	44°54'S,65°32'W	9,067	1995		6
18	Isla Moreno	44°54'S,65°32'W	242	1994		6
19	Isla Arce	45°00'S,65°29'W	3,500	1995		6
20	Isla Leones	45°03'S,65°36'W	96,287	1995	Decreasing	7, 2
21	Península Lanaud	45°03'S,65°35'W	5,460	1995	Decreasing	7, 2
22	Isla Buque	45°03'S,65°37'W	174	1994	Decreasing	7, 2
23	Isla Sudoeste	45°03'S,65°36'W	867	1995	Decreasing	7, 2
24	Isla Tova	45°06'S,66°00'W	57,174	1995		7
25	Isla Tovita	45°07'S,65°57'W	31,906	2001		1, 2, 7
26	Isla Gaviota	45°06'S,65°58'W	939	2001		1, 2, 7
27	Isla Este	45°07'S,65°56'W	28	2001		1, 2, 7
28	Isla Vernaci Este	45°11'S,66°29'W	2,503	2003	Decreasing	1, 2, 7
29	Isla Vernaci Norte 1	45°11'S,66°30'W	24,105	2003	Decreasing	1, 2, 7
30	Isla Vernaci Norte 2	45°11'S,66°30'W	5,183	2003	Increasing	1, 2, 7
31	Isla Vernaci Sudoeste	45°11'S,66°31'W	52	2003	Increasing	1, 2, 7
32	Isla Vernaci Noroeste	45°10'S,66°31'W	275	2003	Decreasing	1, 2, 7
33	Isla Vernaci Fondo 1	45°10'S,66°31'W	162	1998	Increasing	1, 2, 7
34	Isla Vernaci Fondo 2	45°10'S,66°31'W	219	2003	Increasing	1, 2, 7
35	Isla Viana Mayor	45°11'S,66°24'W	3,165	1993		1, 7

TABLE 14.17 (cont.)

#	COLONY	LOCATION	PAIRS	YEAR	TREND	SOURCE
36	Punta Pájaros	46°57S,66°51'W	300	1994	Increasing	8
37	Isla Chaffers	47°46'S,65°52'W	13,700	1992	Increasing	8
38	Isla Larga	47°45'S,65°56'W	50	1994	Increasing	8
39	Isla Quiroga	47°45'S,65°56'W	760	1992	Increasing	8
40	Islote Burlotti	47°46'S,65°57'W	225	1993	Increasing	8
41	Isla de los Pájaros	47°45'S,65°58'W	8,650	1993	Increasing	8
42	Islote Cañadón del Puerto	47°45'S,66°00'W	580	1992	Increasing	8
43	Isla del Rey	47°46'S,66°03'W	1,100	1993	Increasing	8
44	Isla Pingüino	47°54'S,65°43'W	15,000	1992	Increasing	8
45	Isla Chata	47°56S,65°44W	120	1994	Increasing	8
46	Isla Schwarz	48°04'S,65°54'W	11,000	1994	Increasing	8
47	Islote Burgos	48°05'S,65°54'W	800	1994	Increasing	8
48	Isla Liebres	48°06'S,65°54'W	170	1994	Increasing	8
49	Punta Medanosa (Faro)	48°06'S,65°55'W	22,000	1994	Increasing	8
50	Punta Sur	48°07'S,65°56'W	312	1994	Increasing	8
51	Estancia 8 de Julio	48°07'S,66°08'W	3,900	1993	Increasing	8
52	Cabo Guardián	48°21'S,66°21'W	7,000	1994	Increasing	9
53	Islote del Bajío	48°21'S,66°21'W	175	1994	Increasing	9
54	Isla Rasa Chica	48°22'S,66°20'W	133	1995	Increasing	9
55	Islote Sin Nombre	48°22'S,66°21'W	400	1994	Increasing	9
56	Banco Cormorán	49°16'S,67°40'W	37,150	1993	Increasing	9
57	Banco Justicia I	49°17'S,67°41'W	30	1994	Increasing	9
58	Isla Leones	50°04'S,68°26'W	19,200	1994	Increasing	9
59	Punta Entrada	50°08'S,68°22'W	48,000	1994	Increasing	9
60	Monte León	50°22'S,68°53'W	32,000	1994	Increasing	9
61	Isla Deseada	51°34'S,69°02'W	3,560	1995	Increasing	9
62	Cabo Vírgenes	52°22'S,68°24'W	89,200	1994	Increasing	9
63	Isla Martillo	54°54'S,67°23'W	3,064	2009	Increasing	1, 10
64	Bahía Franklin	54°53'S,64°39'W	1,633	2010	Increasing	1, 10
65	Isla Observatorio	54°39'S,64°08'W	105,534	1995		11
66	Isla Goffré	54°42'S,64°14'W	14,849	1995		11

Sources: 1 Schiavini et al 2005; 2 Pozzi and Garcia Borboroglu unpub.; 3 Yorio et al. 1998a; 4 Boersma 2008; 5 Boersma unpubl. data.; 6 Yorio et al. 1998b; 7 Garcia Borboroglu et al 2002; ; 8 Gandini and Frere 1998; 9 Frere and Gandini 1998; 10 Raya Rey and Schiavini unpubl. data; 11 Schiavini et al. 1998.

TABLE 14.18 Magellanic population in Chile

COLONY	LOCATION	NUMBER OF PAIRS	YEAR OF ESTIMATE	SOURCE
Isla Cachagua	32°35'S, 71°27'W		1919	Philippi 1937
Isla Pájaro Niño	33°21'S, 71°41'W	8	1998	Simeone & Bernal 2000
Isla Pupuya	33°58'S; 71°53'W			Johnson 1965
Isla Santa María	37°01'S; 73°31'W		1936	Housse 1936
Isla Mocha	38°22'S; 73°56'W		1932	Bullock 1935
Isla Maiquillahue	39°27'S; 73°15'W			Johnson 1965
Roca Huenteyao	40°32'S; 73°43'W	102	2008	Cursach et al. 2009
Islote Pingüinos	40°56'S; 73°54'W	349	2008	Cursach et al. 2009
Puñihuil	41°55'S; 74°02'W	458	2004	Simeone 2005
Ahuenco	42°05'S; 74°03'W	200	1991	Soria-Galvarro 1991
Metalqui	42°12'S; 74°10'W	203	2008	Hiriart-Bertrand et al. 2010
Isla Guafo	43°36'S; 74°43'W	1,706	2005	Reyes-Arriagada et al. 2009
Isla sin nombre	45°22'S; 73°35'W		2008	R. Matus, pers. comm.
Isla Tenquehuén	45°39'S;74148'W		1984	Clark 2008
Isla Rugged	47°38'S; 75°08'W		2008	R. Matus, pers. comm.
Isla Buenaventura	50°45'S; 75°08'W		1984	Clark 2008
Seno Otway	52°58'S; 71°12'W	2,216	2007	Téllez 2007
Isla Marta	52°51'S; 70°34'W		1970	Pisano 1971
Isla Contramaestre	52°51'S; 70°21'W	25,000	2002	Bingham & Herrmann 2008
Isla Magdalena	52°55'S; 70°35'W	62,460	2007	Bingham & Herrmann 2008
Isla Rupert	53°46'S; 72°13'W	11,250	2005	Soto 1990, Miranda et al. 2009
Islote Poroto	54°03'S; 71°03'W			R. Matus, pers. comm.
Isla Noir	54°28'S; 73°00'W	>35,000	1999	Venegas 1999, Kusch et al. 2007
Isla Treble	55°04'S; 71°00'W		1984	Clark 2008
Isla Carolina	55°27'S; 69°34'W		1914	Murphy 1936
Isla Morton	55°32'S; 69°17'W		1984	Clark 2008
Isla Bayly	55°37'S; 67°33'W	300	1980	Venegas 1981
Isla Barnevelt	55°49'S; 66°47'W	5,000	1932	Reynolds 1935
Isla Idelfonso	55°50'S; 69°20'W			Venegas 1978
Isla Hornos	55°57'S; 67°16'W		1984	Clark 2008
Isla Diego Ramírez				Schlatter and Riveros 1997
		144,252		

REFERENCES

Akst, E. P., P. D. Boersma, and R. C. Fleischer. 2002. A comparison of genetic diversity between the Galapagos penguin and the Magellanic penguin. *Conservation Genetics* 3:375–83.

Alegre, M. B., S. Ferrari, M. Perroni, P. Gandini, and E. Frere. 2004. Captura incidental de aves acuáticas por redes de enmalle en el estuario del Río Gallegos—Chico (Santa Cruz). II Jornadas Patagónicas sobre Mallines y Humedales. Rio Gallegos, Santa Cruz, Argentina: Ediciones Universidad Nacional de la Patagonia Austral.

Barbosa, A., L. M. Ortega-Mora, F. T. García-Moreno, F. Valera, and M. J. Palacios. 2007. Southernmost record of the Magellanic penguin *Spheniscus magellanicus* in Antarctica. *Marine Ornithology* 35:79.

Bellati, J., and J. von Thungen. 1990. Lamb predation in Patagonian ranches. In *Proceedings of the 14th Vertebrate Pest Conference*, ed. L. R. Davis and R. E. Marsh, 263–68. Davis: University of California.

Bingham, M. 2002. The decline of Falkland Islands penguins in the presence of a commercial fishing industry. *Revista Chilena de Historia Natural* 75:805–18.

Bingham, M., and T. Herrmann. 2008. Magellanic penguin (Spheniscidae) monitoring results for Magdalena Island (Chile), 2000–2008. *Anales Instituto Patagonia (Chile)* 36(2):19–32.

BirdLife International. 2009. Species factsheet: Spheniscus magellanicus. http://www.birdlife.org (accessed 1 December 2010).

Blanco, D. E., P. Yorio, and P. D. Boersma. 1996. Feeding behavior, size asymmetries, and food distribution in Magellanic penguin (*Spheniscus magellanicus*) chicks. *The Auk* 113:496–98.

Boersma, P. D. 1987. Penguin deaths in the South Atlantic. *Nature* 327:96.

———. 1992. Asynchronous hatching and food allocation in the Magellanic penguin *Spheniscus magellanicus*. In *ACTA XX Congressus Interna-*

tionalis Ornithologici, 961–73. International Ornithological Congress, Christchurch, New Zealand, 2–9 December 1990.

———. 2008. Penguins as marine sentinels. *BioScience* 58:597–607.

———. 2012. Penguins and petroleum: Lessons in conservation ecology. *Frontiers in Ecology and the Environment* 10:218–19.

Boersma, P. D., and E. M. Davies. 1987. Sexing monomorphic birds by vent measurements. *The Auk* 104:779–83.

Boersma, P. D., and J. K. Parrish. 1999. Limiting abuse: Marine protected areas, a limited solution. *Ecological Economics* 31:287–304.

Boersma, P. D., and G. A. Rebstock. 2009a. Intraclutch egg-size dimorphism in Magellanic penguins (*Spheniscus magellanicus*): Adaptation, constraints, or noise? *The Auk* 126: 335–40.

———. 2009b. Magellanic penguin eggshell pores: Does number matter? *Ibis* 151:535–40.

———. 2009c. Foraging distance affects reproductive success in Magellanic penguins. *Marine Ecology Progress Series* 375:263–75.

———. 2009d. Flipper bands do not affect foraging-trip duration of Magellanic penguins. *Journal of Field Ornithology* 80:408–18.

———. 2010a. Calculating egg volume when shape differs: When are equations appropriate? *Journal of Field Ornithology* 81:442–48.

———. 2010b. Effects of double bands on Magellanic penguins. *Journal of Field Ornithology* 81:195–205.

Boersma, P. D., G. A. Rebstock, E. Frere, and S. E. Moore. 2009. Following the fish: Penguins and productivity in the South Atlantic. *Ecological Monographs* 79:59–76.

Boersma, P. D., G. A. Rebstock, and D. L. Stokes. 2004. Why penguin eggshells are thick. *The Auk* 121:148–55.

Boersma, P. D., G. A. Rebstock, D. L. Stokes, and P. Majluf. 2007. Oceans apart: Conservation models for two temperate penguin species shaped by the marine environment. *Marine Ecology Progress Series* 335:217–25.

Boersma, P. D., and D. L. Stokes. 1995 Mortality patterns, hatching asynchrony, and size asymmetry in Magellanic penguin (*Spheniscus magellanicus*) chicks. In *The Penguins: Ecology and Management*, ed. P. Dann, I. Norman, and P. Reilley, 3–25. Chipping Norton, NSW, Australia: Surrey Beatty and Sons.

Boersma, P. D., D. L. Stokes, and I. Strange. 2002. Applying ecology to conservation: Tracking breeding penguins at New Island South Reserve, Falkland Islands. *Aquatic Conservation: Marine and Freshwater Ecosystems* 12:1–11.

Boersma, P. D., D. L. Stokes, and P. M. Yorio. 1990. Reproductive variability and historical change of Magellanic penguins (*Spheniscus magellanicus*) at Punta Tombo, Argentina. In *Penguin Biology*, ed. L. Davis and J. Darby, 15–43. San Diego, CA: Academic Press.

Boswall, J., and D. MacIver. 1975. The Magellanic penguin *Spheniscus magellanicus*. In *The Biology of Penguins*, ed. B. Stonehouse, 271–305. London: Macmillan.

Bouzat, J. L., B. G. Walker, and P. D. Boersma. 2009. Regional genetic structure in the Magellanic penguin (*Spheniscus magellanicus*) suggests metapopulation dynamics. *The Auk* 126:326–34.

Brandão, M. L., K. M. Braga, J. L. Luque. 2011. Marine debris ingestion by Magellanic penguins *Spheniscus magellanicus* (Aves: Sphenisciformes), from the Brazilian coastal zone. *Marine Pollution Bulletin* 62(10):2246–49.

Bullock, D. S. 1935. Las aves de la isla de la Mocha. *Revista Chilena de Historia Natural* (Chile) 39:232–53.

Capurro, A., E. Frere, M. Gandini, P. Gandini, T. Holik, V. Lichtschein, and P. D. Boersma. 1988. Nest density and population size of Magellanic penguins (*Spheniscus magellanicus*) at Cabo Dos Bahias, Argentina. *The Auk* 105:585–88.

Cardoso, L. G., L. Bugoni, P. L. Mancini, and M. Haimovici. 2011. Gillnet fisheries as a major mortality factor of Magellanic penguins in wintering areas. *Marine Pollution Bulletin* 62:840–44.

Carrara, S. 1952. Lobos marinos, pinguinos y guaneras del litoral marítimo e islas adyacentes de la República Argentina. Facultad de Ciencias Veterinarias, Universidad Nacional de la Plata, La Plata.

Carribero, A., D. Pérez, and P. Yorio. 1995. Actualización en la distribución y abundancia del pingüino de Magallanes en Península Valdés, Chubut. *El Hornero* 14:33–37.

Cevasco, C., E. Frere, and P. Gandini. 2001. El valor reproductivo de la nidada y la intensidad de visitas como condicionantes de la respuesta del pingüino de Magallanes (*Spheniscus magellanicus*) al disturbio humano. *Ornitología Neotropical* 12:75–81.

Ciancio, J., D. A. Beauchamp, and M. Pascual. 2010. Marine effect of introduced salmonids: Prey consumption of exotic steelhead and anadromous brown trout in the Patagonian Continental Shelf. *Limnology and Oceanography* 55:2181–92.

Clark, G. 2008. *La Travesía del Totorore.* Santiago, Chile: Editorial El Mercurio-Aguilar S.A.

Clark, J. A., and P. D. Boersma. 2006. Southern elephant seal (*Mirounga leonina*) kills Magellanic penguins (*Spheniscus magellanicus*) on land. *Marine Mammal Science* 22(1):222–25.

Clark, J. A., P. D. Boersma, and D. O. Olmsted. 2006. Name that tune: Call discrimination and individual recognition in Magellanic penguins. *Animal Behavior* 72:1141–48.

Clausen, A. P., and K. Pütz. 2002. Recent trends in diet composition and productivity of gentoo, Magellanic, and rockhopper penguins in the Falkland Islands. *Aquatic Conservation in Marine and Freshwater Ecosystems* 12:51–61.

Conway, W. G. 1971. Predation on penguins at Punta Tombo. *Animal Kingdom* 74:2–6.

Copello, S., F. Quintana, and F. Perez. 2008. The diet of the southern giant petrel in Patagonia: Fishery-related items and natural prey. *Endangered Species Research* 6: 15–23.

Crawford, R. J. M., L. G. Underhill, L. Upfold, and B. M. Dyer. 2007. An altered carrying capacity of the Benguela upwelling ecosystem for African penguins (*Spheniscus demersus*). *ICES Journal of Marine Science* 64:570–76.

Croxall, J. P., S. J. McInnes, and P. A. Prince. 1984. The status and conservation of seabirds at the Falkland Islands. In *Status and Conservation of the World's Seabirds*, ed. J. P. Croxall, P. G. H. Evans, and R. W. Schreiber, 271–91. ICBP Technical Publication No. 2. Cambridge: International Council for Bird Preservation.

Cursach, J., J. Vilugrón, C. Tobar, J. Ojeda, J. Rau, C. Oyarzún, and O. Soto. 2009. Nuevos sitios de nidificación para cuatro especies de aves marinas en la provincia de Osorno, centro-sur de Chile. *Boletín Chileno de Ornitología* 15:17–22.

Daciuk, J. 1976. Notas faunísticas y bioecológicas de Península Valdés y Patagonia XV. Estudio bioecológico inicial de los esfeníscidos visitantes y colonizadores de Península Valdés y costas aledañas (Prov. De Chubut, Argentina). *Physis Seccion C* 35:43–56.

Darby, J. T. 1991. A second Magellanic penguin in New Zealand. *Notornis* 38:36.

Davis, L. S., and M. Renner. 2003. *Penguins.* London: T & A D Poyser.

Dillingham, P. W., and D. Fletcher. 2008. Estimating the ability of birds to sustain additional human-caused mortalities using a simple decision rule and allometric relationships. *Biological Conservation* 141:1783–92.

Field, J. C., A. D. MacCall, R. W. Bradley, and W. J. Sydeman. 2010. Estimating the impacts of fishing on dependent predators: A case study in the California Current. *Ecological Applications* 20:2223–36.

Fonseca, V. S da S., M. V. Petry, and A. H. Jost. 2001. Diet of the Magellanic penguin on the coast of Rio Grande so Sul, Brazil. *Waterbirds* 24:290–93.

Forero, M. G., K. A. Hobson, G. R. Bortolotti, J. A. Donázar, M. Bertellotti, and G. Blanco. 2002. Food resource utilization by the Magellanic penguin evaluated through stable-isotope analysis: Segregation by sex and age and influence on offspring quality. *Marine Ecology Progress Series* 234:289–99.

Fowler, G. S. 1993. Ecological and endocrinological aspects of long-term pair bonds in the Magellanic penguin (*Spheniscus magellanicus*). PhD diss., University of Washington.

———. 1999. Behavioral and hormonal response of Magellanic penguins (*Spheniscus magellanicus*) to tourism and nest visitation. *Biological Conservation* 90:143–49.

Fowler, G. S., J. C. Wingfield, and P. D. Boersma. 1995. Hormonal and reproductive effects of low levels of petroleum fouling in the Magellanic penguin (*Spheniscus magellanicus*). *The Auk* 112:382–89.

Fowler, G. S., J. C. Wingfield, P. D. Boersma, and A. R. Sosa. 1994. Reproductive endocrinology and weight change in relation to reproductive success in the Magellanic penguin (*Spheniscus magellanicus*). *General and Comparative Endocrinology* 94:305–15.

Frere, E., and P. Gandini. 1998. Distribución reproductiva y abundancia de las aves marinas de Santa Cruz. Parte 2: de Bahía Laura a Punta Dungeness. In *Atlas de la distribución reproductiva de aves marinas en el litoral Patagónico Argentino*, ed. P. Yorio, E. Frere, P. Gandini, and G. Harris, 153–77. Buenos Aires, Argentina: Society Instituto Salesiano de Artes Gráficas.

Frere, E., P. Gandini, and P. D. Boersma. 1996a. Aspectos particulares de la biologia de reproduccion y tendencia poblacional del pinguino de magellanes (*Spheniscus magellanicus*) en la colonia de Cabo Virgenes, Santa Cruz, Argentina. *Hornero* 14: 79–88.

———. 1998. Breeding ecology of the Magellanic penguin at Cabo Virgenes, Argentina: What determines reproductive success? *Colonial Waterbirds* 21:205–10.

Frere, E., P. Gandini, and V. Lichtschein. 1996b. Variación latitudinal en la dieta del pinguino de Magallanes (*Spheniscus magellanicus*) en la costa patagonica, Argentina. *Ornitologia Neotropical* 7:35–41.

Frere, E., P. A. Gandini, and P. D. Boersma. 1992. Effects of nest type and location on reproductive success of the Magellanic penguin (*Spheniscus magellanicus*). *Marine Ornithology* 20:1–6.

Frere, E., A. Millones, A. Morgenthaler, A. Travaini, and P. Gandini. 2010. High predation rates by pumas on Magellanic penguin adults: New conflicts in coastal protected areas in Argentina. 1st World Seabird Conference, Victoria, Canada, 7–11 September 2010.

Gandini, P., P. D. Boersma, E. Frere, M. Gandini, T. Holik, and V. Lichtschein. 1994. Magellanic penguins (*Spheniscus magellanicus*) are affected by chronic petroleum pollution along the coast of Chubut, Argentina. *The Auk* 111:20–27.

Gandini, P., and E. Frere. 1998. Distribución y abundancia de las aves marinas de Santa Cruz. Parte 1: La Loberia a Islote del Cabo. In *Atlas de la distribución reproductiva de aves marinas en el litoral Patagónico Argentino*, ed. P. Yorio, E. Frere, P. Gandini, and G. Harris, 119–51. Buenos Aires, Argentina: Instituto Salesiano de Artes Gráficas.

Gandini, P., E. Frere, and P. D. Boersma. 1996. Status and conservation of Magellanic penguins *Spheniscus magellanicus* in Patagonia, Argentina. *Bird Conservation International* 6:307–16.

———. 1997. Efectos de la calidad de habitat sobre el éxito reproductivo del pinguino de magellanes (*Spheniscus magellanicus*) en Cabo Virgenes Santa Cruz, Argentina. *Ornitologia Neotropical* 8:37–48.

———. 1999a. Nest concealment and its relationship to predation and reproductive success in the Magellanic penguin at its southern-most continental colony. *Ornitologia Neotripical* 10:145–50.

Gandini, P., E. Frere, S. Ferrari, and M. Perroni. 2000. Magellanic penguin mortality in a gillnet fishery of Southern Patagonia, Argentina. 4th International Penguin Conference, Coquimbo, Chile. 22 pp.

Gandini, P., E. Frere, and T. Holik. 1992. Implicancias de las diferencias en el tamaño corporal entre colonias para el uso de medidas morfométricas como método de sexado en *Spheniscus magellanicus*. *Hornero* 13:211–14.

Gandini, P., E. Frere, A. Petovello, and P. Cedrola. 1999b. Interactions between seabirds and shrimp fishery at Golfo San Jorge. *Condor* 101:783–89.

Garcia Borboroglu, P. 1998. Distribución, abundancia y requerimientos del hábitat de nidificación del pingüino de Magallanes (*Spheniscus magellanicus*) en colonias del Chubut. Licenciatura thesis, Universidad Nacional de la Patagonia San Juan Bosco, Argentina.

———. 2010. Plan de Manejo de colonia de pinguinos de Magallanes del Pedral, Chubut. Elevado a la Dirección de Fauna y Flora de la Provincia del Chubut. 23 pp.

Garcia Borboroglu, P., P. D. Boersma, L. Reyes, and E. Skewgar. 2008. Petroleum pollution and penguins: Marine conservation tools to reduce the problem. In *Marine Pollution: New Research*, ed. T. N. Hofer, 339–56. New York: Nova Science Publishers.

Garcia Borboroglu, P., P. D. Boersma, V. Ruoppolo, R. Pinho-da-Silva-Filho, A. Corrado-Adornes, D. Conte-Sena, R. Velozo, C. Myiaji-Kolesnikovas, G. Dutra, P. Maracini, C. Carvalho-do-Nascimento, V. Ramos-Júnior, L. Barbosa, and S. Serra. 2010. Magellanic penguin mortality in 2008 along the SW Atlantic coast. *Marine Pollution Bulletin* 60:1652–57.

Garcia Borboroglu, P., P. D. Boersma, V. Ruoppolo, L. Reyes, G. A. Rebstock, K. Griot, S. Rodrigues Heredia, A. Corrado Adornes, and R. Pinho da Silva. 2006. Chronic oil pollution harms Magellanic penguins in the southwest Atlantic. *Marine Pollution Bulletin* 52:193–98.

Garcia Borboroglu, P., P. Yorio, P. D. Boersma, H. D. Del Valle, and M. Bertellotti. 2002. Habitat use and breeding distribution of Magellanic penguins in northern San Jorge Gulf, Patagonia, Argentina. *The Auk* 119(1):233–39.

Godoy, J. C. 1963. *Fauna Silvestre: Evaluación de los Recursos Naturales de la Argentina*. Book 8, vols. 1 and 2. Buenos Aires: Consejo Federal de Inversiones.

González Zevallos, D., and P. Yorio. 2006. Seabird use of discards and incidental captures at the Argentine hake trawl fishery in Golfo San Jorge, Argentina. *Marine Ecology Progress Series* 316:175–83.

González-Zevallos, D., P. Yorio, and G. Caille. 2007. Seabird mortality at trawler warp cables and a proposed mitigation measure: A case of study in Golfo San Jorge, Patagonia, Argentina. *Biological Conservation* 136:108–16.

González-Zevallos, D., P. Yorio, and W. S. Svagelj. 2011. Seabird attendance and incidental mortality at shrimp fisheries in Golfo San Jorge, Argentina. *Marine Ecology Progress Series* 432:125–35.

Hahn, I., U. Römer, P. Vergara, and H. Walter. 2009. Biogeography, diversity, and conservation of the birds of the Juan Fernández Islands, Chile. *Vertebrate Zoology* 59:103–14.

Hiriart-Bertrand, L., A. Simeone, R. Reyes-Arriagada, V. Riquelme, K. Pütz, and B. Lüthi. 2010. Descripción de una colonia mixta de pingüino de Humboldt (*Spheniscus humboldti*) y de Magallanes (*S. magellanicus*) en isla Metalqui, Chiloé, sur de Chile. *Boletín Chileno de Ornitología* 16(1):42–47.

Hood, L. C. 1996. Adrenocortical response to stress in incubating Magellanic penguins (*Spheniscus magellanicus*) and mate switching in Magellanic penguins at Punta Tombo, Argentina. MSc thesis, University of Washington.

Hood, L. C., P. D. Boersma, and J. C. Wingfield. 1998. The adrenocortical response to stress in incubating Magellanic penguins (*Spheniscus magellanicus*). *The Auk* 115:76–84.

Housse, R. 1936. Avifauna de la isla Santa María. *Revista Chilena de Historia Natural* 40:63–69.

Howarth, R. W. 2008. Coastal nitrogen pollution: A review of sources and trends globally and regionally. *Harmful Algae* 8:14–20.

IUCN (International Union for Conservation of Nature). 2009. IUCN Red List of Threatened Species. Version 2009.2. http://www.iucnredlist.org (accessed 11 January 2010).

Jahncke, J., D. M. Checkley Jr., and G. L. Hunt Jr. 2004. Trends in carbon flux to seabirds in the Peruvian upwelling system: Effects of wind and fisheries on population regulation. *Fisheries Oceanography* 13:208–23.

Jehl, J. 1980. Mortality of magellanic penguins in Argentina. *The Auk* 92:596–98.

Johnson, A. W. 1965. *The Birds of Chile and Adjacent Regions of Argentina, Bolivia and Peru.* Vol. 1. Buenos Aires: Platt Establecimientos Gráficos.

Kane, O. J., J. R. Smith, P. D. Boersma, N. J. Parson, V. Strauss, P. Garcia Borboroglu, and C. Villanueva. 2010. Feather-loss disorder in African and Magellanic penguins. *Waterbirds* 33:415–21.

Kane, O. J., M. M. Uhart, V. Rago, A. J. Pereda, J. R. Smith, A. Van Buren, J. A. Clark, and P. D. Boersma. 2012. Avian pox in Magellanic penguins (*Spheniscus magellanicus*). *Journal of Wildlife Disease* 48:790–94.

Kikkawa, F. F., T. T. Tsuda, T. K. Naruse, D. Sumiyama, M. Fukuda, M. Kurita, K. Murata, R. P. Wilson, Y. LeMaho, M. Tsuda, J. K. Kulski, and H. Kurita. 2005. Analysis of the sequence variations in the Mhc DRB1-like gene of the endangered Humboldt penguin (*Spheniscus humboldti*). *Immunogenetics* 57(1–2):99–107.

Kusch, A., M. Marín, D. Oheler, and S. Drieschman. 2007. Notas sobre la avifauna de isla Noir (54°28´S-73°00´W). *Anales Instituto de la Patagonia* (Chile) 35:61–66.

Lluch-Belda, D., R. A. Schwartzlose, R. Serra, R. Parrish, T. Kawasaki, D. Hedgecock, and R. J. M. Crawford. 1992. Sardine and anchovy regime fluctuations of abundance in four regions of the world oceans: A workshop report. *Fisheries Oceanography* 1:339–47.

Losano, P., and A. Tagliorette. 2009. Situación actual del turismo en las localidades de la costa patagónica. Puerto Madryn, Argentina: Fundación Patagonia Natural.

Marinao, M. J., and P. Yorio. 2011. Fishery discards and incidental mortality of seabirds attending coastal shrimp trawlers at Isla Escondida, Patagonia, Argentina. *The Wilson Journal of Ornithology* 123(4): 709–19.

Martinez, E., D. Antoine, F. D'Ortenzio, and B. Gentili. 2009. Climate-driven basin-scale decadal oscillations of oceanic phytoplankton. *Science* 326:1253–56.

Martínez, J. I. Z., A. Travaini, S. Zapata, D. Procopio, and M. Á. Santillán. 2012. The ecological role of native and introduced species in the diet of the puma *Puma concolor* in Southern Patagonia. *Oryx* 46:106–11.

Matus, R., and O. Blank. 2008. Habilitación de una estación de rehabilitación para la atención de pinguinos de Magallanes contaminados por Petróleo en Punta Arenas. 9th Congreso Chileno de Ornitología, El Tabo, 27–30 August 2008. *Boletín Chileno de Ornitología*, 22.

Miranda, M., J. Gibbons, J. Cárcamo, and Y. Vilina. 2009. Breeding habitat and estimation of population size of the Magellanic penguin (*Spheniscus magellanicus*) on Rupert Island, Francisco Coloane Marine Park, Strait of Magellan. *Anales Instituto Patagonia (Chile)* 37(1):103–11.

Moreno, J., J. Potti, P. Yorio, and P. Garcia Borboroglu. 2001. Sex differences in cell-mediated immunity in the Magellanic penguins *Spheniscus magellanicus*. *Annales Zoologici Fennici* 38:111–16.

Moreno, J., P. Yorio, P. Garcia Borboroglu, and J. Potti. 2002. Health state and reproductive output in Magellanic penguins (*Spheniscus magellanicus*). *Ethology, Ecology and Evolution* 14:19–28.

McEvey, A. R. 1980. Australian specimen of a Magellanic penguin. *Emu* 80:40.

Murphy, R. C. 1936. *Oceanic Birds of South America.* Vol. 1. New York: Macmillan.

Novaro, A. J., M. C. Funes, and R. S. Walker. 2005. An empirical test of source-sink dynamics induced by hunting. *Journal of Applied Ecology* 42:910–20.

Otley, H. M. 2005. Nature-based tourism: Experiences at the Volunteer Point penguin colony in the Falkland Islands. *Marine Ornithology* 33:181–87.

Otley, H. M., A. P. Clausen, D. J. Christie, and K. Pütz. 2004. Aspects of the breeding biology of the Magellanic penguin in the Falkland Islands. *Waterbirds* 27:396–405.

Pascual, M. A., J. L. Lancelotti, B. Ernst, J. E. Ciancio, E. Aedo, and M. Garcia-Asorey. 2009. Scale, connectivity, and incentives in the introduction and management of non-native species: The case of exotic salmonids in Patagonia. *Frontiers in Ecology and the Environment* 7:533–40.

Pennisi, E. 2010. How a little fish keeps overfished ecosystem productive. *Science* 329:268.

Perkins, J. S. 1984. Breeding ecology of Magellanic penguins *Spheniscus magellanicus* at Caleta Valdes, Argentina. *Cormorant* 12:3–13.

Peters, G., R. P. Wilson, J. A. Scolaro, S. Laurenti, J. Upton, and H. Galleli. 1998. The diving behavior of Magellanic penguins at Punta Norte, Peninsula Valdés, Argentina. *Colonial Waterbirds.* 21:1–10.

Petry, M. V., and V. S. da S. Fonseca. 2002. Effects of human activities on the marine environment on seabirds along the coast of Rio Grande do Sul, Brazil. *Ornitología Neotropical* 13:137–42.

Philippi, R. A. 1937. Aves de la Región de Zapallar. *Revista Chilena de Historia Natural* 41:28–38.

Pikitch, E., P. D. Boersma, I. L. Boyd, D. O. Conover, P. Cury, T. Essington, S. S. Heppell, E. D. Houde, M. Mangel, D. Pauly, É. Plagányi, K. Sainsbury, and R. S. Steneck. 2012. *Little Fish, Big Impact: Managing a Crucial Link in Ocean Food Webs.* Washington, DC: Lenfest Ocean Program. 108 pp.

Pisano, E. 1971. Estudio ecológico preliminar del parque nacional "Los pingüinos" (Estrecho de Magallanes). *Anales Instituto de la Patagonia* 2:76–92.

Pistorius, P. 2009. Falkland Island Seabird Monitoring Programme annual report 2008/2009. Falklands Conservation, Stanley. 48 pp.

Potti, J., J. Moreno, P. Yorio, V. Briones, P. Garcia Borboroglu, S. Villar, and C. Ballesteros. 2002. Bacteria divert resources from growth for Magellanic penguin chicks. *Ecology Letters* 5:709–14.

Prince, P. A., and M. R. Payne. 1979. Current status of birds at South Georgia. *British Antarctic Survey Bulletin* 48:103–18.

Punta Tombo Management Plan. 2006. Subsecretary of Tourism of Chubut Province, Argentina. 72 pp.

Pütz, K., R. J. Ingham, and J. G. Smith. 2000. Satellite tracking of the winter migration of Magellanic penguins *Spheniscus magellanicus* breeding in the Falkland Islands. *Ibis* 142:614–22.

———. 2002. Foraging movements of Magellanic penguins *Spheniscus magellanicus* during the breeding season in the Falkland Islands. *Aquatic Conservation: Marine and Freshwater Ecosystems* 12:75–87.

Pütz, K., R. J. Ingham, J. G. Smith, and J. P. Croxall. 2001. Population trends, breeding success and diet composition of gentoo (*Pygoscelis papua*), Magellanic (*Spheniscus magellanicus*), and rockhopper (*Eudyptes chrysocome*) penguins in the Falkland Islands. *Polar Biology* 24:793–807.

Pütz, K., L. Hiriart-Bertrand, A. Simeone, V. Riquelme, R. Reyes-Arriagada, and B. Lüthi. 2011. Entanglement and drowning of a Magellanic penguin (*Spheniscus magellanicus*) in a gill net recorded by a time-depth recorder in southern Chile. *Waterbirds* 34(1):121–25.

Pütz, K., A. Schiavini, A. Raya Rey, and B. Lüthi. 2007. Winter migration of Magellanic penguins (*Spheniscus magellanicus*) from the southernmost distributional range. *Marine Biology* 152:1227–35.

Radl, A., and B. M. Culik. 1999. Foraging behavior and reproductive success in Magellanic penguins (*Spheniscus magellanicus*): A comparative study of two colonies in southern Chile. *Marine Biology* 133:381–93.

Rafferty, N. E., P. D. Boersma, and G. A. Rebstock. 2005. Intraclutch egg-size variation in Magellanic penguins. *Condor* 107:921–26.

Raya Rey, A., C. A. Bost, A. Schiavini, and K. Pütz. 2010. Foraging movements of Magellanic penguins *Spheniscus magellanicus* in the Beagle Channel, Argentina, related to tide and tidal currents. *Journal of Ornithology* 151:933–43.

Rebstock, G. A., and P. D. Boersma. 2011. Parental behavior controls incubation period and asynchrony of hatching in Magellanic penguins. *Condor* 113:316–25.

Renison, D., P. D. Boersma, and M. Martella. 2002. Winning and losing: Causes for variability in outcome of fights in male Magellanic penguins (*Spheniscus magellanicus*). *Behavioral Ecology* 13:462–66.

———. 2003. Fighting in female Magellanic penguins: When, why and who wins? *Wilson Bulletin* 115:58–63.

Renison, D., P. D. Boersma, A. N. Van Buren, and M. Martella. 2006. Agonistic interactions in wild male Magellanic penguins: When and how do they interact? *Journal of Ethology* 24(2):189–93.

Reyes-Arriagada, R., P. Campos-Ellwanger, and R. P. Schlatter. 2009. Avifauna de isla Guafo. *Boletín Chileno de Ornitología (Chile)* 15:35–43.

Reynolds, P. 1935. Notes on the birds of Cape Horn. *Ibis* 5:65–101.

Robertson, C. J. R., R. S. Abel, and F. C. Kinsky. 1972. First New Zealand record of the Magellanic penguins (*Spheniscus magellanicus*). *Notornis* 19:111–13.

Sæther, B. E., and Ø. Bakke. 2000. Avian life history variation and contribution of demographic traits to the population growth rate. *Ecology* 81:642–53.

Sánchez, R. P., and J. D. de Ciechomski. 1995. Spawning and nursery grounds of pelagic fish species in the sea-shelf off Argentina and adjacent areas. *Scientia Marina* 59:455–78.

Schiavini, A. C. M., and A. Raya Rey. 2001. Aves y mamíferos marinos en Tierra del Fuego. Estado de situación, interacción con actividades humanas y recomendaciones para su manejo. Fundación Patagonia Natural, Puerto Madryn.

Schiavini, A., and P. Yorio. 1995. Distribution and abundance of seabird colonies in the Argentine sector of the Beagle Channel, Tierra del Fuego. *Marine Ornithology* 23:39–46.

Schiavini, A., P. Yorio, and E. Frere. 1998. Distribución reproductiva y abundancia de las aves marinas de la Isla Grande de Tierra del Fuego, Isla de los Estados e Islas de Año Nuevo (Provincia de Tierra del Fuego, Antártida e Islas del Atlántico Sur). In *Atlas de la distribución reproductiva de aves marinas en el litoral Patagónico Argentino*, ed. P. Yorio, E. Frere, P. Gandini, and G. Harris, 179–221. Buenos Aires, Argentina: Instituto Salesiano de Artes Gráficas.

Schiavini, A. P., P. Yorio, P. Gandini, A. Rey, and P. D. Boersma. 2005. Los pinguinos de las costas argentinas: Estado poblacional y conservacion. *Hornero* 20:5–23.

Schlatter, R. P. 1984. The status and conservation of seabirds in Chile. In *Status and Conservation of the World's Seabirds*, ed. J. P. Croxall, P. G. H. Evans, and R. W. Schreiber, 261–69. ICBP Technical Publication No. 2. Cambridge: International Council for Bird Preservation.

Schlatter, R. P., E. Paredes, J. Ulloa, J. Harris, A. Romero, J. Vásquez, A. Lizama, C. Hernández, and A. Simeone. 2009. Mortandad de pingüino de Magallanes (*Spheniscus magellanicus*) en Queule, Región de la Araucanía, Chile. *Boletín Chileno de Ornitología* 15:78–86.

Schlatter, R., and G. Riveros. 1997. Historia Natural del Archipiélago Diego Ramírez, Chile. *Serie Científica INACH (Chile)* 47:87–112.

Scioscia, G. 2011. Ecología trófica del Pingüino de Magallanes (*Spheniscus magellanicus*) y sus implicancias en la ecología reproductiva en el Canal Beagle, Tierra del Fuego. PhD diss., Universidad de Mar del Plata, Argentina.

Scioscia, G., A. Raya Rey, M. Favero, and A. Schiavini. 2009. Patrón de asistencia a la colonia en el pingüino de Magallanes (*Spheniscus magellanicus*) de Isla Martillo, Canal Beagle, Tierra del Fuego: Implicancias para minimizar el disturbio humano. *Ornitología Neotropical* 20: 27–34.

———. 2010. Factores que afectan el éxito reproductivo y la calidad de la nidada en el pingüino de Magallanes (*Spheniscus magellanicus*) en el Canal Beagle, Tierra del Fuego, Argentina. *El Hornero* 25:17–25.

Scolaro, J. A. 1978. El pingüino de Magallanes (*Spheniscus magellanicus*). 4: Notas biológicas y de comportamiento. *Publicaciones Ocasionales del Instituto de Biología Animal, Serie científica* 10:1–6.

———. Revisión sobre biología de la reproducción del pingüino de Magallanes (*Spheniscus magellanicus*): El ciclo biológico anual. *Contribuciones Centro Nacional Patagónico* 91:1–26.

Scolaro, J. A., M. A. Hall, and I. M. Ximénez. 1983. The Magellanic penguin (*Spheniscus magellanicus*): Sexing adults by discriminant analysis of morphometric characters. *The Auk* 100:221–24.

Scolaro, J. A., and O. Kovacs. 1978. El pingüino de Magallanes (*Spheniscus magellanicus*) 3: Nota sobre una nueva colonia de reproducción. *Publicaciones Ocasionales del Instituto de Biología Animal, Serie científica* (Mendoza, Argentina) 19:10–16.

Scolaro, J. A., R. P. Wilson, S. Laurenti, M. A. Kierspel, H. Gallelli, and J. A. Upton. 1999. Feeding preferences of the Magellanic penguin *Spheniscus magellanicus* over its breeding range in Argentina. *Waterbirds* 22:104–10.

Shumway, S. E., S. M. Allen, and P. D. Boersma. 2003. Marine birds and harmful algal blooms: Sporadic victims or under-reported events? *Harmful Algae* 2:1–17.

Simeone, A. 2005. Evaluación de la población reproductiva del pingüino de Magallanes y del pingüino de Humboldt en los islotes Puñihuil, Chiloé. Informe final. Estudio financiado por la Fundación Otway (Chile) y Zoo Landau in der Pfalz (Alemania). Viña del Mar. 49 pp.

Simeone, A., and M. Bernal. 2000. Effects of habitat modification on breeding seabirds: A case study in central Chile. *Waterbirds* 23:449–56.

Simeone, A., M. Bernal, and J. Meza. 1999. Incidental mortality of Humboldt penguins *Spheniscus humboldti* in gill nets, central Chile. *Marine Ornithology* 27:157–61.

Simeone, A., L. Hiriart-Bertrand, R. Reyes-Arriagada, M. Halpern, J. Dubach, R. Wallace, K. Pütz, and B. Lüthi. 2009. Heterospecific pairing and hybridization between wild Humboldt and Magellanic penguins in southern Chile. *Condor* 111:544–50.

Simeone, A., and R. P. Wilson. 2003. In-depth studies of Magellanic penguin (*Spheniscus magellanicus*) foraging: Can we estimate prey consumption by perturbations in the dive profile? *Marine Biology* 143:825–31.

Skewgar, E., P. D. Boersma, G. Harris, and G. Caille. 2007. Anchovy fishery threat to Patagonian ecosystem. *Science* 315:45.

Skewgar, E., A. Simeone, and P. D. Boersma. 2009. Marine reserve in Chile would benefit penguins and ecotourism. *Ocean & Coastal Management* 52:487–91.

Soria-Galvarro, Y. 1991. Pingüineras de Ahuenco y Puñihuil (Parque Nacional Chiloé), un ejemplo sobre la importancia de crear reservas marinas en Chile. *Resúmenes I Congreso Chileno de Ornitología* 21.

Soto, N. 1990. Proyecto de protección y manejo de las colonias de pingüinos presentes en Isla Rupert e Isla Recalada, Reserva Nacional Alacalufes. Informe de temporada 1989–1990. CONAF (Corporacion Nacional Forestal de Chile). 29 pp.

Stokes, D. L., and P. D. Boersma. 1991. Effects of substrate on the distribution of Magellanic penguins (*Spheniscus magellanicus*). *The Auk* 108:923–33.

———. 1998. Nest-site characteristics and reproductive success in Magellanic penguins (*Spheniscus magellanicus*). *The Auk* 115:34–49.

———. 2000. Nesting density and reproductive success in a colonial seabird, the Magellanic penguin. *Ecology* 81:2878–891.

Stokes, D. L., P. D. Boersma, and L. S. Davis. 1998. Satellite tracking of Magellanic penguin (*Spheniscus magellanicus*) migration. *Condor* 100:376–81.

Strange, I. J. 1992. *A field guide to the wildlife of the Falkland Islands and South Georgia*. London: HarperCollins.

Tamini, L. L., J. E. Perez, G. E. Chiaramonte, and H. L. Cappozzo. 2002. Magellanic penguins *Spheniscus magellanicus* and fish as bycatch in the cornalito *Sorgentinia incisa* fishery at Puerto Quequén, Argentina. *Atlantic Seabirds* 4:109–14.

Téllez, L. 2007. Monitoreo de la población de pingüino de Magallanes, *Spheniscus magellanicus*, del Seno Otway, temporada 2006–2007. Informe Turis Otway Limitada, Punta Arenas.

Thompson, K. R. 1993. Variation in Magellanic penguin *Spheniscus magellanicus* diet in the Falkland Islands. *Marine Ornithology* 21:57–67.

Venegas, C. 1978. Pingüinos de barbijo (*Pygoscelis antarctica*) y macaroni (*Eudyptes chrysolophus*) en Magallanes. *Anales Instituto de la Patagonia (Chile)* 9:180–83.

———. 1981. Aves de las islas Wollaston y Bayly, archipiélago del Cabo de Hornos. *Anales Instituto de la Patagonia (Chile)* 12:213–19.

———. 1999. Estado de conservación de las especies de pingüinos en la región de Magallanes, Chile. *Estudios Oceanológicos (Chile)* 18:45–56.

Villanueva, C., B. Walker, and M. Bertellotti. 2012. A matter of history: Effects of tourism on physiology, behaviour and breeding parameters in Magellanic penguins (*Spheniscus magellanicus*) at two colonies in Argentina. *Journal of Ornithology* 153:219–28.

Wagner, E. L., and P. D. Boersma. 2011. Effects of fisheries on seabird community ecology. *Reviews in Fisheries Sciences* 19:157–67.

Wagner, E. L., E. J. Lee, and P. D. Boersma. In revision. Patterns of acceptance of artificial eggs and chicks by Magellanic penguins (*Spheniscus magellanicus*). *Journal of Ornithology*. DOI: 10.1007/s 10336-012-0875-6.

Walker, B. G., and P. D. Boersma. 2003. Diving behavior of Magellanic penguins (*Spheniscus magellanicus*) at Punta Tombo, Argentina. *Canadian Journal of Zoology* 81:1471–83.

Walker, B. G., P. D. Boersma, and J. C. Wingfield. 2005a. Age and food deprivation affects expression of the glucocorticosteroid stress response in Magellanic penguin (*Spheniscus magellanicus*) chicks. *Physiological and Biochemical Zoology* 78:78–89.

———. 2005b. Physiological and behavioral differences in Magellanic penguin chicks in undisturbed and tourist-visited locations of a colony. *Conservation Biology* 19(5):1571–1577.

———. 2006. Habituation of adult Magellanic penguins to human visitation as expressed through behavior and corticosterone secretion. *Conservation Biology* 20:146–54.

Williams, T. D. 1995. *The Penguins*: Spheniscidae. Oxford: Oxford University Press.

Wilson, R. P. 2003. Penguins predict their performance. *Marine Ecology Progress Series* 249:305–10.

Wilson, R. P., D. C. Duffy, M. P. Wilson, and B. Araya. 1995. Aspects of the ecology of species replacement in Humboldt and Magellanic penguins in Chile. *Le Gerfaut* 85:49–61.

Wilson, R. P., S. Jackson, and M. Thor Straten. 2007. Rates of food consumption in free-living Magellanic penguins *Spheniscus magellanicus*. *Marine Ornithology* 35:109–11.

Wilson, R. P., P. G. Ryan, A. James, and M. P. T. Wilson. 1987. Conspicuous coloration may enhance prey capture in some piscivores. *Animal Behaviour* 35:1558–60.

Wilson, R. P., J. A. Scolaro, D. Grémillet, M. A. M. Kierspel, S. Laurenti, J. Upton, H. Gallelli, F. Quintana, E. Frere, G. Müller, M. T. Straten, and I. Zimmer. 2005. How do Magellanic penguins cope with variability in their access to prey? *Ecological Monographs* 75:379–401.

Wilson, R. P., E. L. C. Shepard, A. Gómez Laich, E. Frere, and F. Quintana. 2010. Pedaling downhill and freewheeling up: A penguin perspective on foraging. *Aquatic Biology* 8:193–202.

Wilson, R. P., and I. Zimmer. 2004. Inspiration by Magellanic penguins: Reduced swimming effort when under pressure. *Marine Ecology Progress Series* 278:303–7.

Woods, R. W., and A. Woods. 1997. *Atlas of Breeding Birds of the Falkland Islands.* Shropshire, United Kingdom: Anthony Nelson.

Yorio, P. 2009. Marine protected areas, spatial scales and governance: Implications for the conservation of breeding seabirds. *Conservation Letters* 2:171–78.

Yorio, P., M. Bertellotti, P. Garcia Borboroglu, A. Carribero, M. Giaccardi, M. E. Lizurume, P. D. Boersma, and F. Quintana. 1998b. Distribución reproductiva y abundancia de las aves marinas de Chubut. Parte 1: de Península Valdés a Islas Blancas. In *Atlas de la distribución reproductiva de aves marinas en el litoral Patagónico Argentino*, ed. P. Yorio, E. Frere, P. Gandini, and G. Harris, 39–73. Buenos Aires, Argentina: Instituto Salesiano de Artes Gráficas.

Yorio, P., and P. D. Boersma. 1992. The effects of human disturbance on Magellanic penguin behavior and breeding success. *Bird Conservation International* 2: 161–73.

———. 1994a. Causes of nest desertion during incubation in the Magellanic penguin (*Spheniscus Magellanicus*). *Condor* 96:1076–83.

———. 1994b. Consequences of nest desertion and inattendance for Magellanic penguin hatching success. *The Auk* 111:215–18.

Yorio, P., and G. Caille. 1999. Seabird interactions with coastal fisheries in northern Patagonia: Use of discards and incidental captures in nets. *Waterbirds* 22:207–16.

Yorio, P., E. Frere, P. Gandini, and G. Harris, eds. 1998a. *Atlas de la distribución reproductiva de aves marinas en el litoral Patagónico Argentino. Plan de Manejo Integrado de la Zona Costera Patagónica*. Fundación Patagonia Natural y Wildlife Conservation Society. Buenos Aires, Argentina: Instituto Salesiano de Artes Gráficas.

Yorio, P., E. Frere, P. Gandini, and A. Schiavini. 2001a. Tourism and recreation at seabird breeding sites in Patagonia, Argentina: Current concerns and future prospects. *Bird Conservation International* 11:231–45.

Yorio, P., P. Garcia Borboroglu, M. Bertellotti, M. E. Lizurume, M. Giaccardi, G. Punta, J. Saravia, G. Herrera, S. Sollazzo, and P. D. Boersma. 1998c. Distribución reproductiva y abundancia de las aves marinas de Chubut. Parte 2: Norte del Golfo San Jorge, de Cabo Dos Bahías a Comodoro Rivadavia. In *Atlas de la distribución reproductiva de aves marinas en el litoral Patagónico Argentino*, ed. P. Yorio, E. Frere, P. Gandini, and G. Harris, 76–117. Buenos Aires: Instituto Salesiano de Artes Gráficas.

Yorio, P., P. Garcia Borboroglu, J. Potti and J. Moreno. 2001b. Breeding biology of Magellanic penguins *Spheniscus magellanicus* at Golfo San Jorge, Patagonia, Argentina. *Marine Ornithology* 29:75–79.

Yorio, P., F. Quintana, P. Dell'Arciprete, and D. González Zevallos. 2010. Spatial overlap between foraging seabirds and trawl fisheries: Implications for the effectiveness of a marine protected area at Golfo San Jorge, Argentina. *Bird Conservation International* 20:320–34.

Zavalaga, C., and R. Paredes. 2009. Records of Magellanic penguins *Spheniscus magellanicus* in Peru. *Marine Ornithology* 37:281–82.

15

Humboldt Penguin

(Spheniscus humboldti)

SANTIAGO DE LA PUENTE, ALONSO BUSSALLEU, MARCO CARDEÑA,
ARMANDO VALDÉS-VELÁSQUEZ, PATRICIA MAJLUF, AND ALEJANDRO SIMEONE

1. SPECIES (COMMON AND SCIENTIFIC NAMES)

Humboldt penguin, *Spheniscus humboldti* (Meyen, 1834)

The Humboldt penguin is also known as the Peruvian penguin and *pingüino de Humboldt, pájaro bobo, pájaro niño, pachanca,* and *patranca* (Spanish).

2. DESCRIPTION OF THE SPECIES

The Humboldt is a medium-size penguin, with body length of 67–72 centimeters and body mass of approximately 4.2–5.0 kilograms, depending on breeding condition, sex, and prey availability (Coker 1919; Murphy 1936; Zavalaga and Paredes 1997a) (table 15.1). Although similar to the other *Spheniscus* penguins, Humboldts have large fleshy margins at the base of the bill (UNEP 2003).

Sexes are similar in plumage, but males tend to be larger than females (Zavalaga and Paredes 1997a; Wallace et al. 2008) (table 15.2). The head is mostly blackish gray, with a white chin and a narrow white stripe extending from the bill on each side of the crown, looping over the eyes and broadening at the junction with the white upper breast. The upperparts, flippers, legs, feet, and tail tend to be blackish gray, while most of the underparts are white (fig. 2). Adults have an inverted black horseshoe-shaped band extending down the flank to the thigh, which is lacking on juveniles and chicks. Black feathers on the breast appear as spots and are individually distinct. The Humboldt's bill is black with a gray

FIG. 1 (*FACING PAGE*) On Isla Chañaral, Chile, a Humboldt penguin stands in the desert vegetation. (T. Mattern)

FIG. 2 Adult male Humboldt penguin at Punta San Juan, Peru. (P. D. Boersma)

transverse bar and a fleshy pink area at the base that is especially prominent during the breeding season. The irises of chicks and juveniles are dark gray, becoming pale and then darkening with age, turning reddish brown as they become adults (Scholten 1999). Juveniles are easily distinguished from adults by the brownish plumage on their heads, grayer cheeks, and lack of the white head stripe and black horseshoe-shaped breast-band (fig. 3).

Gender determination of free-ranging penguins using discriminant functions of morphometric characters is accurate (88–95%) but is hindered by geographic

FIG. 3 Adult Humboldt penguins and a juvenile Humboldt penguin (lower right) rest on a guano island in Peru. (P. D. Boersma)

variation and effects such as abnormal bill growth associated with captivity (Zavalaga and Paredes 1997a; Wallace et al. 2008). The main characteristics used for sex determination were bill length (BL), measured from the point in the V where feathers start to the tip of the culmen, which is hooked on the upper mandible; bill depth (BD), measured dorso-ventrally at the nares or nostrils; and the width of the head (WH), measured from the crevice just posterior to the bulge behind the eyes.

3. TAXONOMIC STATUS

The genus *Spheniscus* has four closely related species (Baker et al. 2006). An ancestral population of *Spheniscus* on the Pacific coast may have split, forming the Humboldt and the Galápagos penguins (*Spheniscus mendiculus*). This dispersal event is consistent with the flow of the Humboldt Current. Consensus on the evolutionary history and dispersal of the Spheniscidae across the globe has not been reached (Bertelli and Giannini 2005; Baker et al. 2006; Ksepka et al. 2006).

The Humboldt penguin is monotypic (UNEP 2003), although its wide latitudinal distribution (5–42° S) should expose the species to different selective pressures over its breeding range (Simeone et al. 2002). Hybridization between Humboldt and Magellanic (*S. magellanicus*) penguins in mixed colonies occurs in southern Chile (Simeone et al. 2009).

Although there is evidence of strong colony and nest fidelity from field observations (Teare et al. 1998), penguins are also capable of dispersion (Culik and Luna-Jorquera 1997b; Wallace et al. 1999; Taylor et al. 2004).

Satellite tracking of individuals in periods of scarce prey availability showed that penguins moved more than 600 kilometers away from their home colonies (Culik and Luna-Jorquera 1997a). Genetic evidence suggests that sharp episodic fluctuations in climate, such as El Niño events, or human pressure can change the relationship between breeding locations and demography, making it impossible to view colonies as separate entities (Schlosser et al. 2009). This connectivity is perhaps the main reason that microsatellite data show no evidence of bottlenecks or strong population fragmentation, although the number of Humboldt penguins has decreased dramatically over the past two centuries (Schlosser et al. 2009).

4. RANGE AND DISTRIBUTION

At least 60 colonies have been described between Foca (5°12′ S) and Metalqui (42°12′ S) (41 in Peru and 19 in Chile) (table 15.3, fig. 4). Most of the breeding population is located in Chile, but recent estimates suggest an increase in the overall population (McGill et al., unpubl. data). Great uncertainty resulting from inaccurate counting methods and fragmented and uncoordinated counts throughout the Humboldt's distribution make a population estimate problematic.

In Peru, most of the population was concentrated in five sites in 1999–2000 (Paredes et al. 2003) (table 15.3a), and in 2003–4 (Majluf, unpubl. data). The 1997–98 El Niño, with high surface water temperatures that altered prey distribution, and increased human presence and activities in the central coastal areas of Peru may have caused the penguins' clustered distribution (Paredes et

FIG. 4 Distribution and abundance of the Humboldt penguin, with counts based on individuals.

al. 2003). In Chile, the most important breeding colony, with about 22,000 penguins, is at Isla Chañaral (table 15.3b) (Mattern et al. 2004).

Since the early 1960s, colonies between 05–15° south are stable, with 1,000 birds. At the best-studied penguin-breeding colony in Peru, Punta San Juan (15°22′ S), El Niño events were followed by drastic reductions in Humboldt populations (Majluf et al. 2002); penguins slowly recovered (table 15.3). At present, Punta San Juan is the largest colony in Peru, with a total of 4,000 adult birds (fig. 5) (Cardeña and Bussalleu, unpubl. data).

In Chile (25–35° S), declines following El Niño events were not as great as in Peru, and populations likewise recovered. The increased number of penguins in the time series in Chile is likely due not to more penguins but to a change in census methodology that allowed greater accuracy in describing the Humboldt population at Isla Chañaral (29°01′ S) (Mattern et al. 2004). The southernmost colonies (35–45° S) are small and, since the 1980s, stable.

Although restricted to Peru and Chile, Humboldt penguins have been found in the wild in the Northern Hemisphere on several occasions. One was caught by a herring fisher in Alaska and released, but it likely had been transported by boat (Van Buren and Boersma 2007).

5. SUMMARY OF POPULATION TRENDS

Humboldts have declined since the 19th century (Murphy 1936), with a marked reduction after the 1982–83 El Niño event, leaving no more than 10,000 penguins

(Duffy et al. 1984; Hays 1986; Araya and Todd 1987). A decade later, the population was estimated at 10,000–13,000 (Boersma et al. 1992).

Increases in numbers continued in the 1990s but were regarded skeptically by some researchers. Paredes et al. (2003) suggest that increases in Peruvian counts were masking a process of population clustering, leaving the species vulnerable to catastrophes (78% of the Peruvian population was found in San Juan, San Juanito, Hornillos, Pachacamac, and Tres Puertas).

In Chile, numbers also increased (Simeone et al. 2003). However, the confirmation of 22,000 penguins in Isla Chañaral in 2003 was a surprise (Mattern et al. 2004), as it was more than the total estimated population. This number was the product of inconsistent census methods rather than a real population increase.

Later estimates range between 30,000 to 40,000 breeding birds (Boersma 2008). Other recent estimates describe a total population of 48,000 individuals (Schlosser et al. 2009); however, this number is based on an expert's opinion and not on field data, highlighting the absence, obsolescence, and fragmented nature of accurate census data and the difficulty of counting this species along the coast.

The Isla Chañaral counts stress that estimated population size is closely related to the census methodology used (Mattern et al. 2004). An efficient method for counting these birds is a pressing need (Araya et al. 2000; Luna-Jorquera et al. 2000; UNEP 2003; Mattern et al. 2004; BirdLife International 2008).

Recent improvements in census methodologies have

FIG. 5 Humboldt penguins at Punta San Juan, Peru, rest on the coast. (P. D. Boersma)

led to more accurate counts (Mattern et al. 2004) but may not indicate increases in the global population (BirdLife International, 2008). After all, the magnitude of possible positive population trends in recent years will be masked by decades of underestimations (Mattern et al. 2004).

In Punta San Juan since the early 1990s, the penguins have been counted twice during the breeding season and once during the molt. The largest count is during the molt and at the peak of breeding in May (Cardeña et al., unpubl. data). For large-scale counts across the Humboldt's entire distribution, Luna-Jorquera et al. (2000) suggests counting penguins on the beach just above where the waves are breaking between 11:00 AM and 4:00 PM.

Ideally, as Boersma (2008) suggests, each penguin colony should be visited at least annually.

6. IUCN STATUS

The International Union for Conservation of Nature lists the Humboldt penguin as Vulnerable on its Red List of Threatened Species because the species has undergone extreme population size fluctuations and decline (IUCN 2009). Local estimates and expert opinions suggest that the Humboldt is slowly bouncing back. The status of Vulnerable is justified, as the population has not reached historical numbers of "hundreds of thousands" (Murphy 1936), it is clustered (Paredes et al. 2003; Simeone et al. 2003), and major threats to the species (e.g., overfishing, entanglement in gill nets, habitat destruction) have not been ameliorated.

Nevertheless, actions have been taken toward increasing the protection of Humboldt breeding colonies in Peru through the Decreto Supremo 024–2009-MINAM and in Chile (Simeone et al. 2003; UNEP 2003) and reducing the impact of guano-harvesting methods at the penguins' largest colony in Peru at Punta San Juan (Majluf, unpubl. data). Both countries have implemented the Convention on International Trade in Endangered Species of Wild Flora and Fauna (CITES) under national law (Paredes et al. 2003; Iriarte 1999).

The Humboldt penguin is legally protected throughout its distribution (table 15.4). Peruvian legislation, Decreto Supremo 013–99-AG, prohibits the hunting,

possession, capture, transportation, and export of the birds for commercial purposes and categorizes the species as Endangered in Decreto Supremo 034–2004-AG. Chile implemented a 30-year hunting ban, Decreto Supremo 225, in 1995, and since 1967, Decreto Supremo 506 forbids hunting, transport, possession, and commercialization of penguins (Iriarte 1999). Chile adopted a new species-categorization system for conservation, granting legal status to international categories but interpreting them locally; thus, Decreto Supremo 50/2008 ranks the Humboldt penguin as "vulnerable."

Most of the penguins breed within protected areas in Peru and Chile: Pingüino de Humboldt National Reserve, Isla Cachagua Natural Monument, Islotes de Puñihuil National Monument, Guano System National Reserve (officially established in January 2010), the Paracas National Reserve, and the San Fernando National Reserve (UNEP 2003; SPIJ 2011).

7. NATURAL HISTORY

BREEDING BIOLOGY. The breeding period starts right after the molt (Zavalaga and Paredes 1997b). Nest selection and occupation begins when penguins return to their breeding colonies after an extended period of foraging and recovery at sea (Zavalaga and Paredes 1997b; Simeone et al. 2002).

Humboldt penguins nest on cliff tops, beaches, and scrapes covered by vegetation and in sea caves, rock crevices (fig. 7), and burrows dug into dirt or guano (Simeone and Schlatter 1998; Battistini and Paredes 1999; Simeone and Bernal 2000; Paredes and Zavalaga 2001).

Surface nests are common in areas without terrestrial predators and with little human disturbance (fig. 6) (Paredes and Zavalaga 2001). In Punta San Juan, pen-

FIG. 6 Humboldt penguins breeding on Isla Chañaral, Chile. (T. Mattern)

FIG. 7 Adult Humboldt penguin on a rock crevice with two eggs at Algarrobo Island, Chile. (P. D. Boersma).

guins in surface nests endure persistent cool winds, sea breezes, and shade through the year, as these counteract the high solar radiation on hot days. Penguins at Pájaro Niño nest in dirt burrows and rock crevices that offer protection from solar radiation and egg predators (Simeone and Bernal 2000).

Humboldts breed throughout the year when resources are abundant (Battistini 1998; Paredes and Zavalaga 2001; Paredes et al. 2002; Simeone et al. 2002). In Punta San Juan, reproduction occurs from March to December (Paredes and Zavalaga 2001), with two prominent reproductive peaks in April–May and August–September (Paredes and Zavalaga 2001; Cardeña et al., unpubl. data). In central Chile, there are also two reproductive peaks, but offset by a month (May, October), and the number of nests depends on prey availability and climatic factors such as rainfall (Simeone et al. 2002).

Females lay two eggs 4 days apart and of similar size (see table 15.5). Incubation lasts for six weeks (approx. 40–42 days). Hatching, like egg laying, is 2–4 days apart (Deeming et al. 1991; Paredes et al. 2002).

Based on recovery of banded birds at Pajaro Niño in central Chile, age at first breeding ranges from 3.6 to

FIG. 8 Humboldt penguin chick waits at its nest for its parents to return and is nearly ready to fledge. (T. Mattern)

FIG. 9 Adult Humboldt penguin in a nest site with two chicks marked with fiber bands at Punta San Juan, Peru. (P. D. Boersma)

6.1 years (mean = 5 ± 1 year, n = 7 [A. Simeone, unpubl. data]). Araya et al. (2000) suggest that first reproduction takes place 3 or 4 years after hatching, as fledglings are not sexually mature. Humboldt penguins on Pajaro Niño tend to breed in nests within 10–80 meters of their natal nests, and most use the same nest type as their natal nests when they first breed (A. Simeone, unpubl. data).

Chick rearing takes 10–12 weeks (approx. 75 days) until fledging (Paredes et al. 2002). The whole process from incubation to fledging lasts for approximately 121 days (Paredes and Zavalaga 2001). Both parents take turns foraging at sea and feeding the chicks (fig. 8) (Luna-Jorquera and Culik 1999; Taylor et al. 2002; Hennicke and Culik 2005).

In southern Peru, double brooding is common (approx. 73% [Paredes et al. 2002]) (fig. 9). In northern-central Chile (32–36° S), parents rarely raise both chicks (Simeone et al. 2002). Breeding success (fledglings per nest) varies within and among years and sites. At Punta San Juan, however, breeding success averaged around one chick: 0.78 ± 0.04 in 1993 (n = 535) and 1.05 ± 0.04 in 1996 (n = 507) (Paredes and Zavalaga 2001).

Although this species is monogamous, extra-pair mating and copulation occurred but failed to result in young (Schwartz et al. 1999).

FORAGING BEHAVIOR AND PREY. Humboldts are pelagic predators (Luna-Jorquera and Culik 1999) and probably visual hunters (Martin and Young 1984). They may use smell to recognize areas of high productivity and conspecifics (Coffin et al. 2012). Nonetheless, their foraging rhythm depends on light intensity, and dive depths in excess of 30 meters are uncommon (Taylor et al. 2002).

Humboldts feed close to their breeding colonies during reproduction, within a radius of about 35 kilometers (Culik et al. 1998; Luna-Jorquera and Culik 1999; Boersma et al 2007). As for other penguins, their foraging distance is estimated by how long they are gone (Boersma et al. 2007). They follow pelagic shoals or are found foraging in loose aggregations, even engaging in solitary foraging, which is rare among Peruvian seabirds (Duffy 1983). Their main target species vary depending on location and prey availability (Wilson et al. 1989; Herling et al. 2005). Frequent prey include anchovies, silversides, and jack mackerel, among others (table 15.6). Cephalopods such as Patagonian squid (*Loligo gahi*), Southern Ocean squid (*Todarodes filippovae*), and Humboldt squid (*Dosidiscus gigas*), and crustaceans (isopods and stomatopods) have also been described as parts of their diet (Wilson et al. 1989; Herling et al. 2005).

The Humboldt's foraging effort is directly proportional to foraging distance traveled and average dive depth. Energy investment during foraging varies with food scarcity. During El Niño events, energy investment doubles to 4300 kilojoules spent per day, compared to 2800 kilojoules spent per day before the El Niño event (Culik et al. 2000). Penguins must eat at least 340–600 grams of anchovies per day just to cope with the energy costs of foraging (Luna-Jorquera and Culik 2000). As cephalopods have a lower energy content than fish, the greater the proportion of cephalopods in the penguins' diet, the lower the energy per stomach load (Herling et al. 2005).

PREDATORS. Little is known about predators of the Humboldt. Williams (1995) states that killer whales (*Orcinus orca*), great white sharks (*Charcarodon charcarias*), and South American fur seals (*Arctocephalus australis*) prey on Humboldt penguins at sea. The South American sea lion (*Otaria flavescens*) occasionally kills penguins at sea and on the shore (Marco Cardeña, pers. obs.). Desert foxes (*Lycalopex sechurae*) are the main predators of adults, juveniles, and chicks on land (Williams 1995). Without the walls that exclude foxes from Punta San Juan, few penguins would survive; for example, a single desert fox killed more than 25 adult penguins in 2010 (Cardeña, pers. obs.). Peruvian gulls (*Larus belcheri*) and kelp gulls (*Larus dominicanus*) prey on the eggs (Wilson et al. 1995; Paredes and Zavalaga 2001). Vampire bats (*Desmodus rotundus*) are also micro-predators of this species, biting the feet and tarsometatarsi of chicks and juveniles (Luna-Jorquera and Culik 1995). Introduced predators such as dogs (*Canis familiaris*) kill adults; rats (*Rattus rattus*) prey on eggs, but their impact is not uniform and depends on location and year (Simeone and Bernal 2000).

MOLT. The penguins' feathers protect their skin against water and work as an insulator. During molt, Humboldt penguins do not enter the sea to forage (Paredes et al. 2002; Otsuka et al. 2004). To allow for an energy-demanding molt and a period of forced fasting, they become hyperphagic during the pre-molting period and then return to the colonies to replace their entire plumage over a short period of time (Zavalaga and Paredes 1997b; Paredes et al. 2002). This process is closely related to circulating thyroxine, testosterone, and estradiol levels; increases of thyroxine in conjunction with low levels of testosterone and estradiol might trigger feather growth (Otsuka et al. 2004).

Molting initiation and duration vary with latitude (Scholten 1987; Zavalaga and Paredes 1997b; Luna-Jorquera et al. 2000; Paredes et al. 2002; Simeone et al. 2002). Adult molting in southern Peru (14–16° S) lasts for approximately 21 days, peaking between January and February (Paredes et al. 2002), while in north-central Chile, the peak of molting tends to be one month later (Simeone et al. 2002). Captive penguins in the Northern Hemisphere start molting in July, reflecting that hemisphere's solar radiation and photoperiod (Schloten 1987), and have shown shorter molting periods (10 days), perhaps because of more constant food.

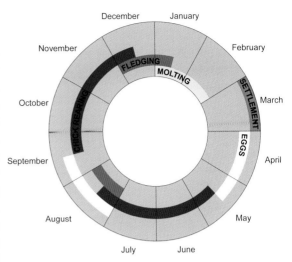

FIG. 10 Annual cycle of the Humboldt penguin.

ANNUAL CYCLE. Humboldts have two breeding peaks, but breeding depends on prey availability and can cease, with penguins deserting their nests during strong El Niño periods (fig. 10).

8. POPULATION SIZES AND TRENDS

Although the Humboldt population has severely declined since the 19th century and showed marked reductions after the 1982–83 and 1997–98 El Niños, the population is now increasing slowly. Some authors estimate a population of 48,000 breeding adults (Schlosser et al. 2009); however, these are based not on actual counts but on the opinions of researchers.

As population numbers are prone to fluctuate in relation to changes in prey availability (El Niño events) and interactions with fisheries, recent improvements in Peruvian counts could be a result of a fixed escapement policy of 5 million tons of spawning biomass (Guevarra-Carrasco et al. 2010) for Peruvian anchoveta (*Engraulis ringens*) and the recent implementation of an individual transferable quota system for fisheries established by Decreto Legislativo 1084. Nevertheless, it is a priority to assess the impact of these policies on the conservation of penguins and other anchovy-dependent fauna. In terms of actual population numbers, field data on species abundance are scarce and of limited accuracy. Counting methodologies are inconsistent (perhaps masking the Chilean trends), and species monitoring is concentrated in just a few sites, neglecting newly discovered colonies (Bussalleu, pers. obs.).

9. MAIN THREATS

Threats to Humboldt populations are shown in table 15.7. Most of the threats described in the 1930s are still major concerns today (Murphy 1936; Boersma and Stokes 1995; BirdLife International 2008).

EL NIÑO AND CLIMATE VARIATION. The Humboldt Current upwelling system has alternating blooms and depletions of productivity triggered by Kelvin waves coming into contact with the South American Pacific coast (Bertrand et al. 2008b). El Niño events, recurring periods of reduced upwelling intensity, increase the average depth of pelagic schools of Peruvian anchoveta (Bertrand et al. 2008a) and thus reduce prey availability (Culik et al. 2000; Taylor et al. 2002) and increase adult penguin mortality.

Together with other environmental changes associated with Kelvin waves (such as increased rainfall leading to the collapse of nests), increases in nest abandonment and chick mortality during El Niño events have alarmed researchers in Ecuador, Peru, and Chile (Boersma 1978; Paredes and Zavalaga 1998; Simeone et al. 2002).

Species inhabiting the Humboldt Current have survived periods of food scarcity, but the increased frequency of El Niño events likely will harm the Humboldt penguin as it has the Galápagos penguin (Boersma 1998; Vargas et al. 2006). Dramatic increases in fishing effort (ongoing since the 1950s) modify energetic and nutritional flows across the trophic webs, reducing the carbon flux toward top predators, such as seabirds (Jahncke et al. 2004). The introduction of large-scale industrial fisheries increases the ecosystems' vulnerability and can lead to losses in biodiversity and even ecological collapse (Pauly et al. 1998; Cury et al. 2000; Pauly and Palomares 2005). As the commercial fisheries of Chile and Peru intensively exploit the Humboldt's main prey species, it is thought that this competition is the most pressing threat to the species' conservation, hindering its ability to recover population sizes similar to those of the 19th century (Duffy et al. 1984; Herling et al. 2005; BirdLife International 2008).

FISHING. Bycatch in the small-scale fisheries in Peru and Chile is an ongoing problem (Duffy 1983; Simeone et al. 1999; Majluf et al. 2002; Melly et al. 2006; Boersma et al. 2007). Most of the penguins' foraging activities take place within a radius of 30 kilometers around their colonies and in depths of less than 30 meters (Culik et al. 1998; Luna-Jorquera and Culik 1999; Taylor et al. 2002; Boersma et al. 2007); thus, they are vulnerable to entanglement in fishing nets that target pelagic commercial species (Boersma and Stokes 1995; Simeone et al. 1999; Majluf et al. 2002; Taylor et al. 2004; Melly et al. 2006). For example, at least 1,500 penguins (about 10% of the overall estimated population size at the time) were bycatch in gill nets of small-scale fisheries between 1991 and 1996 in Chile and Peru (Simeone et al. 1999; Majluf et al. 2002). Their accidental drownings were associated mostly with drift gill nets in fisheries targeting palm ruffs (*Seriolella violacea*) and corvina drums (*Cilus gilberti*).

Mesh size and fishing methods may change rates of penguin mortality. It seems that penguins are entangled while resting at sea at night after foraging or while they are approaching anchovy schools during the day (Duffy 1983; Simeone et al. 1999; Majluf et al. 2002; Taylor et al. 2002; Boersma et al. 2007). Some authors have suggested banning drift gill nets at night and establishing no-take zones in areas within 30 kilometers of breeding colonies (Taylor et al. 2002; Boersma et al. 2007), but there has been little response from Chilean and Peruvian authorities. Other authors are deeply concerned with this issue, as local depletion of valuable fish continues to force artisanal fishermen to change target fish and shellfish species and seek new fishing grounds, often around penguin rookeries (Paredes et al. 2003).

GUANO HARVESTING. Guano harvesting is detrimental to the conservation of penguins (Boersma and Stokes 1995). It is important to note that there are nevertheless examples of sustainable, less harmful harvests in Punta San Juan (Majluf et al., unpubl. data). Peruvian guano is a valued organic fertilizer derived from the excrement of the Guanay cormorant (*Phalacrocorax bougainvillii*), Peruvian booby (*Sula variegata*), and Peruvian pelican (*Pelecanus thagus*), and the fecal deposits of these birds are the nesting substrates for Humboldt penguins in Peru (fig. 11) (Boersma and Stokes 1995; Battistini 1998; Battistini and Paredes 1999; Paredes and Zavalaga 2001).

Thus, guano harvesting reduces nesting habitat for penguins (Murphy 1936; Duffy et al. 1984; Boersma and Stokes 1995; Paredes et al. 2003), but guano miners also increase adult and egg mortality through direct harvest and by the introduction of alien species such as dogs

FIG. 11 At Punta San Juan, Peru,
Humboldt penguins dig burrows in
the guano to nest. (P. D. Boersma)

and rats (Duffy et al. 1984). Even without guano harvest, human presence has harmed colonies (Simeone and Bernal 2000) through the introduction of species such as goats that trample nests (Simeone and Schlatter 1998).

HABITAT LOSS. Coastal development reduces breeding sites (Duffy et al. 1984; Araya et al. 2000). The increased human presence near Humboldt colonies is of concern because these penguins are among the timid species of the Sphenisciformes (Ellenberg et al. 2006). Habituation is possible, but visual contact with humans increases their heart rate and can result in nest abandonment (Taylor et al. 2002; Ellenberg et al. 2006). Unregulated tourism is problematic for the species (Simeone and Schlatte 1998; Ellenberg et al. 2006; Skewgar et al. 2009); however, its impacts on breeding success, attendance, and mortality are not quantified.

DISEASE. Changes in coastal use (e.g., increased chicken aviculture) are a possible driver for new diseases that could further endanger the species (Duffy et al. 1984; Adkesson, pers. comm.), but impacts on penguin health are not quantified (Smith et al. 2008). A team of veterinary researchers led by Michael Adkesson is quantifying changes in pathogens, diseases, viruses, and pollutants that were previously described at Punta San Juan in Smith et al. (2008).

10. RECOMMENDED PRIORITY RESEARCH ACTIONS FOR CONSERVATION

Research can aid in conserving this species. The following actions address the most pressing needs:

1. Define a consolidated census methodology. What are the optimum survey times and methods for determining the population size of the species across its distribution?
2. Monitor distribution and abundance. How many penguins are there? How do colonies behave in periods of food abundance and scarcity?
3. Quantify the impact of human activities (urban growth, tourism, industrial fishing, small-scale fishing, mining, farming, animal husbandry) on distribution, abundance, and breeding success.
4. Identify and quantify the impacts of climate change on population size, distribution, and breeding success.
5. Identify critical areas for conservation. Which colonies are facing the worst pressures? Where are penguins most prolific? What areas do penguins use for transit and migration? Where should marine protected areas (MPAs) or zoning for conservation be located?
6. Generate relevant information for industrial fishery management and policy (define catch quotas and fishery bans based on ecosystem parameters). Monitor targeted prey species and their relative

contribution (numbers, mass, and energy) to the penguins' diet.

7. Generate a baseline of health parameters across the Humboldt penguin's distribution. What are the most common pathogens, parasites, and diseases? How do they affect the penguins' mortality and reproductive success? Are there areas where or times when penguins are more prone to diseases?

8. Assess the impacts of MPAs on penguin conservation. Does the current MPA system protect the species? Have population sizes and breeding success improved after the establishment of protected areas? Do reductions in harvests of fish result in increases in numbers of penguins and other seabirds?

9. Develop innovative educational programs on forage fish and seabird conservation for adults and children to better understand economic and conservation trades-offs with human well-being. Conservation programs should include interactive websites, talks in schools, TV presentations, and community-based projects.

11. CURRENT CONSERVATION EFFORTS

Humboldt penguins are protected by MPAs (marine reserves, marine parks, natural monuments, national reserves, and reserved zones). Some protect marine extensions (foraging habitats), others land (nesting habitats), and some both. Their effectiveness is related to the ability of the Peruvian and Chilean governments to properly enforce conservation measures against urgent threats and short-term economic gain. This capacity is limited and perhaps insufficient. Moreover, although some breeding sites are protected, many areas lack appropriate participatory management plans that include Humboldts as an object of conservation (Skewgar et al. 2009) (fig. 12).

In January 2010, the Peruvian government established the Guano System National Reserve (Decreto Supremo 024–2009-MINAM) (Ministerio del Ambiente 2012). This network of headlands, capes, and islands harbors Humboldt nesting sites and protects major foraging areas around them. The San Fernando National Reserve (established in July 2011 by the Decreto Supremo 017–2011-MINAM) is also a major site for penguins.

It is also important to highlight that Isla Chañaral, the main breeding colony for the species, is protected (UNEP 2003) and that improvements in MPA cover-

age and management are significant in Chile (Pizarro 2004).

Establishing sustainable guano harvest methodologies, implementing protective tourism routes with camouflaged observation points, removing alien species from breeding sites, enforcing fishery no-take zones around breeding colonies, and applying the ecosystem approach to fisheries (Garcia et al. 2003) are all major challenges for the future of the species.

12. RECOMMENDED PRIORITY CONSERVATION ACTIONS

In general, information on species distribution and abundance is needed. Census methodologies for Chile and Peru should be unified and a complete penguin count made over the range of the species (Araya et al. 2000). Additional recommendations include the following specific actions:

1. Acquire information on abundance, breeding phenology, reproductive success, and adult mortality for the most representative reproductive colonies at

FIG. 12 Adult Humboldt penguin at its nest on Isla Choros, Chile. (T. Mattern)

least once a year and perhaps conduct a full census across the species' range once every three to four years (Boersma 2008).

2. Determine migratory and transit corridors between colonies by means of ecological and genetic studies (Schlosser et al. 2009).

3. Implement a sustainable guano harvest method, to minimize disturbance at the breeding colonies and better preserve nesting habitat, as was done for two guano harvests at Punta San Juan (Majluf, unpubl. data). The guano sites in Peru are part of a national reserve, and tools developed at Punta San Juan should be incorporated into guano management plans for all sites. Promote compliance with protective guano harvesting guidelines as a part of national and regional regulations.

4. The industrial anchovy fishery is a threat to Humboldt penguins and other seabirds, so total allowable catches should be set based on trophic and oceanographic models that include ecological parameters (such as prey dependence) (Christensen and Walters 2004; Cury et al. 2005). Include a precautionary approach to uncertainty in fishery policy (Ward et al. 2002), reducing fishing pressure during El Niño events.

5. Monitoring of bycatch and fishing efforts is critical for penguin conservation (Majluf et al. 2002; Melly et al. 2006; Skewgar et al. 2009). Monitor fishing areas and landings of small-scale and artisanal fisheries for areas and gear that render the greatest bycatch. Vessel monitoring and landing data would also provide a description of spatial and resource use patterns for these small-scale fisheries around breeding colonies.

6. Use these data to inform zoning proposals (fishing areas, no-take zone, areas where only hook and line are allowed, etc.) and fishing policies (fishing bans, mesh sizes, catch shares, etc.) that could enhance conservation of commercially targeted species and reduce disturbance to Humboldt breeding colonies (Araya et al. 2000). Local stakeholders need to be involved in these activities, which should be coupled with educational and outreach programs that will enhance their ability (Olsen et al. 2009) to reduce bycatch, improve self-governance, and convince local stakeholders to accept MPAs and comply with their regulations.

7. Implement on-board observer programs for obtaining data on bycatch. Integrating this information with satellite data from vessel monitoring programs and GPS tracking and telemetry data from projects studying the Humboldt's foraging areas will provide a better sense of how, if, and when industrial fisheries cause local depletions of prey. This information could then be used for rapid adaptive management so that fishing policies and activities can minimize their impact in key areas during the species' breeding seasons.

8. Many MPAs have been established across this species' distribution and include important breeding colonies. Enforcement of these protections should be coupled with management plans that have clearly defined, measurable objectives so that conservation practitioners can monitor the efficiency of MPAs in relation to Humboldt penguins (Skewgar et al. 2009) and adapt to changes as necessary (e.g., assess the effectiveness of a no-take zone after a predetermined period of time).

9. Include capacity building and local involvement of stakeholders from civil society when discussing sustainable practices and policies intended to help conserve marine biodiversity (Olsen et al. 2009), including especially the Humboldt penguin. Initiate efforts to communicate science in more novel ways that can better reach target stakeholders like fishermen and policy makers.

Other conservation activities include developing clearly defined regulations on tourism in MPAs and near penguin breeding sites and including these regulations in regional conservation action plans, to reduce negative impacts on the species; enhancing waste treatment efforts in coastal regions where mining, industries, and urban areas are located near breeding colonies; and implementing health-monitoring programs at breeding colonies.

ACKNOWLEDGMENTS

We thank the Chicago Zoological Society, the Saint Louis Zoo, and the Philadelphia Zoo for funding our research and conservation efforts at Punta San Juan during the past decade. The Chicago Zoological Society and the Saint Louis Zoo are also currently sponsoring Humboldt penguin health-monitoring programs for wild populations, Humboldt colony identification, and population size surveys across the species distribution. Moreover, we would like to thank P. D. Boersma and P. Garcia Borboroglu for entrusting this task to our team and for their patience.

TABLE 15.1 Mean measurements of adult Humboldts

SITE	GENDER	BM	WH	BL	BD	FL	DATA SOURCE
Wild penguins at Punta San Juan Colony 1992–1994	Males (n=165)	4711.06 ± 31.55 (g)					Zavalaga and Paredes (1997a)
	Females (n=123)	4047.39 ± 35.15 (g)					
	Pooled (n=288)	4427.62 ± 30.42 (g)					
Captive penguins at Washington Park Zoo	Males (n=19)	4802.6 ± 100.5 (g)	5.49 ± 0.045 (cm)	6.44 ± 0.075 (cm)	2.97 ± 0.045 (cm)		Zavalaga and Paredes (1997a)
	Females (n=16)	4328.0 ± 92.6 (g)	5.22 ± 0.055 (cm)	6.03 ± 0.057 (cm)	2.71 ± 0.004 (cm)		
	Pooled (n=35)	4585.7± 79.2 (g)	5.36 ± 0.042 (cm)	6.25 ± 0.059 (cm)	2.85 ± 0.037 (cm)		
Dead penguins recovered at San Pedro Port 1992–1993	Males (n=112)*	4931.08 ± 56.05 (g)	5.13 ± 0.02 (cm)	6.54 ± 0.02 (cm)	2.56 ± 0.01 (cm)	15.63 ± 0.06 (cm)	Zavalaga and Paredes (1997a)
	Females (n=111)*	4317.85 ± 52.15 (g)	4.75 ± 0.02 (cm)	6.08 ± 0.02 (cm)	2.27 ± 0.01 (cm)	14.93 ± 0.05 (cm)	
	Pooled (n=223)	4643.54 ± 43.90 (g)	4.92 ± 0.02 (cm)	6.31 ± 0.02 (cm)	2.42 ± 0.01 (cm)		
Wild penguins at Pájaro Niño Colony 1994–1995	Males (n=388)			65.21 ± 2.61 (mm)	27.59 ± 1.38 (mm)	216.67 ± 6.92 (mm)	Wallace et al. (2008)
	Females (n=368)			61.42 ± 2.45 (mm)	24.68 ± 1.48 (mm)	209.07 ± 7.63 (mm)	

Note: The measurements taken where Body mass (BM), Length of head (LH), Width of head (WH), Bill length (BL), Bill depth (DB) and Flipper length (FL). Data from Zavalaga and Paredes (1997 a) is presented as mean value in centimetres ± the standard error, whilst data from Wallace et al. (2008) is presented as mean value in millimetres ± the standard deviation. Strong differences in FL measurements are due to methodological differences. For Wallace et al. (2008) FL was measured from the tip of the flipper to its junction with the thoracic wall, when the flipper was held perpendicular to the sagittal plane of the body. For Zavalaga and Paredes (1997 a) FL was measured from the humero-radial joint to the tip of the flipper (maximum flattened chord). *Only 201 penguins (98 males and 103 females) were measured for the FL parameter.

TABLE 15.2 Sex determination of Humboldts

ZAVALAGA AND PAREDES (1997 A)	WALLACE ET AL. (2008)
D = 3.16(WH) + 3.69(BL)–38.98	Ln $(P/(1-P))$ = 42.889–0.244(BL)–1–052(BD)
If D >0 the penguin was classified as male.	$P = \dfrac{e^{(42.889-0.244\,BL-1-052\,BD)}}{1+e^{(42.889-0.244\,BL-1-052\,BD)}}$
	P is the probability that the penguin is female (if $P \geq 0.5$, the penguin is female).

Note: Bill length (BL), bill depth (BD), and width of the head (WH).

TABLE 15.3 (A) Humboldt colonies* in Peru

SITES	LAT. °S	1980	1981	1996	1997	1999	2000	2001	2002	2003	2004	2005	2006	2007	2008	2009	REFERENCES
Foca	05°12					20	20			25	0						8, 12
Aguja	05°41'			10		20	20										1, 12
Lobos de Tierra	06°25'	1000	900	100		0	0			35	0						1, 4, 8
Macabi	07°48'			15													1
Guañape Norte	08°32'					20	20			43	0						8, 12
Huarmey	10°04'					0	0			300	0						8
Mazorca	11°22'	150	120	100													1, 4
Pachacamac	12°18'	600	750	800		230	575			959	120						1, 4, 8, 12
Santa María	12°18'					0	3										12
Asia	12°47'					0	0			227	206						8
Chincha Norte	13°38'			50		0	146			145	168						1, 8, 12
Ballestas	13°44'		9	60		79	153			157	98						1, 4, 8, 12
Culebras	13°50'					46	0										12
San Gallán	13°51'		111	60		191	259			192	101						1, 4, 8, 12
Tambillo	13°51'					3	3										12
Arquillo	13°52'			50													1
Mendieta-Paracas	14°03'					15	0			0	0						8, 12
Independencia	14°10'					0	10										12
Tres Puertas	14°10'					220	900			145	110						8, 12
La vieja	14°17'		8			0	0										1, 4, 12
Santa Rosa	14°19'		0			0	1			309	131						1, 8, 12
Gallinazo	15°09'					0	4										12
San Fernando	15°09'			500		3	25			22	67						1, 8, 12
San Juanito	15°16'					505	538			812	917						8, 12
San Juan	15°22'	3680	2220	3650	3686	1631	1774	801	1388	2305	1407	2267	3214	2545	3199	4421	1, 4, 8, 10–12, 19
Sombrerillo	15°30'					113	0			49	80						8, 12
Pampa Redonda	15°50'	1800	300														1, 4
La Chira	16°29'			300		6	0										1, 12
Caleta	16°31'					93	133			120	79						8, 12
La Norte	16°31'					4	0										12
Quilca	16°43'					15	0										12
Honoratos	16°52'					4	17										12
Hornillos	16°53'			60		512	502			298	492						1, 8, 12
Carrizales	16°55'					23	0			16	14						8, 12
Tarpuy	16°58'					6	0										12
Islay	17°01'					0	0			13	37						8
Cocotea	17°15'					103	160										12
Cordel	17°15'					21	160			68	163						8
Corio	17°15'					9	0										12
Platanales	17°24'					0	26										12
Coles	17°42'					178	91			289	146						8, 12

* The Maximum Recorded Values (individuals) for each year were extracted for all of Table 15.3 (a and b) from: (1) Araya et al.,2000; (2) Culik and Luna-Jorquera, 1997b; (3) Cursach et al., 2009; (4) Duffy et al., 1984; (5) Hertel et al., 2005; (6) Hiriart-Bertrand et al,. 2010, (7) Luna-Jorquera and Cortes, 2007; (8) Majluf*unpublished data*; (9) Mattern et al., 2004; (10) Paredes and Zavalaga, 2001; (11) Paredes et al., 2002; (12) Paredes et al., 2003; (13) Simeone, 2005; (14) Simeone and Schlatter, 1998; (15) Simeone et al., 1999; (16) Simeone et al., 2003; (17) Simeone et al., 2009; (18) Thiel et al., 2007; (19) UNEP, 2003.

TABLE 15.3 (B) Humboldt colonies in Chile

SITES	LAT. °S	1981	1982	1983	1984	1985	1986	1987	1988	1989	1990	1991	1995	1996	1997	2000	2002	2003	2005	2008	2009	REF.'S
Cueva del Caballo	20°12'							16														1
Pan de Azúcar	26°09'		6000		131	2500	2570		4000	600	1000		5900	1750								1, 2
Grande	27°15'	58					34						40			300			4270		4570	1, 7, 16, 18
Cima Cuadrada	27°41'			180																		1
Chañaral	29°01'		750		146	6000	1000			788	1500			2500				22021				1, 9, 16
Damas	29°13'																20	20				9, 16
Choros	29°15'		96		32		14							50			720	720				1, 9, 16
Chungungo	29°24'																	400				5
Tilgo	29°32'																	1200				5
Pájaros (1)	29°35'		624				54			880				1000			1200	1200				1, 9, 16
De los Huevos	31°55'		60		64	274	34							120								1
Papudo	32°30'			100																		1
Cachagua	32°35'		1000		1055						2030			2000		1200						1, 14, 16
Concón	32°52'		500		12		100	46		12	10			20		10						1, 16
Pájaro Niño	33°21'				530	1000	2000							1600		250						1, 16, 18
Papuya	34°00'			14																		1
Islote Pingüinos	40°56'																			4		3
Puñihuil	41°55'						12					450		50	315				152			1, 13, 14, 15
Metalqui	42°12'																				56	6, 15

TABLE 15.4 Legal actions undertaken by Chile and Peru for the protection of Humboldt penguins in their national territories

YEAR	LEGAL MEASURE	COUNTRY	DESCRIPTION	REFERENCE
1952	DS 797	Chile	Bans penguin hunting	Iriarte 1999
1963	DFL RRA 25	Chile	Regulates guano trade, banning extraction of guano in areas with active bird nests (including those of penguins).	Iriarte 1999
1966	DS 531	Chile	Bans penguin-hunting indefinitely.	Iriarte 1999
1967	SM 506	Chile	Bans hunting, transportation, possession and commercialization of penguins in Chile.	Iriarte 1999
1977	RM 01710–77AG-GFFF	Peru	Lists the Humboldt penguin as a Vulnerable species.	SPIJ 2011
1981	DFL 3.557	Chile	Allows for extraction of guano in marine birds' reproductive sites only with Governmental clearance.	Iriarte 1999
1993	DS 133	Chile	Bans penguin hunting and capturing.	Iriarte 1999
1995	DS 225	Chile	Implements a 30-year long extractive ban for marine vertebrates-including Humboldt penguins.	Iriarte 1999
1999	DS 013–99-AG	Peru	Bans hunting, extraction and exportation of Humboldt penguins for commercial purposes.	SPIJ 2011
2004	DS 034–2004-AG	Peru	Lists the Humboldt penguin as an Endangered Species and bans its hunting, capturing, possession, transportation and exportation for commercial purposes.	SPIJ 2011
2008	DS 50/2008	Chile	Lists the Humboldt penguin as Vulnerable in Chile	CONAMA 2010
2009	DS 024–2009-MINAM	Peru	Implements the guano system as a National Protected Area protecting most of the Peruvian breeding sites for the species.	SPIJ 2011

Note: DFL = Decreto con Fuerza de Ley; DS = Decreto Supremo; RM =Resolución Ministerial

TABLE 15.5 Sizes and masses of captive and wild Humboldt penguins' eggs

LOCATION	CAPTIVE POPULATION-BIRDWORLD GREAT BRITAIN		WILD POPULATION-PUNTA SAN JUAN	
Year (sample size)	1990 (n=13)		2002 (n=50)	
Measurements	Mean ± SD	Range	Mean ± SD	Range
Length (cm)	7.200 ± 0.179	6.997–7.448	7.35 ± 0.33	6.67–8.10
Breadth (cm)	5.559 ± 0.216	5.056–5.890	5.54 ± 0.32	4.80–7.10
Initial Mass (g)	122.11 ± 10.81	95.18–134.5	119.28 ± 12.97	84.00–143.00

Sources: Deerling et al. 1991; Cardeña et al., unpublished data

TABLE 15.6 Main constituents of the Humboldt's diet

LOCATION	MAIN DIETARY CONSTITUENTS		REFERENCE
Punta San Juan 15°22' S	Peruvian anchovy	*Engraulis ringens*	Zavalaga and Paredes 1997b
	Silversides	*Odontesthesregia*	
Pan de Azucar 26°09' S	Garfishes	*Scomberesox saurus*	Herling et al. (2005)
	Peruvian anchovy	*Engraulis ringens*	
	Jack Mackerel	*Trachurus murphyi*	
	South American pilchard	*Sardinops sagax*	
Chañaral 29°01' S	Garfishes	*Scomberesox* spp.	Wilson et al. (1989)
	Peruvian anchovy	*Engraulis ringens*	
Algarrobo 33°30' S	Peruvian anchovy	*Engraulis ringens*	Wilson et al. (1989)
	South American pilchard	*Sardinops sagax*	
Puñihuil 45° 55' S	Peruvian anchovy	*Engraulis ringens*	Herling et al. (2005)
	Araucanian herring	*Strangomera bentincki*	
	Silversides	*Odontesthes regia*	
	Common hake	*Merluccius gayi*	
	Jack Mackerel	*Trachurus murphyi*	

TABLE 15.7 Main threats for wild Humboldt populations

REFERENCE	EN	GW	EP	HA	P AND Z	B	CF	HD	HL	IAS	TH
Murphy (1936)	x		x (h)	x (h)	x (h)	x	x	x			
Duffy (1983)			x (h)	x (h)		x	x				
Duffy et al. (1984)			x (h)	x (h)		x	x	x	x	x	
Hays (1984)											
Hays (1986)							x				
Culik and Luna-Jorquera (1997a)							x				
Battistini (1998)											
Paredes and Zavalaga (1998)	x										
Araya et al. (2000)	x	x				x	x	x	x		
Simeone and Schlatter (1998)			x	x						x	x
Simeone et al. (1999)						x					
Wallace et al. (1999)						x					
Culik et al. (2000)	x										
Simeone and Bernal (2000)										x	
Majluf et al. (2002)	x	x				x					
Simeone et al (2002)	x										
Taylor et al. (2002)						x	x				x
Cushman (2003)			x (h)	x (h)				x			x
Paredes et al. (2003)	x		x (r)	x (r)	x (r)	x	x	x			
Simeone et al. 2003				x						x	x
Herling et al (2005)							x				
Ellenberg et al. (2006)								x			x
Boersma et al. (2007)						x	x				
BirdLife International (2008)	x	x		x		x	x	x			
Skewgar et al. (2009)						x				x	x

TABLE 15.7 (cont.)

Note: Threats include EN: El Niño; GW: global warming; EP: Egg poaching; HA: Hunting of adult birds; P and Z: Capturing birds for pets and zoos; B: By-catch and drowning by net entanglement; CF: Competition with commercial fisheries; HD: Habitat degradation and reproductive failure from Guano harvests; HL: Loss of nesting sites and reproductive habitat due to coastal development; IAS: Introduction of alien species; TH: Tourism and human presence. "x" denotes the mentioning or discussing of the threat in the references, (h) stands for historical threat and (r) for recent threat.

REFERENCES

Araya, B., D. Garland, G. Espinoza, A. Sanhuesa, A. Simeone, A. Teare, C. Zavalaga, R. Lacy, and S. Ellis, eds. 2000. Population and habitat viability analysis for the Humboldt penguin (*Spheniscus humboldti*). Final report. Apple Valley, MN: IUCN/SSC Conservation Breeding Specialist Group.

Araya, B., and F. S. Todd. 1987. Status of the Humboldt penguin in Chile following the 1982–83 El Niño. Proceedings of the Jean Delacour/IFCB Symposium, Los Angeles, 148–57.

Baker, A. J., S. L. Pereira, O. P. Haddrath, and K. A. Edge. 2006. Multiple gene evidence for expansion of extant penguins out of Antarctica due to global cooling. *Proceedings of the Royal Society B* 273:11–17.

Battistini, G. 1998. El nido del pingüino de Humboldt (*Spheniscus humboldti*) y su relación con el éxito reproductivo. Tesis para optar el título Licenciado en Biología, Universidad Peruana Cayetano Heredia, Lima, Perú.

Battistini, G., and R. Paredes. 1999. Nesting habits and nest characteristics of Humboldt penguins at Punta San Juan, Peru. *Penguin Conservation* 12:12–19.

Bertelli, S., and N. P. Giannini. 2005. A phylogeny of extant penguins (Aves: Sphenisciformes) combining morphology and mitochondrial sequences. *Cladistics*, 21:209–39.

Bertrand, A., F. Gerlotto, S. Bertrand, M. Gutierrez, L. Alza, A. Chipollini, E. Diaz, P. Espinoza, J. Ledesma, R. Quesquen, S. Peraltilla, and F. Chavez. 2008a. Schooling behaviour and environmental forcing in relation to anchoveta distribution: An analysis across multiple spatial scales. *Progress in Oceanography* 79:264–77.

Bertrand, S., B. Dewitte, J. Tam, E. Díaz, and A. Bertrand. 2008b. Impacts of Kelvin wave forcing in the Peru Humboldt Current system: Scenarios of spatial reorganizations from physics to fishers. *Progress in Oceanography* 79:278–89.

BirdLife International 2008. BirdLife International/American Bird Conservancy Workshop on Seabirds and Seabird-Fishery Interactions in Peru. RSPB (Royal Society for the Protection of Birds), Sandy, United Kingdom.

Boersma, P. D. 1978. Breeding patterns of Galápagos penguins as an indicator of oceanographic conditions. *Science* 200:1481–83.

———. 1998. Population trends of the Galápagos penguin: Impacts of El Niño and La Niña. *Condor* 100:245–53.

———. 2008. Penguins as marine sentinels. *BioSience* 58(7):597–607.

Boersma, P. D., S. Branch, D. Butler, S. Ellis-Joseph, P. Garland, P. McGill, G. Phipps, U. Seal, and P. Stockdale. 1992. Penguin Conservation Assessment and Management Plan. Apple Valley, MN: IUCN/SSC Captive Breeding Specialist Group.

Boersma, P. D., G. A. Rebstock, D. L. Stokes, and P. Majluf. 2007. Oceans apart: Conservation models for two temperate penguin species shaped by the marine environment. *Marine Ecology Progress Series* 335:217–25.

Boersma, P. D., and D. L. Stokes. 1995. Conservation of penguins: Threats to penguin populations. In *Bird Families of the World: The Penguins*, ed. T. D. Williams, 127–39. Oxford: Oxford University Press.

Christensen, V., and C. J. Walters. 2004. Ecopath with Ecosim: Methods, capabilities and limitations. *Ecological Modelling* 172:109–39.

Coffin, H. R., J. V. Watters, and J. M. Mateo. 2011. Odor-based recognition of familiar and related conspecifics: A first test conducted on captive Humboldt penguins (*Spheniscus humboldti*). *PLoS ONE* 6(9):e25002.

Coker, R. E. 1919. Habits and economic relations of the guano birds of Peru. *Proceedings of the United States National Museum* 56:449–511.

CONAMA (Comisión Nacional del Medio Ambiente). 2010. Decreto Supremo No. 50/2008. http://www.conama.cl/clasificacionespecies/Anexos_segundo_proceso/DS_50_2008_2doProcesoClasif_completo.pdf.

Culik, B., J. Hennicke, and T. Martin. 2000. Humboldt penguins outmanoeuvring El Niño. *Journal of Experimental Biology* 203:2311–22.

Culik, B. M., and G. Luna-Jorquera. 1997a. Satellite tracking of Humboldt penguins (*Spheniscus humboldti*) in northern Chile. *Marine Biology* 128:547–56.

———. 1997b. The Humboldt penguin *Spheniscus humboldti*: A migratory bird? *Journal of Ornithology* 138:325–30.

Culik, B. M., G. Luna-Jorquera, H. Oyarzo, and H. Correa. 1998. Humboldt penguins monitored via VHF telemetry. *Marine Ecology Progress Series* 162:279–86.

Cursach, J., J. Vilugrón, C. Tobar, J. Ojeda, J. Rau, C. Oyarzún, and O. Soto. 2009. Nuevos sitios de nidificación para cuatro especies de aves marinas en la provincia de Osorno, centro-sur de Chile. *Boletín Chileno de Ornitología* 15:17–22.

Cury, P., A. Bakun, R. J. M. Crawford, A. Jarre, R. A. Quiñones, L. J. Shannon, and H. M. Verheye. 2000. Small pelagics in upwelling systems: Patterns of interaction and structural changes in "wasp-waist" ecosystems. *ICES Journal of Marine Science* 57:603–18.

Cury, P. M., L. J. Shannon, J-P. Roux, G. M. Daskalov, A. Jarre, C. L. Moloney, and D. Pauly. 2005. Trophodynamic indicators for an ecosystem approach to fisheries. *ICES Journal of Marine Science* 62:430–42.

Cushman, G. T. 2003. The lords of guano: Science and the management of Peru's marine environment, 1800–1973. PhD diss., University of Texas at Austin.

Deeming, D. C., R. L. Harvey, L. Harvey, S. Carey, and D. Leuchars. 1991. Artificial incubation and hand rearing of Humboldt penguins (*Spheniscus humboldti*) at Birdworld, Franham. *International Zoo Yearbook* 30:165–73.

Duffy, D. C. 1983. The foraging ecology of Peruvian seabirds. *The Auk* 100(4):800–810.

Duffy, D. C., C. Hays, and M. A. Plengue. 1984. The conservation status of Peruvian seabirds. In *Status and Conservation of the World's Seabirds*, ed. J. P. Croxall, P. G. H. Evans, and R. W. Schreiber, 245–59. ICBP Technical Publication No. 2. Cambridge: International Council for Bird Preservation.

Ellenberg, U., T. Mattern, P. J. Seddon, and G. Luna-Jorquera. 2006. Physiological and reproductive consequences of human disturbance in Humboldt penguins: The need for species-specific visitor management. *Biological Conservation* 133:95–106.

Garcia, S. M., A. Zerbi, C. Aiaume, T. Do Chi, and G. Lasserre. 2003. The ecosystem approach to fisheries: Issues, terminology, principles, institutional foundations, implementation and outlook. Food and Agricultural Organization Fisheries Technical Paper No. 443. Rome: Food and Agricultural Organization.

Guevara-Carrasco, R., M. Niquen, and M. Bouchon. 2010. A summary on monitoring, assessment and criteria to estimate the total allowable catch of the Peruvian anchoveta (northern-central stock). Document submitted by the Instituto del Mar del Perú (IMARPE) at the Lenfest Forage Fish Task Force Meeting, Paracas, Peru.

Hays, C. 1984. The Humboldt penguin (*Spheniscus humboldti*) in Peru and the effects of the 1982–1983 El Niño. MSc thesis, University of Florida.

———. 1986. Effects of the El Niño 1982–83 on Humboldt penguin colonies in Perú. *Biological Conservation* 36:169–80.

Hennicke, J. C., and B. M. Culik. 2005. Foraging performance and reproductive success of Humboldt penguins in relation to prey availability. *Marine Ecology Progress Series* 296:173–81.

Herling, C., B. M. Culik, and J. C. Hennicke. 2005. Diet of the Humboldt penguin (*Spheniscus humboldti*) in northern and southern Chile. *Marine Biology* 147:13–25.

Hertel, F., D. Martinez, M. Lemus, and J. C. Torres-Mura. 2005. Birds from Chungungo, Tilgo, and Pájaros Islands in north-central Chile. *Journal of Field Ornithology* 76(2): 197–203.

Hiriart-Bertrand, L., A. Simeone, R. Reyes-Arriagada, V. Riquelme, K. Pütz, and B. Lüthi. 2010. Descripción de una colonia mixta de pingüino de Humboldt (*Spheniscus humboldti*) y de Magallanes (*S. magellanicus*) en Isla Metalqui, Chiloé, sur de Chile. *Boletín Chileno de Ornitología* 16:42–47.

Iriarte, A. 1999. Marco legal relativo a la conservación y usosustentable de aves, mamíferos y reptiles en Chile. *Estudios Oceanológicos* 18:5–12.

IUCN (International Union for Conservation of Nature). 2009. IUCN red list of threatened species. Version 2009. http://www.iucnredlist.org.

Jahncke, J., D. M. Checkley, and G. L. Hunt. 2004. Trends in carbon flux of seabirds in the Peruvian upwelling system: Effects of wind and fisheries on population regulation. *Fisheries Oceanography* 13(3):208–23.

Ksepka, D. T., S. Bertelli, and N. P. Giannini. 2006. The phylogeny of living and fossil Sphenisciformes (penguins). *Cladistics* 22:412–41.

Luna-Jorquera, G., and M. Cortes. 2007. Estudio del ensamble de aves y mamíferos marinos al interior del Área Marina y Costera Protegida de Múltiples Usos Isla Grande de Atacama. In *Conservación de la Biodiversidad de Importancia Mundial a lo largo de la Costa Chilena*. Gobierno de Chile.

Luna-Jorquera, G., and B. M. Culik. 1995. Penguins bled by vampires. *Journal of Ornithology* 136:471–72.

———. 1999. Diving behaviour of Humboldt penguins *Spheniscus humboldti* in northern Chile. *Marine Ornithology* 27:67–76.

———. 2000. Metabolic rates of swimming Humboldt penguins. *Marine Ecology Progress Series* 203:301–9.

Luna-Jorquera, G., S. Garthe, F. G. Sepulveda, T. Weichler, and J. A. Vásquez. 2000. Population size of Humboldt penguins assessed by combined terrestrial and at-sea counts. *Waterbirds: The International Journal of Waterbird Biology* 23(3):506–10.

Majluf, P., E. A. Babcock, J. C. Riveros, M. Arias-Schreiber, and W. Alderete. 2002. Catch and by-catch of sea birds and marine mammals in the small-scale fishery of Punta San Juan, Peru. *Conservation Biology* 16(5):1333–43.

Martin, G. R., and S. R. Young. 1984. The eye of the Humboldt penguin, *Spheniscus humboldti*: Visual fields and schematic optics. *Proceedings of the Royal Society of London B* 223:197–222.

Mattern, T., U. Ellenberg, G. Luna-Jorquera, and L. Davis. 2004. Humboldt penguin census in Isla Chañaral, Chile: Recent increase or past underestimate of penguin numbers? *Waterbirds: The International Journal of Waterbird Biology* 27(3):368–76.

Melly, P., J. Alfaro-Shigueto, J. Mangel, M. Pajuelo, C. M. Cáceres, L. Santillán-Corrales, D. Montes-Iturrizaga, K. Baella, and J. Janhcke. 2006. Assessment of seabird by-catch in Peruvian artisanal fisheries. Final report to the British Petroleum Conservation Programme. Pro Delphinus, Lima.

Ministerio del Ambiente. 2012. Servicio Nacional de Áreas Naturales Protegidas por el Estado. www.sernanp.gob.pe/sernanp/bmapas.jsp?NroPag=1%20and%20ID=428 (accessed 9 April 2012).

Murphy, R. C. 1936. *Oceanic Birds of South America*. Vol. 1. New York: American Museum of Natural History.

Olsen, S. B., G. G. Page, and E. Ochoa. 2009. The analysis of governance responses to ecosystem change: A handbook for assembling a baseline. LOICZ Reports and Studies No. 34. Geesthacht, Germany: GKSS Research Center. 87 pp.

Otsuka, R., T. Machida, and M. Wada. 2004. Hormonal correlations at transition from reproduction to molting in an annual life cycle of Humboldt penguins (*Spheniscus humboldti*). *General and Comparative Endocrinology* 135:175–85.

Paredes, R., and C. B. Zavalaga. 1998. Overview of the effects of El Niño 1997–98 on Humboldt penguins and other seabirds at Punta San Juan, Perú. *Penguin Conservation* 11:5–7.

———. 2001. Nesting sites and nest types as important factors for the conservation of Humboldt penguins (*Spheniscus humboldti*). *Biological Conservation* 100:199–205.

Paredes, R., C. B. Zavalaga, G. Battistini, P. Majluf, and P. McGill. 2003. Status of the Humboldt penguin in Peru, 1999–2000. *Waterbirds: The International Journal of Waterbird Biology* 26(2):129–38.

Paredes, R., C. B. Zavalaga, and D. J. Boness. 2002. Patterns of egg laying and breeding success in Humboldt penguins (*Spheniscus humboldti*) at Punta San Juan. *The Auk* 119(1):244–50.

Pauly, D., V. Christensen, J. Dalsgaard, R. Froese, and F. C. Torres. 1998. Fishing down marine food webs. *Science* 279:860–63.

Pauly, D., and M. L. Palomares. 2005. Fishing down marine food web: It is far more pervasive than we thought. *Bulletin of Marine Science* 76 (2):197–211.

Pizarro, C. A. 2004. Áreas marinas protegidas y usutilidad en la conservación de las aves marinas en Chile. Tesis de pregrado, Facultad de Ciencias, Universidad de Chile.

Schlosser, J. A., J. M. Dubach, T. W. J. Garner, B. Araya, M. Bernal, A. Simeone, K. A. Smith, and R. S. Wallace. 2009. Evidence for gene flow differs from observed dispersal patterns in Humboldt penguin, *Spheniscus humboldti*. *Conservation Genetics* 10:839–49.

Scholten, C. J. 1987. Breeding biology of the Humboldt penguin (*Spheniscus humboldti*) at Emmen Zoo. *International Zoo Yearbook* 26:198–204.

———. 1999. Iris colour of Humboldt penguins *Spheniscus humboldti*. *Marine Ornithology* 27:187–94.

Schwartz, M. K., D. J. Boness, C. M. Schaeff, P. Majluf, E. A. Perry, and R. C. Fleischer. 1999. Female-solicited extrapair matings in Humboldt penguins fail to produce extrapair fertilizations. *Behavioral Ecology* 10(3):242–50.

Simeone, A. 2005. Evaluación de la población reproductiva del pingüino de Magallanes y del pingüino de Humboldt en los islotes Puñihuil, Chiloé. Informe final. Estudio financiadopor la Fundación Otway (Chile) y Zoo Landau in der Pfalz (Alemania). Viña del Mar. 49 pp.

Simeone, A., B. Araya, M. Bernal, E. N. Diebold, K. Grzybowski, M. Michaels, J. A. Tare, R. C. Wallace, and M. J. Willis. 2002. Oceanographic and climatic factors influencing breeding and colony attendance patterns of Humboldt penguins *Spheniscus humboldti*. *Marine Ecology Progress Series* 227:43–50.

Simeone, A., and M. Bernal. 2000. Effects of habitat modification on breeding seabirds: A case study in central Chile. *Waterbirds: The International Journal of Waterbird Biology* 23(3):449–56.

Simeone, A., M. Bernal, and J. Meza. 1999. Incidental mortality of Humboldt penguins *Spheniscus humboldti* in gill nets, central Chile. *Marine Ornithology* 27:157–61.

Simeone, A., L. Hiriart-Bertrand, R. Reyes-Arriagada, M. Halpern, J. Dubach, R. Wallace, K. Pütz, and B. Lüthi. 2009. Heterospecific pairing and hybridization between wild Humboldt and Magellanic penguins in southern Chile. *Condor* 111(3):544–50.

Simeone, A., G. Luna-Jorquera, M. Bernal, S. Garthe, F. Sepúlveda, R. Villablanca, U. Ellenberg, M. Contreras, J. Muñoz, and T. Ponce. 2003. Breeding distribution and abundance of seabirds on islands off north-central Chile. *Revista Chilena de Historia Natural* 76:323–33.

Simeone, A., and R. P. Schlatter. 1998. Threats to a mixed-species colony of Spheniscus penguins in southern Chile. *Colonial Waterbirds* 21(3):418–21.

Skewgar, E., A. Simeone, and P. D. Boersma. 2009. Marine reserve in Chile would benefit penguins and ecotourism. *Ocean and Coastal Management* 52:487–91.

Smith, K. M., W. B. Karesh, P. Majluf, R. Paredes, C. Zavalaga, A. Hoo-gesteijn-Reul, M. Stetter, W. E. Braselton, H. Puche, and R. A. Cook. 2008. Health evaluation of free-ranging Humboldt penguins (*Spheniscus humboldti*) in Peru. *Avian Diseases* 52:130–35.

SPIJ (Sistema Peruano de Información Jurídica). 2011. Sistema Peruano de Información Jurídica. http://spij.minjus.gob.pe.

Taylor, S. S., M. L. Leonard, D. J. Boness, and P. Majluf. 2002. Foraging by Humboldt penguins (*Spheniscus humboldti*) during the chick-rearing period: General patterns, sex differences, and recommendations to reduce incidental catches in fishing nets. *Canadian Journal of Zoology* 80:700–707.

———. 2004. Humboldt penguins *Spheniscus humboldti* change their foraging behaviour following breeding failure. *Marine Ornithology* 32:63–67.

Teare, J. A., E. N. Diebold, K. Grzybowski, M. G. Michaels, R. S. Wallace, and M. J. Willis. 1998. Nest site fidelity in Humboldt penguins (*Spheniscus humboldti*) at Algarrobo, Chile. *Penguin Conservation* 11:22–23.

Thiel, M., E. C. Macaya, E. Acuña, W. E. Arntz, H. Bastias, K. Brokordt, P. A. Camus, J. C. Castilla, L. R. Castro, M. Cortés, C. P. Dumont, R. Escribano, M. Fernandez, J. A. Gajardo, C. F. Gaymer, I. Gomez, A. E. Gonzales, H. E. Gonzalez, P. A. Haye, J. E. Illanes, J. L. Iriarte, D. A. Lancellotti, G. Luna-Jorquera, C. Luzoro, P. H. Manriquez, V. Marín, P. Muñoz, S. A. Navarrete, E. Perez, E. Poulin, J. Sellanes, H. Sepúlveda, W. Stotz, F. Tala, A. Thomas, C. A. Vargas, J. A. Vasquez, and J. M. A. Vega. 2007. The Humboldt Current System of northern and central Chile: Oceanographic processes, ecological interactions and socioeconomic feedback. *Ocean-ography and Marine Biology: An Annual Review* 45:195–344

UNEP (United Nations Environment Programme). 2003. World Conservation Monitoring Centre report on the status and conservation of the Humboldt penguin *Spheniscus humboldti*. United Nations Environment Programme World Conservation Monitoring Centre, Cambridge.

Van Buren, A. N., and P. D. Boersma. 2007. Humboldt penguins (*Spheniscus humboldti*) in the Northern Hemisphere. *The Wilson Journal of Ornithology* 119(2):284–88.

Vargas, F. H., S. Harrison, S. Rea, and D. W. Macdonald. 2006. Biological effects of El Niño on the Galápagos penguin. *Biological Conservation* 127:107–14.

Wallace, R. S., J. Dubach, M. G. Michaels, N. S. Keuler, E. D. Diebold, K. Grzybowski, J. A. Teare, and M. J. Willis. 2008. Morphometric determination of gender in adult Humboldt penguins (*Spheniscus humboldti*). *Waterbirds* 31(3):448–53.

Wallace, R. S., K. Grzybowski, E. Diebold, M. G. Michaels, J. A. Teare, and M. J. Wills. 1999. Movements of Humboldt penguins from a breeding colony in Chile. *Waterbirds: The International Journal of Waterbird Biology* 22(3):441–44.

Ward, T., D. Tarte, E. Hegerl, and K. Short. 2002. Policy proposals and operational guidance for ecosystem-based management of marine capture fisheries. World Wide Fund for Nature, Australia.

Williams, T. D. 1995. *The Penguins.* Oxford: Oxford University Press.

Wilson, R. P., D. C. Duffy, M. P. Wilson, and B. Araya. 1995. The ecology of species replacement of Humboldt and Magellanic penguins in Chile. *Le Gerfaut* 85:49–61.

Wilson, R. P., M. P. Wilson, D. C. Duffy, B. M. Araya, and N. Klages. 1989. Diving behavior and prey of Humboldt penguin (*Spheniscus humboldti*). *Journal of Ornithology* 130:75–79.

Zavalaga, C. B., and R. Paredes. 1997a. Sex determination of adult Humboldt penguins using morphometric characters. *Journal of Field Ornithology* 68(1):102–12.

———. 1997b. Humboldt penguins in Punta San Juan, Peru. *Penguin Conservation* 10(1):6–8.

16

Galápagos Penguin

(Spheniscus mendiculus)

P. Dee Boersma, Antje Steinfurth, Godfrey Merlen, Gustavo Jiménez-Uzcátegui, F. Hernan Vargas, and Patricia G. Parker

1. SPECIES (COMMON AND SCIENTIFIC NAMES)

Galápagos penguin (*Spheniscus mendiculus*) (C. J. Sundevall, 1871)

The Galápagos penguin is also known as *pingüino de Galápagos* (Spanish), *manchot des Galápagos* (French), and *Galápagospinguin* (German).

2. DESCRIPTION OF THE SPECIES

ADULT. The Galápagos is the smallest of the *Spheniscus* penguins, with males generally larger and slightly heavier than females (table 16.1). The adult has a dark blackish to brown back and a white breast flecked with dark feather spots that are individually distinct (fig. 1). On the side of the head, a narrow white line of feathers extends from behind the eyes, around the ear coverts, and down to join at the throat. Galápagos penguins have a dark throat band and a second dark band on the breast that extends down both sides of the upper white breast and along the flanks to the legs (Williams 1995). Their markings are similar to those of the other *Spheniscus* penguins but are finer and subtler. Generally, the white feathers on the chin beneath the bill are more pronounced with less mottling in males than in females (Boersma 1977); the white chin is absent in the other *Spheniscus* penguins. Flippers are generally dark except for a pale central pattern on the ventral side. The bill is not as deep but nearly the same length as in its congeners, with males having deeper bills than females

FIG. 1 (*FACING PAGE*) Adult Galápagos penguin after the molt, when fully feathered around the bill and eyes. (P. D. Boersma)

FIG. 2 Male (*left*) and female (*right*) Galápagos penguin, showing the deeper bill depth and generally wider chin feathers of males. Both sexes lose feathers around the base of the bill and the eyes when in breeding condition. (P. D. Boersma)

(fig. 2). The maxilla is black and hooked at the tip, fitting into the groove of the lower mandible. Usually the distal third of the mandible is black, shading from white to yellowish white to pink at the base. The feet of adults are black with some light shading on the web and are less mottled than juvenile feet (fig. 3).

During the breeding season, Galápagos penguins shed feathers around the bill and eye, exposing bare skin, which is better for losing heat. When pigmented, the skin is black and individually distinct (fig. 4a). The unpigmented skin turns pink or red from blood flow

FIG. 3 The juvenile Galápagos penguin's foot (left) is more mottled and less pigmented than the adult's foot (right). (P. D. Boersma)

FIG. 5 Juvenile Galápagos penguin showing the gray plumage and lack of facial bands. (P. D. Boersma)

FIG. 4 When penguins are not breeding and spend more time in the water, they retain the feathers around the bill and eyes. (a) A Galápagos penguin that is about half defeathered; it has lost feathers around the base of the bill and the eyes and is ready to breed. (b) A Galápagos penguin fully feathered around the base of the bill, after the molt. (P. D. Boersma)

when the penguin is panting to reduce its body temperature. Body temperature can be estimated by counting the number of pants per minute (Boersma 1975). After the molt, when the penguins are not tied to a site, they spend more time in the water, and the skin around the bill and eyes is covered in white or black feathers (Boersma 1977) (fig. 4b). Before the molt, penguins stop oiling their feathers, so feathers become brown; after the molt, feathers are gray-black (Boersma, per. obs.). Unlike their congeners, Galápagos penguins always lack a white tail spot (Boersma, unpubl. data).

IMMATURE. The juvenile Galápagos lacks the white feathers on both sides of the head that outline the cheeks and the dark feather band around the breast to the legs (fig. 5). Like the adult, the juvenile has individually distinct dark feather spots on the breast. The back and head are dark, and the breast is white. The chin and lower throat are gray, and the face can be white to gray. Fledglings have more blue-gray plumage that becomes grayer as they age and eventually turns brown before the molt. They lose the feathers around the bill, and by the time they molt,

their facial skin is bare and their plumage brownish. Feet are pale with black mottling in the webbing that becomes blacker with age (see fig. 3). The coloration of the lower bill is variable but often black at the tip and whitish or pinkish at the base. Juveniles molt into adult plumage at about six months of age, often when adults are breeding (Boersma 1977).

CHICK. A newly hatched chick is covered in grayish-brown protoptile down (fig. 6a) that is eventually worn away, exposing the second coat of gray-brown down (clover down) on the back and white down on the belly (fig. 6b). The juvenile plumage pushes out the second coat of down and is exposed as the down is worn away. Recent fledglings may retain down on their heads and necks for several days.

3. TAXONOMIC STATUS

There are no subspecies recognized.

The oldest penguin fossil is about 55 million years old (Fordyce and Jones 1990). DNA evidence suggests that the genus *Spheniscus* split from the genus *Eudyp-*

FIG. 6 (a) Galápagos penguin chick, one or two days old, showing the first coat of down. (b) Galápagos penguin chicks approximately three to four weeks old, in clover down. (P. D. Boersma)

tula around 25 million years ago, diversifying less than 4 million years ago into the four Spheniscus species: Magellanic (*S. magellanicus*), Humboldt (*S. humboldti*), African (*S. demsersus*), and Galápagos penguins (Baker et al. 2006; Goehlich 2007). DNA analysis suggests that the Humboldt is the closest relative to the Galápagos penguin (Duffy 1991; Thumser and Karron 1994; Baker et al. 2006; Bollmer et al. 2007). The heterozygosity measured at five microsatellites of the Galápagos penguin was 3%, significantly lower than the 46% found for Magellanic penguins, and reflects serial bottlenecks for the species (Akst et al. 2002).

Penguins are restricted to the Galápagos Archipelago but do move among islands (Harris 1973; Boersma 1977; Vargas et al. 2005a; Steinfurth 2007; Vargas et al. 2007). Nims et al. (2008) shows a symmetrical degree of gene flow between island populations and considers the population a single panmictic unit, which is reasonable, as the penguins move among the islands.

4. RANGE AND DISTRIBUTION

The Galápagos penguin, the most northerly species, is endemic to the Galápagos Archipelago, breeding on Isabela, Fernandina, Bartolomé, Santiago, and Floreana Islands (fig. 7a) (Boersma 1977; Vargas et al. 2007). About 95% are found in the westernmost islands, Isabela (including Rocas Marielas, #7 in fig. 7b) and Fernandina (fig. 7b) (Boersma 1977; Vargas et al. 2006). The population's stronghold is along the coasts of northern and eastern Fernandina Island and the southwestern areas of Isabela Island (Boersma 1977; Vargas et al. 2006). The Marielas Islands in Elizabeth Bay were the most important breeding area in the 1970s (Boersma 1977). Chick growth, reproductive success, and nest density were higher in the Marielas Islands than at any other site (Boersma 1977). The center of breeding activity shifted to southwest Isabela Island by 2000, with Caleta Iguana now supporting the largest breeding colony (Vargas 2006; Steinfurth 2007). The change likely occurred because of the introduction of rats on the Marielas Islands in the 1990s; the rats were eradicated, and as of 2011, the Marielas Islands remain rat free (Boersma and Merlen, unpubl. data). At the same time, the removal of dogs from the southwest coast of Isabela Island in the 1980s likely made sites more suitable for the penguins. The penguins' range (fig. 8) also includes the coasts of

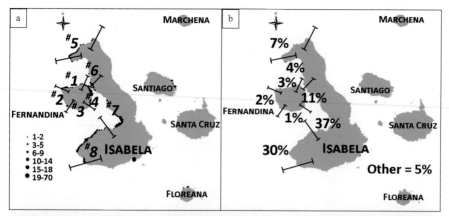

FIG. 7 (a) Distribution of Galápagos penguins (black dots) within eight census areas during the 2005 count. (b) Percentage of the Galápagos penguin population in each of the eight census areas, based on the 2005 count (Vargas et al. 2005).

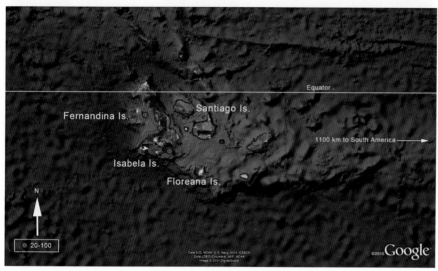

FIG. 8 Distribution and abundance of the Galápagos penguin, with counts based on individuals.

Santiago Island from Bartolomé to Sombrero Chino and the west coast of Floreana Island to at least the village of Velasco Ibarra, where about 5% of the population resides (Vargas et al. 2006). Vagrants are seen throughout the archipelago, including at Baltra, Pinzon, Rábida, Santa Cruz, and San Crístobal Islands (Boersma 1977; Vargas et al. 2005a; Naturalist Guide records at the Charles Darwin Research Station).

5. SUMMARY OF POPULATION TRENDS

The Galápagos population has declined substantially since the early 1970s because of increased frequency of El Niños, decreased frequency of La Niñas, and introduced predators. See section 8.

6. IUCN STATUS

The Galápagos penguin's small population size and restricted geographic range warrant its listing as Endangered on the International Union for Conservation of Nature's Red List of Threatened Species (BirdLife International 2011; IUCN 2010). This species was first listed as Endangered in 2000 on the IUCN's Red List (Stattersfield et al. 2000), is classified as Endangered in the current Red Book of the Birds of Ecuador (Granizo 2002), listed as Endangered under the U.S. Endangered Species Act, and classified as Near Threatened in *Birds to Watch 2* (Collar et al. 1994). The species, although adapted to the unpredictable fluctuations between productive La Niña and unproductive El Niño events (Boersma 1974, 1978), faces new challenges with the increasing intensity and frequency of these events (Boersma 1998a; Vargas et al. 2007). Disease and parasites may be synergistic in combination with climate variation (Levin et al. 2009; Vargas et al. 2007). Introduced alien species (especially rats and cats) remain a problem, and their spread within the archipelago would further endanger the penguins.

Potential threats such as oil transport, fisheries, shipping (including shipwrecks), and tourism are also increasing in the islands.

7. NATURAL HISTORY

BREEDING BIOLOGY. Galápagos penguins breed in lava tubes, caves, and crevices formed by lava plates or fallen basaltic boulders, where they find shade (fig. 9) (Boersma 1975, 1977; Steinfurth 2007). They can breed in any month of the year, depending on food availability, up to three times a year, and in strong El Niño years, they may skip breeding or fail to breed (Boersma 1977, 1978). Steinfurth (2007) notes that 1 of 54 pairs (2%) laid eggs three times in a year, 31 bred twice (57%), and 22 (41%) bred once. Boersma (1977) reports that, in a 15-month period, 74% of 74 pairs laid eggs twice, 24% laid eggs three times, and 11 pairs skipped breeding because they were molting when others were laying eggs.

The penguins may bring sticks, feathers, bones, and leaves to cover the rough bottom of the nest (Boersma 1975, 1977). The eggshells are extremely thick and, even without any nesting material, are unlikely to break (Boersma et al. 2004). Males occupy the nest first, and the pair may stay in the nest until both eggs are laid. The two eggs are laid 3–4 days apart but hatch usually within 2 days of each other (table 16.2). Incubation of the eggs takes 38–42 days and is shared equally between the male and the female (Boersma 1977). The difference in the incubation period for the two eggs is likely a result of parental inattention to the first egg, as has been shown for the Magellanic penguin (Rebstock and Boersma

FIG. 9 Adult Galápagos penguin in a shady lava crevice nest. White guano is in front of the nest site's opening, and a chick is behind the adult. (P. D. Boersma)

2011). Upon hatching, chicks are often brooded for a few days and guarded for a few weeks. When chicks are 25–30 days old, they are left unattended while both parents forage for food. Adults usually return in the late afternoon to feed their chicks (Boersma 1977; Steinfurth et al. 2008). Chicks do not form crèches and do not leave their nest sites until they are ready to fledge. They venture out of their nests only when the nests are located in the most protected sites and close to the water (Boersma, pers. obs.). They fledge at approximately 60 days of age (Boersma 1974, 1977) (table 16.2). After fledging, chicks often return to the shore near the nest, beg for food, and are fed by their parents (Boersma, pers. obs.).

The Galápagos penguin is monogamous, faithful within and often between breeding attempts, and can live for more than 10 years. Of 75 penguins that nested two or more times, 93% stayed with the same mate in successive breeding seasons and 4% of the females switched mates, though their former partners were available, and moved into their new mates' nests (Boersma 1977). Three females divorced their mates, and one of the divorced males remained unmated for the next two breeding periods (Boersma 1977). Five banded males remained unmated for five breeding seasons; no female, once she laid eggs, remained unpaired (Boersma 1977). Galápagos penguins have high rates of nest fidelity. Steinfurth (2007) observes that 16 of 17 pairs returned to the same nest the next time they bred. At least 20 of the same nest sites have been used for 40 years (Boersma, pers. obs.).

Mate switching occurred when a pair failed to raise its brood, a mate was molting, or a partner disappeared and presumably died. Of 79 adults, 4 died over a 15-month period, suggesting survival is near 95% in some years. The recapture rate was 89% for banded individuals at Punta Espinosa, Fernandina Island, over a 10-month period from June 1971 to March 1972, which included part of the 1972 El Niño (Boersma 1977).

PREY AND FORAGING BEHAVIOR. The Galápagos is a generalist and opportunist that, within limits, eats what is available close to shore (Boersma 1977; Steinfurth 2007). These birds will herd fish toward obstacles such as rocks, beaches, or boats where they are trapped or will catch crustaceans and eat small fish held in the edges of a pelican's (*Pelecanus occidentalis urinator*) beak as the pelican strains out the water (Boersma and Merlen, pers.

FIG. 10 (a) Galápagos penguins (circled in red) feeding with pelicans and noddy terns at Punta Mangle, Fernandina Island. (b) The crustaceans that the birds are eating are mysids. (P. D. Boersma)

obs.) (fig. 10a). They normally swallow their prey underwater, catching it as they travel upward.

The penguins prey mainly on nearshore schooling fish typical of upwelling systems, dominated by Engraulidae (likely represented by Pacific anchoveta [*Cetengraulis mysticetus*]), followed by South American pilchard (*Sardinops sagax*) and mullet (*Mugil* spp.) (Boersma 1977; Mills 1998; Vargas et al. 2006; Steinfurth 2007; Steinfurth et al. 2008). Other prey species include *Salema*, an endemic schooling fish (Boersma and Merlen, pers. obs.), the striped herring or Pacific piquitingas (*Lile stolifera*) (Fitter et al. 2000; Vargas et al. 2006; Merlen, pers. obs.), fish in the families Sphyraenidae and Carangidae (Allain, pers. comm.), as well as crustaceans like mysids (fig. 10b) and cephalopods (Steinfurth 2007). In 2005, nonbreeding adults at Caleta Iguana in the southwest of Isabela Island fed almost exclusively on very young pelagic fish, as all prey in stomach samples were 30 millimeters or less in size (Steinfurth 2007). When customary prey are rare, penguins will take damselfish (Pomacentridae) and blennies (Blenniidae) (Boersma 1977, pers. obs.). In April 2007, Boersma saw a penguin come to the surface with a 20-by-6-centimeter large-banded blenny (*Ophioblennius steindachneri*) and try to position and swallow it twice before a sea lion stole the fish. In 1997, Romero Davila (pers. comm.) photographed a Galápagos penguin trying unsuccessfully to swallow an 80-centimeter tiger snake-eel (*Myrichthys tigrinus*) (fig. 11).

The Galápagos forages in shallow water alone or in small groups, following close to the shore, eating small fish 10–150 millimeters in length (Boersma 1977; Stein-

furth et al. 2008). They are a coast-hugging species and shallow divers, spending 90% of the time at depths of less than 6 meters with the deepest known dive being 52 meters (Steinfurth et al. 2008). Dives are short—generally less than a minute—but can last for more than three minutes (Boersma 1977; Steinfurth et al. 2008).

When prey is more plentiful and water temperatures are low, penguins are more likely to be seen in multispecies feeding flocks (fig. 10a) (Boersma 1977, 1978; Mills 1998). When feeding in association with species such as tuna (Scombridae), mackerel (Scombridae), sharks (Carcharhinidae), pelicans, boobies (*Sula* spp.), flightless cormorants (*Phalacrocorax harrisi*), brown noddy terns (*Anous stolidus*), and Galápagos shearwaters (*Puffinus subalaris*), penguins can be well off shore and sometimes

FIG. 11 Galápagos penguin at Tagus Cove in 1997 trying to swallow an approximately 80-centimeter-long tiger snake-eel. Penguins normally swallow their prey underwater. This penguin tried several times to swallow the eel, allowing Fabian Romero Davila time to get the picture. (Romero Davila)

in groups of 200 (Boersma 1977). The presence of penguins increases the duration of feeding by mixed flocks (Mills 1998). Penguins sometimes have slashes on their flippers, legs, and feet, which probably are inflicted while they are feeding in these mixed foraging assemblages with large fish such as tuna, Sierra mackerel (*Scomberomus sierra*), or sharks, dolphins, and whales (Boersma, pers. obs.).

Galápagos penguins are daytime foragers. They generally leave at sunrise and return to land around sunset to rest or to spend the night, although weather and prey availability can modify this pattern (Boersma 1977; Mills 2000; Steinfurth et al. 2008). On southwestern Isabela Island, data loggers recorded penguins foraging during daylight, leaving their nests between 5:11 and 5:48 AM and returning between 11:04 and 5:00 PM (Mills 2000). Individuals may relieve their mates in the late morning or early afternoon when food is available close to their breeding site (Boersma and Steinfurth, pers. obs.). The mean distance traveled from their nest site was 5.2 kilometers, with one penguin traveling 24 kilometers, and foraging trips lasted on average eight hours. The foraging pattern of males and females were similar at two sites (Steinfurth et al. 2008).

MOVEMENT AND MIGRATION. The movement patterns of males and females are similar (Boersma 1977; Steinfurth et al. 2008). Boersma (1977) reports that males moved more than females and were less likely than females to be recaptured. Galápagos penguins move more often when they are not breeding and under poor food conditions. Juveniles wandered more and had higher mortality than adults. Further, adults were twice as likely to be recaptured compared to juveniles. Over a 16-month period, less than 10% of recaptured penguins were found in sites other than the ones where they were banded, indicating high site fidelity.

When not breeding, Galápagos penguins move among islands; no organized migration pattern is known. One nonbreeding penguin moved more than 120 kilometers in less than a month at the beginning of the 1972 El Niño (Boersma, unpubl. data). When not breeding, penguins may travel hundreds of kilometers, for example, from Cabo Douglas, Fernandina Island, to Elizabeth Bay, Isabela Island (Boersma 1977). Vargas et al. (2005a) note that penguins traveled from eastern Fernandina Island to Cabo Douglas, on northwestern Fernandina Island. Stein-

furth (2007) documents a nonbreeding penguin that went 64 kilometers. Two nonbreeding penguins marked by Boersma in May during the 1998 El Niño were captured a month later 27 kilometers and 59 kilometers, respectively, from the place where they had been marked (Vargas et al. 2005a). Penguins would likely die from lack of food if they left the islands. A swimmer caught one juvenile Galápagos, in the company of a larger (probably adult) penguin, in Panama. The other penguin escaped, but both likely came to Panama on a boat (Eisenmann 1956).

PREDATORS. Predation is rare. The Galápagos hawk (*Buteo galapagoensis*) preys on young penguins (Boersma 1977), and the short-eared owl (*Asio flammeus*) and barn owl (*Tyto alba punctatissima*) may prey on adults (Harris 1970, 1974). The Galápagos snake (*Alsophis* and *Antillophis* spp.) and the Sally Lightfoot crab (*Grapsus grapsus*) are native terrestrial predators and act as scavengers, feeding on starving and unattended penguin chicks and eggs (Boersma 1977). Hatchlings are killed mainly by introduced mammals such as black rats (*Rattus rattus*), domestic cats (*Felis catus*), and domestic dogs (*Canis familiaris*) (Boersma 1977). Cats and dogs can also kill adult penguins (Barnett 1986; Steinfurth and Merlen 2005; Steinfurth 2007). Gashes around the tails and on the sides of penguins observed in the 1970s suggest shark attacks (Boersma 1977; Vargas 2009). These marks were absent in 2010 and 2011 (Boersma, pers. obs.), probably because shark populations have declined substantially (Stevens et al. 2005).

MOLT. Following a 7- to 28-day period at sea, during which Galápagos penguins increase their body weight by about a third, molting occurs a week to months before the onset of breeding (Boersma 1975, 1977). During the approximately 10–15 days when penguins lose their old feathers and grow new ones, they fast and avoid the water, often molting in the shade at their nest sites. Many penguins, particularly juveniles and young adults, molt in small aggregations on the coast or in shady spots near water. The interval between molting is a minimum of 5 months and a maximum of 12 months with an average of just over 6 months. Unlike any other penguin species, juveniles molt into adult plumage at about 6 months of age and adults molt twice a year, before they breed, although the timing of molt and breeding is flexible (Boersma 1974, 1977).

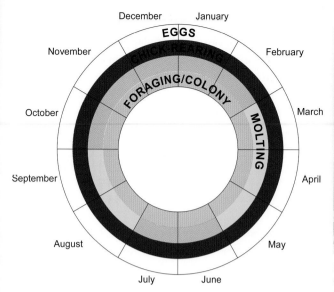

FIG. 12 Annual cycle of the Galápagos penguin.

FIG. 13 Galápagos penguins mating on land. Note the upturned tail and distended cloaca of the receptive female as the male vibrates his bill and treads on her back. The two have lost the feathers around the base of the bill and around the eyes, displaying the individual pigmentation of their skin. (F. Romero Davila)

ANNUAL CYCLE. The life history of the Galápagos penguin is well adapted to the unpredictable availability of food that characterizes the Galápagos marine environment (Boersma 1974, 1977, 1978). They molt as often as twice a year and then breed opportunistically when oceanographic conditions are favorable (fig. 12) (Boersma 1977, 1978). There is no strict breeding season, but sea surface temperatures are usually between 15° and 22°C when breeding occurs; there is no breeding when surface waters are above 25°C (Boersma 1978). Breeding is most frequent in the cool, dry season, generally May through December (Boersma 1978; Vargas et al. 2006; Steinfurth 2007). Galápagos penguins mate both on land and in the water (fig. 13) (Boersma 1977).

8. POPULATION SIZES AND TRENDS

The Galápagos is likely the rarest penguin species. Various estimates of the population in the 1960s were based on opinion and ranged between 500 and 5,000 (references in Boersma 1977). Boersma (1977) places the population at 6,000–15,000 individuals in 1972. Vargas and colleagues (2005a) estimate the 1999 population at 1,054–1,403 individuals and 600–4,000 individuals from 1970 to 2003. The current population is likely between 1,500 and 4,700 individuals. The population has declined and is likely less than half of what it was in the 1970s.

Boersma (1974, 1977) discusses a mark-recapture study conducted in 1972 to determine detection probability and population size. The author searched for banded penguins after 4:00 PM from a Zodiac around Punta Espinosa, Fernandina Island, where she knew how many resident penguins were banded. From 25 June to 23 September 1972, she did 22 searches and saw on average 22% of the banded penguins. The highest resighting rate for a survey was 46% and the lowest was 7% (Boersma 1977). On five surveys performed when the penguins were breeding, the author saw between 19% and 46% of the banded penguins, for a recapture rate of 30%. She did two additional counts around the Marielas Islands and found a 13% resighting rate. Population estimates of Galápagos penguins based on mark-recapture methods may be biased for two reasons. First, the population is not a closed population, and, second, bands and other marks are difficult to detect in the water, so marked penguins are counted as unmarked.

Vargas et al. (2005a) report on a 1999 mark-recapture population estimate in which the authors marked 141 penguins on five islands with picric acid and ethanol. Two weeks later, from 10 to 17 September, they resighted 80 of the marked penguins (57%). The variance among islands was 67%. The recapture rate was lowest on Santiago and Bartolomé Islands (8%) and highest on Floreana Island (75%). On Fernandina Island, the recapture rate was 57%, the same as when all the islands were pooled. Vargas et al. (2005a) consider the population closed and states that marked individuals were not likely to have been resighted more than once because the small boat traveled at 12 kilometers an hour and *Spheniscus* penguins usually travel at slower speeds. Counting penguins, particularly in the water, is challenging, and without individually identifiable marks, it is hard to assess whether penguins were sighted more than once. The authors found one marked penguin that had moved from the site where it had been marked to another island.

Recapture and resighting depend on a number of factors, especially whether the penguin is in the water or on land. A penguin is harder to see in the water, and on land is harder to see when lying with its dark back to the observer than when standing with its white chest facing the observer. Marks are also harder to see in the water. A penguin is most likely to stand on shore late in the afternoon or when its stomach is full or it is molting. Thus, Galápagos penguins are more likely to be seen on land resting when prey is abundant during La Niña events than when it is rare in El Niño periods. Census methods, type of mark, areas surveyed, time of day, weather conditions, and breeding status also influence the number of penguins seen. Detection of both penguins and marks is unlikely to be the same among areas, years, or even days, and a correction factor based on one set of conditions probably will not be constant. Unlike most other penguins, the Galápagos is not a strong seasonal breeder (Boersma 1978).

The two annual cycles in sea surface temperature have opposing forces that obscure a season pattern, and the western archipelago has a quasipermanent pool of cold productive water on the west side of Isabela and Fernandina Islands (Houvenaghel 1978; Palacios 2004). Maximum sea surface temperature occurs in March, when the southeast trade winds are weakest (Palacios

2004). Although penguins may be less likely to breed in March when surface water temperatures are higher and productivity likely lower, there is no best time of the year for surveying the population. The number of penguins seen was smaller during El Niño events (Boersma 1977, 1978, 1998a; Vargas 2006; Vargas et al. 2007) (fig. 14), probably because more penguins were at sea and harder to detect and also because the population crashed. The Galápagos penguin was in much poorer body condition during El Niño events (Boersma 1998b). Major population crashes likely occurred in 1972–73, 1982–83, and 1997–98 (Boersma 1978; Valle and Coulter 1987; Vargas et al. 2006, 2007). From 1970 to 2009, the population may have decreased by about half (Vargas et al. 2005a) (table 16.3).

Mills and Vargas (1997) try to standardize methods for counts and cover penguins around Floreana Island, Santiago Island, and other areas not surveyed in Boersma (1977). The number of counters and type of boat used, weather, time in the breeding cycle (breeding, molting, or not breeding), and time of day all create variation among counts. Most of the population is around Isabela and Fernandina Islands, which were counted 25 times between 1970 and 2009 (see fig. 7a; table16.3). Before 2005, the counts were inconsistent in areas covered, number of observers, time of year, time of day, breeding status, size of vessel used, travel speed, the vessel's distance from shore, weather, sea state, and availability of prey.

The area counted may be the most important factor, as some areas always have more penguins than other areas (see fig. 7b). We divided the area into eight zones,

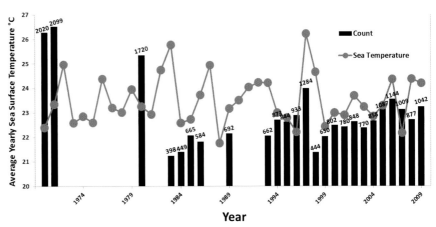

FIG. 14 Counts of Galápagos penguins around Isabela and Fernandina Islands, with the circles showing average yearly sea surface temperature, taken at Punta Ayora, Santa Cruz, Galápagos. Strong El Niño years were 1982–83 and 1997–98.

and for each count, we estimated what the count might have been had all the same areas around Fernandina and Isabela Islands been counted. For counts done before 2005, we determined how much of the census area was not counted or was poorly counted because of bad conditions and extrapolated a count based on that area (table 16.3). Although more area was surveyed starting in 2005, fewer than 200 penguins were found on these other islands (Vargas et al. 2005a). We increased early counts to estimate what would have been counted had all the census zones on Isabela and Fernandina been included but did not subtract the number of penguins counted at other islands in later counts as the number was small, always fewer than 50 (Vargas et al. 2005a).

In summary, the penguin population is small, though the exact population size is unknown. It is likely that counts underestimate the population in El Niño periods, as more penguins are likely to be in the water hunting for food and more should be counted when penguins are molting and standing on shore. The number of penguins missed is likely to vary among days, years, and areas. The correction estimate of 57% in Vargas et al. (2005a) and Boersma's average correction estimate of 22% suggest that the population has ranged from a low of 700 individuals in 1983 to a high of nearly 10,000 in 1971. Using the two correction factors, the 2009 population was likely between 1,800 and 4,700 penguins. Using the lower confidence interval of 1,500 in Vargas et al. (2005a) and our upper figure, we estimate the current population to be between 1,500 and 4,700 penguins.

9. MAIN THREATS

CLIMATE VARIATION. The breeding biology of the Galápagos penguin is adapted to the unpredictable oceanic productivity fluctuations (Boersma 1974, 1977) between productive La Niña and unproductive El Niño events. These climatic events influence the upwelling system and availability of prey for penguins and other seabirds (Boersma 1978). During El Niño events, the Galápagos's body condition declines and mortality increases (Boersma 1998b; Vargas et al. 2007); in response to strong El Niño events, breeding success and the population crash dramatically (Boersma 1978; Vargas et al. 2007). The penguins are a marine sentinel of climate variability in the Galápagos (Boersma 2008). If, as predicted, El Niño events become more frequent and intense, the Galápagos penguin will likely face a higher probability of extinction, which

has been estimated to be 30% within the next century (Boersma 1998a; Boersma et al. 2005; Vargas et al. 2007).

FISHERIES. Development of large-scale fisheries for small, schooling species that penguins depend on could indirectly affect the population. Removal of predatory fish that drive small fish toward the surface where they are available to penguins and other seabirds may also have indirect impacts. Direct effects of fisheries, such as incidental catch, will be high if fishing gear such as gill nets, which kill Galápagos penguins (Merlen, pers. obs.), are allowed in the penguins' range.

INTRODUCED PREDATORS. Introduced predators kill adult penguins or eat their eggs and young, decreasing the species' reproductive success (Boersma 1977; Hoeck 1984; Valle 1986; Cruz and Cruz 1987; Loope et al. 1988; Steinfurth and Merlen 2005; Steinfurth 2007; Vargas 2009). Many of the problems currently faced by seabirds in the Galápagos started in the 1600s–1800s, when pirates, whalers, and fur sealers introduced black rats and house mice (*Mus musculus*) (Hoeck 1984). In 1832, when Ecuador officially claimed the islands, alien plants and animals, including domestic pigs (*Sus scrofa*), goats (*Capra hircus*), dogs, and cats, were introduced with human colonization (MacFarland and Cifuentes 1996; Snell et al. 2002; Jiménez-Uzcátegui et al. 2007). Steinfurth (2007) records cat predation on penguins at Caleta Iguana, Isabela Island, in 2005 and calculates that a single cat could increase adult penguin mortality by 49% per year at this site.

POLLUTION. As economic and political interest in the islands accelerates, demand for resources increases. Human impacts will remain a major problem and threaten not only penguins but the islands' biodiversity; for this reason, UNESCO added the Galápagos to its List of World Heritage in Danger (UNESCO 2007). The islands were removed from the list in 2010 (UNESCO 2010, 2011) without fully addressing the issue. Waste management and growing infrastructure problems put considerable pressure on the managers of Galápagos National Park (Boersma et al. 2005). More than 173,000 tourists visited the islands in 2008, and as land-based tourism increases, visitor impact is hard to control. Growing numbers of visitors and colonists traveling to and from the islands mean more boats, flights, and supply ships (FCD et al. 2007; Naula, pers. comm.). Contamination from oil spills poses a severe threat to many marine organisms, including penguins (Edgar et al. 2003), but few, if any, spill-prevention measures are in place in the islands.

DISEASE. The naive nature of most island populations, due to their evolution in a pathogen-scarce environment, is thought to make them more susceptible to introduced exotic diseases and parasites and thus prone to extinction (Diamond 1984; Groombridge 1992; Dobson and Foufopoulus 2001; Wikelski et al. 2004; Parker et al. 2006). Small populations in particular, such as Galápagos penguins, may be compromised by the loss of genetic diversity (Lyles and Dobson 1993), which often results in reduced ability to react to new pathogens. In addition to extremely low estimates of nuclear genetic diversity (Akst et al. 2002; Nims et al. 2008), this species has extremely low MHC (major histocompatibility complex) diversity (Bollmer et al. 2007), leaving it likely more susceptible to the arrival of new pathogens.

A health assessment of the Galápagos penguin found antibodies to *Chlamydophila psittaci* and *Toxoplasmosis gondii*, although the penguins showed no symptoms of infection (Travis et al. 2006; Deem et al. 2010). The Galápagos has the hemoparasite extraerythrocytic microfilaria (*Nematoda*) (Parker et al. 2006; Travis et al. 2006; Merkel et al. 2007) and the protozoan parasite of the genus *Plasmodium* (Levin et al. 2009). The latter raised serious concerns, as parasites of this genus can cause avian malaria and are known to result in severe mortality in other birds (Miller et al. 2001; Levin et al. 2009). The mosquito *Culex quinquefasciatus*, a vector of avian malaria, has been established in the Galápagos Islands since the 1980s (Fonseca et al. 1998; Whiteman and Parker 2005; Parker et al. 2006), but to date the most pathogenic strain known of the malarial parasite *Plasmodium relictum* has not been found in the Galápagos (Parker et al. 2006; Levin et al. 2009). The parasite identified in Galápagos penguins is most similar to *Plasmodium elongatum* (Levin et al. 2009), which is also highly pathogenic in some avian hosts. Since evidence of all of these pathogenic agents was found during relatively benign times, concern remains that any of them, but particularly *Plasmodium*, may have greater impact when the penguins are under physiological stress such as during an El Niño event. In 1972 and 1997, both El Niño

years, penguins were seen with avian pox–like symptoms, and in 1972, an El Niño period, about three times as many juveniles had the disease as adults (Boersma 1977, pers. obs.).

TOURISM. The Galápagos penguin is a major tourist attraction in the Galápagos Islands (Vargas 2009). Although tourists per se do not cause damage to the islands, the touristic infrastructure creates avenues by which degrading influences such as diseases, alien species, and habitat degradation reach the islands. Tourist sites are generally well controlled, with defined paths and boardwalks. Penguins at Bartolomé and Floreana Islands can be viewed from passing Zodiacs or by snorkeling. No tourism-induced negative effects on the Galápagos penguin are documented, but penguins on shore may enter the water when approached too closely. Generally, other species of penguins tolerate intense tourism, but stress hormones are elevated (Walker et al. 2005). The Galápagos's adrenocortical response to stress depended on its body condition and was higher in 1998, an El Niño year, than in 1999 (Wingfield, unpubl. data).

10. RECOMMENDED PRIORITY RESEARCH ACTIONS FOR CONSERVATION

Recommendations for research and conservation actions should focus on increasing persistence and reducing the risk of extinction. The following are the most important recommendations:

1. Monitor for threats including disease, disease vectors, and invasive alien species and analyze human activities that may indirectly and directly harm penguins.
2. Develop a reliable and cost-effective monitoring program with the aim of determining population trends, demographic parameters, disease, availability of food, reproductive success, and health of the penguin population.
3. If the population declines because of climate variation, disease, fishing, or volcanic activity, consider removing eggs and starting *ex situ* breeding programs.
4. Study the physical and ecological requirements for successful nesting to evaluate and mitigate the impacts of extreme climate variability. Consider the impacts of temperature, solar radiation, humidity, substrate, and presence and absence of predators on penguin survival and success.

11. CURRENT CONSERVATION EFFORTS

The Galápagos penguin is protected under laws of the Republic of Ecuador that established the Galápagos National Park and the Galápagos Marine Reserve, both of which are administered by the staff of the national park. Access to breeding sites is strictly regulated, collection of adults, juveniles, chicks, and eggs is prohibited, and the national park regulates research. The Galápagos Islands were also designated a UNESCO World Heritage site in 1978 and a Man and the Biosphere Reserve in 1984.

Since 1970, Galápagos penguins were counted from small boats around their main breeding areas of Fernandina and Isabela Islands, and counts were conducted annually from 1993 to 2009. The Galápagos population declined by perhaps more than half from its numbers in the early 1970s, and the fewest penguins are seen in El Niño years. An inexpensive and robust system for monitoring the population is needed. About 900 penguins were marked with microchips between 2001 and 2010 (Vargas 2009).

Regulations are in place to reduce the introduction of disease vectors (especially flying insects, including mosquitoes) to areas where penguins occur through the use of lights that are less attractive to insects and installation of insect-elimination equipment on boats.

In response to the growing risk of invasive disease transmission via airplanes and vessels (Wikelski et al. 2004; Travis et al. 2006), the Saint Louis Zoo and the University of Missouri–St. Louis, in cooperation with the Charles Darwin Foundation and Galápagos National Park, initiated an avian-disease surveillance program in 2001 (e.g., Parker et al. 2006; Parker 2009; Gottdenker et al. 2005). In addition, all flights to the Galápagos from continental Ecuador should be fumigated to minimize ongoing introduction of arthropod disease vectors.

Efforts to eradicate or control invasive alien species and predators such as rats are ongoing.

Since 2010, the use of two-stroke engines for tourist boats was banned within the Galápagos Marine Reserve, eliminating discharges that could harm penguins.

The infrastructure that supports a growing human population increases the risk of threats such as oil spills, introduction of alien species, and diseases that threaten penguins. Immigration is strictly controlled through the Special Law for Galápagos, approved by the Ecuadorian Congress in 1998.

12. RECOMMENDED PRIORITY CONSERVATION ACTIONS FOR INCREASING POPULATION RESILIENCE AND MINIMIZING THREATS AND IMPACTS

The substantial decline of Galápagos penguins requires a number of immediate conservation actions.

1. Pay greater attention to the impact of introduced species at the penguins' main breeding sites. Increase efforts to control and/or remove alien species such as feral cats, rats, and mice on islands where penguins breed and make such efforts a high priority.

2. Continue the experiments building shaded nests initiated by the Galápagos National Park and Boersma in 2010 until there are results (Boersma 2010). Penguin breeding success may be limited by a lack of quality nest sites. Breeding attempts may be increased by building shady nest sites or grouping nests. If most of the population can be induced to breed and successfully raise chicks when conditions are favorable, these fledglings, learning to forage when food is abundant, may have a better chance of survival. Penguins able to nest close to other penguins may benefit from group foraging and have higher reproductive success.

3. Initiate efforts to prevent the introduction of new diseases and parasites to the Galápagos ecosystem, especially those that have insect vectors, such as West Nile virus and avian malaria. In addition to preying on native fauna, alien species can introduce infectious diseases and parasites, and people can augment their spread. It is imperative that direct international flights to the Galápagos continue to be prohibited and fumigation of marine vessels, including cargo holds, visiting the islands from the mainland be reinforced. Continue the avian health and baseline study in the Galápagos Archipelago, which will enable early detection of novel avian diseases or parasites, and make timely recommendations to the Galápagos National Park to prevent further spread of diseases and parasites that have already been detected in the islands.

4. Protecting the species where it is most successful would benefit the population and reduce the penguins' risk of extinction. The Galápagos penguin is an icon that attracts tourists and an endangered species. The highest-density breeding area for the penguins (nests per square meter) was on the Mari-

elas Islands until rats were introduced. By mid-2005, more penguins were breeding in Caleta Iguana on Isabela Island than in the Marielas, perhaps because dogs were removed from the southwest coast of Isabela Island in the 1980s and 1990s. The shallow, productive bay around the Marielas Islands has supported the penguin population even during El Niño events. Elizabeth Bay has higher reproductive success, and chicks grew faster there than at sites on Fernandina Island (Boersma 1977). Protection of southwestern Isabela Island is critical for the penguins. The Galápagos National Park has already taken two important steps by removing black rats and restricting tourism activities around the Marielas Islands, with protection zones extending about 3.5 kilometers from the shoreline. Restricting tourist or fishing vessels from anchoring near the island would reduce the chance of harmful rat introduction and petroleum leakage. Protecting the penguins' feeding and breeding area in southwestern Isabela by making the area from Punta Essex, Isabela Island, to Elizabeth Bay, including the Marielas Islands, a no-take zone will help protect the population. The seabirds, including penguins, shearwaters, pelicans, boobies, and terns, depend on large fish, sharks, and dolphins to concentrate their prey. Making this area a no-take zone would be an important step toward maintaining the productive and dynamic nature of this system, benefiting not only penguins but many other species of seabirds, marine mammals, reptiles, sharks, and fish.

5. Maintain regular control and vigilance of feeding and breeding areas to ensure strict compliance with fishery regulations, especially regarding fishing techniques that are responsible for penguin deaths, such as the use of gill nets.

6. Develop a contingency plan for responding rapidly to the appearance of new threats (e.g., oil spills, introduced animals, diseases, shipwrecks, and volcanic eruptions).

7. Reduce human conflicts with wildlife and biodiversity. Reducing human impact on the islands and the surrounding ocean should be woven into all aspects of governance and development.

8. Continued marking and recapturing penguins provides information on population dynamics and helps assess the health of the population. An external mark, like a web-tag, would aid in measuring the failure rate of microchips (Boersma and Rebstock 2010).

9. Conservation in the Galápagos, particularly of the inhabited islands, depends on the local community. Comprehensive environmental education and awareness programs are critical for success. The Galápagos is one of the few places where the natural world should be favored over the human world. The islands are a treasure for all humans and deserve stronger political protection than they currently receive.

ACKNOWLEDGMENTS

We thank the staff at the Galápagos National Park for their encouragement and interest in the protection of penguins. The Charles Darwin Foundation facilitated our research. The David and Lucile Packard Foundation, Commonweal, the Wadsworth Endowed Chair, the German Academic Exchange Service, and the Leiden Conservation Fund provided support for this chapter. We thank David Duffy, Burr Heneman, Ginger Rebstock, Tom Leiden, Amy Van Buren, Eleanor Lee, Rebecca Sheridan, and Sue Moore for reviewing and improving the manuscript and Fabian Romero Davila for his photographs.

TABLE 16.1 Biometric data for the Galápagos penguin (showing mean, standard deviation, and sample size)

VARIABLE	MALES	FEMALES	UNSEXED	REFERENCES
Body mass (kg)[1]	2.2±0.2 (n=61)[10]	1.9±0.2 (n=50)[10]		Boersma 1977
	2.3±0.3 (n=103)	1.9±0.3 (n=62)		Travis et al. 2006
	2.1±0.2 (n=11)[11]	1.8±0.3 (n=12)[11]		Steinfurth et al. 2008
	2.2±0.3 (n=141)	1.7±0.3 (n=90)		Boersma unpublished data
Body length (mm)[2]	526.7±23.56 (n=107)	489.4±28.40 (n=66)		Boersma unpublished data
Body height (mm)[3]			360±11.4 (n=5)	Boersma unpublished data
Body girth (mm)[4]	352.4±17.2 (n=105)	335.1±12.7 (n=63)		Boersma unpublished data
Head length (mm)[5]	122.8±3.1 (n=108)	114.8±3.31 (n=64)		Boersma unpublished data
Bill length (mm)[6]	58.2±2.89 (n=93)	53.9±1.82 (n=83)		Boersma 1977
	58.8±2.40 (n=136)	54.7±2.03 (n=86)		Boersma unpublished data
Bill depth (mm)	19.84±2.77 (n=158)	16.68±0.89 (n=161)		Boersma 1977
	18.67±0.93 (n=136)	16.10±0.76 (n=86)		Boersma unpublished data
Bill gape (mm)[7]	23.1±2.15 (n=108)	20.1±2.44 (n=65)		Boersma unpublished data
Flipper length (mm)[8]	118.66±6.17 (n=126)	113.25±6.33 (n=119)		Boersma 1977
	128.46±4.64 (n=135)	121.0±4.84 (n=85)		Boersma unpublished data
Toenail (mm)[9]	16.41±1.16 (n=134)	15.09±0.96 (n=114)		Boersma 1977
	17.09±0.96 (n=81)	15.60±1.15 (n=52)		Boersma unpublished data
Egg length (mm)			First 62.51±2.32 (n=74) Second 61.74±2.42 (n=65)	Boersma 1974
Egg width (mm)			First 47.88±1.20 (n=74) Second 48.62±1.61 (n=65)	Boersma 1974

1 Weight depends on life cycle stage (see Boersma 1977).
2 Body length is taken from tip of beak to end of the tail when bird is held outstretched horizontally.
3 Body height is taken from standing penguin, base of their feet to the top of their flat head.
4 Body girth is circumference of penguin taken underneath flippers.
5 Head length is tip of bill to the back of the base of skull.
6 Bill length is taken from tip of bill to the base of the "V," where feathering starts on forehead.
7 Bill gape is mouth slit (rimaoris).
8 Flipper length measurements are taken from elbow joint to tip of flipper.
9 Length of the middle toenail on the right foot.
10 Nonreproductive, nonmolting period.
11 Brooding chicks.

TABLE 16.2 Key life-history parameters of the Galápagos penguin

PARAMETER	VALUE	REFERENCE
Age at first breeding	unknown	
Sex ratio	1.1 males:females (n=212)	Boersma unpublished data
Eggs	1st longer, 2nd wider, 2nd more likely to be addled	Boersma 1977
Incubation period	38–42 days	Boersma 1977
Hatching interval	2 to 3 days	Boersma 1977
Guard period	2 weeks	Boersma 1977
Crèche behavior	nest density too low for crèching	
Chick period	50–60 days	Boersma 1977
Molt	before breeding, 2x/year	Boersma 1977
Molt duration	17–43 days	Boersma 1974, 1977
Distance between nests	>1.8m, often >10m	Boersma unpublished data
Max dive depth	52m (most dives less than 6m)	Steinfurth et al. 2008
Juvenile survival rate	variable (0–30%, possibly higher)	Boersma 1977
Adult survival rate	variable (2x higher than juveniles in El Niño events 23–35%); 89% in 10 month period	Boersma 1977 Vargas et al. 2006
Maximum lifespan	unknown (>11 years, n=3)	Boersma 1977

TABLE 16.3 Counts of Galápagos penguins around Isabela, Fernandina, Floreana, Rabida, Santiago (including Logie and Sombrero Chino) and Bartolomé islands made from a small boat

YEAR	COUNT	REFERENCE
1970	1584 (2020)*	Boersma 1974, 1977
1971	1931 (2099)**	Boersma 1974, 1977
1980	1720	Harcourt 1980, Rosenberg et al. 1990
1983 (Sep)	398	Valle and Coulter 1987, Vargas et al. 2005
1984 (Jan)	463	Valle and Coulter 1987, Vargas et al. 2005
1984 (Sep)	435	Valle and Coulter 1987, Vargas et al. 2005
1985	665	Rosenberg et al. 1990
1986	584	Rosenberg and Harcourt 1987
1989	692	Mills and Vargas 1997
1993	662	Mills and Vargas 1997
1994	873	Mills and Vargas 1997
1995	844	Mills and Vargas 1997
1996	933	Vargas et al. 2005b
1997	1284	Vargas et al. 2005b
1998	444	Vargas et al. 2005b
1999	650	Vargas et al. 2005b
2000	802	Vargas et al. 2005b
2001	780	Vargas et al. 2005b
2002	848	Wiedenfeld and Vargas 2002, unpublished
2003	770	Vargas and Wiedenfeld 2003, unpublished.
2004	858	Wiedenfield and Vargas 2004, unpublished
2005	1087	Vargas et al. 2005b, unpublished
2006	1144	Jiménez-Uzcátegui et al. 2006, unpublished
2007	1009	Jiménez-Uzcátegui and Vargas 2007, unpublished
2008	877	Jiménez-Uzcátegui and Vargas 2008, unpublished
2009	1042	Jiménez-Uzcátegui and Devineau 2009, unpublished

Note: Counts for the eight areas around Fernandina and Isabela were estimated for 1970 and 1971 by ranking of the % of penguins counted in the eight areas (fig 7b). From 1970 to 1986 counts were of Fernandina and Isabela islands. Floreana, Rabida, Santiago, or Bartolomé were not counted from 1970 to 1989. When they were counted, less than 50 penguins were seen.

* About 45% of the coastline was not counted. Actual count is 1584, but using the % of penguins in each zone (Fig. 1, Vargas et al. 2005), the count was adjusted to 2020.

** About 20% of the coastline was not counted. Actual count is 1931, but using the % of penguins in each zone (Fig. 1, Vargas et al. 2005), the count was adjusted to 2099.

REFERENCES

Akst, E. P., P. D. Boersma, and R. C. Fleischer. 2002. A comparison of genetic diversity between the Galápagos penguin and the Magellanic penguin. *Conservation Genetics* 3:375–83.

Baker, A. J., S. L. Pereira, O. P. Haddrath, and K. A. Edge. 2006. *Proceedings of the Royal Society B* 273:11–17.

Barnett, B. D. 1986. Eradication and control of feral and free-ranging dogs in the Galápagos Islands. Proceedings of the Twelfth Vertebrate Pest Conference, Nebraska, 358–68.

BirdLife International. 2011. Species factsheet: *Spheniscus mendiculus*. http://www.birdlife.org (accessed 4 May 2011).

Boersma, P. D. 1974. The Galápagos penguin: A study of adaptations for life in an unpredictable environment. PhD diss., Ohio State University.

———. 1975. Adaptations of Galápagos penguins for life in two different environments. In *The Biology of Penguins*, ed. B. Stonehouse, 101–14. London: Macmillan.

———. 1977. An ecological and behavioural study of the Galápagos penguin. *Living Bird* 15:43–93.

———. 1978. Breeding patterns of Galápagos penguins as an indicator of oceanographic conditions. *Science* 200:1481–83.

———. 1998a. Population trends of the Galápagos penguin: Impacts of El Niño and La Niña. *Condor* 100:245–53.

———. 1998b. The 1997–1998 El Niño: Impacts on penguins. *Penguin Conservation*, 10–11.

———. 2008. Penguins as marine sentinels. *BioScience* 58:597–607.

———. 2010. Increasing the population of the Galápagos penguins. Unpublished interim report, 1 October, to the Galápagos National Park Service, Puerto Ayora, Ecuador. 6 pp.

Boersma, P. D., and G. Rebstock. 2010. Effects of double bands on Magellanic penguins. *Journal of Field Ornithology* 81:195–205.

Boersma, P. D., G. A. Rebstock, and D. L. Stokes. 2004. Why penguin eggshells are thick. *The Auk* 121:148–55.

Boersma, P. D., H. Vargas, and G. Merlen. 2005. Living laboratory in peril. *Science* 308:925.

Bollmer, L., F. H. Vargas, and P. G. Parker. 2007. Low MHC variation in the endangered Galápagos penguin (*Spheniscus mendiculus*). *Immunogenetics* 59:593–602.

Collar, N. J., M. J. Crosby, and A. J. Stattersfield. 1994. Birds to watch 2: The world list of threatened birds. Cambridge: BirdLife International.

Cruz, J. B., and F. Cruz. 1987. Conservation of the dark-rumped petrel *Pterodroma phaeopygia* in the Galápagos Islands, Ecuador. *Biological Conservation* 42:303–11.

Deem, S. L., J. Merkel, L. Ballweber, F. H. Vargas, M. B. Cruz, and P. G. Parker. 2010. Exposure to Toxoplasma gondii in Galápagos penguins (*Spheniscus mendiculus*) and flightless cormorants (*Phalacrocorax harrisi*) in the Galápagos Islands, Ecuador. *Journal of Wildlife Diseases* 46(3):1005–11.

Diamond, J. 1984. Historic extinctions: A Rosetta stone for understanding prehistoric extinctions. In *Quaternary Extinctions: A Prehistoric Revolution*, ed. P. S. Martin and R. G. Klein. Tucson: University of Arizona Press.

Dobson, A. P., and J. Foufopoulos. 2001. Emerging infectious pathogens of wildlife. *Philosophical Transactions of the Royal Society of London B* 356:1001–12.

Duffy, D. C. 1991. Field studies of *Spheniscus* penguins. *Spheniscus Penguin Newsletter* 4:10–15.

Edgar, G. J., H. L. Snell, and L. W. Lougheed. 2003. Impacts of the *Jessica* oil spill: An introduction. *Marine Pollution Bulletin* 47:273–75.

Eisenmann, E. 1956. Galápagos penguin in Panama. *Condor* 58:74.

Fitter, J., D. Fitter, and D. Hosking. 2000. *Wildlife of the Galápagos*. Princeton, NJ: Princeton University Press.

Fonseca, D. M., C. T. Atkinson, and R. C. Fleischer. 1998. Microsatellite primers for *Culex pipiens quinquefasciatus*, the vector of avian malaria in Hawaii. *Molecular Ecology* 7:1617–19.

Fordyce, R. E., and C. M. Jones. 1990. Penguin history and new fossil material from New Zealand. In *Penguin Biology*, ed. L. S. Davis and J. T. Darby, 419–46. San Diego, CA: Academic Press.

Galápagos National Park Service. 2009. *Informe de Ingreso de Turistas 2008*. Puerto Ayora, Santa Cruz, Ecuador: Galápagos National Park Service.

Goehlich, U. B. 2007. The oldest fossil record of the extant penguin genus *Spheniscus*: A new species from the Miocene of Peru. *Acta Palaeontologica Polonica* 52(2):285–98.

Gottdenker, N. L., T. Walsh, H. Vargas, J. Merkel, G. Jiménez, R. Eric Miller, M. Dailey, and P. G. Parker. 2005. Assessing the risks of introduced chickens and their pathogens to native birds in the Galápagos Archipelago. *Biological Conservation* 126:429–39.

Granizo, T. 2002. Galápagos penguin. In *Red Book of the Birds of Ecuador*, ed. T. Granizo, 100–101. Red Books Series of Ecuador. Book 2. Quito, Ecuador: SIMBIOE, Conservation International, Ecociencia, Ministerio del Ambiente, UICN.

Groombridge, B. 1992. Global biodiversity: Status of the earth's living resources; A report compiled by the World Conservation Monitoring Centre. London: Chapman and Hall.

Harcourt, S. 1980. Report on a census of the flightless cormorant and the Galápagos penguin. *Noticias de Galápagos* 32:7–11.

Harris, M. P. 1970. The biology of an endangered species, the dark-rumped petrel (*Pterodroma phaeopygia*) in the Galápagos Islands. *Condor* 72:72–84.

———. 1973. The Galápagos avifauna. *Condor* 75:265–78.

———. 1974. *A Field Guide to the Birds of the Galápagos Islands*. London: Collins.

Hoeck, H. N. 1984. Introduced fauna. In *Key Environments: Galápagos*, ed. R. Perry, 223–45. Oxford: Pergamon Press.

Houvenaghel, G. T. 1978. Oceanographic conditions in the Galápagos Archipelago and their relationships with life on the islands. In *Upwelling Ecosystems*, ed. R. Boje and M. Tomczak, 181–200. Berlin: Springer Verlag.

Jiménez-Uzcátegui, G., V. Carrión, J. Zabala, P. Buitrón, and B. Milstead. 2007. Status of introduced vertebrates in Galápagos. In *Galápagos Report, 2006–2007*, 136–41. FCD, PNG, and INGALA. Puerto Ayora, Ecuador.

Jiménez-Uzcátegui, G., and O. Devineau. 2009. Censo del pingüino de Galápagos y cormorán no volador. Informe técnico para la Fundación Charles Darwin y Parque Nacional Galápagos. Manuscript. Puerto Ayora, Ecuador. 15 pp.

Jiménez-Uzcátegui, G., and H. F. Vargas. 2008. Censo del pingüino de Galápagos y cormorán no volador. Informe técnico para la Fundación Charles Darwin y Parque Nacional Galápagos. Manuscript. Puerto Ayora, Ecuador. 20 pp.

Jiménez-Uzcátegui, G., H. F. Vargas, C. Larrea, B. Milstead, and W. Llerena. 2006. Censo del pingüino de Galápagos y cormorán no volador. Informe técnico para la Fundación Charles Darwin, Parque Nacional Galápagos y Sea World & Busch Gardens Conservation Fund. Manuscript. Puerto Ayora, Ecuador. 24 pp.

Levin, I. I., D. C. Outlaw, F. H. Vargas, and P. G. Parker. 2009. *Plasmodium* blood parasite found in endangered Galápagos penguins (*Spheniscus mendiculus*). *Biological Conservation* 142(12):3191–95.

Loope, L. L., O. Hamann, and C. P. Stone. 1988. Comparative conservation biology of oceanic archipelagos. *BioScience* 38:272–82.

Lyles, A. M., and A. P. Dobson. 1993. Infectious disease and intensive management: Population dynamics, threatened hosts and their parasites. *Journal of Zoo and Wildlife Medicine* 24:315–26.

MacFarland, C., and M. Cifuentes. 1996. Case study: Galápagos, Ecuador. In *Human Population, Biodiversity and Protected Areas: Science and Policy Issues*, ed. V. Dompka, 135–88. Washington, DC: American Association for the Advancement of Science.

Merkel, J., H. I. Jones, N. K. Whiteman, N. Gottdenker, H. Vargas, E. K.

Travis, R. E. Miller, and P. G. Parker. 2007. Microfilariae in Galápagos penguins (*Spheniscus mendiculus*) and flightless cormorants (*Phalacrocorax harrisi*): Genetics, morphology, and prevalence. *Journal of Parasitology* 93:495–503.

Miller, G. D., B. V. Hofkin, H. Snell, A. Hahn, and R. D. Miller. 2001. Avian malaria and Marek's disease: Potential threats to Galápagos penguins *Spheniscus mendiculus. Marine Ornithology* 29:43–46.

Mills, K. L. 1998. Multispecies seabird feeding flocks in the Galápagos Islands. *Condor* 100:277–85.

———. 2000. Diving behaviour of two Galápagos penguins *Spheniscus mendiculus. Marine Ornithology* 28:75–79.

Mills, K. L., and H. Vargas. 1997. Current status, analysis of census methodology, and conservation of the Galápagos penguin, *Spheniscus mendiculus. Noticias de Galápagos* 58:8–15.

Nims, B. D., F. H. Vargas, J. Merkel, and P. G. Parker. 2008. Low genetic diversity and lack of population structure in the endangered Galápagos penguin (*Spheniscus mendiculus*). *Conservation Genetics* 9:1413–20.

Palacio, D. M. 2004. Seasonal patterns of sea-surface temperature and ocean color around the Galápagos: Regional and local influences. *Deep Sea Research Part II* 51:43–57.

Parker, P. G. 2009. Parasites and pathogens: Threats to native birds. *Galápagos: Preserving Darwin's Legacy*, ed. Tui de Roi, 177–83. Richmond Hill, Ontario: Firefly Books.

Parker, P. G., N. Kerness Whiteman, and R. E. Miller. 2006. Conservation medicine on the Galápagos Islands: Partnerships among behavioural, population, and veterinary scientists. *The Auk* 123 (3):625–38.

Rebstock, G. A., and P. D. Boersma. 2011. Parental behavior controls incubation period and hatching asynchrony in Magellanic penguins. *Condor* 113(2):316–25.

Rosenberg, D. K., and S. A. Harcourt. 1987. Population sizes and potential conservation problems of the endemic Galápagos penguin and flightless cormorant. *Noticias de Galápagos* 45:24–25.

Rosenberg, D. K., C. A. Valle, M. C. Coulter, and S. A. Harcourt. 1990. Monitoring Galápagos penguins and flightless cormorants in the Galápagos Islands. *Wilson Bull* 102(3):525–32.

Snell, H. L., A. Tye, C. E. Causton, and R. Bensted-Smith. 2002. Current status of and threats to the terrestrial biodiversity of Galápagos. In *A Biodiversity Vision for the Galápagos Islands*, 30–47. Charles Darwin Foundation and World Wildlife Fund. Manuscript. Puerto Ayora, Ecuador.

Stattersfield, A. J., D. R. Capper, and G. C. L. Dutson. 2000. *Threatened Birds of the World: The Official Source for Birds on the IUCN Red List*. Cambridge: BirdLife International.

Steinfurth, A. 2007. Marine ecology and conservation of the Galápagos penguin, *Spheniscus mendiculus*. PhD diss., University of Kiel, Germany.

Steinfurth, A., and G. Merlen. 2005. Predación de gatos salvajes (*Felis catus*) sobre el pingüino de Galápagos (*Spheniscus mendiculus*) en Caleta Iguana, Isla Isabela. Unpublished report to the Galápagos National Park Service and the Charles Darwin Foundation, Puerto Ayora, Ecuador.

Steinfurth, A., F. H. Vargas, R. P. Wilson, D. W. Macdonald, and M. Spindler. 2008. Space use by foraging Galápagos penguins during chick rearing. *Endangered Species Research* 4:105–12.

Stevens, J. D., T. I. Walker, S. F. Cook, and S. V. Fordham. 2005. Threats faced by chondrichthyan fish. In *Sharks, Rays and Chimaeras: The Status of the Chondrichthyan Fishes*, ed. S. L. Fowler, R. D. Cavanagh, M. Camhi, G. H. Burgess, G. M. Cailliet, S. V. Fordham, C. A. Simpfendorfer, and J. A. Musick, 48–57. Gland, Switzerland, and Cambridge: IUCN/SSC Shark Specialist Group.

Sundevall, C. J. 1871. On birds from the Galápagos Islands. *Proceedings of the Zoological Society of London* 1871:124–29.

Thumser, N., and J. D. Karron. 1994. Patterns of genetic polymorphism in five species of penguins. *The Auk* 111(4):1018–22.

Travis, E. K., F. H. Vargas, J. Merkel, N. Gottdenker, G. Jiménez-Uzcátegui, E. Miller, and P G. Parker. 2006. Hematology, serum chemistry, and disease surveillance of the Galápagos penguin (*Spheniscus mendiculus*) in the Galápagos Islands, Ecuador. *Journal of Wildlife Diseases* 42(3):625–32.

UNESCO (United Nations Educational, Scientific and Cultural Organization). 2007. Galápagos and Niokolo-Koba National Park inscribed on UNESCO's List of World Heritage in Danger. http//:whc.unesco.org (accessed 3 April 2011).

———. 2010. List of World Heritage in Danger: World Heritage Committee inscribes the Tombs of Buganda Kings (Uganda) and removes Galápagos Islands (Ecuador). http//:whc.unesco.org (accessed 3 April 2011).

———. 2011. Galápagos Islands. http//: whc.unesco.org (accessed 3 April 2011).

Valle, C. A. 1986. Status of the Galápagos penguin and flightless cormorant populations in 1985. *Noticias de Galápagos* 43:16–17.

Valle, C. A., and M. C. Coulter. 1987. Present status of the flightless cormorant, Galápagos penguin and greater flamingo populations in the Galápagos Islands, Ecuador, after the 1982–83 El Niño. *Condor* 89:276–89.

Valle, C. A., F. Cruz, J. B. Cruz, G. Merlen, and M. C. Coulter. 1987. The impact of the 1982–83 El Niño–Southern Oscillation on seabirds in the Galápagos Islands, Ecuador. *Journal of Geophysical Research* 92(C13):14437–44.

Vargas, F. H. 2006. The ecology of small population of birds in a changing climate. PhD diss., Oxford University.

Vargas, F. H., S. Harrison, S. Rea, and D. W. Macdonald. 2006. Biological effects of El Niño on the Galápagos penguin. *Biological Conservation* 127:107–14.

Vargas, F. H., R. C. Lacy, P. J. Johnson, A. Steinfurth, R. J. M. Crawford, P. D. Boersma, and D. W. Macdonald. 2007. Modelling the effects of El Niño on the persistence of small populations: The Galápagos penguin as a case study. *Biological Conservation* 137:138–48.

Vargas, H. 2009. Penguins on the equator, hanging on by a thread. In *Galápagos: Preserving Darwin's Legacy*, ed. T. De Roy, 154–61. Richmond Hill, Ontario, Canada: Firefly Books.

Vargas, H., C. Lougheed, and H. Snell. 2005a. Population size and trends of the Galápagos penguin *Spheniscus mendiculus. Ibis* 147:367–74.

Vargas, H., A. Steinfurth, C. Larrea, G. Jiménez-Uzcátegui, and W. Llerena. 2005b. Penguin and cormorant census. Unpublished report to the Charles Darwin Foundation and the Galápagos National Park Service, Puerto Ayora, Ecuador.

Walker, B. G., P. D. Boersma, and J. C. Wingfield. 2005. Physiological and behavioural differences in Magellanic penguin chicks in undisturbed and tourist-visited locations of a colony. *Conservation Biology* 137:138–48.

Whiteman, N. K., and P. G. Parker. 2005. Using parasites to interfere host population history: A new rationale for parasite conservation. *Animal Conservation* 8:175–81.

Wiedenfeld, D., and H. Vargas. 2002. Penguin and cormorant survey 2002. Unpublished report to the Charles Darwin Foundation and the Galápagos National Park Service, Puerto Ayora, Ecuador.

———. 2003. Penguin and cormorant survey 2003. Unpublished report to the Charles Darwin Foundation and the Galápagos National Park Service, Puerto Ayora, Ecuador.

———. 2004. Penguin and cormorant survey 2004. Unpublished report to the Charles Darwin Research Station and the Galápagos National Park Service, Puerto Ayora, Ecuador.

Wikelski, M., J. Foufopoulus, H. Vargas, and H. Snell. 2004. Galápagos birds and diseases: Invasive pathogens as threats for island species. *Ecology and Society* 9(1):5.

Williams, T. D. 1995. *The Penguins*. Oxford: Oxford University Press.

VI

Little (or Blue) Penguin

Genus *Eudyptula*

17

Little Penguin

(Eudyptula minor)

Peter Dann

1. SPECIES (COMMON AND SCIENTIFIC NAMES)

Little (or fairy) penguin (Australia), *Eudyptula minor*

The little penguin is also known as the blue penguin and white-flippered penguin in New Zealand.

2. DESCRIPTION OF THE SPECIES

The little penguin is the smallest of extant species, white below and blue above, approximately 33 centimeters tall, with pinkish-white feet, a dark bill, and neotenous plumage that lacks conspicuous coloration or head markings (fig. 1). The plumage wears to a duller bluish gray toward molt. Males are usually slightly larger and heavier, and the bill is more hooked. Juveniles are distinguished by their smaller bills and fresher or bluer plumage, except for post-molt adults, which can be distinguished from juveniles only by bill length (fig. 2). Little penguins from Penguin Island, Western Australia, are consistently heavier and have larger bills than conspecifics elsewhere in Australia or New Zealand (Klomp and Wooller 1988a).

MORPHOMETRIC DATA. See table 17.1.

3. TAXONOMIC STATUS

This species consists of two clades—one occurs across southern Australia and the southeastern part of the South Island, New Zealand, and the other is in the northern parts of the South Island and around the North

FIG. 1 (*FACING PAGE*) Little penguins are the smallest of penguins coming ashore normally under cover of darkness to avoid predators. (J. Harrison)

FIG. 2 Little penguin in a nest burrow lined with grass at Penguin Place, South Island, New Zealand. (H. Ratz)

Island, New Zealand (Banks et al. 2002, 2008; Peucker et al. 2009).

This taxon exhibits regional variation in morphology, coloration, and breeding phenology, which has resulted in six geographically partitioned subspecies being described in the past (Kinsky and Falla 1976). Meredith and Sin (1988a, 1988b) discuss genetic and

FIG. 3 Little penguins often rear two chicks. The younger chick (*foreground*) has more down than its sibling, which is nearly ready to fledge. (H. Ratz)

morphometric studies of four populations within New Zealand. Although individuals of all morphological types were found to be present in most colonies, the authors confirmed three subspecies of those proposed by Kinsky and Falla (1976): *E. m. iredalei*, *E. m. albosignata*, and *E. m. variabilis*. Banks et al. (2002) investigate the division of the six subspecies described in Kinsky and Falla (1976) using morphometric characters, vocalizations, and mitochondrial DNA analysis and finds that *E. minor* comprised two deeply divergent mitochondrial lineages, one restricted to New Zealand and the other shared between Australia and Otago in southeastern New Zealand.

Overeem et al. (2008) examine phylogeographic structuring in a limited area of southeastern Australia, revealing only haplotypes that group within the Australia-Otago clade identified by Banks et al. (2002). There was no phylogeographic structuring within this Australian region (Overeem et al. 2008). In a distribution-wide survey of mitochondrial DNA variation in little penguins, Peucker et al. (2009) also observes that phylogeographic structuring is absent among Australian colonies. In keeping with findings in Banks et al., Peucker et al. note that some Australian individuals exhibit close phylogenetic relationships with a subset of New Zealand individuals, while the remaining New Zealand individuals are phylogenetically distinct. These patterns suggest an origin in New Zealand, followed by recent colonization of Australia and back-dispersal to New Zealand (Banks et al. 2002; Peucker et al. 2009).

The taxonomic status of the distinct white-flippered form described in Kinsky and Falla (1976), *Eudyptula m. albosignata*, as a subspecies is not currently supported by genetic analyses because it is present in both clades. It is critical, however, that this distinct form retains the status required for its conservation (Challies and Burleigh 2004). Individuals from the Chatham Islands in eastern New Zealand (described as *E. m. chathamensis* in Kinsky and Falla) have not been examined genetically in sufficient detail to confirm their subspecific status.

4. RANGE AND DISTRIBUTION

BREEDING DISTRIBUTION. In New Zealand, breeding occurs around the coasts of the North, South, Stewart, and Chatham Islands (fig. 4) (Kinsky and Falla 1976; Robertson et al. 2007).

In Australia, breeding distribution extends from the

FIG. 4 Distribution of the little penguin, with counts based on individuals.

FIG. 5 Little penguin out of its element on Snares Island is surrounded by Snares penguins interested in pecking him. (T. Mattern)

Shoalwater Island Group (Penguin, Garden, and Carnac Islands), near Perth, Western Australia, across the southern coast (including Bass Strait and Tasmania), and up the east coast as far as South Solitary Island, New South Wales (near Coffs Harbour) (Dunlop et al. 1988; Marchant and Higgins 1990; Barrett et al. 2003; Cannell, pers. comm.). There are few colonies on the Australian mainland. The northern limit of the penguins' breeding distribution on both sides of the Australian mainland coincides with the summer isotherms of 20°C (Kinsky and Falla 1976; Marchant and Higgins 1990).

NONBREEDING DISTRIBUTION. Nonbreeding distribution generally is similar to breeding distribution, with occasional records as far south as the Snares Islands in New Zealand (fig. 5) (Miskelly et al. 2001) and occasional Australian records as far north as Brisbane on the east coast and Dirk Hartog Island on the west coast (Marchant and Higgins 1990). Tracking and flipper-banding studies have suggested that there is some separation in the nonbreeding distribution at sea of adults and juveniles (Dann et al. 1992; Collins et al. 1999).

5. SUMMARY OF POPULATION TRENDS

The little penguin population is stable over most of its range, with increases and decreases reported at some sites (less than 5%) in both New Zealand and Australia. It is, however, decreasing overall in several regions, notably mainland southeastern Tasmania and in parts of Australia and the South Island in New Zealand.

6. IUCN STATUS

The International Union for Conservation of Nature lists the little penguin as "Least Concern" on its Red List of Threatened Species because the population is numerous and widespread across southern Australia and New Zealand (IUCN 2011). Some quantitative as well as circumstantial evidence indicates declines in numbers at some sites (less than 5%) and increases at a lesser number of sites over the past 100 years. No change has been recorded at the majority of breeding sites, but most are not monitored systematically.

Overall numbers have declined slightly in Australia, particularly on mainland southeastern Tasmania and, in New Zealand, on some parts of the South Island. In both

New Zealand and Australia, there has been little apparent change on most offshore islands, although quantitative data are lacking. Recent declines have been reported at four islands in southern Australia (table 17.3).

LEGAL STATUS. This species is protected by both federal and state wildlife laws in Australia and federal laws in New Zealand. Destructive use of the penguin is prohibited in both Australia and New Zealand. In New Zealand, under the Wildlife Act 1953, it is an offense to take, capture, kill, or disturb the nest of any protected species.

In New Zealand, littles are listed as a *taonga* species, one of significant cultural and spiritual value, under the Deed of Settlement with Ngāi Tahu. This is not a right for cultural take but a partnership in the protection and management of the species (D. Houston, pers. comm.). The use of parts from dead birds (e.g., feathers) would be permitted for a cultural purpose.

Breeding habitat is not protected in either country under wildlife protection laws, but land is reserved under national park, reserves, or conservation acts. Most breeding sites in Australia (74%) are in reserves of some kind under either federal or state jurisdictions (Dann et al. 1996). These reserves include only those providing some protection for breeding habitat (e.g., national parks, state parks, wildlife reserves, wildlife sanctuaries, wildlife management cooperatives, nature reserves, coastal reserves, and conservation parks). The remaining 26% of breeding sites are on private land, unreserved crown land, leased crown land, harbor or lighthouse reserves, and recreation reserves. It is not known how many penguin breeding sites in New Zealand are currently protected, but it is considered to be around 75% (D. Houston, pers. comm.).

By contrast, less than 5% of the feeding areas of most little penguin colonies in Australia and New Zealand have any form of protection either as marine protected areas (marine national parks, special management areas) or Ramsar Sites.

7. NATURAL HISTORY

BREEDING BIOLOGY. The main life-history characteristics for little penguins are well known (table 17.2). In Australia, this species usually breeds on offshore islands or, less commonly, along parts of mainland coasts (often on talus at the base of cliffs) that are inaccessible to mammalian predators. In New Zealand, many colonies are established on the mainland coasts as well as offshore islands. Most breeding sites are adjacent to the sea with burrows in sand or soil or under vegetation, but in some areas, the birds nest in caves or crevices in rock falls or in human-made nesting boxes (fig. 6). A few sites are in urban areas, and some are on anthropogenic structures such as breakwaters (Preston et al. 2008). Little penguins are territorial about their nest sites (Mouterde et al. 2012). The type and structure of vegetation in breeding areas vary from sparsely vegetated caves and rock screes, to grasslands, herblands, and scrublands, to woodlands and forests (Marchant and Higgins 1990; Dann 1994a; Fortescue 1995).

Breeding activity begins when both parents are in the nest together for about 5 days approximately 30 days before laying occurs (Chiaradia and Kerry 1999). Both parents then go on a pre-laying exodus of about 9 days for males and about 11 days for females (Chiaradia and

FIG. 6 Little penguin in a rock crevice nest on the Otago Peninsula, New Zealand. (P. D. Boersma)

FIG. 7 Two little penguin chicks in their characteristic dark down plumage at Banks Peninsula, South Island, New Zealand. (P. D. Boersma)

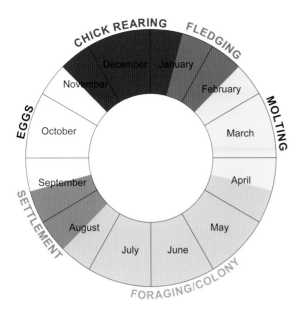

FIG. 8 Annual cycle of the little penguin in Australia.

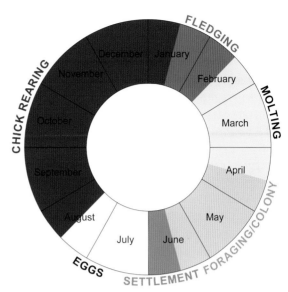

FIG. 9 Annual cycle of the little penguin in New Zealand.

Kerry 1999). Males tend to return earlier (60%) or on the same day (25%). Once at the nest, males and females remain ashore for 5 days until the first egg is laid (Chiaradia and Kerry 1999). Parents alternate stints of 1–10 days incubating the usual clutch of two eggs (Stahel and Gales 1987). The chicks hatch after 35 days (range 33–44) (Kemp and Dann 2001), and both parents feed them. During the guard stage, the parents brood chicks alternately, and from when chicks are about two weeks old (later at some sites) both parents go to sea usually for 1–2 days (Chiaradia and Kerry 1999). Chicks fledge at seven to nine weeks (fig. 7) (Chiaradia and Nisbet 2006). If the eggs or chicks are lost, the pair may initiate a second nesting, and this may also happen after a successful first clutch (Reilly and Cullen 1981; Gales 1985; Fortescue 1995; Perriman and Steen 2000). Third clutches have been reported (Reilly and Cullen 1981; Johannesen et al. 2003).

Young birds do not usually breed until they are two or three years old (Dann and Cullen 1990), although breeding at less than 16 months of age has been reported (Perriman and Steen 2000). Recruits usually return for breeding to the colony where they were hatched (Reilly and Cullen 1981; Dann 1992). Breeding dates and breeding performance vary markedly from year to year, and measures of individual breeding performance (clutch size, hatching success, chick masses, and productivity) are related to early laying, parental age, and duration of the pair-bond (Nisbet and Dann 2009). Early breeding is significantly related to age and pair-bond duration.

High-quality birds, meaning those that produce more chicks, are less prone than low-quality birds to change mates and burrows (Nisbet and Dann 2009).

MATE AND NEST-SITE FIDELITY. On Lion Island, New South Wales, the average number of female and male birds faithful to their nests each year was 76% and 79%, respectively (Rogers and Knight 2006). Across the range of the species, the divorce rate varies widely between years, with a range of 0–40% (Reilly and Cullen 1981; Bull 2000; Rogers and Knight 2006). Nisbet and Dann (2009) report that birds on Phillip Island had 1.8 mates on average in their lifetimes, and pair-bonds lasted 1–13 years. The success of a clutch in one year and fidelity to a mate in the following year were correlated in studies in Otago and New South Wales (Johannesen et al. 2002; Rogers and Knight 2006), and breeding performance is strongly related to pair-bond duration (Nisbet and Dann 2009).

PREY. Diets vary significantly among colonies and between years at the same colony (Klomp and Wooller 1988b; Gales and Pemberton 1990; Cullen et al. 1992; Chiaradia et al. 2003, 2012; Fraser and Lalas 2004). Little penguins feed mainly on clupeids such as anchovy (*Engraulis* spp.), pilchards (*Sardinops sagax*), and sandy sprat (*Hyperlophus vittatus*) when feeding chicks but may also consume krill (*Nyctiphanes australis*) and several species of cephalopods during breeding (Gales and Pemberton 1990; Cullen et al. 1992).

PREDATORS. Introduced mammalian predators in Australia (foxes [*Vulpes vulpes*], dogs [*Canis familiaris*], and cats [*Felis catus*]) and in New Zealand (ferrets [*Mustela furo*], stoats [*Mustela erminea*], cats, and dogs) have been reported taking eggs, chicks, or adult penguins (Stahel and Gales 1987; Dann 1992; Taylor 2000; Challies and Burleigh 2004).

There is no direct evidence of rats (*Rattus* spp.) taking eggs or chicks, but *rakali*, or water rats (*Hydromys chrysogaster*), have been implicated in chick deaths at St. Kilda, Victoria (Preston 2008), and king's skink (*Egernia kingii*) were thought to take eggs in Western Australia (Meathrel and Klomp 1990). Pacific gulls (*Larus pacificus*) and white-bellied sea eagles (*Haliaeetus leucogaster*) take chicks and adults in Australia (mainly at sea or at water's edge) (Wiebkin 2012; Dann, pers. obs.), and brown skuas (*Catharcta skua lonnbergi*) also take chicks and adults crossing wave platforms in the Chatham Islands (Houston, pers. comm.). Tiger snakes (*Notechis ater*) take eggs and chicks (Serventy et al. 1971), and little ravens (*Corvus mellori*) take eggs (Dann, pers. obs.) and chicks (Renwick, pers. comm.). New Zealand fur seals take adults in New Zealand and Australia (Notman 1985; Page et al. 2005; Clemens et al. 2011), and a leopard seal (*Hydrurga leptonyx*) also took adults and one newly fledged bird with a PIT (passive integrated transponder) tag during a stay of some months on Phillip Island at the northern limit of its range (Renwick and Kirkwood 2004).

MOLT. Littles often return to their natal colonies to molt when they are around a year old. Molt for adults occurs once a year after each breeding season and is often done in the same burrow as breeding (Reilly and Cullen 1983). Before the molt, they put on a considerable amount of fat that will sustain them during the three-week molt period (Stahel and Gales 1987). The mean date of molting varies little from year to year compared with egg laying, and birds are more likely to use the same burrow or remain in the vicinity following successful breeding (Reilly and Cullen 1983).

ANNUAL CYCLE. The breeding cycle varies with location. See figure 8 for Australia and figure 9 for New Zealand.

8. POPULATION SIZES AND TRENDS
Overall numbers have declined slightly in Australia,

particularly on mainland southeastern Tasmania, and in New Zealand, on the South Island (table 17.3). In both New Zealand and Australia, there has been little apparent change on most offshore islands, although quantitative data are lacking. However, declines have been reported at four islands in southern Australia. Evidence indicates a decrease in breeding sites in Australia, particularly on mainland Tasmania; penguin populations at several other sites have declined, while two new mainland sites have become established. At present, declines have not been reported where the majority of the species occurs, so a conservative assessment is that numbers are relatively stable, with declines and increases recorded at a few (less than 5%) sites, or that numbers have declined slightly.

In New Zealand, the number of breeding sites on the South Island has been reduced, although three South Island colonies have increased substantially in recent times under active management, suggesting that numbers may have declined slightly or are relatively stable.

The world breeding population is thought to be approximately 600,000 birds, comprising about 500,000 in Australia (Ross et al. 1995; Dann et al. 1996) and about 50,000–100,000 in New Zealand (estimated from Robertson and Bell 1984). These figures are underestimates, as new colonies are still being found and many have not been adequately surveyed. Numbers of pre-breeding individuals are unknown.

9. MAIN THREATS
In an analysis of little colonies along the Victorian coast in Australia, Dann and Norman (2006) conclude that both available breeding area and food supply during breeding may be involved in the regulation of little populations locally. The relationship between population sizes and available breeding area suggests that little penguin numbers may be limited by area on smaller islands but clearly are not on larger islands. Breeding success is related to the foraging area available around breeding colonies (Chiaradia et al. 2007) and colony size (Dann and Norman 2006). Changes in adult survival have a much greater impact on the size of populations than do changes in juvenile survival or breeding productivity (Dann 1992), and therefore factors affecting adult survival are likely to be more significant threats. The main threats to little penguins vary among colonies and the following is a brief review of known and potential threats.

CLIMATE VARIATION. Overall, penguins are likely to be affected in a number of aspects, some positively and some negatively, by predicted climate change over the next 100 years (Dann and Chambers 2009). Breeding productivity and juvenile survival seem likely to improve with increasing sea temperatures in southeastern Australia (Cullen et al. 2009; Sidhu et al. 2012), although this may not be the case for Penguin Island in Western Australia (at the northern limit of the range), where chick mass has decreased with increasing sea temperatures (Cannell and Chambers, in prep.). By contrast, adult mass appears to vary in size between months in response to increasing ocean temperature (Sidhu 2007), while climate change may have both negative and positive effects on feeding behavior (Dann and Chambers 2009; Ropert-Coudert et al. 2009). Some of the negative impacts, particularly those resulting from expected changes to the terrestrial environment, can be addressed on a small scale in the short-term by habitat management, particularly fire and vegetation management (Dann and Chambers 2009).

HISTORICAL OR CURRENT HARVEST. Latham (1802) notes that New Zealand Maori killed little penguins with sticks and ate them, considering them a delicacy. However, there is not much indication that humans in Australia have ever taken adults in numbers for food; eggs and chicks are presumed to have been taken in both countries, but evidence is lacking. Generally, littles cohabit breeding areas with more abundant, and possibly more palatable, shearwaters (*Ardenna* spp.).

FISHERIES. The use of penguins for crayfish bait is thought to have reduced the population of De Witt Island in Tasmania "alarmingly" during the 1950s and 1960s (White 1980). Although this practice may once have been widespread, it now appears rare. Inshore fishing nets have been a source of some mortality for littles in New Zealand (A. Tennyson, pers. comm.), at Victor Harbour and Nepean Bay in South Australia (Copley 1996), in Corner Inlet and Corio Bay in Victoria (Norman 2000), and in southeastern Tasmania (Stevenson and Woehler 2007). The extent of this type of mortality is not known, as reporting rates are likely to be low. This use of nets is more widespread in New Zealand than in Australia.

A number of important prey species of little penguins are taken commercially in Australia and New Zealand, including pilchards and Australian anchovy (*Engraulis australis*) (Klomp and Wooller 1988b; Gales and Pemberton 1990; Cullen et al. 1992; Chiaradia et al. 2010, 2012). However, the direct impacts of these fisheries on penguin breeding success and survival are unknown.

HABITAT DEGRADATION. Introduced mammalian predators such as foxes and dogs are the most significant threat to penguins on land in Australia (Dann 1992), and in New Zealand, ferrets, stoats (Hocken 2000; Challies and Burleigh 2004), and dogs are implicated in a number of colony extinctions or declines. For example, an estimated 500 littles were killed by foxes over a period of six years in a colony in western Victoria (Overeem and Wallis 2007). The role of cats in determining the distribution and abundance of penguins varies, being relatively unimportant on Phillip Island (Dann 1992) but significant on Wedge Island in Tasmania (Stahel and Gales 1987) until the cats were eradicated (C. Vertigan, pers. comm.).

Penguins are or have been killed by cars at a number of places where they cross coastal roads at night to reach their burrows, notably Phillip Island and Portland in Victoria, Bruny Island and Lillico Beach in Tasmania, and Oamaru and Wellington and on the west coast of the South Island in New Zealand (see Hodgson 1975; Dann 1992; Hocken 2000; Heber et al. 2008). The effects on population sizes at each site vary, and, in some cases, road mortality has contributed to local declines in breeding numbers. At its worst, traffic mortality killed an estimated 180 adult penguins per year on Phillip Island before traffic management measures eliminated this cause of death. Penguins have also been killed by trains in Oamaru in New Zealand (Hocken 2000) and in Penguin in Tasmania (Dann, pers. obs.).

Deliberately lit fires are believed to have caused declines in the numbers of penguins breeding on De Witt Island between 1975 and 1977 (White 1980). Fortunately, the deliberate burning of islands along the Australian coast seems to be decreasing but is likely to have contributed to declines in numbers at some sites. Little penguins are particularly susceptible to being killed or injured by fires in their breeding areas (Chambers et al. 2009).

Indirect threats, such as habitat loss through weed invasion, erosion, grazing, and housing developments, remain a concern (Harris and Bode 1981; Dann 1992; Fortescue 1995; Priddel et al. 2008) and have an impact on the distribution and abundance of penguins in some areas. Habitat loss is particularly a problem at mainland sites or on islands that are intensely settled. Trampling of burrows by humans and stock contributes to habitat loss, particularly where erosion develops as a consequence. Fire and rabbits destroy the above-ground vegetation where nests are situated, reducing soil support and fostering burrow collapse.

Some introduced weeds cause severe loss of breeding area, such as kikuyu grass (*Pennisetum clandestinum*), which forms dense patches that penguins find difficult to penetrate. On Bowen and Montagu Islands, the grass is recognized as a serious pest for seabirds, and measures are in place to eliminate it (Fortescue 1995; Weerheim et al. 2003).

COASTAL DEVELOPMENTS. Coastal developments have the potential to affect the marine and terrestrial environments through disturbance at fish nursery sites or increased watercraft access to marine areas used by penguins. For example, the penguins on Penguin Island, Western Australia, feed mainly on sandy sprat when raising chicks (Wooler et al, unpubl. data). The sandy sprat originate from a highly productive nursery (Lenanton et al. 2003), adjacent to which a boat ramp was constructed in 2010. Its impact on both the abundance of sandy sprat and the reproductive success of the penguins is the subject of a 2010–13 study (Cannell, pers. comm.). Increasing human access to marine areas used by little penguins could also result in more injuries and deaths associated with collisions with watercraft.

POLLUTION. Littles are the most likely seabirds in southern Australia to come into contact with oil spills at sea (Dann 1994b). Approximately 10,000–20,000 were affected in the *Iron Baron* oil spill in northern Tasmania in 1995 (Goldsworthy et al. 2000). Oil spills have the potential to reduce populations significantly or, if frequent enough, entirely. Recent advances in less invasive and potentially more effective techniques for cleaning oiled penguins may improve post-oiling survival (Orbell et al. 2005; Van Dao et al. 2006), but cleaning is an ineffective and costly mitigation tool. On October 5, 2011, the *Rena* went aground on a reef in the Bay of Plenty, North Island, New Zealand, oiling several hundred little penguins and thousands of other seabirds (D. Houston, pers. comm.).

Organochlorine and heavy metal accumulations at levels typical for temperate seabirds are found in little penguins living near Sydney and at Phillip Island (Gibbs 1995), but it is not known whether these levels interfere with health or breeding success.

DISEASE. The immunological penalty for malnutrition is poorly understood but is reflected in severe internal helminth parasitic disease in starved birds (Harrigan 1992). Similar burdens of helminth parasites may not be as pathogenic in adult birds in good condition, suggesting that helminths must be regarded as opportunistic pathogens (Norman 2006). Ectoparasites, protozoa, bacteria, and fungi are primary causes of death or agents that contribute to multifactorial deaths in individuals (Harrigan 1988; Norman 2006). There is no indication that anthropogenic sources are contributing to penguin diseases. The role of known bacterial and viral pathogens in birds in the epidemiology of penguin mortality and the costs of morbidity or reproductive wastage have not been investigated or have been only superficially and unsystematically examined.

Disease has affected little penguins indirectly through a viral outbreak among pilchard, one of their main prey species (Murray et al. 2003). In 1995 and, to a lesser extent, in 1998, widespread die-offs of pilchards were first reported at Port Lincoln in South Australia and spread east and west around southern Australia and, in 1995, to the North Island of New Zealand (Griffin et al. 1997). These die-offs were associated with reduced survival and breeding success of little penguins at a number of colonies in Victoria (Dann et al. 2000) and caused fundamental changes in their diet (Chiaradia et al. 2003) that have remained until the present (Chiaradia et al. 2010). The cause of the viral outbreak was undetermined, but according to one hypothesis, the viral disease came from infected pilchard, imported to supply tuna farms.

TOURISM. There are many kinds of human disturbance, and the commercial pressure to allow public viewing without the resulting detrimental effects on the population presents a major challenge. There have been at least 15 sites in Australia and New Zealand where organized viewing of wild little penguins occurs for commercial or community purposes. The most popular of these viewing sites, known as the Penguin Parade, is on Phillip Island in southeastern Australia and currently attracts almost 500,000 tourists each year. Breeding productivity, philopatry, and the movements of adults between areas were similar for birds breeding inside and outside the tourism area (Dann 1992). While tourism is often suspected of having detrimental effects on penguins at viewing sites, in reality, penguin numbers have increased substantially at four monitored and well-managed tourism sites in the past ten years (Perriman and Steen 2000; Houston and Russell 2001; Preston et al. 2008; Dann, pers. obs.). Numbers declined at one viewing site, which instigated changes to viewing behavior of visitors to reduce disturbance (Shaughnessy and Briggs 2009).

10. RECOMMENDED PRIORITY RESEARCH ACTIONS FOR CONSERVATION

Little penguins have a wide distribution; consequently, research priorities are directed toward understanding variations among populations.

1. Monitor distribution of breeding sites and estimate population size in areas where data are scarce, particularly New Zealand and southwest Western Australia.
2. Estimate population in areas where data are more than 15–25 years old (most Australian sites, particularly important in Bass Strait).
3. Establish optimal census methods.
4. Assess the potential competition between fisheries and penguins in Australia and New Zealand.
5. Evaluate gillnetting as a potential source of significant mortality in New Zealand and Tasmania.
6. Develop more effective methods of controlling or eradicating mammalian predators, particularly ferrets and stoats in New Zealand and foxes in Tasmania.
7. Determine why ferrets have not become established in the wild in Australia and take steps to ensure that they never do.
8. Assess and mitigate the impacts of coastal developments and human recreational activities on penguin demography.
9. Continue investigations into effective habitat management, in particular of fire and weeds.
10. Determine interactions with New Zealand and Australian fur seals, both of which are increasing in parts of the littles' range and may have negative impacts on penguin numbers at some sites.
11. Model the effects of changes in sea surface temperature on the size of little penguin populations and their prey.
12. Determine the effects of wind and thermoclines on

foraging success, which is predicted to be affected by climate change.
13. Develop ways to determine prey abundance and availability more readily.

11. CURRENT CONSERVATION EFFORTS

The eradication or control of introduced mammalian predators (i.e., foxes, dogs, cats, ferrets, and stoats) would benefit this species and is the single most important terrestrial management action needed. Dogs have been effectively controlled at Oamaru, New Zealand, and on Phillip Island, Australia; cats have been eradicated from Wedge Island, Australia; foxes have been controlled and, it is hoped, will be eradicated from Phillip Island soon; and ferrets are being controlled on the Banks Peninsula, Taiaroa Head, Nelson, Kaikoura, and at several other small breeding sites on the west coast in New Zealand (D. Houston, pers. comm.).

Availability of habitat appears to be a limiting resource for little penguins at some sites, but penguin habitat can be improved to encourage adult survival and breeding success. Vegetation management for controlling weeds that reduce survival or inhibit burrowing activity and the provision of local fire-resistant and insulative types of vegetation (particularly succulents) increase resilience to climate change. Nest boxes placed in shallow soils in New Zealand and Australia attract penguins to previously unused areas. Traffic and penguin-access management reduce penguin deaths. A spectacular example of habitat protection for little penguins is the Summerland Estate on Phillip Island, where 600 commercial and residential blocks and 180 houses were purchased for AUD$25 million and removed for the safety of the penguins.

Penguin numbers increase or remain stable when tourism is managed by limiting tourists to boardwalks through the colony and allowing penguins unrestricted or undisturbed access; managing levels of light and noise, including banning flash photography; supervising visitation; and investing in predator control.

Threats to little penguins at sea are difficult to detect and manage. The impacts of some (e.g., competition with commercial fishing, climate change) on the population are being assessed, and some conspicuous threats such as oil spills have been reduced through enforcement of pollution regulations, advocacy, and improvements in methods of treating oiled penguins.

12. RECOMMENDED PRIORITY CONSERVATION ACTIONS FOR INCREASING POPULATION RESILIENCE AND MINIMIZING THREATS AND IMPACTS

1. Eradicate foxes from Tasmania and Phillip Island in Australia and cats from any breeding islands uninhabited by humans; develop effective controls for ferrets, stoats, dogs, and cats wherever they are having an impact on penguins.

2. Reduce oil released into the oceans and develop more effective methods of cleaning oiled penguins.

3. Determine the likely changes to inshore marine productivity from predicted changes in climate

4. Develop a thorough understanding of the role of food availability in determining population size (and trends) and the processes involved.

5. Increase long-term studies of foraging ecology to understand natural and human-induced factors affecting the penguins' foraging success over time.

6. Remove harmful weeds, provide fire-resistant and insulated breeding sites (natural and artificial), and establish traffic management systems and fire-fighting strategies designed to reduce penguin mortality where appropriate.

7. Advocate for robust environmental management plans for large-scale human-induced activities (e.g., dredging or long-term discharge of pollutants) that are likely to affect penguins' foraging ecology.

8. Actively manage tourist activities in penguin areas and ensure that they follow established protocols that minimize impacts on penguins.

9. Given the widespread human affection for penguins, we must make it patently clear that the future of penguins is in our hands; managing our consumption and population growth is critical to the future of many, if not all, penguin species.

ACKNOWLEDGMENTS

I would like to acknowledge the assistance of David Houston, Andre Chiaradia, Martin Fortescue, Roz Jessop, Richard Norman, Martin Renner, Chris Challies, Eric Woehler, Nick Holmes, Perviz Marker, Nic Dunlop, Roger Kirkwood, Belinda Cannell, Leanne Renwick, Paula Wasiak, and Kerry-Jayne Wilson, all of whom have made important contributions to or commented on drafts or sections of this chapter. I'm grateful to the Australian Bird and Bat Banding Scheme (Department of Environment and Heritage) for the supply of flipper bands and coordination of the information on banded birds and would also like to acknowledge the considerable efforts of Penguin Study Group members over the past 42 years in assisting with data collection on Phillip Island. The research team at the Phillip Island Nature Park has contributed enormously in many ways to our body of knowledge on little penguins, and the enthusiasm and assistance of its members are inspirational. Finally, I would like to thank my long-suffering family, Moragh, Chloe, and Tarkyn, for their interest in, support of, and tolerance for my addiction to Sphenisciformes.

TABLE 17.1 Measurements for the little penguin (showing mean; standard error for head length, egg mass, egg length and egg breadth; and sample size)

PARAMETER*	MALES	FEMALES	REFERENCE
Mass (g)	1,172 ±164 (17,452)	1,048 ±159 (16,245)	Dann et al. 1995
Flipper length (mm)	120.6 ±2.49 (22)	117.6±3.52 (22)	Phillips 1975 in Marchant & Higgins 1990
Bill length (mm)	39.1±0.1 (207)	36.9 ± 0.1 (193)	Arnould et al. 2004
Head length (mm)	58.0 ± 0.2 (207)	56.2 ± 0.2 (193)	Arnould et al. 2004
Tarsus length (mm)	34.8±1.3 (11)	34.2 ± 01.6 (8)	Phillips 1960 in Marchant & Higgins 1990
Tarsus-toe length (mm)	49.0±1.4 (22)	47.2±1.9 (22)	Phillips 1960 in Marchant & Higgins 1990
Eggs	A-egg	B-egg	
Mass (g)	53.7±0.4 (94)	53.5±0.4 (94)	Kemp & Dann 2001
Length (mm)	55.8±0.2 (95)	54.6±0.2 (95)	
Breadth (mm)	42.0±0.1 (95)	42.2±0.1 (95)	

Note: Additional values can be found in Williams (1995).

* These parameters are for south-eastern Australian birds

TABLE 17.2 Main life-history characteristics of the little penguin

	MEAN AND/OR RANGE	REFERENCE
Incubation period	35.4 days (range 33–44)	Kemp & Dann 2001
Chick-rearing period	59–66 days	Chiaradia & Nisbet 2006
Reproductive (egg) success		
Phillip Island, Aust.	26% (range 3–43%)	Reilly & Cullen 1981
Otago Peninsula, NZ	23–78%	Perriman & Steen 2000
Hatching		
Phillip Island, Aust.	67% (range 47–87%)	Kemp & Dann 2001
Otago Peninsula, NZ	40–81%	Perriman & Steen 2000
Fledging		
Phillip Island, Aust.	40% (range 6- 69%)	Reilly & Cullen 1981
Otago Peninsula, NZ	58–95%	Perriman & Steen 2000
Age of first breeding		
Phillip Island, Aust.	2 years- 50%	Dann & Cullen 1990
	3 years- 50%	
Otago Peninsula, NZ	2 years- 57%	Perriman & Steen 2000
	3 years- 19%	
Survival	17% first year, 71% second year, 78% third year, and 83% thereafter	Sidhu et al. 2007
Maximum lifespan	25.7 years	Dann et al. 2005

TABLE 17.3 Population trends for the little penguin

SITE	YEARS	TREND	SOURCE
New Zealand			
Oamaru	1993–2006	Increasing	Agnew & Houston 2008
Tairoa Head	1993–1999	Increasing	Dann 1994a, Perriman & Steen 2000
Flea Bay, Banks Peninsula	2000–2008	Increasing	T. Stracke & F. Helps unpubl. data
Otago region	<1990	Decreased	Dann 1994a
Westland region	1998–2008	Decreasing	Heber et al. 2008
Banks Peninsula	Mid 1900s-2004	Decreasing	Challies & Burleigh 2004
Australia			
St Kilda, Victoria	1974–2007	Recent site (post 1956) & increasing	Preston et al. 2008
Manly, New South Wales		Declined in the past, now increasing	Priddel et al. 2008
Phillip Island, Victoria	1968–2010	Has decreased & increased, now relatively stable	Dann 1992, Dann et al. 2000, Sutherland and Dann 2012
North-western Tasmania	1985–2010 (c. 24 colonies)	Some declines but overall numbers increasing slightly	P. Marker pers. comm.
Garden Island, Western Australia	2001–2009	Increased	B. Cannell unpubl. data
South-eastern Tasmania (12 colonies)	<1996, 2002–03	Decreasing	Stevenson & Woehler 2007
Granite & West Islands, South Australia	1990–2006	Decreasing	Bool et al. 2007
Middle Island, Victoria	1999–2005	Decreasing	Overeem & Wallis 2007
Bruny Island, Tasmania	<1975	Decreased	Hodgson 1975
Griffiths Island, Victoria	1971–1980	Decreased	Bowker 1980
De Witt Island, Tasmania	1975–77	Decreased	White 1980
Penguin Island, Western Australia	2007–2008	Decreased	B. Cannell unpubl. data
Eden, New South Wales	1935–1972	Decreased (now extinct)	Barton 1978, Priddel et al. 2008
Portland, Victoria	<1990	Decreased (probably now extinct)	P. Dann unpubl. data
Various sites around Sydney, New South Wales	<1960	Now extinct	Rogers et al. 1995, Priddel et al. 2008.

REFERENCES

Agnew, P., and D. Houston. 2008. Oamaru's blue penguins continue to thrive. *New Zealand Journal of Zoology* 35:302.

Arnould, J. P. Y., P. Dann, and J. M. Cullen. 2004. Determining the sex of little penguins (*Eudyptula minor*) in northern Bass Strait using morphometric measurements. *Emu* 104:261–65.

Banks, J. C., R. H. Cruickshank, G. M. Drayton, and A. M. Paterson. 2008. Few genetic differences between Victorian and Western Australian blue penguins, *Eudyptula minor*. *New Zealand Journal of Zoology* 35:265–70.

Banks, J. C., A. D. Mitchell, J. R. Waas, and A. M. Paterson. 2002. An unexpected pattern of molecular divergence within the blue penguin (*Eudyptula minor*) complex. *Notornis* 49:29–38.

Barrett, G., A. Silcocks, S. Barry, R. Cunningham, and R. Poulter. 2003. *The New Atlas of Australian Birds.* Melbourne: Birds Australia.

Barton, D. 1978. Breeding fairy penguins at Twofold Bay, NSW. *Corella* 2:71–72.

Bool, N. M., B. Page, and S. D. Goldsworthy. 2007. What is causing the decline of little penguins *Eudyptula minor* on Granite Island, South Australia? Report to the South Australian Department for Environment and Heritage, Wildlife Conservation Fund and the Nature Foundation SA.

Bowker, G. 1980. Seabird Island no. 99: Griffiths Island. *Corella* 4:104–6.

Bull, L. 2000. Fidelity and breeding success of the blue penguin *Eudyptula minor* on Matiu-Somes Island, Wellington, New Zealand. *New Zealand Journal of Zoology* 27:299–304.

Challies, C. N., and R. R. Burleigh. 2004. Abundance and breeding distribution of the white-flippered penguin (*Eudyptula minor albosignata*) on Banks Peninsula, New Zealand. *Notornis* 51:1–6.

Chambers, L. E., L. Renwick, and P. Dann. 2009. Climate, fire and the little penguin. In *Australia's Biodiversity and Climate Change: A Strategic Assessment of the Vulnerability of Australia's Biodiversity to Climate Change*, ed. W. Steffen, A. A. Burbidge, L. Hughes, R. Kitching, D. Lindenmayer, W. Musgrave, M. Stafford Smith, and P. Werner. Melbourne: CSIRO Publishing.

Chiaradia, A., A. Costalunga, and K. Kerry. 2003. The diet of little penguins *Eudyptula minor* at Phillip Island, Victoria, following the 1995 mass mortality of one of their main prey, the pilchard *Sardinops sagax*. *Emu* 103:43–48.

Chiaradia, A., M. G. Forero, K. A. Hobson, and J. M. Cullen. 2010. Changes in diet and trophic position of a top predator 10 years after a mass mortality of a key prey. *ICES Journal of Marine Science* 67:1710–20.

Chiaradia, A., M. G. Forero, K. Hobson, S. Swearer, F. Hume, L. Renwick, and P. Dann. 2011. Diet segregation between two colonies of little penguins *Eudyptula minor* in southeast Australia. *Austral Ecology* 37:610–19.

Chiaradia, A. F., and K. R. Kerry. 1999. Nest attendance and breeding success in the little penguins *Eudyptula minor* at Phillip Island, Australia. *Marine Ornithology* 27:13–20.

Chiaradia, A., and I. C. T. Nisbet. 2006. Plasticity in parental provisioning and chick growth in little penguins *Eudyptula minor* in years of high and low breeding success. *Ardea* 94:257–70.

Chiaradia, A., Y. Ropert-Coudert, A. Kato, T. Mattern, and J. Yorke. 2007. Diving behaviour of little penguins from four colonies across their whole distribution range: Bathymetry affecting diving effort and fledging success. *Marine Biology* 151:1535–42.

Clemens, S. R. Boss, A. Light, and K. A. Stockin. Attack on blue penguin by a New Zealand fur seal. *New Zealand Journal of Zoology*. 38:33–36.

Collins, M., J. M. Cullen, and P. Dann. 1999. Seasonal and annual foraging movements of little penguins. *Wildlife Research* 26:705–21.

Copley, P. 1996. A review of the status of seabirds in South Australia. In *The Status of Australia's Seabirds*, ed. G. J. B. Ross, K. Weaver, and J. C. Greig. Proceedings of the National Seabird Workshop, Canberra, 1–2 November 1993. Biodiversity Group, Environment Australia. Canberra: Australian Nature Conservation Agency.

Cullen, J. M., L. E. Chambers, P. C. Coutin, and P. Dann. 2009. Predicting onset and success of breeding of little penguins *Eudyptula minor* from ocean temperatures off south-eastern Australia. *Marine Ecology Progress Series* 378:269–78.

Cullen, J. M., T. L. Montague, and C. Hull. 1992. Food of little penguins *Eudyptula minor* in Victoria: Comparison of three localities between 1985 and 1988. *Emu* 91:318–41.

Dann, P. 1992. Distribution, population trends and factors influencing the population size of little penguins *Eudyptula minor* on Phillip Island, Victoria. *Emu* 91:263–72.

———. 1994a. The abundance, breeding distribution and nest sites of blue penguins in Otago, New Zealand. *Notornis* 41:157–66.

———. 1994b. The vulnerability of Australian seabirds to oil spills. Proceedings of the RAOU Seabird Congress, Hobart, 1993. *Australasian Seabird Group Bulletin* 27:16.

Dann, P., M. Carron, B. Chambers, L. Chambers, T. Dornom, A. McLaughlin, B. Sharp, M. E. Talmage, R. Thoday, and S. Unthank. 2005. Longevity in little penguins *Eudyptula minor*. *Marine Ornithology* 33:71–72.

Dann, P., and L. Chambers. 2009. Climate change and little penguins. Report to the Department of Sustainability and Environment, Melbourne, Australia.

Dann, P., and J. M. Cullen. 1990. Survival, patterns of reproduction and lifetime reproductive success in little blue penguins (*Eudyptula minor*) in Victoria, Australia. In *Penguin Biology*, ed. L. Davis and J. Darby, 63–84. San Diego, CA: Academic Press.

Dann, P., J. M. Cullen, and R. Jessop. 1995. Cost of reproduction in little penguins. In *The Penguins: Ecology and management*, ed. P. Dann, I. Norman, and P. Reilly, 39–55. Sydney: Surrey Beatty and Sons.

Dann, P., J. M. Cullen, R. Thoday, and R. Jessop. 1992. The movements and patterns of mortality at sea of little penguins *Eudyptula minor* from Phillip Island, Victoria. *Emu* 91: 278–86.

Dann, P., J. M. Cullen, and I. Weir. 1996. National review of the conservation status and management of Australian little penguin colonies. Canberra: Australian Nature Conservation Agency.

Dann, P., and F. I. Norman. 2006. Population regulation in little penguins *Eudyptula minor*: The role of intraspecific competition for nesting sites and food during breeding. *Emu* 106:289–96.

Dann, P., F. I. Norman, J. M. Cullen, F. Neira, and A. Chiaradia. 2000. Mortality and breeding failure of little penguins in 1995 following a widespread mortality of pilchard *Sardinops sagax*. *Marine and Freshwater Research* 51:355–62.

Dunlop, J. N., N. I. Klomp, and R. Wooller. 1988. Seabird Island no. 188: Penguin Island, Western Australia. *Corella* 12:93–98.

Fortescue, M. E. 1995. Biology of the little penguin *Eudyptula minor* on Bowen Island and at other Australian colonies. In *The Penguins: Ecology and Management*, ed. P. Dann, I. Norman, and P. Reilly, 364–92. Sydney: Surrey Beatty and Sons.

Fraser, M., and C. Lalas. 2004. Seasonal variation in the diet of blue penguins *Eudyptula minor* at Oamaru, New Zealand. *Notornis* 51:7–15.

Gales, R. P. 1985. Breeding seasons and double brooding of the little penguin *Eudyptula minor* in New Zealand. *Emu* 85:127–30.

Gales, R. P., and D. Pemberton. 1990. Seasonal and local variation in the diet of the little penguin, *Eudyptula minor*, in Tasmania. *Australian Journal of Wildlife Research* 17:231–59.

Gibbs, P. J. 1995. Heavy metal and organochlorine in tissues of the little penguin *Eudyptula minor*. In *The Penguins: Ecology and Management*, ed. P. Dann, I. Norman, and P. Reilly, 364–92. Sydney: Surrey Beatty and Sons.

Goldsworthy, S. D, R. P. Gales, M. Giese, and N. Brothers. 2000. Effects of the *Iron Baron* oil spill on little penguins (*Eudyptula minor*). 1: Estimates of mortality. *Wildlife Research* 27:559–71.

Griffin, D. A., P. A. Thompson, N. J. Bax, R. W. Bradford, and G. M. Hallegraeff. 1997. The 1995 mass mortality of pilchard: No role found for physical or biological oceanographic factors in Australia. *Marine and Freshwater Research* 48:27–42.

Harrigan, K. E. 1988. Causes of mortality of little penguins. In *Australian Wildlife, Proceedings*, 705–16. Post-graduate Committee on Veterinary Science, University of Sydney.

———. 1992. Causes of mortality of little penguins *Eudyptula minor* in Victoria. *Emu* 91: 273–77.

Harris, M. P., and K. G. Bode. 1981. Populations of short-tailed shearwaters, little penguins, and other seabirds on Phillip Island, 1978. *Emu* 81:20–28.

Heber, S., K. J. Wilson, and L. Molles. 2008. Breeding biology and breeding success of the blue penguin (*Eudyptula minor*) on the West Coast of New Zealand's South Island. *New Zealand Journal of Zoology* 35:63–71.

Hocken, A. G. 2000. Cause of death in blue penguins (*Eudyptula m. minor*) in North Otago, New Zealand. *New Zealand Journal of Zoology* 27:305–10.

Hodgson, A. 1975. Some aspects of the ecology of the fairy penguin *Eudyptula minor novaehollandiae* (Forster) in southern Tasmania. PhD diss., University of Tasmania.

Houston, D. M., and J. J. Russell. 2001. The impact of tourism on blue penguins (*Eudyptula minor*). *New Zealand Journal of Zoology* 28:440.

IUCN (International Union for Conservation of Nature). 2011. IUCN Red List of Threatened Species. Version 2011.2. www.iucnredlist.org (accessed 9 April 2012).

Johannesen, E., D. Houston, and J. Russell. 2003. Increased survival and breeding performance of double breeders in little penguins *Eudyptula minor*, New Zealand: Evidence for individual bird quality? *Journal of Avian Biology* 34:198–210.

Johannesen, E., L. Perriman, and H. Steen. 2002. The effect of breeding success on nest and colony fidelity in the little penguin *Eudyptula minor* in Otago, New Zealand. *Emu* 102:241–47.

Kemp, A., and P. Dann. 2001. Egg size, incubation periods and hatching success of little penguins *Eudyptula minor*. *Emu* 101:249–53.

Kinsky, F. C., and R. A. Falla. 1976. A subspecific revision of the Australian blue penguin (*Eudyptula minor*) in the New Zealand region. *National Museum of New Zealand Records* 1:105–26.

Klomp, N. I., and R. D. Wooller. 1988a. The size of little penguins *Eudyptula minor*, on Penguin Island, Western Australia. *Records of the Western Australian Museum* 14:211–15.

———. 1988b. Diet of little penguins, *Eudyptula minor*, from Penguin Island, Western Australia. *Australian Journal of Marine and Freshwater Research* 39: 633–39.

Latham, J. 1802. *A General Synopsis of Birds, 1781–1787*. London: Benjamin White and Leigh and Sotheby.

Lenanton, R. C. J., F. Valesini, T. P. Bastow, G. B. Nowara, J. S. Edmonds, and M. N. Connard. 2003. The use of stable isotope ratios in whitebait otolith carbonate to identify the source of prey for Western Australian penguins. *Journal of Experimental Marine Biology and Ecology* 291:17–27.

Marchant, S., and P. J. Higgins, eds. 1990. *Handbook of Australian, New Zealand and Antarctic Birds*. Vol. 1, part A. Melbourne: Oxford University Press.

Meathrel, C. E., and N. T. Klomp. 1990. Predation of little penguins by king's skinks on Penguin Island, Western Australia. *Corella* 14:129–30.

Meredith, M., and F. Sin. 1988a. Genetic variation of four populations of the little blue penguin, *Eudyptula minor*. *Heredity* 60:69–76.

———. 1988b. Morphometrical analysis of four populations of the little blue penguin, *Eudyptula minor*. *Journal of Natural History* 22:801–9.

Miskelly, C., P. Sagar, A. Tennyson, and R. Scofield. 2001. Birds of the Snares Islands, New Zealand. *Notornis* 48:1–42.

Mouterde, S. C., D. M. Duganzich, L. E. Molles, S. Helps, F. Helps, and J. R. Waas. 2012. Triumph displays inform eavesdropping little blue penguins of new dominance asymmetries. *Animal Behaviour* 83:605–11.

Murray, A. G., M. O. O'Callaghan, and B. Jones. 2003. A model of spatially evolving herpesvirus epidemics causing mass mortality in Australian pilchard *Sardinops sagax*. *Diseases of Aquatic Organisms* 54:1–14.

Nisbet, I. C. T., and P. Dann. 2009. Reproductive performance in little penguins in relation to year, age, pair-bond duration, breeding date and individual quality. *Journal of Avian Biology* 40:296–308.

Norman, F. I. 2000. Preliminary investigation of the bycatch of marine birds and mammals in inshore commercial fisheries, Victoria, Australia. *Biological Conservation* 92:217–26.

Norman, R. 2006. Gastric parasitism in the little penguin, *Eudyptula minor*, by nematodes of the genus *Contracaecum* (Anisakidae). PhD diss., University of Melbourne.

Notman, P. 1985. Blue penguin attacked by fur seal. *Notornis* 32:260.

Orbell, J. D., H. Van Dao, L. N. Ngeh, S. W. Bigger, M. Healy, R. Jessop, and P. Dann. 2005. Acute temperature dependency in the cleansing of tarry feathers utilizing magnetic particles. *Environmental Chemistry Letters* 3:25–27.

Overeem, R. L., A. J. Peucker, C. M. Austin, P. Dann, and C. P. Burridge. 2008. Contrasting genetic structuring between colonies of the world's smallest penguin, *Eudyptula minor* (Aves: Spheniscidae). *Conservation Genetics* 9:893–905.

Overeem, R. L., and R. L. Wallis. 2007. Decline in numbers of little penguin *Eudyptula minor* at Middle Island, Warrnambool, Victoria. *Victorian Naturalist* 124:19–22.

Page, B., J. McKenzie, and S. D. Goldsworthy. 2005. Dietary resource partitioning among sympatric New Zealand and Australian fur seals. *Marine Ecology Progress Series* 293:283–302.

Peucker, A. J., P. Dann, and C. R. Burridge. 2009. Range-wide phylogeography of the little penguin (*Eudyptula minor*): Evidence of long-distance dispersal. *The Auk* 126:397–408.

Perriman, L., and H. Steen. 2000. Blue penguin (*Eudyptula minor*) nest distribution and breeding success on Otago Peninsula, 1992 to 1998. *New Zealand Journal of Zoology* 27:269–75.

Phillips, A. 1960. A note on the ecology of the fairy penguin *Eudyptula minor novaehollandiae* (Forster) in southern Tasmania. *Proceedings of the Royal Society of Tasmania* 94:63–67.

Preston, T. 2008. Water-rats as predators of little penguins. *Victorian Naturalist* 125:165–68.

Preston, T. J., Y. Ropert-Coudert, A. Kato, A. Chiaradia, R. Kirkwood, P. Dann, and R. D. Reina. 2008. Foraging behaviour of little penguins *Eudyptula minor* in an artificially modified environment. *Endangered Species Research* 4:95–103.

Priddel, D., N. Carlile, and R. Wheeler. 2008. Population size, breeding success and provenance of a mainland colony of little penguins (*Eudyptula minor*). *Emu* 108:35–41.

Reilly, P. N., and J. M. Cullen. 1981. The little penguin *Eudyptula minor* in Victoria. 2: Breeding. *Emu* 81:1–19.

———. 1983. The little penguin *Eudyptula minor* in Victoria. 4: Moult. *Emu* 83:94–98.

Renwick, L., and R. Kirkwood. 2004. An extended visit by a leopard seal to Phillip Island, Victoria. *Victorian Naturalist* 121:55–59.

Rogers, T., G. Eldershaw, and E. Walraven. 1995. Reproductive success of little penguins, *Eudyptula minor*, on Lion Island, New South Wales. *Wildlife Research* 22:709–15.

Rogers, T., and C. Knight. 2006. Burrow and mate fidelity in the little penguin *Eudyptula minor* at Lion Island, New South Wales, Australia. *Ibis* 148:801–6.

Robertson, C. J. R., and B. D. Bell. 1984. Seabird status and conservation in the New Zealand region. In *Status and Conservation of the World's Seabirds*, ed. J. P. Croxall, P. G. H. Evans, and R. W. Schreiber, 573–86. ICBP Technical Publication No. 2. Cambridge: International Council for Bird Preservation.

Robertson, C. J. R., P. Hyvonen, M. J. Fraser, and C. R. Pickard. 2007. *Atlas *of Bird Distribution in New Zealand, 1999–2004.* Wellington: The Ornithological Society of New Zealand.

Ropert-Coudert, Y., A. Kato, and A. Chiaradia. 2009. Impact of small-scale environmental perturbations on local marine food resources: A case study of a predator, the little penguin. *Proceedings of the Royal Society B* 276:4105–9.

Ross, G. J. B., A. A. Burbidge, N. Brothers, P. Canty, P. Dann, P. J. Fuller, K. R. Kerry, F. I. Norman, P. W. Menkhorst, D. Pemberton, G. Shaughnessy, P. D. Shaughnessy, G. C. Smith, T. Stokes, and J. Tranter. 1995. The status of Australia's seabirds. In *The State of the Marine Environment Report for Australia: Technical Annex 1; The Marine Environment*, ed. L. P. Zann and P. Kailola, 167–82. Townsville: Great Barrier Reef Marine Park Authority; Canberra: Department of the Environment, Sport, and Territories.

Serventy, D., V. Serventy, and J. Warham. 1971. *Handbook of Australian Seabirds.* Sydney: A. H and A. W. Reed.

Shaughnessy, P. D., and S. V. Briggs. 2009. Tourists and little penguins *Eudyptula minor* at Montague Island, New South Wales. *Corella* 33:25–29.

Sidhu, L. 2007. Analysis of recovery-recapture data for little penguins. PhD diss., University of New South Wales.

Sidhu, L. A., E. A. Catchpole, and P. Dann. 2007. Mark-recapture-recovery modelling and age-related survival in little penguins *Eudyptula minor*. *The Auk* 124:815–27.

Sidhu, L. A., P. Dann, L. Chambers, and E. A. Catchpole. 2012. Seasonal ocean temperatures and the survival of first-year little penguins *Eudyptula minor* in south-eastern Australia. *Marine Ecology Progress Series* 454:263–72.

Stahel, C., and R. Gales. 1987. Little penguins: Fairy penguins in Australia. Kensington, Australia: New South Wales University Press.

Stevenson, C., and E. J. Woehler. 2007. Population decreases in little penguins *Eudyptula minor* in south-eastern Tasmania, Australia, over the past 45 years. *Marine Ornithology* 35:71–76.

Sutherland, D. R., and P. Dann. 2012. Improving the accuracy of population size estimates for burrow-nesting seabirds. *Ibis* 154:488–98.

Taylor, G. A. 2000. Action plan for seabird conservation in New Zealand. Department of Conservation, Biodiversity Recovery Unit, Wellington, New Zealand.

Van Dao, H., L. N. Ngeh, S. W. Bigger, J. D. Orbell, M. Healy, R. Jessop, and P. Dann. 2006. Magnetic cleansing of weathered/tarry oiled feathers: The role of pre-conditioners. *Marine Pollution Bulletin* 52:1591–94.

Weerheim, M. S., N. I. Klomp, A. M. H. Brunsting, and J. Komdeur. 2003. Population size, breeding habitat and nest-site distribution of little penguins (*Eudyptula minor*) on Montague Island, New South Wales. *Wildlife Research* 30:151–57.

White, C. 1980. Islands of south-west Tasmania. Sydney: published by the author.

Wiebkin, A. S. 2012. Feeding and foraging ecology of little penguins *Eudyptula* in the eastern Great Australian Bight. PhD diss., University of Adelaide.

Williams, T. 1995. Little penguin. In *Bird Families of the World: The Penguins*, 230–38. London: Oxford University Press.

Species	Population (PAIRS)	Population Trends	THREATS: Climate Change	Fisheries	Pollution	Habitat Degradation	Introduced predators	Human Disturbance	Disease	IUCN Status
King	1.6 million	Stable (Regionally variable)	Potential	Major	Minor	Minor, increasing	None	Minor	Minor	LC
Emperor	238,000	Unknown (Decreasing or stable)	Major	None	Minor	Minor	None	Minor	Unknown/ None	NT
Adélie	4-5 million	Stable (Regionally variable)	Moderate	Minor, potential increase	Minor	Minor	None	None	None, potential	NT
Chinstrap	4 million	Declining	Major	Minor, potential increase	Minor	Minor	None	Minor	None, potential	LC
Gentoo	387,000	Stable/ Increasing	None/Minor	Minor/ Moderate	Minor	Minor	Minor	Minor	Minor	NT
Yellow-eyed	1,700	Stable/Highly variable	Minor, potential increase	Minor	Minor, potential increase	Major	Major	Moderate	Moderate/ Major	EN
Southern Rockhopper	1.2 million	Declining	Moderate	Minor/ Moderate	Moderate, potential increase	Moderate	Minor	Minor	Minor, potential increase	VU
Northern Rockhopper	190,000-230,000	Declining	Minor, potential increase	Minor	Minor, potential increase	Moderate	Minor	Minor	Minor	EN
Erect-crested	80,000	Declining	Unknown, potential	Minor	Minor	Minor	None	None	None	EN
Fiordland	2,500-3,000	Declining/ Uncertain	Minor, potential increase	Minor/ Moderate	Minor, potential Increase	Minor, potential increase	Major	Minor	Unknown/ None	VU
Snares	26,000-31,000	Stable	Minor, potential increase	Minor, potential increase	Minor, potential increase	None	None	None	Unknown/ None	VU
Macaroni	6.3 million	Declining	Moderate	Minor, potential increase	Minor, potential increase	Minor	Moderate/ Major	Minor	Moderate, potential increase	VU
Royal	500,000	Declining	Moderate	Minor, potential increase	Moderate	Minor	Moderate/ Major	Minor	Minor, potential increase	VU
African	25,000	Declining	Major	Major	Major	Moderate	Minor	Minor	Minor, increasing	EN
Magellanic	1.2-1.6 million	Uncertain (Variable)	Moderate	Moderate, potential increase	Moderate, potential increase	Minor, potential increase	Minor	Minor	Minor, increasing	NT
Humboldt	15,000-20,000	Declined/ Slow Recovery	Moderate	Major	Minor	Moderate, potential increase	Minor	Minor	Minor/ unknown	VU
Galápagos	750-2,300	Declining	Major	Minor	Minor, potential	Minor	Moderate/ Major	Minor, potential increase	Minor, potential increase	EN
Little blue	300,000	Stable/ Declining regionally	Minor/ Variable	Minor/ Unknown	Moderate, potential	Moderate, potential increase	Major	Minor	Minor	LC

TABLE 1 Penguin species population size, trends, threats and IUCN conservation status. Declining populations and endangered and vulnerable species have bold red letters. For threats, the boxes are colored to indicate their impacts: red = major, orange = moderate, and yellow = minor.

Conclusion

PABLO GARCIA BORBOROGLU AND P. DEE BOERSMA

The previous chapters clearly point out that most penguin species are in peril—about half are declining, very few are increasing, and the rest are stable or have uncertain population trends. The variation in abundance among species covers more than three orders of magnitude, reflecting factors such as their distribution range. The rarest penguin is the Galápagos; the most abundant are the Adélies and chinstraps, with 4 million or more pairs. The ones in the most trouble have restricted distributions and are threatened by human alterations to their environment. The penguins with the largest populations and the largest distribution ranges are in Antarctica, where human contact is rare, and these make up the majority of the species of least concern. Most of the temperate penguins that live close to people have the largest problems, although the little penguin is an exception to that rule.

The threats facing penguins vary in intensity and frequency among the species. Table 1 shows population size and trends, along with the major threats for 18 species. Major threats are highlighted in red, moderate threats in orange, and minor threats in yellow. Dark red lettering indicates declining population trends and species listed as Endangered and Vulnerable by the International Union for Conservation of Nature (IUCN).

Some threats have historically been particularly damaging. The harvesting of penguins was a severe problem from the 1800s until the late 1970s, when penguins were killed for their oil, but this is no longer causing populations to decline. Introduced predators continue to hamper penguins that breed on islands that were formerly predator-free but that now swarm with new mammalian predators. Petroleum pollution seriously affects temperate penguin species along tanker routes or where oil is discharged. Species classified as endangered or vulnerable are found mainly around New Zealand, in the Eastern Pacific, at the southeast Atlantic islands, and along the south coast of Africa (fig. 1).

NATURAL HISTORY

The natural-history traits penguins share with most other seabirds tell us that there are very few ways of surviving and reproducing successfully in a variable oceanic environment. Success depends on sufficient food, and the birds' breeding times are often flexible because of ocean variability. The annual cycle presented for each penguin species masks the variability within species and among locations; none shows the exact timing for each location but is instead a general representation for the species. Breeding location is the variable that determines the timing of events within a species' annual cycle.

Many avian traits contribute to success in rearing young. For most birds, reproduction would be unsuccessful without both parents caring for and feeding offspring; monogamy and biparental care overwhelmingly dominate among birds. Penguins follow the expected pattern in being monogamous and having both parents incubate the eggs and feed the offspring. Other traits such as delayed maturity, longevity, slow growth, small clutch size, and flexibility in the timing and duration of

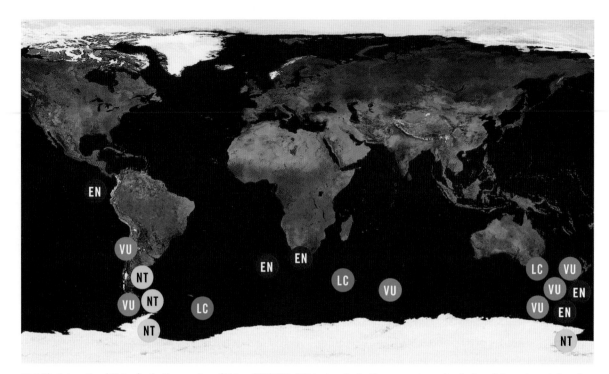

FIG. 1 The International Union for the Conservation of Nature (IUCN) Red List categories for the 17 penguin species. Each oval shows the main breeding concentration for each species with its red list category. EN=Endangered, VU=Vulnerable, NT=Near threatened, LC=Least concern

replacing feathers have evolved. These traits foster survival when food is variable and supply is unpredictable, features that dominate the oceanic environment. Nearly all seabirds have similar life-history traits. One fundamental difference between other seabirds and penguins is that only penguins cannot fly and thus they cannot rapidly traverse large distances. Penguins are efficient swimmers, but swimming is much slower than flying; therefore, their foraging range is restricted, particularly when they must return to a fixed location to incubate and rear chicks.

Penguins depend on highly productive waters near their breeding locations. Little, yellow-eyed, and Galápagos penguins stay close to their breeding colonies throughout the year because food is always available. They are close to their nest sites and can be ready to breed when conditions are right.

When food is not predicable at one location throughout the year, penguins must move to a place where it is dependably available. Some species, such as Magellanic, Adélie, and king penguins, migrate thousands of miles between their breeding and wintering areas. Humboldt penguins in El Niño years may migrate south to Peru, where anchovies are more reliable.

Galápagos penguins can lay eggs three times a year, the African and Humboldt penguins twice a year, and littles may sometimes lay two clutches. All of these species live in relatively more productive environments and usually remain near their breeding sites throughout the year. All other penguins lay eggs only once a year because of seasonality.

All birds replace their feathers, typically at least once a year, because feathers wear with age. Growing new feathers is energetically costly and nutrient intensive, and during the molt, mobility is usually impaired, making flight costlier. The frequency, timing, and pattern of molt are highly variable because molting's cost imposes tradeoffs. Many birds molt only a few flight feathers simultaneously and can always fly. At the opposite extreme, most waterfowl shed all their flight feathers at once, losing their ability to fly, so they molt at sites where they can feed while flightless. Flying seabirds, however, do not have such places, and not being able to fly means certain starvation. In the most extreme cases, some albatrosses take two years to complete their molt.

Unlike any other seabird, penguins fast, molt, and grow new feathers while resting on land or ice. Only after their new feathers are grown can they return to the ocean to forage. Before the molt, they gorge, nearly doubling their weight, and then come ashore to fast for about three weeks while they push out their old feathers and grow new ones. Penguins have individual flexibility in when, where, and how long they molt, and the timing and location of their molts vary greatly. A penguin may delay or even skip a molt if it is in poor condition, but there are costs. One Magellanic penguin kept its juvenile plumage for more than two years, but its back and tail were nearly bare, increasing its rate of heat loss. Penguins in poor condition take longer to molt, which makes them more likely to die. The Galápagos penguin molts twice a year before breeding; all other penguins molt once a year. The timing of the Galápagos's molt implies that the birds take care of themselves first because productivity is highly variable. They may need new feathers so that they will be in top condition and able to grow chicks rapidly, as living on the equator likely increases feather wear. Galápagos penguins do not migrate far, so they can feed their chicks after the chicks fledge, the same way gentoo penguins do, which do not have to migrate.

Adélies, like most penguins, molt after the young fledge, but they are so constrained by the seasonality of their breeding sites that they must desert their chicks to fatten up for the molt. If they continue to feed their chicks, they lose the opportunity to fatten up when food is abundant, and if they do not gain enough weight for the molt, they die. A decline in their foraging time or prey availability forces Adélies to migrate and thereby shortens their breeding period. Molting more than once a year would be so costly for most penguins that it cannot be done. The energy, time constraints, and foraging costs must be too high, except in a very productive ocean. The natural-history traits of a species reflect the adaptations it has made so that it can cope with its dynamic environment, but these may not be adaptive in the current human-dominated, rapidly changing world.

CONSERVATION

What insights from penguin natural history can be applied to conservation? First, because penguins cannot cover long distances quickly, they are restricted to rich, productive waters. Changes in spatial and temporal patterns of food impact their breeding success and winter survival. Many penguins depend on pelagic prey—krill, squid, anchovy, and herring. Marine protected areas, marine zoning, and other conservation tools implemented around breeding colonies or in wintering areas are likely to benefit penguins. Few such areas have been designated or established effectively. This important conservation gap needs to be filled.

Climate change is creating penguin winners and losers. Two species, the king and the gentoo, have benefited by expanding their range south. Others, like the Galápagos, African, and Humboldt penguins, are losers because of the increased frequency or intensity of environmental events such as El Niño. Chinstraps and Adélies in the Antarctic Peninsula depend on krill, which winter under ice where they can feed on algae. The melting of ice decreases krill survival, and their abundance plummets. Climate warming is predicted to be highest at high latitudes, where it has already caused chinstrap and Adélie penguins in the Antarctic Peninsula to decline. Warming will also likely remove breeding habitat for emperor penguins. Early breakup of shore-fast ice where emperor penguins breed can cause complete reproductive failure for the colony. Lack of ice may also reduce molting sites for ice-associated penguins.

The penguin's longevity is likely to mask some of the negative impacts of climate variation for several decades. Increased rainfall coupled with cold winds can kill chicks that historically grew up in a dry desert climate on the coasts of Peru, Argentina, South Africa, or a sub-Antarctic island. From the Galápagos to the Antarctic, climate variability will likely have a negative impact on breeding success. Increased rainfall in deserts and rain instead of snow in the Antarctic, depending on when it occurs, lower reproductive success.

Fishers can catch and kill penguins. Most worrisome are fisheries that target penguin prey. The Humboldt penguin and other Peruvian seabirds never recovered from the development of the anchovy fishery. Fisheries that target forage fish, particularly in productive waters, will have a negative impact on penguins whether they are located on the coast of Peru, Chile, Argentina, or South Africa, on islands like the Falkland/Malvinas Islands or South Georgia, or in Antarctica. Fisheries are already harming king, Humboldt, and African penguins. The development or expansion of fisheries is likely to be damaging to all penguins.

Petroleum pollution kills penguins. It caused the African penguin population to collapse and is a moderate problem for four other temperate penguin species (table

1). Pollution is likely to grow as risky petroleum development and oil transport increase around the world.

Habitat degradation and coastal development are certain to continue in many places where penguins live. Penguins and seabirds must lay their eggs on land, but they are extremely vulnerable to mammalian predators when they are breeding. Unfortunately, rats, cats, and a variety of other predators, deliberately or accidentally introduced to islands, have decimated populations and endangered many seabird species. Shortly after the Maoris arrived in New Zealand, one species of penguin that was bigger than the yellow-eyed became extinct. Mitigation and restoration are likely the only means of improving penguin habitats. The challenges of removing introduced predators such as rats, stoats, and cats from New Zealand and other islands are formidable.

Human presence on breeding grounds can be a problem. Unless visits are regulated and managed as they are in the Snares Island, human disturbance can harm penguins. Guidelines for minimizing human impacts are important for penguins as well as other seabirds.

Disease is thought to be a moderate or minor factor for most species, but our knowledge is very limited. Globalization of trade and rapid transportation increase connections and lower barriers, facilitating the introduction and spread of disease. The spread of the featherless chick condition from Africa to Argentina and now found in zoos in Asia shows how easily disease can be spread between species that normally do not come in contact with one another. The threat of disease may be higher than currently recognized.

In the past twenty years, the conservation status of 14 species has been upgraded. In 1988, the IUCN listed 3 species of penguins as Threatened, Vulnerable, or Endangered; in 1994 it listed 5, and in 2004, 11. This increase reflects less a change in conservation status than the pooling of information by scientists. It highlights how long it took to recognize the threats and the need for scientific discussion to draw attention to upgrading the IUCN Red List.

This book has already started the process of protecting penguins. When Australian scientists and contributors to the book realized that the emperor penguin was not protected under their national laws, they started discussions about adding the bird to the Environmental and Biodiversity Conservation Act. We gladly allowed the government of South Africa to use the chapter on African penguins in developing its biodiversity management plan.

Each chapter in this volume provides a list of recommended actions that need implementation if they are to be useful. Larger-scale ecosystem-based planning and protection are needed. Success requires focused local efforts for conservation. In addition to these actions, protection for penguins starts with the decisions we make as individuals—from what we eat, to what we do, to what we conserve. Our actions matter for people and for penguins.

Acknowledgments

Books are labors of love and can take years to produce. This one is no different, and it would never have happened if Popi Garcia Borboroglu hadn't insisted that "we" bring together all the available information on every species of penguin to help preserve them and conserve their environment. We are both glad we did. The University of Washington Press encouraged us every step of the way, and we thrived under then-director Pat Soden's encouragement. We are also grateful to Thomas Eykemans for his masterful handling of the illustrations and for his fine design, and to editors Laura Iwasaki and Marilyn Trueblood for their attention to detail. Laura Marina Reyes and Sue Moore advised us on aspects of the book throughout its preparation. The maps were provided by Jeffrey Ryan Smith and the annual cycles by Courtney Wenneborg. El Lee read many drafts of the manuscript. Rebecca Sheridan's early work on the book was invaluable: she researched and edited and was instrumental in the book's structure and layout. Funds from the Leiden Foundation helped with the preparation of the book, and the Foundation hosted a workshop in South Africa, using the material in the book for conservation planning. Boersma's Wadsworth Endowed Chair in Conservation Science and Borboroglu's Pew Fellowship also provided support to enhance the book's content.

We are grateful to the volume contributors, all of whom devote themselves to researching penguins and who joined us in compiling the natural history, conservation issues, and research needs for each of eighteen species. This book was made possible because of the research community's enthusiasm for the project. Many of the scholars were writing their chapters even while collecting penguin data on remote islands, bobbing around the oceans of the world, or living in research stations with limited Internet access.

Our biggest debt is to our long-term research funders. We are grateful to all the researchers who conducted a detailed review of the chapters and to all the people who offered their beautiful images to illustrate the book.

PABLO GARCIA BORBOROGLU: I am deeply grateful to Luc Hoffmann for his sustained interest and support for more than two decades. Rufford Small Grants Foundation was a tremendous partner in my work with penguins in Argentina, allowing me to think and act on a large scale. The Pew Fellows Program in Marine Conservation, the Whitley Fund for Nature, and George and Natasha Duffield provided the tipping point to make the effort international and allow dreams to come true. The Centro Nacional Patagonico (CONICET), the University of Washington, Duke University, Wildlife Conservation Network, Disney Interactive Studios, and individuals supported my research and conservation work. I am also grateful to my parents and my grandmother, Melania, who used to tell me stories about penguins in Patagonia—stories I have now told my sons, Alejo and German. And special thanks to my wife, Laura Reyes, who gives a vast dimension of light to my life.

P. DEE BOERSMA: I am grateful to Dr. William Conway, who as director general of the Wildlife Conservation Society saw the importance of a long-term research project at Punta Tombo and helped it happen for thirty years;

and to G. Harris, who continues to foster the project. Institutions that helped fund the research include the University of Washington, the Wildlife Conservation Society, Pew's Fellows Program in Marine Conservation, and the foundations of CGKM, Chase, Disney, Exxon/Mobil, W. K. Kellogg Foundation National Fellowship Program, National Geographic, Leiden, Offield, and Tortuga. The Wadsworth Endowed Chair at the University of Washington gave me the freedom to pursue my passion for understanding penguins. Both Argentine and American students associated with the University and numerous volunteers helped collect data on Magellanic penguins. If it hadn't been for the Punta Tombo project, I would not have met many of the people who have so enriched my life. I hope you know who you are. I also thank my mother and my sister, who visited to work at Punta Tombo. Most importantly, I thank Dr. Sue Moore for her help and because she believed that the best thing I've done was to keep Punta Tombo in my life.

Finally, both of us owe the penguin. Penguins teach us that we only live once, that life isn't easy, and that human existence should make a positive difference to the other creatures with whom we share the planet.

PDB & PGB
January 2013

GRANT BALLARD
PRBO Conservation Science
3820 Cypress Drive #11
Petaluma, CA 94954
USA

CHRISTOPHE BARBRAUD
Centre d'Etudes Biologiques de Chizé
UPR 1934 CNRS
79360 Villiers-en-bois
France

P. DEE BOERSMA
Department of Biology
University of Washington
Seattle, Washington 98195-1800
USA;
Global Penguin Society
Marcos Zar 2716, U9120ACD
Puerto Madryn
Argentina;
Wildlife Conservation Society
2300 Southern Boulevard
Bronx, NY 10460
USA
E-mail: boersma@u.washington.edu

CHARLES-ANDRÉ BOST
Centre d'Etudes Biologiques de Chizé
UPR 1934 CNRS
79360 Villiers-en-bois
France
E-mail: bost@cebc.cnrs.fr

ALONSO BUSSALLEU
Centro para la Sostenibilidad Ambiental
Universidad Cayetano Heredia
Av. Armendariz 445
Miraflores, Lima 18
Peru

MARCO CARDEÑA
Centro para la Sostenibilidad Ambiental
Universidad Cayetano Heredia
Av. Armendariz 445
Miraflores, Lima 18
Peru

YVES CHEREL
Centre d'Etudes Biologiques de Chizé
UPR 1934 CNRS
79360 Villiers-en-bois
France

CÉDRIC COTTÉ
Centre d'Etudes Biologiques de Chizé
UPR 1934 CNRS
79360 Villiers-en-bois
France

ROBERT JM CRAWFORD
Oceans and Coasts
Department of Environmental Affairs
PO Box 52126
Cape Town 8000
South Africa;
Animal Demography Unit
University of Cape Town
Rondebosch 7700
South Africa
E-mail: crawford@environment.gov.za

GLENN T. CROSSIN
Department of Biology
Dalhousie University
Halifax
Nova Scotia, B3H 4R2
Canada
E-mail: gtc@dal.ca

RICHARD J. CUTHBERT
Royal Society for the Protection of Birds
The Lodge, Sandy
Bedfordshire, SG192DL
United Kingdom

PETER DANN
Research Department
Phillip Island Nature Parks
P.O. Box 97
Cowes, Phillip Island
Victoria, 3922
Australia
E-mail: pdann@penguins.org.au

LLOYD SPENCER DAVIS
Centre for Science Communication
University of Otago
P.O. Box 56, Dunedin
New Zealand
E-mail: lloyd.davis@otago.ac.nz

SANTIAGO DE LA PUENTE
Centro para la Sostenibilidad Ambiental
Universidad Cayetano Heredia
Av. Armendariz 445, Miraflores
Lima 18
Peru
E-mail: sdelapuente@csa-upch.org

KARINE DELORD
Centre d'Etudes Biologiques de Chizé
UPR 1934 CNRS
79360 Villiers-en-bois
France

URSULA ELLENBERG
Department of Zoology
University of Otago
P.O. Box 56, Dunedin
New Zealand

ESTEBAN FRERE
Wildlife Conservation Society
2300 Southern Boulevard
Bronx, NY 10460
USA;
Centro de Investigaciones Puerto
    Deseado
Universidad Nacional de la Patagonia
    Austral
CONICET

PABLO GARCIA BORBOROGLU
Department of Biology
University of Washington
Seattle, WA 98195-1800
USA;
Global Penguin Society
Marcos Zar 2716
U9120ACD Puerto Madryn
Argentina;
Centro Nacional Patagónico (CONICET)
Blvd. Brown 2915
U9120ACD
Puerto Madryn, Chubut
Argentina
E-mail: pgborbor@cenpat.edu.ar

GUSTAVO JIMÉNEZ-UZCÁTEGUI
Charles Darwin Foundation
Isla Santa Cruz
Puerto Ayora, Galápagos
Ecuador

OLIVIA KANE
Department of Biology
University of Washington
Seattle, WA 98195-1800
USA

JESSICA KEMPER
African Penguin Conservation Project
P.O. Box 583, Lüderitz
Namibia

GERALD KOOYMAN
Scripps Institution of Oceanography
Scholander Hall 0204
9500 Gilman Drive
La Jolla, CA 92093
USA

YVON LE MAHO
Département Ecologie, Physiologie et
    Ethologie
Institut Pluridisciplinaire Hubert Curien
UMR 7178 CNRS-UDS
23 rue Becquerel
67087 Strasbourg Cedex 2
France

H. J. Lynch
University of Maryland
College Park, MD 20742
USA;
Oceanites, Inc.
Chevy Chase, MD 20825
USA;
SUNY at Stony Brook
Stony Brook, NY 11794
USA
E-mail: hlynch@life.bio.sunysb.cdu

Patricia Majluf
Centro para la Sostenibilidad Ambiental
Universidad Cayetano Heredia
Av. Armendariz 445
Miraflores, Lima 18
Peru

Thomas Mattern
Department of Zoology
University of Otago
Dunedin
New Zealand
E-mail: t.mattern@eudyptes.net

Godfrey Merlen
Galápagos National Park Service
Isla Santa Cruz, Puerto Ayora
Galápagos
Ecuador

Helen Otley
West Coast Tai Poutini Conservancy
Department of Conservation
Hokitika
New Zealand

Patricia G. Parker
Department of Biology
University of Missouri–St. Louis
Saint Louis, MO 63121
USA

Clara Péron
Centre d'Etudes Biologiques de Chizé
UPR 1934 CNRS
79360 Villiers-en-bois
France

Luciana M. Pozzi
Centro Nacional Patagónico (CONICET)
Blvd. Brown 2915, U9120ACD
Puerto Madryn, Chubut
Argentina

Klemens Pütz
Antarctic Research Trust
Am Oste-Hamme-Kanal 10
D-27432 Bremervoerde
Germany
E-mail: puetz@antarctic-research.de

Ginger A. Rebstock
Department of Biology
University of Washington
Seattle, WA 98195-1800
USA

Andrea Raya Rey
Centro Austral de Investigaciones Cientí-
ficas (CONICET)
Houssay 200 (9410)
Ushuaia, Tierra del Fuego
Argentina
E-mail: arayarey@cadic-conicet.gob.ar

Philip J. Seddon
Department of Zoology
University of Otago
P.O. Box 56, Dunedin
New Zealand
E-mail: philip.seddon@otago.ac.nz

Alejandro Simeone
Departamento de Ecología y Biodiversidad
Facultad de Ecología y Recursos Naturales
Universidad Andres Bello
República 470, Santiago
Chile

Jeffrey Smith
Department of Biology
University of Washington
Seattle, WA 98195-1800
USA

Antje Steinfurth
Animal Demography Unit
Department of Zoology
University of Cape Town
Rondebosch 7700
South Africa
E-mail: antje.steinfurth@uct.ac

Phil N. Trathan
British Antarctic Survey
High Cross, Madingley Road
Cambridge CB3 0ET
United Kingdom
E-mail: p.trathan@bas.ac.uk

Susan Trivelpiece
Antarctic Ecosystem Research Division
Southwest Fisheries Science Center
La Jolla, CA 92065
USA

Wayne Trivelpiece
Antarctic Ecosystem Research Division
Southwest Fisheries Science Center
La Jolla, CA 92065
USA
E-mail: wayne.trivelpiece@noaa.gov

Les G. Underhill
Animal Demography Unit
Department of Zoology
University of Cape Town
Rondebosch 7700
South Africa
E-mail: les.underhill@uct.ac.za

Armando Valdés-Velázques
Facultad de Ciencias y Filosofía
Universidad Peruana Cayetano Heredia
Av. Honorio Delgado 430
Urb. Ingeniaría S. M. P. Lima, 31
Perú

Amy Van Buren
Department of Biology
University of Washington
Seattle, WA 98195
USA

Yolanda Van Heezik
Department of Zoology
University of Otago
P.O. Box 56, Dunedin
New Zealand

F. Hernan Vargas
The Peregrine Fund
5668 W. Flying Hawk Lane
Boise, Idaho 83709
USA

H. Weimerskirch
Centre d'Etudes Biologiques de Chizé
UPR 1934 CNRS
79360 Villiers-en-bois
France

Barbara Wienecke
Australian Antarctic Division
203 Channel Highway
Kingston 7050
Australia
E-mail: Barbara.Wienecke@aad.gov.au

Pablo Yorio
Wildlife Conservation Society
2300 Southern Boulevard
Bronx, NY 10460
USA;
Centro Nacional Patagónico (CONICET)
Blvd. Brown 2915
U9120ACD Puerto Madryn, Chubut
Argentina